Mathematical Analysis in Fluid Mechanics

Selected Recent Results

CONTEMPORARY MATHEMATICS

710

Mathematical Analysis in Fluid Mechanics

Selected Recent Results

International Conference on
Vorticity, Rotation and Symmetry (IV)—Complex Fluids
and the Issue of Regularity
May 8–12, 2017
Centre International de Rencontres Mathématiques
(CIRM), Luminy, Marseille, France

Raphaël Danchin
Reinhard Farwig
Jiří Neustupa
Patrick Penel
Editors

AMERICAN MATHEMATICAL SOCIETY

2010 *Mathematics Subject Classification.* Primary 35B40, 35B65, 35Q30, 35Q35, 76D05, 76D07, 76N10, 76W05.

Library of Congress Cataloging-in-Publication Data

Names: Danchin, Raphaël, editor. | International Conference on Vorticity, Rotation, and Symmetry (4th : 2017 : Luminy, Marseille, France)

Title: Mathematical analysis in fluid mechanics : selected recent results : International Conference on Vorticity, Rotation, and Symmetry (IV) : complex fluids and the issue of regularity : May 8-12, 2017, Centre International de Rencontres Mathematiques (CIRM), Luminy, Marseille, France / Raphaël Danchin [and three others], editors.

Description: Providence, Rhode Island : American Mathematical Society, [2018] | Series: Contemporary mathematics ; volume 710 | Includes bibliographical references.

Identifiers: LCCN 2018003589 | ISBN 9781470436469 (alk. paper)

Subjects: LCSH: Fluid mechanics–Mathematics–Congresses. | AMS: Partial differential equations – Qualitative properties of solutions – Asymptotic behavior of solutions. msc | Partial differential equations – Qualitative properties of solutions – Smoothness and regularity of solutions. msc | Partial differential equations – Equations of mathematical physics and other areas of application – Navier-Stokes equations. msc | Partial differential equations – Equations of mathematical physics and other areas of application – PDEs in connection with fluid mechanics. msc | Fluid mechanics – Incompressible viscous fluids – Navier-Stokes equations. msc | Fluid mechanics – Incompressible viscous fluids – Stokes and related (Oseen, etc.) flows. msc | Fluid mechanics – Compressible fluids and gas dynamics, general – Existence, uniqueness, and regularity theory. msc | Fluid mechanics – Magnetohydrodynamics and electrohydrodynamics – Magnetohydrodynamics and electrohydrodynamics. msc

Classification: LCC QA901 .M3577 2018 | DDC 532.001/515353–dc23

LC record available at https://lccn.loc.gov/2018003589

DOI: `http://dx.doi.org/10.1090/conm/710`

∞ The paper used in this book is acid-free and falls within the guidelines
established to ensure permanence and durability.
Visit the AMS home page at `http://www.ams.org/`

10 9 8 7 6 5 4 3 2 1 23 22 21 20 19 18

Contents

Preface

Since the seminal paper by Jean Leray in 1934 on global in time existence of weak (turbulent) solutions to the Navier-Stokes system mathematical fluid mechanics has been a fast growing field based on the theory of partial differential equations, linear and nonlinear functional analysis and harmonic analysis. Some of the open problems of J. Leray's time could be solved because of his paper, but most of them are still open and led to one of the seven Millennium Prize problems of the Clay Mathematics Institute in 2000. Moreover, models describing the behaviour of non-Newtonian fluids such as blood and polymeric flows, mixtures of fluids, compressible and inviscid fluids and interaction with other physical impacts opened a fast field of problems requiring new analytical tools.

We selected 13 out of more than 35 talks that are representative of the great variety of topics that have been tackled during the international conference *Vorticity, Rotation and Symmetry (IV)—Complex Fluids and the Issue of Regularity* held at the Centre International de Rencontres Mathématiques (CIRM) in Luminy (Marseille, France), May 8–12, 2017. Four articles deal with the classical problem of regularity of solutions to the (Navier-)Stokes system, e.g., see the articles by K. Nakao & Y. Taniuchi (Serrin type regularity criterion using logarithmic Brezis-Gallouet-Wainger type inequalities), D. Chamorro, P.G. Lemarié–Rieusset & K. Mayoufi discussing an extension of the Caffarelli–Kohn–Nirenberg theory for weak solutions to the case with no control on the pressure, P. Maremonti proving an interpolation estimate for the L^∞-norm for the Stokes problem, as well as H. Kozono & S. Shimizu on strong solutions and singular data in Lorentz/Besov spaces. Periodic solutions and their asymptotic profile are in the focus of an article by M. Kyed & G.P. Galdi, and D. Maity & M. Tucsnak extend the notion of maximal regularity to fluid-rigid body interaction problems. Two-phase flows with sharp interface governed by a convective mean curvature flow equation are analyzed by H. Abels & M. Moser; another important auxiliary regularity problem on elliptic equations with drift is investigated by N. Filonov and T. Shilkin. Two articles deal with compressible fluids. The first is by T. Piasecki and M. Pokorný on chemically reacting heat conducting compressible mixtures in the steady case and the second is by H. J. Choe and M. Yang on a blow up criterion for unsteady solutions involving the density and the divergence of the velocity field. A class of complex fluids with a simplified viscoelastic rate-type model with a stress-diffusion term is discussed by M. Bulíček, J. Málek, V. Průša and E. Süli; the authors prove the existence of a global in time weak solution. Electrical interation, magnetic and gravitational forces are crucial ingredients in the articles by Th. Holding & E. Miot on uniqueness and stability for the Vlasov-Poisson system, and by W. Deng & P. Zhang on global smooth solutions to the incompressible MHD system with initial data close to an equilibrium state.

We are grateful to the financial support of Agence Nationale de la Recherche, l'Université Paris-Est Créteil, and Technische Universität Darmstadt as well as the hospitality and support by CIRM in Luminy. Moreover, we thank all of the authors for their contribution and all of the reviewers for their reports. Finally, we would like to express our deep thanks to all of the people behind *Contemporary Mathematics*.

The Editors

Well-posedness of a Navier-Stokes/mean curvature flow system

Helmut Abels and Maximilian Moser

ABSTRACT. We consider a two-phase flow of two incompressible, viscous and immiscible fluids which are separated by a sharp interface in the case of a simple phase transition. In this model the interface is no longer material and its evolution is governed by a convective mean curvature flow equation, which is coupled to a two-phase Navier-Stokes equation with Young-Laplace law. The problem arises as a sharp interface limit of a diffuse interface model, which consists of a Navier-Stokes system coupled with an Allen-Cahn equation. We prove existence of strong solutions for sufficiently small times and regular initial data.

1. Introduction

We consider the flow of two immiscible, incompressible Newtonian fluids with phase transition in a bounded, smooth domain $\Omega \subseteq \mathbb{R}^n, n = 2, 3$. At time $t \in (0, T)$ the fluids fill domains $\Omega^{\pm}(t)$ separated by the interface $\Gamma(t) := \partial\Omega^+(t)$. For simplicity we set the densities to one. Moreover, we assume that the fluids have constant viscosities $\mu^{\pm} > 0$. Then the stress tensor has the form $T(v, p) = 2\mu^{\pm}\text{sym}(\nabla v) - pI$, where $\text{sym}(\nabla v) := \frac{1}{2}(\nabla v + \nabla v^{\top})$, $v : \Omega \times (0, T) \to \mathbb{R}^n$ is the velocity and $p : \Omega \times (0, T) \to \mathbb{R}$ is the pressure in Eulerian coordinates.

To formulate the model, we need some notation. We denote by $\nu_{\Gamma(t)}$ the outer unit normal to $\Gamma(t) = \partial\Omega^+(t)$ and by $V_{\Gamma(t)}, H_{\Gamma(t)}$ the normal velocity and mean curvature (for convenience the sum of the principal curvatures) with respect to $\nu_{\Gamma(t)}$. Furthermore the jump of a quantity f defined on $\Omega^{\pm}(t)$ across $\Gamma(t)$ (with respect to $\nu_{\Gamma(t)}$) is defined as

$$[\![f]\!](x) := \lim_{r\to 0} f(x + r\nu_{\Gamma(t)}(x)) - f(x - r\nu_{\Gamma(t)}(x))$$

for all $x \in \Gamma(t)$.

2010 *Mathematics Subject Classification.* Primary 35R35; Secondary 35Q30, 76D27, 76D45, 76T99.

As model we consider the following Navier-Stokes/mean curvature flow system

$$\partial_t v + v \cdot \nabla v - \mu^{\pm}\Delta v + \nabla p = 0 \qquad \text{in } \Omega^{\pm}(t), t \in (0,T), \tag{1.1}$$

$$\operatorname{div} v = 0 \qquad \text{in } \Omega^{\pm}(t), t \in (0,T), \tag{1.2}$$

$$-[\![T(v,p)]\!]\, \nu_{\Gamma(t)} = \sigma H_{\Gamma(t)}\nu_{\Gamma(t)} \qquad \text{on } \Gamma(t), t \in (0,T), \tag{1.3}$$

$$[\![v]\!] = 0 \qquad \text{on } \Gamma(t), t \in (0,T), \tag{1.4}$$

$$v|_{\partial\Omega} = 0 \qquad \text{on } \partial\Omega \times (0,T), \tag{1.5}$$

$$v|_{t=0} = v_0 \qquad \text{in } \Omega, \tag{1.6}$$

$$V_{\Gamma(t)} - \nu_{\Gamma(t)} \cdot v|_{\Gamma(t)} = H_{\Gamma(t)} \qquad \text{on } \Gamma(t), t \in (0,T), \tag{1.7}$$

$$\Gamma(0) = \Gamma_0, \tag{1.8}$$

where $\sigma > 0$ is a surface tension constant and Γ_0 is the initial interface. Here the velocity v, the pressure p and the family of interfaces $\{\Gamma(t)\}_{t\in(0,T)}$ are to be determined. Equations (1.1)-(1.2) describe the (local) mass and momentum conservation of the fluids and (1.3) is the balance of forces at the interface, see Landau and Lifshitz [**19**], §61. Moreover we require continuity of v on $\Gamma(t)$ in (1.4) and the no-slip condition for v on $\partial\Omega$ in (1.5). (1.6) and (1.8) are the initial conditions for v and the family of hypersurfaces $\{\Gamma(t)\}_{t\in(0,T)}$, respectively. Equation (1.7) without the curvature term $H_{\Gamma(t)}$ means that the interface is just transported by the fluids. Then (1.1)-(1.8) reduces to the classical two-phase Navier-Stokes system with surface tension, which was studied e.g. by Denisova [**8**], [**9**], Denisova and Solonnikov [**10**], Köhne, Prüss and Wilke [**18**] and Solonnikov [**26**]. For this system the total mass of each fluid is preserved since the densities are constant and the enclosed volume of $\Gamma(t)$ stays the same. Formally this follows from (references in) Escher and Simonett [**11**], p. 623, the divergence theorem and (1.2):

$$\frac{d}{dt}|\Omega^+(t)| = \int_{\Gamma(t)} V_{\Gamma(t)}\, d\mathcal{H}^{n-1} = \int_{\Gamma(t)} v \cdot \nu_{\Gamma(t)}\, d\mathcal{H}^{n-1} = \int_{\Omega^+(t)} \operatorname{div} v\, dx = 0.$$

The additional curvature term $H_{\Gamma(t)}$ on the right hand side in (1.7) allows for a change of total masses and couples the Navier-Stokes system to the mean curvature flow

$$V_{\Gamma(t)} = H_{\Gamma(t)} \qquad \text{for } t > 0.$$

Therefore we call (1.1)-(1.8) the Navier-Stokes/mean curvature flow system. The system is obtained as a sharp interface limit of a Navier-Stokes-Allen-Cahn system. Another motivation is the regularizing effect of the curvature term in (1.7). But we can also view it as a simple model for a two-phase flow with phase transition. In the case that the Navier-Stokes equations are modified for shear thickening non-Newtonian fluid of power-law type Liu, Sato, and Tonegawa in [**20**] have shown existence of weak solutions. In the present contribution we prove local existence of strong solutions to (1.1)-(1.8) in an L^p-setting. The result is part of the second author's Master thesis, which was supervised by the first author.

We apply a similar strategy as A. and Wilke [**3**], who study local well-posedness and qualitative behaviour of solutions for a two-phase Navier-Stokes-Mullins-Sekerka system. The equations are transformed with a variant $\tilde{\Theta}_h$ of the Hanzawa transform to a reference surface Σ and disjoint domains $\Omega^{\pm}$, where $\Omega = \Omega^+ \cup \Omega^- \cup \Sigma$ is similar as above. Here $h : \Sigma \times (0,T) \to \mathbb{R}$ is the "height function" of $\{\Gamma(t)\}_{t\in(0,T)}$ with respect to Σ. The coupling with the Navier-Stokes part in (1.7) is of lower order.

Therefore we solve the Navier-Stokes-part (1.1)-(1.6) for a given appropriate family of hypersurfaces $\{\Gamma(t)\}_{t\in(0,T)}$ or equivalently for a given suitable time-dependent height function h. Then we insert the obtained velocity field $u(h)$ into equation (1.7) and (1.7)-(1.8) turns into an abstract evolution equation for h. The latter is solved by using the theory of maximal L^p-regularity.

The paper is organized as follows. In Section 2 we describe the transformation to a fixed reference hypersurface (and related domains) and analyse the occurring terms, especially the transformed mean curvature. In Section 3 we consider the Navier-Stokes part for a given family of hypersurfaces $\{\Gamma_h(t)\}_{t\in(0,T)}$ and in Section 4 we show the local well-posedness for the transformed system. Definitions and some properties of the used Banach-space valued function spaces are summarized in Appendix A.

We always assume $\Omega \subseteq \mathbb{R}^n, n = 2, 3$, to be a bounded, connected and smooth domain and throughout let $\Sigma \subseteq \Omega$ be a compact, connected and smooth hypersurface, that separates Ω in two disjoint, connected domains $\Omega^\pm$ with $\Sigma = \partial\Omega^+$ and outer unit normal ν_Σ.

2. Preliminaries

2.1. Notation and scalar-valued function spaces. Let $\mathbb{R}_+ := [0, \infty)$ and $\mathbb{R}^n_+ := \mathbb{R}^{n-1} \times (0, \infty)$ for $n \in \mathbb{N}, n \geq 2$. The Euclidean norm in $\mathbb{R}^n$ is denoted by $|.|$. For matrices $A, B \in \mathbb{R}^{n\times n}$ let $A : B$ be the matrix-product and $|A| := \sqrt{A : A}$ be the induced norm. The set of invertible matrices in $\mathbb{R}^{n\times n}$ is $GL(n, \mathbb{R})$. For $A \in GL(n, \mathbb{R})$ we set $A^{-\top} := (A^{-1})^\top$.

For metric spaces X denote by $B_X(x, r)$ the ball with radius $r > 0$ around $x \in X$. For normed spaces X, Y over $\mathbb{K} = \mathbb{R}$ or $\mathbb{C}$ the set of bounded, linear operators $T : X \to Y$ is $\mathcal{L}(X, Y)$ and $\mathcal{L}(X) := \mathcal{L}(X, X)$. If X, Y are additionally complete and $U \subseteq X$ is open, for $k \in \mathbb{N}$ the set of all k-times continuously differentiable functions $f : U \to Y$ is denoted by $C^k(U; Y)$. The set of all bounded $f \in C^k(U; Y)$ with bounded derivatives is $BC^k(U; Y)$ with usual norm. Let X, Y, Z be Banach spaces over $\mathbb{K} = \mathbb{R}$ or $\mathbb{C}$. Then $B : X \times Y \to Z$ is called *product* if B is bilinear and $\|B(f, g)\|_Z \leq C_0\|f\|_X\|g\|_Y$ holds for all $(f, g) \in X \times Y$ and some $C_0 > 0$.

If (X_0, X_1) is admissible and $\theta \in (0, 1), 1 \leq p \leq \infty$, we denote by $(X_0, X_1)_{\theta,p}$ the real interpolation spaces and with $\|.\|_{(X_0,X_1)_{\theta,p}} \equiv \|.\|_{\theta,p}$ the norm, see Lunardi [**21**], Bergh and Löfström [**6**] and Triebel [**27**].

Let $\Omega \subseteq \mathbb{R}^n, n \in \mathbb{N}$, be open and nonempty. Spaces of continuous and continuously differentiable functions $f : \overline{\Omega} \to \mathbb{R}$ are defined as usual. Now let $1 \leq p \leq \infty$ and $k \in \mathbb{N}$. Then the Lebesgue and Sobolev spaces are denoted by $L^p(\Omega) =: W^0_p(\Omega)$ and $W^k_p(\Omega)$, respectively. Moreover, we set

$$W^k_{p,0}(\Omega) := \overline{C^\infty_0(\Omega)}^{W^k_p(\Omega)} \quad \text{and} \quad W^{-k}_p(\Omega) := W^k_{p',0}(\Omega)',$$

where p' is the conjugate exponent, as well as $L^p_{(0)}(\Omega) := \{f \in L^p(\Omega) : \int_\Omega f\,dx = 0\}$,

$$W^k_{p,(0)}(\Omega) := W^k_p(\Omega) \cap L^p_{(0)}(\Omega) \quad \text{and} \quad W^{-k}_{p,(0)}(\Omega) := W^k_{p',(0)}(\Omega)'.$$

The closure of the divergence-free $\phi \in C^\infty_0(\Omega)^n$ in $L^p(\Omega)^n$ is denoted by $L^p_\sigma(\Omega)$. Besides, we need the Besov spaces $B^s_{p,q}(\mathbb{R}^n)$ for $0 < s < \infty$ and $1 \leq p, q \leq \infty$. We set $W^s_p(\mathbb{R}^n) := B^s_{p,p}(\mathbb{R}^n)$ for all $s \in (0, \infty) \setminus \mathbb{N}, 1 \leq p \leq \infty$. Respective spaces can also be defined on bounded, smooth domains $\Omega \subseteq \mathbb{R}^n$ and, if $n \geq 2$, on their boundaries $\partial\Omega$. For embeddings, interpolation results and trace theorems, compare e.g. Triebel [**27**], [**28**].

2.2. Hanzawa transformation Θ_h. For the construction of Θ_h we use that there is an $a > 0$, such that

$$\Phi : \Sigma \times (-a, a) \to \mathbb{R}^n : (s, r) \mapsto s + r\,\nu_\Sigma(s)$$

is a C^∞-diffeomorphism onto $\Sigma_a := B_a(\Sigma)$. We set $(\Pi, d_\Sigma) := \Phi^{-1}$. Then it holds $\nabla d_\Sigma = \nu_\Sigma \circ \Pi$. Here Π is called projection onto Σ and d_Σ signed distance function. For the proof cf. Hildebrandt [**15**], Chapter 4.6, Theorems 1-3.

Let $a < \operatorname{dist}(\Sigma, \partial\Omega)$ and $a_0 < \frac{a}{4}$ fixed, $\chi \in C^\infty(\mathbb{R})$ be a cutoff function such that $|\chi'| \leq 4$ and $\chi(s) = 1$ for $|s| \leq \frac{1}{3}$, as well as $\chi(s) = 0$ for $|s| \geq \frac{2}{3}$. For $h \in C^2(\Sigma)$ with $\|h\|_\infty < a_0$ we define the *Hanzawa transformation* Θ_h by

$$\Theta_h(x) := x + \chi(\frac{d_\Sigma(x)}{a})\, h(\Pi(x))\, \nu_\Sigma(\Pi(x)) \qquad \text{for all } x \in \mathbb{R}^n.$$

Essential properties of Θ_h are listed in

LEMMA 2.1. *The Hanzawa transformation $\Theta_h : \Omega \to \Omega$ is a C^2-diffeomorphism and Θ_h is the identity on $\mathbb{R}^n \setminus \Sigma_{2a/3}$, in particular nearby $\partial\Omega$. Moreover,*

$$\det D\Theta_h \geq c > 0 \quad \text{and} \quad \|D\Theta_h\|_\infty + \|D\Theta_h^{-1}\|_\infty \leq C(1 + \|h\|_{C^1(\Sigma)})$$

with $c, C > 0$ independent of h.

PROOF. The first part can be shown as in Kneisel [**16**], Lemma 2.2.1 and one also obtains a local representation for $D\Theta_h$. The latter yields the estimates by compactness of Σ. □

REMARK 2.2. Let $h \in C^2(\Sigma)$ such that $\|h\|_\infty < a_0$. Then $\Gamma_h := \Theta_h(\Sigma)$ is a connected and compact C^2-hypersurface as zero level set of $\phi_h : \Sigma_a \to \mathbb{R} : x \mapsto d_\Sigma(x) - h(\Pi(x))$, where $|\nabla\phi_h| \geq 1$ as one can show $\nabla\phi_h|_x \cdot \nu_\Sigma|_{\Pi(x)} = 1$ for all $x \in \Sigma_a$ similar to Kneisel [**16**], p. 13. Moreover, let ν_{Γ_h} be the outer unit normal to $\Gamma_h = \partial(\Theta_h(\Omega^+))$.

2.3. Function spaces for the height function and modification $\tilde{\Theta}_h$ of Θ_h. In the following we specify, in which function spaces we consider height functions h. Let $p > n + 2$ be fixed. We set

$$X_0 := W_p^{1-\frac{1}{p}}(\Sigma) \quad \text{and} \quad X_1 := W_p^{3-\frac{1}{p}}(\Sigma).$$

Since $3 - \frac{3}{p} - \frac{n-1}{p} > 2$ we have $X_\gamma := (X_0, X_1)_{1-\frac{1}{p},p} = W_p^{3-\frac{3}{p}}(\Sigma) \hookrightarrow C^2(\Sigma)$. Moreover, we define $\mathbb{E}_1(T) := L^p(0, T; X_1) \cap W_p^1(0, T; X_0)$. Theorem A.3 yields $\mathbb{E}_1(T) \hookrightarrow C^0([0, T]; X_\gamma)$. Hence

$$\Theta_h(x, t) := \Theta_{h(.,t)}(x) \qquad \text{for } (x, t) \in \mathbb{R}^n \times [0, T]$$

is well-defined for $h \in \mathbb{E}_1(T) \cap C^0([0, T]; U)$ where $U := \{h \in X_\gamma : \|h\|_\infty < a_0\}$. Let $R_0 > 0$ arbitrary and fixed from now on. It suffices to consider height functions in the space

$$V_T := \{h \in \mathbb{E}_1(T) \cap C^0([0, T]; U) : \|h\|_{\mathbb{E}_1(T)} + \|h(0)\|_{X_\gamma} < R_0\}.$$

The restriction due to the estimate is negligible as we are interested in local well-posedness, but it is needed to show suitable properties of a modification of the Hanzawa transformation Θ_h which we need for technical reasons: if we transform differential operators in space, the term $D\Theta_h^{-\top}$ always appears. In order to show appropriate regularity properties in space of $D\Theta_h^{-\top}$, it is complicated (but possible) to work with the Hanzawa transformation directly. However, it seems difficult to obtain suitable regularity in time for $D\Theta_h^{-\top}$ since derivatives of h in space are present. It is much simpler to replace $h \circ \Pi$ in Θ_h by an extension Eh of h, where

$$E : X_0 = W_p^{1-\frac{1}{p}}(\Sigma) \to W_p^1(\Sigma_{\overline{a}}) \text{ with } E \in \mathcal{L}(W_p^{k-\frac{1}{p}}(\Sigma), W_p^k(\Sigma_{\overline{a}})), k = 1, 2, 3 \tag{2.1}$$

for $\overline{a} := \frac{3a}{4}$ is a suitable extension operator. Then we have $E \in \mathcal{L}(X_\gamma, \tilde{X}_\gamma)$, where we have set $\tilde{X}_\gamma := (W_p^1(\Sigma_{\overline{a}}), W_p^3(\Sigma_{\overline{a}}))_{1-\frac{1}{p},p}$. Here because of $p > n+2$ it holds

$$\tilde{X}_\gamma = W_p^{3-\frac{2}{p}}(\Sigma_{\overline{a}}) \hookrightarrow C^2(\overline{\Sigma_{\overline{a}}}). \tag{2.2}$$

So we define the modification $\tilde{\Theta}_h$ of the Hanzawa transformation Θ_h as follows:

DEFINITION 2.3. Let $p > n+2, R_0 > 0$ as above and $T > 0$. For $h_0 \in U$ we set

$$\tilde{\Theta}_h(x) := x + \chi\Big(\frac{d_\Sigma(x)}{a}\Big)Eh(x)\nu_\Sigma(\Pi(x)) \quad \text{for all } x \in \mathbb{R}^n$$

and for $h \in V_T$ let $\tilde{\Theta}_h(x,t) := \tilde{\Theta}_{h(.,t)}(x)$ for all $(x,t) \in \mathbb{R}^n \times [0,T]$.

The following lemma shows that similar properties as for Θ_h also hold for the modification $\tilde{\Theta}_h$ if the extension operator E is chosen properly.

LEMMA 2.4. *Let* $\overline{a} = \frac{3a}{4}$ *and* $\varepsilon > 0$. *Then* E *can be chosen such that for all* $h \in X_\gamma$

$$\|h \circ \Pi - Eh\|_{C^1(\overline{\Sigma_{\overline{a}}})} \le \varepsilon \|h\|_{X_\gamma}. \tag{2.3}$$

Moreover, for $R > 0$ *there is an* $\varepsilon = \varepsilon(R) > 0$ *such that with the extension operator* E *with respect to* ε *the following hold:* $\tilde{\Theta}_h : \Omega \to \Omega$ *is for all* $h \in U$ *with* $\|h\|_{X_\gamma} \le R$ *a* C^2*-diffeomorphism, the identity on* $\mathbb{R}^n \setminus \Sigma_{2a/3}$ *and we have*

$$\det D\tilde{\Theta}_h \ge \tilde{c} > 0 \quad \text{and} \quad D\tilde{\Theta}_h^{-\top}|_x \nu_\Sigma|_{\Pi(x)} \cdot \nu_\Sigma|_{\Pi(x)} \ge \frac{1}{2} \quad \text{for all } x \in \overline{\Sigma_{a/3}}.$$

REMARK 2.5. Theorem A.3 implies $\mathbb{E}_1(T) \hookrightarrow C^0([0,T]; X_\gamma)$ and the embedding constant is bounded independently of $T > 0$ if we add the term $\|h(0)\|_{X_\gamma}$ in the $\mathbb{E}_1(T)$-norm. Now we can do the following: we choose $R = \overline{R_0} > R_0$ in Lemma 2.4 (and so E) such that

$$h(t) \in U_0 := \{h_0 \in U : \|h_0\|_{X_\gamma} < \overline{R_0}\}$$

holds for all $h \in V_T$ and $0 \le t \le T$. In particular for $h \in V_T$ the last part in Lemma 2.4 is valid for $\tilde{\Theta}_h(.,t)$ instead of $\tilde{\Theta}_h$. Here $0 \le t \le T$ and $0 < T < \infty$ are arbitrary.

PROOF OF LEMMA 2.4. Similar to A. and Wilke [**3**] one can construct a new operator E that fulfils (2.3) by scaling E in normal direction of Σ. Then for $R > 0$ and $h \in U$ with $\|h\|_{X_\gamma} \le R$ it holds $\|D\Theta_h - D\tilde{\Theta}_h\|_\infty \le C\varepsilon R$. From this the second part follows using the properties of Θ_h. For the details we refer to the preprint [**2**]. □

2.4. The transformed equations. We transform the Navier-Stokes/mean curvature flow system (1.1)-(1.8) with $\tilde{\Theta}_h$ to Σ and $\Omega^\pm$, respectively. Therefore we fix the height function $h \in V_T$ and let $v : \Omega \times (0,T) \to \mathbb{R}^n$ be a vector-valued and $\tilde{p} : \Omega \times (0,T) \to \mathbb{R}$ be a scalar-valued function. Then we define $u(x,t) := v(\tilde{\Theta}_h(x,t),t)$ and $\tilde{q}(x,t) := \tilde{p}(\tilde{\Theta}_h(x,t),t)$ for $(x,t) \in \Omega \times (0,T)$. Moreover, let

$$A(h) := D\tilde{\Theta}_h^{-\top}, \quad \nabla_h := A(h)\nabla, \quad \nabla_h u := (\nabla_h u_k^\top)_{k=1}^n \text{ and } \operatorname{div}_h u := \operatorname{Tr}\nabla_h u.$$

Let $t \in [0,T]$. We set $\Gamma_h(t) := \Gamma_{h(.,t)}$ and denote the outer unit normal by $\nu_{\Gamma_h(t)}$ and by $H_{\Gamma_h(t)}$ the corresponding mean curvature. Furthermore, we define

$$K(h) : \Sigma \times [0,T] \to \mathbb{R} : (x,t) \mapsto H_{\Gamma_h(t)}|_{\Theta_h(x,t)}.$$

Applying the chain rule to $\phi_{h(.,t)} \circ \tilde{\Theta}_h(.,t)$ with $\phi_{h(.,t)}$ as in Remark 2.2 implies

$$\nu_h|_{(x,t)} := \nu_{\Gamma_h(t)}|_{\tilde{\Theta}_h(x,t)} = \frac{A(h)|_{(x,t)}\nu_\Sigma|_x}{|A(h)|_{(x,t)}\nu_\Sigma|_x|} \quad \text{for all } (x,t) \in \Sigma \times (0,T).$$

Moreover, we have $V_{\Gamma_h(t)}|_{\tilde{\Theta}_h(x,t)} = \partial_t(\Theta_h(x,t)) \cdot \nu_{\Gamma_h(t)}|_{\Theta_h(x,t)} = \partial_t h(x,t)\,\nu_\Sigma|_x \cdot \nu_h|_{(x,t)}$ for the normal velocity. Since $\tilde{\Theta}_h$ is the identity in a neighbourhood of $\partial\Omega$, the no-slip condition on v is retained. Therefore (1.1)-(1.8) becomes

$$\partial_t u + u \cdot \nabla_h u - \mu^\pm \operatorname{div}_h \nabla_h u + \nabla_h \tilde{q} = \nabla_h u\, \partial_t \tilde{\Theta}_h \quad \text{in } \Omega^\pm \times (0,T), \tag{2.4}$$

$$\operatorname{div}_h u = 0 \quad \text{in } \Omega^\pm \times (0,T), \tag{2.5}$$

$$-[\![T_h(u,\tilde{q})]\!]\,\nu_h = \sigma K(h)\,\nu_h \quad \text{on } \Sigma \times (0,T), \tag{2.6}$$

$$[\![u]\!] = 0 \quad \text{on } \Sigma \times (0,T), \tag{2.7}$$

$$u|_{\partial\Omega} = 0 \quad \text{on } \partial\Omega \times (0,T), \tag{2.8}$$

$$u|_{t=0} = v_0 \circ \tilde{\Theta}_{h_0} \quad \text{in } \Omega, \tag{2.9}$$

$$\partial_t h - a_1(h) \cdot u|_\Sigma = a_2(h)K(h) \quad \text{on } \Sigma \times (0,T), \tag{2.10}$$

$$h|_{t=0} = h_0 \quad \text{on } \Sigma, \tag{2.11}$$

where we set $T_h(u,\tilde{q}) := 2\mu^\pm \operatorname{sym}(\nabla_h u) - \tilde{q}I$ as well as

$$a_1(h) := \frac{A(h)\nu_\Sigma}{A(h)\nu_\Sigma \cdot \nu_\Sigma} \quad \text{and} \quad a_2(h) := \frac{|A(h)\nu_\Sigma|}{A(h)\nu_\Sigma \cdot \nu_\Sigma}.$$

Moreover $u \cdot \nabla_h u \equiv (u \cdot \nabla_h)u$ and div_h is applied row-wise to $\nabla_h u$ in (2.4). For this system we show local well-posedness. As a preparation, in the next two Subsections 2.5 and 2.6 we show properties of $\tilde{\Theta}_h$, $A(h) = D\tilde{\Theta}_h^{-\top}$ and of the transformed mean curvature $K(h)$, respectively.

2.5. Properties of $\tilde{\Theta}_h$ and $D\tilde{\Theta}_h^{-\top}$. For later use we must know the regularity of $\tilde{\Theta}_h$ and $A(h) = D\tilde{\Theta}_h^{-\top}$. This is provided by

LEMMA 2.6. *Let U_0 as in Remark 2.5, $0 < T \leq T_0$. Then $\tilde{\Theta}_. \in BC^1(U_0; C^2(\overline{\Omega})^n)$, $A \in BC^1(U_0; C^1(\overline{\Omega})^{n\times n})$ and $\tilde{\Theta}_. \in BC^1(V_T; \tilde{X}_T)$, $A \in BC^1(V_T; X_T)$, where*

$$\tilde{X}_T := C^\tau([0,T]; C^2(\overline{\Omega})^n) \times C^{\frac{1}{2}+\tau}([0,T]; C^1(\overline{\Omega})^n) \cap W_p^1(0,T; W_p^1(\Omega)^n),$$

$$X_T := C^\tau([0,T]; C^1(\overline{\Omega})^{n\times n}) \times C^{\frac{1}{2}+\tau}([0,T]; C^0(\overline{\Omega})^{n\times n}) \cap W_p^1(0,T; L^p(\Omega)^{n\times n})$$

for some $\tau > 0$. In the time-dependent case the mappings and their derivatives are bounded by a $C > 0$ independent of $0 < T \leq T_0$ if we add $\|h(0)\|_{X_\gamma}$ in the $\mathbb{E}_1(T)$-norm.

REMARK 2.7. Lemma 2.4 and Remark 2.5 yield $|A(h)\nu_\Sigma|_\Pi| \geq \frac{1}{2}$ on $\overline{\Sigma_{a/3}}$. Thus we can extend ν_Σ by $\nu_\Sigma|_\Pi$ smoothly to $\overline{\Sigma_{a/3}}$ and obtain from Lemma 2.6 that $\nu_., a_1$ and a_2 from Subsection 2.4 have the same regularity properties as A with $\overline{\Omega}$ replaced by Σ, if we omit the Sobolev space in X_T.

PROOF OF LEMMA 2.6. Let $h \in U_0$ or $h \in V_T$. First, we show the properties of $\tilde{\Theta}_h$. It holds

$$\tilde{\Theta}_h = \mathrm{id} + \phi E h \quad \text{with } \phi := \chi(d_\Sigma(.)/a)\, \nu_\Sigma \circ \Pi \in C_0^\infty(\Sigma_{3a/4})^n. \tag{2.12}$$

Here E is the extension operator from Remark 2.5 that induces $E \in \mathcal{L}(X_\gamma, \tilde{X}_\gamma)$ with $\tilde{X}_\gamma \hookrightarrow C^2(\overline{\Sigma_{\overline{a}}})$ by (2.2) and $E \in \mathcal{L}(\mathbb{E}_1(T), \tilde{\mathbb{E}}_1(T))$ with

$$\tilde{\mathbb{E}}_1(T) := L^p(0,T; W_p^3(\Sigma_{\overline{a}})) \cap W_p^1(0,T; W_p^1(\Sigma_{\overline{a}})),$$

where the operator norm is bounded by a $C > 0$ independent of $T > 0$. We need suitable embeddings for $\tilde{\mathbb{E}}_1(T)$: Theorem A.4 yields for $0 < \theta < 1 - \frac{1}{p}$

$$\tilde{\mathbb{E}}_1(T) \hookrightarrow C^{1-\theta-\frac{1}{p}}([0,T]; B_{p,p}^{1+2\theta}(\Sigma_{\overline{a}})) \tag{2.13}$$

and the embedding constant is bounded independently of $T > 0$ if we add the term $\|\tilde{h}(0)\|_{\tilde{X}_\gamma}$ in the $\tilde{\mathbb{E}}_1(T)$-norm. For $\theta = \frac{n}{2p} + \frac{k-1}{2} + \delta, k = 1, 2$ and $\delta > 0$ small we obtain because of $p > n+2$

$$1 + 2\theta - \frac{n}{p} = 1 + k - 1 + 2\delta > k \quad \text{and} \quad 1 - \frac{1}{p} - \theta = 1 - \frac{n+2}{2p} - \frac{k-1}{2} - \delta > 1 - \frac{k}{2}.$$

Embeddings and (2.13) imply $\tilde{\mathbb{E}}_1(T) \hookrightarrow C^\tau([0,T]; C^2(\overline{\Sigma_{\overline{a}}})) \cap C^{\frac{1}{2}+\tau}([0,T]; C^1(\overline{\Sigma_{\overline{a}}}))$ for some $\tau > 0$. Since $\tilde{\Theta}_h$ is affine linear in h, we get the properties of $\tilde{\Theta}_h$.

Now we deduce the assertions for $A(h) = D\tilde{\Theta}_h^{-\top}$. The first part yields analoguous properties for $D\tilde{\Theta}_h$. Moreover, using suitable product estimates, Lemma 2.4 and the inverse formula we obtain $A(h) \in C^1(\overline{\Omega})^{n\times n}$ for all $h \in U_0$ and $A(h) \in X_T$ for all $h \in V_T$. Cf. the preprint [**2**] for more details. Furthermore, we have boundedness as in the lemma. The implicit function theorem applied to $G : Z \times X \to X : (h, B) \mapsto D\tilde{\Theta}_h^\top B - I$ at $(h, A(h))$ for $(Z, X) = (U_0, C^1(\overline{\Omega})^{n\times n})$ or (V_T, X_T) implies that A is C^1 in the respective spaces and gives a formula for the derivative. Now from the properties of $D\tilde{\Theta}_h$ the claim follows. □

2.6. Mean curvature. For $h \in U$ we denote by H_{Γ_h} the mean curvature of $\Gamma_h = \Theta_h(\Sigma)$ with respect to ν_{Γ_h}. This subsection is devoted to the transformed mean curvature $K(h) := H_{\Gamma_h} \circ \Theta_h : \Sigma \to \mathbb{R}$. In particular we show that $K(h)$ has a quasilinear structure in terms of h and the principal part for $h = 0$ is given by the Laplace-Beltrami operator Δ_Σ.

LEMMA 2.8. *Let $p > n+2$. Then there exist $P \in C^1(U; \mathcal{L}(X_1, X_0))$ and $Q \in C^1(U; X_0)$ with $P(0) = \Delta_\Sigma$ and $K(h) = P(h)h + Q(h)$ for all $h \in U \cap X_1$.*

PROOF. Let $\varphi_l : V_l \to U_l \subseteq \Sigma$ for $l = 1, ..., N$ be suitable parametrizations of Σ with $\Sigma = \bigcup_{l=1}^N U_l$ and $V_l \subseteq \mathbb{R}^{n-1}$. Then $\tilde{\varphi}_l := (\varphi_l, \mathrm{id}) : V_l \times (-a, a) \to U_l \times (-a, a)$ are appropriate parametrizations of $\Sigma \times (-a, a)$. Now fix l. The Euclidean metric g^{eucl} on Σ_a induces via the diffeomorphism Φ from Subsection 2.2 a Riemannian metric g^Φ on $\Sigma \times (-a, a)$. We denote by w_{ij}, Γ_{ij}^k for $i, j, k = 1, ..., n$ the local representation of g^Φ and the Christoffel symbols with respect to $\tilde{\varphi}_l$, respectively, and w^{ij} is as usual. Moreover, we set $w_{ij}(h)|_s := w_{ij}|_{(s,h(s))}$ for $s \in U_l$. $\Gamma_{ij}^k(h)$ and

$w^{ij}(h)$ are defined analogously. As in Escher and Simonett [**12**], proof of Lemma 3.1 on p. 274 ff. and Remark 3.2 on p. 277 it follows that

$$K(h)|_{U_l} = P_l(h)h + Q_l(h)$$

where P_l and Q_l are altered by a factor $-(n-1)$ here. Additionally let $p_{jk}, p_j, q, l.$ be as in Escher and Simonett [**12**] modified by the same factor. Then $P_l(0)$ is a local representation of Δ_Σ. One can show that the $w_{ij}|_{(s,r)}$ and $\partial_i\partial_j\Phi \cdot \partial_m\Phi|_{(s,r)}$ for $i,j,m = 1,...,n$ are at most quadratic in $r \in (-a,a)$ with smooth coefficients in $s \in \overline{U_l}$. Because of $\|h\|_\infty < a_0 < \frac{a}{4}$ and $w_{ij}(h) = w_{ij}(.,h(.))$ for $i,j = 1,...,n$, a compactness argument yields

$$\det((w_{ij}(h))_{i,j=1}^n) \geq c > 0 \tag{2.14}$$

for a $c > 0$ independent of h.

Using $X_\gamma \hookrightarrow C^2(\Sigma)$, the product rule for the Fréchet-derivative, the inverse formula as well as properties of $h \mapsto h^\beta$ in certain Hölder spaces (cf. Lemma A.3 in the preprint [**2**]) for the local representations in Escher and Simonett [**12**] implies

$$p_{jk}(.) \circ \varphi_l,\ p_i(.) \circ \varphi_l,\ q(.) \circ \varphi_l \in C^1(U; C^1(\overline{V_l})) \qquad \text{for } i,j,k = 1,...,n-1.$$

Now the claim follows utilizing a suitable partition of unity. □

Furthermore, we need properties of K, when we insert a time-dependent height function $h \in V_T$. This gives us

LEMMA 2.9. *Let $p > n+2$, $1 < q < p$ and $0 < T \leq T_0$. Then it holds*

$$K \in BC^1(V_T; W_q^{1-\frac{1}{q},\frac{1}{2}(1-\frac{1}{q})}(\Sigma_T)).$$

Moreover, K and its derivative are bounded by a constant $C(p,q,T_0) > 0$ when we add the term $\|h(0)\|_{X_\gamma}$ in the $\mathbb{E}_1(T)$-norm.

PROOF. We use the same notation as in the proof of Lemma 2.8. Then

$$K(h)|_{U_l} = \sum_{\alpha \in \mathbb{N}_0^{n-1}, |\alpha| \leq 2} a_\alpha(., h, \nabla_s h)\partial_s^\alpha h, \tag{2.15}$$

where the a_α correspond to p_{ij}, p_j and q, respectively, and ∂_s^α is the α-th derivative with respect to the canonical coordinates.

First we estimate the a_α in suitable Hölder spaces. As in the proof of Lemma 2.6 one can show $\mathbb{E}_1(T) \hookrightarrow C^\tau([0,T]; C^2(\Sigma)) \cap C^{\frac{1}{2}+\tau}([0,T]; C^1(\Sigma))$ for $\tau > 0$ small because of $p > n+2$ and the embedding constant is bounded independently of $T > 0$ if we add the term $\|h(0)\|_{X_\gamma}$ in the $\mathbb{E}_1(T)$-norm. Since the w_{ij} for $i,j = 1,...,n$ are at most quadratic in h with smooth coefficients in $\overline{U_l}$ product estimates in Hölder spaces yield

$$w_{ij}(.) \circ \varphi_l \in BC^1(V_T; C^\tau([0,T]; C^2(\overline{V_l})) \cap C^{\frac{1}{2}+\tau}([0,T]; C^1(\overline{V_l}))) \qquad \text{for } i,j = 1,...,n.$$

Now (2.14) implies[1] using the inverse formula, product estimates and properties of $h \mapsto h^\beta$ in certain Hölder spaces (cf. Lemma A.3 in the preprint [**2**]) that

$$w^{ij}(.) \circ \varphi_l \in BC^1(V_T; C^\tau([0,T]; C^1(\overline{V_l})) \cap C^{\frac{1}{2}+\tau}([0,T]; C^0(\overline{V_l}))) \qquad \text{for } i,j = 1,...,n.$$

[1] In fact, the same regularity as for the $w_{ij}(.) \circ \varphi_l$ should be obtained. But this is not needed.

Hence this also holds for $\Gamma^k_{ij}(.)\circ\varphi_l$ and $(l_.)^r$, where $i,j,k = 1,...,n$ and $r \in \mathbb{R}$. Using the mapping properties in the local representations, we obtain for $\alpha \in \mathbb{N}_0^{n-1}$, $|\alpha| \le 2$

$$a_\alpha \in BC^1(V_T; C^\tau([0,T]; C^1(\overline{V_l})) \cap C^{\frac{1}{2}+\tau}([0,T]; C^0(\overline{V_l}))). \tag{2.16}$$

Furthermore, all above terms and their derivatives are bounded by a $C > 0$ independent of $T > 0$ if we add $\|h(0)\|_{X_\gamma}$ in the $\mathbb{E}_1(T)$-norm. So this also holds for the a_α. The exponent q does not appear here.

To show the desired properties of K we use (2.15). Therefore we need another embedding for $\mathbb{E}_1(T)$ to estimate $\partial_s^\alpha h$ suitably. Theorem A.2 and a well-known interpolation inequality for real interpolation spaces (see Corollary 1.7 in Lunardi [**21**]) yield

$$\mathbb{E}_1(T) \hookrightarrow B^{1-\theta}_{p,\infty}(0,T;(X_0,X_1)_{\theta,p})$$

for $\theta \in (0,1)$ and the embedding constant is bounded independently of $T > 0$. We set $\theta = \frac{1}{2}(1+\frac{1}{p}) + \varepsilon$ for $\varepsilon > 0$ small. Then for $\varepsilon = \varepsilon(p,q) > 0$ small it holds that

$$1 - \frac{1}{p} + 2\theta = 2 + 2\varepsilon \quad \text{and} \quad 1 - \theta = \frac{1}{2}(1-\frac{1}{p}) - \varepsilon > \frac{1}{2}(1-\frac{1}{q}).$$

We infer $\mathbb{E}_1(T) \hookrightarrow B^{\frac{1}{2}(1-\frac{1}{p})-\varepsilon}_{p,\infty}(0,T;W^2_p(\Sigma))$ and, again, the embedding constant is bounded independently of $T > 0$. This implies

$$\partial_{x_i}(.\circ\varphi_l),\ \partial_{x_j}\partial_{x_i}(.\circ\varphi_l) \in BC^1(V_T; B^{\frac{1}{2}(1-\frac{1}{p})-\varepsilon}_{p,\infty}(0,T;L^p(V_l)) \cap L^p(0,T;W_p^{1-\frac{1}{p}}(V_l)))$$

for $i,j = 1,...,n-1$ and the mappings and their derivatives are bounded independently of $T > 0$. Since a product on $C^1(\overline{V_l}) \times W_p^{1-\frac{1}{p}}(V_l) \to W_p^{1-\frac{1}{p}}(V_l)$ is given by pointwise multiplication, using Lemma A.1 and (2.15)-(2.16) yields

$$K(.)\circ\varphi_l \in BC^1(V_T; B^{\frac{1}{2}(1-\frac{1}{p})-\varepsilon}_{p,\infty}(0,T;L^p(V_l)) \cap L^p(0,T;W_p^{1-\frac{1}{p}}(V_l)))$$

where $K(.)\circ\varphi_l$ and the derivative are bounded by a constant $C > 0$ independent of $T > 0$. Now we use the embeddings $L^p(V_l) \hookrightarrow L^q(V_l)$ and $W_p^{1-\frac{1}{p}}(V_l) \hookrightarrow W_q^{1-\frac{1}{q}}(V_l)$ as well as

$$B^{\frac{1}{2}(1-\frac{1}{p})-\varepsilon}_{p,\infty}(0,T;L^p(V_l)) \hookrightarrow B^{\frac{1}{2}(1-\frac{1}{p})-\varepsilon}_{q,\infty}(0,T;L^p(V_l))$$

where the latter can be shown by using Hölder's inequality and the embedding constant is bounded by $C(p,q,T_0) > 0$. We infer from Simon [**25**], Corollary 15 that

$$B^{\frac{1}{2}(1-\frac{1}{p})-\varepsilon}_{q,\infty}(0,T;L^q(V_l)) \hookrightarrow W_q^{\frac{1}{2}(1-\frac{1}{q})}(0,T;L^q(V_l))$$

and the embedding constant is also bounded by a $C(p,q,T_0) > 0$. The claim follows by using a suitable partition of unity. $\square$

3. Two-phase Navier-Stokes system for given time-dependent interface

We consider the Navier-Stokes part (2.4)-(2.9) for a fixed time-dependent height function $h \in V_T$ (or equivalently for a given family of hypersurfaces $\{\Gamma_h(t)\}_{t\in(0,T)}$) and show unique solvability and some properties of the solution operator. Let as before $p > n+2$, $R_0 > 0$ be fixed as well as $X_0, X_1, X_\gamma, U, \mathbb{E}_1(T)$ and V_T for $T > 0$ as in Subsection 2.3. Moreover, let $2 < q < 3$ and $h \in V_T$ with $h_0 := h(0)$. Then we set

$$Y_{\gamma,h_0} := \left\{ v \in L^p(\Omega) : v|_{\tilde{\Theta}_{h_0}(\Omega^\pm)} \in W_q^{2-\frac{2}{q}}(\tilde{\Theta}_{h_0}(\Omega^\pm)) \right\}^n \cap W^1_{q,0}(\Omega)^n, \quad Y_\gamma := Y_{\gamma,0}.$$

Now let $v_0 \in Y_{\gamma,h_0} \cap L^q_\sigma(\Omega)$. In order to write the Navier-Stokes part (2.4)-(2.9) as an abstract fixed-point equation, we rearrange the equations as follows:

$$\partial_t u - \mu^\pm \Delta u + \nabla \tilde{q} = a^\pm(h; D_x)(u, \tilde{q}) - u \cdot \nabla_h u + \nabla_h u\, \partial_t \tilde{\Theta}_h \quad \text{in } \Omega^\pm \times (0,T), \tag{3.1}$$

$$\operatorname{div} u = \operatorname{Tr}(\nabla u - \nabla_h u) =: g(h)u \quad \text{in } \Omega^\pm \times (0,T), \tag{3.2}$$

$$[\![T(u,\tilde{q})]\!]\, \nu_\Sigma = t^\pm(h; D_x)(u,\tilde{q}) - \sigma K(h)\nu_h \quad \text{on } \Sigma \times (0,T), \tag{3.3}$$

$$[\![u]\!] = 0 \quad \text{on } \Sigma \times (0,T), \tag{3.4}$$

$$u|_{\partial\Omega} = 0 \quad \text{on } \partial\Omega \times (0,T), \tag{3.5}$$

$$u|_{t=0} = v_0 \circ \tilde{\Theta}_{h_0} \quad \text{in } \Omega, \tag{3.6}$$

where $A(h) = D\tilde{\Theta}_h^{-\top}, \nabla_h = A(h)\nabla$ and $K(h), \nu_h$ are as in Subsection 2.4 as well as

$$a^\pm(h; D_x)(u,\tilde{q}) := \mu^\pm(\operatorname{div}_h \nabla_h u - \Delta u) + (\nabla - \nabla_h)\tilde{q},$$
$$t^\pm(h; D_x)(u,\tilde{q}) := [\![2\mu^\pm \operatorname{sym}(\nabla u) - \tilde{q} I]\!]\,(\nu_\Sigma - \nu_h) + [\![2\mu^\pm \operatorname{sym}(\nabla u - \nabla_h u)]\!]\, \nu_h.$$

We introduce the space $Y_{T,h_0,v_0} := \{(u,\tilde{q}) \in Y_T : u|_{t=0} = v_0 \circ \tilde{\Theta}_{h_0}\}$ for the transformed velocity and pressure $(u,\tilde{q})$, where $Y_T := Y_T^1 \times Y_T^2$ and with $\Omega_0 := \Omega^+ \cup \Omega^-$ we define

$$\begin{aligned} Y_T^1 &:= W_q^1(0,T; L^q(\Omega))^n \cap L^q(0,T; W_q^2(\Omega_0) \cap W_{q,0}^1(\Omega))^n, \\ Y_T^2 &:= \{\tilde{q} \in L^q(0,T; W^1_{q,(0)}(\Omega_0)) : [\![\tilde{q}]\!] \in W_q^{1-\frac{1}{q},\frac{1}{2}(1-\frac{1}{q})}(\Sigma_T)\}. \end{aligned} \tag{3.7}$$

Then the existence result for the Navier-Stokes part reads as follows:

THEOREM 3.1. *Let $p > n+2$ and $2 < q < 3$ with $1 + \frac{n+2}{p} > \frac{n+2}{q}$. Then for $R > 0$ there are $0 < \varepsilon = \varepsilon(R) < R_0$ and $T_0 = T_0(R) > 0$ such that for all $0 < T \le T_0$ and*

$$h \in V_{T,\varepsilon} := \{h \in V_T : \|h(0)\|_{X_\gamma} \le \varepsilon\}, \qquad v_0 \in Y_{\gamma,h(0)} \cap L^q_\sigma(\Omega) \text{ with } \|v_0\|_{Y_{\gamma,h(0)}} \le R$$

the Navier-Stokes part (3.1)-(3.6) has a unique solution $F_T(h,v_0) := (u,\tilde{q})(h)$ in $Y_{T,h(0),v_0}$ and it holds that $\|F_T(h,v_0)\|_{Y_T} \le C_R$ and $F_T(h,v_0)|_{[0,\tilde{T}]} = F_{\tilde{T}}(h|_{[0,\tilde{T}]}, v_0)$ for all $0 < \tilde{T} \le T$.

Additionally, for $h_j \in V_{T,\varepsilon}$ with $h_0^j := h_j(0)$ as well as $v_0^j \in Y_{\gamma,h_j(0)} \cap L^q_\sigma(\Omega)$ with $\|v_0^j\|_{Y_{\gamma,h_j(0)}} \le R$ and $\tilde{v}_0^j := v_0^j \circ \tilde{\Theta}_{h_0^j}$ for $j = 1,2$ we have the estimate

$$\|F_T(h_1,v_0^1) - F_T(h_2,v_0^2)\|_{Y_T} \le \tilde{C}_R(\|h_1 - h_2\|_{\mathbb{E}_1(T)} + \|h_0^1 - h_0^2\|_{X_\gamma} + \|\tilde{v}_0^1 - \tilde{v}_0^2\|_{Y_\gamma}).$$

For the proof we reformulate (3.1)-(3.6) as an abstract fixed-point equation

$$w = L^{-1}G(w; h, v_0) \quad \text{for } w \in Y_{T,h(0),v_0}, \tag{3.8}$$

where for $w = (u,\tilde{q}) \in Y_T$ we define

$$Lw := \begin{pmatrix} \partial_t u - \mu^\pm \Delta u + \nabla \tilde{q} \\ \operatorname{div} u \\ [\![T(u,\tilde{q})]\!]\, \nu_\Sigma \\ u|_{t=0} \end{pmatrix}, G(w; h, v_0) := \begin{pmatrix} a^\pm(h; D_x)w - u \cdot \nabla_h u + \nabla_h u\, \partial_t \tilde{\Theta}_h \\ g(h)u - \frac{1}{|\Omega|}\int_\Omega g(h)u\, dx \\ t^\pm(h; D_x)w - \sigma K(h)\nu_h \\ v_0 \circ \tilde{\Theta}_{h(0)} \end{pmatrix}.$$

Similar as in A. and Wilke [**3**], Appendix A for $2 \le q < 3$ one can show with the aid of Theorem 2 in Shimizu [**24**][2] that $L : Y_T \to Z_T$ is an isomorphism and that

[2]The results in Shimizu [**24**] were only announced without proof. But the main step is to consider model problems. For $\mathbb{R}^n$ with $\mathbb{R}^{n-1}$ as interface see Shibata and Shimizu [**23**].

for $T_0 > 0$ the operator norm of L^{-1} is bounded uniformly in $0 < T \le T_0$, where

$$Z_T := \{(f, g, a, u_0) \in \tilde{Z}_T \ : \ \operatorname{div} u_0 = g|_{t=0} \text{ in } W^{-1}_{q,(0)}(\Omega)\}$$

with $\tilde{Z}_T := \tilde{Z}_T^1 \times \tilde{Z}_T^2 \times (\tilde{Z}_T^3)^n \times Y_\gamma$ and the spaces

$$\tilde{Z}_T^1 := L^q(0,T; L^q(\Omega))^n, \quad \tilde{Z}_T^3 := W_q^{1-\frac{1}{q},\frac{1}{2}(1-\frac{1}{q})}(\Sigma_T),$$
$$\tilde{Z}_T^2 := \{g \in L^q(0,T; W^1_{q,(0)}(\Omega_0)) \cap W^1_q(0,T; W^{-1}_{q,(0)}(\Omega)) : \operatorname{tr}_\Sigma(g|_{\Omega^+}) \in \tilde{Z}_T^3\}.$$

Here Y_T, Z_T and $\tilde{Z}_T$ are Banach spaces with canonical norms. For the details see Appendix B in the preprint [**2**].

At this point we see why we have to apply[3] the fixed-point argument in the space Y_{T,h_0,v_0} which depends on h_0 and v_0. Without the additional condition for $u|_{t=0}$ in $Y_{T,h(0),v_0}$ the compatibility condition for $G(w;h,v_0)$ in Z_T would not be fulfilled in general. But it is essential that $G(.;h,v_0)$ maps to Z_T so that we can apply L^{-1}. For this reason we also subtracted the mean value in the second component of $G(w;h,v_0)$. In fact, (3.1)-(3.6) and the formulation (3.8) are equivalent because of

REMARK 3.2. Let $2 < q < 3, u \in W^1_{q,0}(\Omega)^n$ and $h \in U_0$ with U_0 as in Remark 2.5. Then

$$\operatorname{div} u = g(h)u - \frac{1}{|\Omega|}\int_\Omega g(h)u \, dx \quad \Leftrightarrow \quad \operatorname{div} u = g(h)u.$$

PROOF. The second equation is equivalent to $\operatorname{div}_h u = 0$. For $u \in W^1_{q,0}(\Omega)^n$ holds $\int_\Omega \operatorname{div} u \, dx = 0$ and this shows the reverse direction. The other part follows as in A. and Wilke [**3**], p. 51. □

The crucial step for showing Theorem 3.1 is to prove the following lemma which states the properties of the right hand side G.

LEMMA 3.3. *Let $p > n+2$ and $2 < q < 3$ with $1 + \frac{n+2}{p} > \frac{n+2}{q}$ as well as $0 < T \le 1$. Then for all $w \in Y_T$, $h \in V_T$ and $v_0 \in Y_{\gamma,h(0)} \cap L^q_\sigma(\Omega)$ we have that $G(w;h,v_0)$ is contained in $\tilde{Z}_T$ and for $w \in Y_{T,h(0),v_0}$ it holds $G(w;h,v_0) \in Z_T$. If $\|v_0\|_{Y_{\gamma,h(0)}}$ is bounded by a constant $R > 0$, then $\|G(0;h,v_0)\|_{\tilde{Z}_T} \le C_R$.*

Moreover, let $R_1, R_2 > 0$. Then for $0 < \varepsilon < R_0$ and $h_j \in V_{T,\varepsilon}$ with $h_0^j := h_j(0)$, $v_0^j \in Y_{\gamma,h_j(0)} \cap L^q_\sigma(\Omega)$ and $\tilde{v}_0^j := v_0^j \circ \tilde{\Theta}_{h_j(0)}$ as well as $w_j = (u_j, \tilde{q}_j) \in Y_T$ with $u_0^j := u_j(0)$, $\|u_0^j\|_{Y_\gamma} \le R_1$ and $\|w_j\|_{Y_T} \le R_2$ for $j = 1,2$ the following estimate holds:

$$\|G(w_1;h_1,v_0^1) - G(w_2;h_2,v_0^2)\|_{\tilde{Z}_T} \le C_{R_1,R_2}(T^\delta + \varepsilon)(\|w_1 - w_2\|_{Y_T} + \|u_0^1 - u_0^2\|_{Y_\gamma}) +$$
$$+ C_{R_1,R_2}(\|h_1 - h_2\|_{\mathbb{E}_1(T)} + \|h_0^1 - h_0^2\|_{X_\gamma}) + \|\tilde{v}_0^1 - \tilde{v}_0^2\|_{Y_\gamma},$$

where $\delta := \min\{\frac{\tau}{2}, \frac{1}{q} - \frac{1}{p}\} > 0$ and $\tau > 0$ is as in Lemma 2.6.

Also as preparation for the proof of Lemma 3.3, in the following we discuss where the conditions for the exponent q come from:

[3] A. and Wilke [**3**] transform with $\tilde{\Theta}_h \circ \tilde{\Theta}_{h_0}^{-1}$. Then for fixed h_0 one could omit the dependence on v_0. But here we have chosen the other strategy for the following reasons: On one hand it is difficult to obtain suitable properties of $\tilde{\Theta}_{h_0}^{-1}$ and on the other hand one needs a priori $C^{2,1}$-regularity for the interface to have the isomorphism result for the linearized Stokes system, see Shimizu [**24**].

REMARK 3.4. First of all, we have the restriction $2 \leq q < 3$ because of the isomorphism result for the linearized Stokes system. Theorem 3.1 gives us the velocity field $u(h) \in Y_T^1$ in dependence of the height function h. Later we will insert it in equation (2.10) which should hold in $L^p(0,T;X_0)$. In order to show local well-posedness afterwards, we need enough regularity for $\operatorname{tr}_\Sigma u$. This means

$$\operatorname{tr}_\Sigma u \in L^s(0,T;X_0)^n \quad \text{for some } s > p. \tag{3.9}$$

Since $Y_\gamma = \{v \in L^p(\Omega) : v|_{\Omega^\pm} \in W_q^{2-\frac{2}{q}}(\Omega^\pm)\}^n \cap W_{q,0}^1(\Omega)^n$, we get from Lemma A.5 below that $Y_T^1 \hookrightarrow C^0([0,T];Y_\gamma) \cap L^q(0,T;W_q^2(\Omega_0)^n)$ and the embedding constant is bounded independently of $T > 0$ if we add the term $\|u(0)\|_{Y_\gamma}$ in the Y_T^1-norm. Now, for $\theta \in (0,1)$ we use the reiteration theorem in Lunardi [**21**], Corollary 1.24 and Lemma A.2 to obtain

$$Y_T^1|_{\Omega^\pm} \hookrightarrow L^s(0,T;B_{q,q}^{2-2(1-\theta)\frac{1}{q}}(\Omega^\pm)^n) \quad \text{with } s = \frac{q}{\theta}.$$

To achieve (3.9) we want to choose $\theta \in (0,1)$ such that $\frac{q}{\theta} = s > p$ and that the Besov-space embeds into $W_p^1(\Omega^\pm)^n$. Because of $q \leq p$ the latter is fulfilled for

$$2 - (1-\theta)\frac{2}{q} - \frac{n}{q} > 1 - \frac{n}{p} \quad \Leftrightarrow \quad \theta > -\frac{q}{2} - \frac{nq}{2p} + \frac{n+2}{2}.$$

Hence (3.9) holds by the trace theorem if

$$\frac{q}{p} > -\frac{q}{2} - \frac{nq}{2p} + \frac{n+2}{2} \quad \Leftrightarrow \quad \frac{q}{2} > -q\frac{n+2}{2p} + \frac{n+2}{2} \quad \Leftrightarrow \quad 1 > \frac{n+2}{q} - \frac{n+2}{p}.$$

Since $p > n+2$ necessary conditions on q are $q > 2$, if $n = 2$, and $q > \frac{5}{2}$, if $n = 3$. So we directly wrote $q > 2$ instead of $q \geq 2$. On the other hand the conditions are fulfilled for many p, q.

Another point is that we have to estimate $u \cdot \nabla_h u$ in the proof of Lemma 3.3 and therefore want to use the embedding $Y_T^1 \hookrightarrow L^r(0,T;C^0(\overline{\Omega}))^n$ for some $r > q$. Embeddings over $\Omega^\pm$ yield $Y_\gamma \hookrightarrow C^0(\overline{\Omega})^n$ for $q > 2$, if $n = 2$ and $q > \frac{5}{2}$, if $n = 3$, since then $2 - \frac{2+n}{q} > 0$ holds. Hence from the above embedding for Y_T we even obtain $Y_T^1 \hookrightarrow C^0([0,T];C^0(\overline{\Omega})^n)$ and the embedding constant is bounded independently of $T > 0$ if we add $\|u(0)\|_{Y_\gamma}$ in the Y_T^1-norm.

Altogether, for simplicity we restrict ourselves in Theorem 3.1 and Lemma 3.3 to exponents $p > n+2$ and $2 < q < 3$ with $1 + \frac{n+2}{p} > \frac{n+2}{q}$.

PROOF OF LEMMA 3.3. For simplicity we consider $0 < T \leq 1$. Properties of $\tilde{\Theta}_h$ and $A(h)$ were shown in Lemma 2.6. The mean value theorem implies for all $h \in V_{T,\varepsilon}$ and $t \in [0,T]$

$$\begin{aligned} &\|A(h)|_t - I\|_{C^1(\overline{\Omega})^{n\times n}} \\ &\leq \|A(h)|_t - A(h)|_0\|_{C^1(\overline{\Omega})^{n\times n}} + \|A(h)|_0 - I\|_{C^1(\overline{\Omega})^{n\times n}} \\ &\leq CT^\tau + c\|h(0)\|_{X_\gamma} \leq C(T^\tau + \varepsilon). \end{aligned} \tag{3.10}$$

Using the latter inequality many of the assertions can be directly shown. The corresponding difference estimate always follows using null additions. See preprint [**2**] for more details. For the first component of G one additionally uses the embedding $Y_T^1 \hookrightarrow C^0([0,T];Y_\gamma) \hookrightarrow C^0([0,T];C^0(\overline{\Omega}))^n$ from Remark 3.4 to treat the convection term and together with $W_p^1(\Omega) \hookrightarrow C^0(\overline{\Omega})$ for $p > n+2$ to get $\nabla_h u\, \partial_t \tilde{\Theta}_h \in \tilde{Z}_T^1$ and the difference estimates.

Now we consider the second component of G. One can directly show $G(w;h,v_0)_2$ in $L^q(0,T;W^1_{q,(0)}(\Omega_0))$. Apart from that $G(w;h,v_0)_2 = g(h)u$ in $L^q(0,T;W^{-1}_{q,(0)}(\Omega))$ since constants drop out when integrating with $\phi \in W^1_{q',(0)}(\Omega)$. In this space we need to show the existence of a weak derivative. To this end let $\eta \in C_0^\infty(0,T)$. We approximate u and $A(h)$ suitably. Using convolution one can show that there are $(u^m)_{m\in\mathbb{N}}$ in $C^\infty([0,T];W^2_q(\Omega_0) \cap W^1_{q,0}(\Omega))^n$ and $(A^m)_{m\in\mathbb{N}}$ in $C^\infty([0,T];C^1(\overline{\Omega})^{n\times n})$ with $u^m \to u$ in Y^1_T and $A^m \to A(h)$ in the space $C^0([0,T];C^1(\overline{\Omega})^{n\times n}) \cap W^1_p(0,T;L^p(\Omega)^{n\times n})$ for $m\to\infty$. Now the product rule yields

$$C^\infty([0,T];W^1_q(\Omega_0)) \ni g^m := (I-A^m) : \nabla u^m \overset{m\to\infty}{\longrightarrow} g(h)u \quad \text{in } L^q(0,T;W^1_q(\Omega_0)).$$

Choose $\phi \in W^1_{q',(0)}(\Omega)$ arbitrary. Then for $m\in\mathbb{N}$ we obtain

$$-\int_0^T \partial_t\eta(t)\int_\Omega g^m(t)\phi\,dx\,dt = \int_0^T \eta(t)\int_\Omega \partial_t g^m(t)\phi\,dx\,dt. \tag{3.11}$$

Here $\partial_t g^m = -\partial_t A^m : \nabla u^m + (I-A^m) : \nabla\partial_t u^m$ and therefore

$$\begin{aligned}\int_\Omega &\partial_t g^m \phi\,dx \\ &= \int_\Omega -\partial_t A^m : \nabla u^m\,\phi + \partial_t u^m \cdot \operatorname{div}A^m\,\phi - ((I-A^m)\partial_t u^m)\cdot\nabla\phi\,dx,\end{aligned} \tag{3.12}$$

where we used integration by parts for the second term. In this equation we want to pass to the limit. Lemma A.5 implies $Y^1_T \hookrightarrow C^0([0,T];Y_\gamma)$ and the embedding constant is bounded independently of $T>0$ if we add $\|u(0)\|_{Y_\gamma}$ in the Y^1_T-norm. This yields

$$\partial_t A^m : \nabla u^m \overset{m\to\infty}{\longrightarrow} \partial_t A(h) : \nabla u \quad \text{in } L^p(0,T;L^{r_1}(\Omega)) \quad \text{with } \frac{1}{r_1} := \frac{1}{p}+\frac{1}{q} \in (0,1).$$

Furthermore we have $\partial_t u^m \cdot \operatorname{div}A^m \overset{m\to\infty}{\longrightarrow} \partial_t u \cdot \operatorname{div} A$ and $(I-A^m)\partial_t u^m \overset{m\to\infty}{\longrightarrow} (I-A(h))\partial_t u$ in $L^q(0,T;L^q(\Omega))$ (vector-valued for the second one). Now it holds $W^1_{q'}(\Omega) \hookrightarrow L^{r_2}(\Omega)$ for $n=2,3$ with $\frac{1}{r_2} := \frac{1}{q'} - \frac{1}{n} \in (0,1)$ and because of

$$\frac{1}{r_1}+\frac{1}{r_2} = 1+\frac{1}{p}-\frac{1}{n} \in (0,1) \quad \text{for } p>n+2, n=2,3$$

the mapping $L^{r_1}(\Omega) \to W^{-1}_{q,(0)}(\Omega) : g \mapsto \langle g,.\rangle$ is bounded and linear. This is also true for

$$L^q(\Omega) \to W^{-1}_{q,(0)}(\Omega) : g \mapsto \langle g,.\rangle \quad \text{and} \quad L^q(\Omega)^n \to W^{-1}_{q,(0)}(\Omega) : g \mapsto [\phi \mapsto \int_\Omega g\cdot\nabla\phi\,dx].$$

Since $\langle .,\phi\rangle : W^{-1}_{q,(0)}(\Omega) \to \mathbb{R}$ is bounded and linear for $\phi \in W^1_{q',(0)}(\Omega)$ we can pass to the limit in (3.11)-(3.12) for fixed ϕ and then we put outside the duality product. Hence $g(h)u \in W^1_q(0,T;W^{-1}_{q,(0)}(\Omega))$ and the derivative is given by

$$\langle \partial_t(g(h)u),\phi\rangle = \int_\Omega (-\partial_t A(h) : \nabla u + \partial_t u \cdot \operatorname{div}A(h))\phi - ((I-A(h))\partial_t u)\cdot\nabla\phi\,dx.$$

For the difference of the derivatives the above observations together with null additions and $\operatorname{div} I = 0$ yield

$$\|\partial_t(g(h_1)u_1) - \partial_t(g(h_2)u_2)\|_{L^q(0,T;W^{-1}_{q,(0)}(\Omega))} \leq$$
$$\leq CT^{\frac{1}{q}-\frac{1}{p}}(C_{R_1,R_2}(\|h_1 - h_2\|_{\mathbb{E}_1(T)} + \|h_0^1 - h_0^2\|_{X_\gamma}) + \|u_1 - u_2\|_{Y_T^1} + \|u_0^1 - u_0^2\|_{Y_\gamma})+$$
$$+ C_{R_2}(\|h_1 - h_2\|_{\mathbb{E}_1(T)} + \|h_0^1 - h_0^2\|_{X_\gamma}) + C(T^\tau + \varepsilon)\|u_1 - u_2\|_{Y_T^1}.$$

Now we show $\operatorname{tr}_\Sigma(G(w; h, v_0)_2|_{\Omega^+}) \in \tilde{Z}_T^3$ and the related difference estimate. To this end let $\operatorname{ext}_T \in \mathcal{L}(W_q^{2,1}(\Omega_T^+), W_q^{2,1}(\Omega^+ \times \mathbb{R}_+))$ for $T > 0$ be extension operators in time such that

$$\|\operatorname{ext}_T(u)\|_{W_q^{2,1}(\Omega^+ \times \mathbb{R}_+)} \leq M(\|u\|_{W_q^{2,1}(\Omega_T^+)} + \|u(0)\|_{W_q^{2-\frac{2}{q}}(\Omega^+)})$$

holds for some $M > 0$. To construct such operators one can e.g. first extend suitably from Ω to $\mathbb{R}^n$ and then proceed as in Amann [**5**], Lemma 7.2 utilizing that the Laplace-operator has maximal L^p-regularity, cf. Prüss and Simonett [**22**], Theorem 6.1.8. Now we can extend $\operatorname{ext}_T u$ by reflection to a $\tilde{u}$ defined on $\Omega^+ \times \mathbb{R}$. From Grubb [**14**], (A.8)-(A.12) we know that $\nabla \tilde{u} \in H_q^{1,\frac{1}{2}}(\Omega^+ \times \mathbb{R})^{n\times n} \hookrightarrow W_q^{1,\frac{1}{2}}(\Omega^+ \times \mathbb{R})^{n\times n}$, where the latter embedding holds because of $q \in (2,3)$. Moreover, we have

$$\|\nabla \tilde{u}\|_{W_q^{1,\frac{1}{2}}(\Omega^+ \times \mathbb{R})^{n\times n}} \leq C\|\tilde{u}\|_{W_q^{2,1}(\Omega^+ \times \mathbb{R})^n} \leq \tilde{C}(\|u\|_{Y_T^1} + \|u(0)\|_{Y_\gamma}).$$

We decompose $A(h) = \tilde{A}(h) + A(h(0))$ with $\tilde{A}(h) := A(h) - A(h(0))$ and extend $\tilde{A}(h)$ by reflection at T to $[0, 2T]$ and then by 0 to all of $\mathbb{R}_+$. Then $\tilde{A}(h)$ is contained in $C^{\frac{1}{2}+\tau}(\mathbb{R}_+; C^0(\overline{\Omega})^{n\times n}) \cap C^\tau(\mathbb{R}_+; C^1(\overline{\Omega})^{n\times n})$ and the norm is estimated by the respective one of $A(h) - A(h(0))$. Since $\tilde{A}(h) = 0$ on $(2T, \infty)$, we can apply Lemma A.1 to $I = (0, 2T+1)$ and $\tilde{A}(h)|_{\Omega^+}$. Afterwards we extend by 0 and obtain

$$(I - A(h(0)) - \tilde{A}(h))|_{\Omega^+} : \nabla \tilde{u} \in W_q^{1,\frac{1}{2}}(\Omega^+ \times \mathbb{R}_+).$$

This then also holds for the mean value and e.g. Denk, Hieber and Prüss [**7**], Lemma 3.5 implies $\operatorname{tr}_\Sigma(G(w; h, v_0)_2|_{\Omega^+}) \in \tilde{Z}_T^3$. Furthermore, a null addition yields

$$\|\operatorname{tr}_\Sigma(G(w_1; h_1, v_0^1)_2|_{\Omega^+}) - \operatorname{tr}_\Sigma(G(w_2; h_2, v_0^2)_2|_{\Omega^+})\|_{\tilde{Z}_T^3} \leq$$
$$\leq C(\|I - A(h_0^1)\|_{C^1(\overline{\Omega})} + \|\tilde{A}(h_1)\|_{C^{\frac{1+\tau}{2}}(\mathbb{R}_+; C^0(\overline{\Omega})) \cap C^0(\mathbb{R}_+; C^1(\overline{\Omega}))}) \cdot \|\tilde{u}_1 - \tilde{u}_2\|_{W_q^{2,1}}+$$
$$+ \|A(h_1) - A(h_2)\|_{C^{\frac{1}{2}+\tau}([0,T]; C^0(\overline{\Omega})) \cap C^0([0,T]; C^1(\overline{\Omega}))^{n\times n}} C_{R_1,R_2} \leq C(T^{\tau/2} + \varepsilon)\cdot$$
$$\cdot (\|u_1 - u_2\|_{Y_T^1} + \|u_0^1 - u_0^2\|_{Y_\gamma}) + C_{R_1,R_2}(\|h_1 - h_2\|_{\mathbb{E}_1(T)} + \|h_0^1 - h_0^2\|_{X_\gamma}),$$

where we used Lemma 2.6, (3.10) and $T \leq 1$.

Next we study the third component of G. Remark 2.7 yields suitable properties of $\nu_.$ and similar to (3.10) we obtain $\|\nu_\Sigma - \nu_h\|_{C^0([0,T]; C^1(\Sigma)^n)} \leq C(T^\tau + \varepsilon)$. Invoking Lemma 2.9 and Remark 2.7 together with Lemma A.1 and the product rule, we obtain $\sigma K(.)\nu_. \in BC^1(V_T; (\tilde{Z}_T^3)^n)$ where $\sigma K(.)\nu_.$ and the derivative is bounded independently of $0 < T \leq 1$ if we add $\|h(0)\|_{X_\gamma}$ in the $\mathbb{E}_1(T)$-norm. In particular we get

$$\|\sigma K(h_1)\nu_{h_1} - \sigma K(h_2)\nu_{h_2}\|_{(\tilde{Z}_T^3)^n} \leq C(\|h_1 - h_2\|_{\mathbb{E}_1(T)} + \|h_0^1 - h_0^2\|_{X_\gamma}).$$

Now consider $t^{\pm}(h; D_x)w = [\![2\mu^{\pm}\mathrm{sym}(\nabla u) - \tilde{q}I]\!]\,(\nu_\Sigma - \nu_h) + [\![2\mu^{\pm}\mathrm{sym}(\nabla u - \nabla_h u)]\!]\,\nu_h$. By definition we have $[\![\tilde{q}]\!] \in \tilde{Z}_T^3$. Similar as before one can show that the jump terms are in $(\tilde{Z}_T^3)^{n\times n}$ together with suitable estimates. Then we use Lemma A.1.

For the fourth component we use that $\tilde{\Theta}_{h(0)} : \Omega \to \Omega$ is a C^2-diffeomorphism and the identity in a neighbourhood of $\partial\Omega$ by Lemma 2.4 and Remark 2.5. Moreover, because of Lemma 2.6 we have that $|\det A(h(0))|$ and the $C^2(\overline{\Omega})$-norm of $\tilde{\Theta}_{h(0)}$ are bounded by a constant independent of $h \in V_T$. Therefore for $k = 0, 1, 2$ we have that $W_q^k(\tilde{\Theta}_{h(0)}(\Omega^{\pm}))^n \to W_q^k(\Omega^{\pm})^n : v \mapsto v \circ \tilde{\Theta}_{h(0)}$ is bounded, linear and the operator norm is bounded independently of $h \in V_T$. The latter follows from Adams and Fournier [**4**], Theorem 3.41 and its proof. By interpolation this also holds for $Y_{\gamma,h(0)} \to Y_\gamma : v \mapsto v \circ \tilde{\Theta}_{h(0)}$. We obtain $G(w; h, v_0)_4 = v_0 \circ \tilde{\Theta}_{h(0)} \in Y_\gamma$ and, if $\|v_0\|_{Y_{\gamma,h(0)}} \leq R$ for some $R > 0$, we have $\|v_0 \circ \tilde{\Theta}_{h(0)}\|_{Y_\gamma} \leq C_R$. The estimate for the difference of $\tilde{v}_0^j$ is trivial.

Altogether we have proven $G(w; h, v_0) \in \tilde{Z}_T$ and the difference estimate in the lemma. Additionally $G(0; h, v_0) = (0,\, 0,\, -\sigma K(h)\nu_h,\, v_0 \circ \tilde{\Theta}_{h(0)})^\top$ and the above implies $\|G(0; h, v_0)\|_{\tilde{Z}_T} \leq C_R$. It remains to show for $w \in Y_{T,h(0),v_0}$ the compatibility condition for $G(w; h, v_0)$ in Z_T which is equivalent to $\operatorname{div}(u(0)) = (g(h)u)|_{t=0}$ in $W^{-1}_{q,(0)}(\Omega)$ since constants drop out when integrating with $\phi \in W^1_{q',(0)}(\Omega)$. Lemma A.5 implies that $Y_T^1 \hookrightarrow C^0([0,T]; Y_\gamma)$ and therefore

$$g(h)u = (I - A(h)) : \nabla u \in C^0([0,T]; L^q(\Omega)).$$

Since $L^q(\Omega) \to W^{-1}_{q,(0)}(\Omega) : g \mapsto \langle g, .\rangle$ is bounded and linear, it suffices to show $\operatorname{div} u(0) = g(h(0))u(0)$ or, equivalently, $\operatorname{div}_{h(0)} u(0) = 0$ in $L^q(\Omega)$. But by chain rule we know $(\operatorname{div} v_0) \circ \tilde{\Theta}_{h(0)} = \operatorname{div}_{h(0)}(u(0))$. Since v_0 is divergence-free, the claim follows. □

Now we can show the existence result for the Navier-Stokes part.

Proof of Theorem 3.1. Let p, q be as in the theorem and $R > 0$ be arbitrary. Then for $h \in V_T$ and $v_0 \in Y_{\gamma,h(0)} \cap L^q_\sigma(\Omega)$ with $\|v_0\|_{Y_{\gamma,h(0)}} \leq R$ Lemma 3.3 yields

$$\|v_0 \circ \tilde{\Theta}_{h(0)}\|_{Y_\gamma} \leq \|G(0; h, v_0)\|_{\tilde{Z}_T} \leq C_R. \tag{3.13}$$

We set $C_L := \sup_{0<T\leq 1} \|L^{-1}\|_{\mathcal{L}(Z_T, Y_T)}$ and choose $R_1 := C_R$ and $R_2 := 2C_R C_L$ in Lemma 3.3. Then (3.13) implies for $0 < \varepsilon < R_0$ and $0 < T \leq 1$ as well as $h \in V_{T,\varepsilon}$ and $w \in Y_{T,h(0),v_0}$ with $\|w\|_{Y_T} \leq R_2$ the estimate

$$\begin{aligned}\|L^{-1}G(w; h, v_0)\|_{Y_T} &\leq \|L^{-1}\|_{\mathcal{L}(Z_T,Y_T)}(\|G(w; h, v_0) - G(0; h, v_0)\|_{\tilde{Z}_T} + \\ &+ \|G(0; h, v_0)\|_{\tilde{Z}_T}) \leq C_L(C_R + C_{R_1,R_2}(T^\delta + \varepsilon)(\|w\|_{Y_T} + \|v_0 \circ \tilde{\Theta}_{h(0)}\|_{Y_\gamma}))\end{aligned}$$

where $\delta > 0$. For $T_0 = T_0(R), \varepsilon = \varepsilon(R) > 0$ small we obtain from (3.13)

$$\|L^{-1}G(w; h, v_0)\|_{Y_T} \leq \frac{1}{2}R_2 + \tilde{C}_R(T^\delta + \varepsilon) \leq \frac{2}{3}R_2 \quad \text{for all } 0 < T \leq T_0. \tag{3.14}$$

Therefore $\|L^{-1}G(w_1; h, v_0) - L^{-1}G(w_2; h, v_0)\|_{Y_T} \leq C_L C_{R_1,R_2}(T_0^\delta + \varepsilon)\|w_1 - w_2\|_{Y_T}$ because of Lemma 3.3 for all $w_j \in Y_{T,h(0),v_0}$ with $\|w_j\|_{Y_T} \leq R_2$ for $j = 1, 2$. We choose $\varepsilon = \varepsilon(R) > 0$ and $T_0 = T_0(R) > 0$ small such that

$$C_L C_{R_1,R_2}(T_0^\delta + \varepsilon) \leq \frac{1}{2} \tag{3.15}$$

holds. Then $L^{-1}G(.;h,v_0) : \overline{B_{Y_{T,h(0),v_0}}(0,R_2)} \to \overline{B_{Y_{T,h(0),v_0}}(0,R_2)}$ is well-defined for all $0 < T \leq T_0$ and a strict contraction. Here $Y_{T,h(0),v_0}$ is a complete metric space as closed subset of a Banach space. Banach's fixed-point theorem yields a unique fixed-point $F_T(h,v_0)$ in $\overline{B_{Y_{T,h(0),v_0}}(0,R_2)}$ and it holds

$$F_T(h,v_0)|_{[0,\tilde{T}]} = F_{\tilde{T}}(h|_{[0,\tilde{T}]}, v_0) \qquad \text{for all } 0 < \tilde{T} \leq T$$

since $h|_{[0,\tilde{T}]} \in V_{\tilde{T},\varepsilon}$ as well as $F_T(h,v_0)|_{[0,\tilde{T}]} \in \overline{B_{Y_{\tilde{T},h(0),v_0}}(0,R_2)}$ and $L^{-1}G(.;h,v_0)$ is compatible with restrictions in time on $[0,\tilde{T}]$ for $0 < \tilde{T} \leq T$. The uniqueness in $Y_{T,h(0),v_0}$ can be shown with the compatibility for restrictions in time and (3.14). The difference estimate follows from Lemma 3.3 and (3.15). □

As preparation for the proof of the local well-posedness for (2.4)-(2.11) we show that the term coming from the Navier-Stokes part in (2.10) is sufficiently regular:

COROLLARY 3.5. *Let p, q and ε, T_0 for $R > 0$ be as in Theorem 3.1. For $0 < T \leq T_0, h \in V_{T,\varepsilon}$ and $v_0 \in Y_{\gamma,h(0)} \cap L^q_\sigma(\Omega)$ with $\|v_0\|_{Y_{\gamma,h(0)}} \leq R$ for $j = 1,2$ let*

$$G_T(h,v_0) := a_1(h) \cdot \mathrm{tr}_\Sigma(F_T(h,v_0)_1),$$

where $a_1(h)$ is defined in Subsection 2.4 and $F_T(h,v_0)$ stems from Theorem 3.1. Then for some $s > p$ it holds $G_T(h,v_0) \in L^s(0,T;X_0)$ and $\|G_T(h,v_0)\|_{L^s(0,T;X_0)} \leq C(R)$ as well as

$$G_T(h,v_0)|_{[0,\tilde{T}]} = G_{\tilde{T}}(h|_{[0,\tilde{T}]}, v_0) \qquad \text{for all } 0 < \tilde{T} \leq T. \tag{3.16}$$

Moreover, for $h_j, h_0^j := h_j(0), v_0^j$ and $\tilde{v}_0^j := v_0^j \circ \tilde{\Theta}_{h_0^j}, j = 1,2$ as in Theorem 3.1

$$\|G_T(h_1,v_0^1) - G_T(h_2,v_0^2)\|_{L^s(0,T;X_0)} \leq C_R(\|h_1 - h_2\|_{\mathbb{E}_1(T)} + \|h_0^1 - h_0^2\|_{X_\gamma} + \|\tilde{v}_0^1 - \tilde{v}_0^2\|_{Y_\gamma}).$$

PROOF. From Remark 3.4 we know that $Y_T^1 \to L^s(0,T;X_0)^n : u \mapsto \mathrm{tr}_\Sigma u$ is bounded, linear for some $s > p$ and the operator norm is bounded independently of $T > 0$ if we add $\|u(0)\|_{Y_\gamma}$ in the Y_T^1-norm. Hence $\mathrm{tr}_\Sigma(F_T(h,v_0)_1) \in L^s(0,T;X_0)^n$ and from Theorem 3.1 and (3.13) we deduce $\|\mathrm{tr}_\Sigma(F_T(h,v_0)_1)\|_{L^s(0,T;X_0)^n} \leq C(R)$ as well as

$$\begin{aligned} &\|\mathrm{tr}_\Sigma(F_T(h_1,v_0^1)_1) - \mathrm{tr}_\Sigma(F_T(h_2,v_0^2)_1)\|_{L^s(0,T;X_0)^n} \leq \\ &\leq C_R(\|h_1 - h_2\|_{\mathbb{E}_1(T)} + \|h_0^1 - h_0^2\|_{X_\gamma} + \|\tilde{v}_0^1 - \tilde{v}_0^2\|_{Y_\gamma}). \end{aligned} \tag{3.17}$$

Since pointwise multiplication is a product on $C^1(\Sigma) \times X_0 \to X_0$, we obtain $G_T(h,v_0) \in L^s(0,T;X_0)$ with $\|G_T(h,v_0)\|_{L^s(0,T;X_0)} \leq \tilde{C}(R)$ by Remark 2.7 and (3.17) together with a null addition yields the estimate in the lemma. Equation (3.16) follows from the respective one for F_T in Theorem 3.1. □

4. Local well-posedness

In this section we show the local well-posedness for the transformed Navier-Stokes/mean curvature flow system (2.4)-(2.11). Let $p > n+2$ and $2 < q < 3$ with $1 + \frac{n+2}{p} > \frac{n+2}{q}$. For $R > 0$ let $\varepsilon, T_0 > 0$ be as in Theorem 3.1 and $0 < T \leq T_0$ as well as $h \in V_T, h_0 \in X_\gamma$ with $\|h_0\|_{X_\gamma} \leq \varepsilon$ and $v_0 \in Y_{\gamma,h(0)} \cap L^q_\sigma(\Omega)$ with $\|v_0\|_{Y_{\gamma,h(0)}} \leq R$ the system is equivalent to the abstract evolution equation

$$\partial_t h - G_T(h,v_0) = a_2(h)\,K(h) \qquad \text{on } \Sigma \times (0,T), \tag{4.1}$$

$$h|_{t=0} = h_0 \qquad \text{on } \Sigma, \tag{4.2}$$

where G_T is defined in Corollary 3.5. Here "equivalent" means that $(u, \tilde{q}, h) \in Y_T \times V_T$ solves (2.4)-(2.11) if and only if h fulfils (4.1)-(4.2) and it holds $(u, \tilde{q}) = F_T(h, v_0)$.

In order to show local well-posedness for (4.1)-(4.2), we use that by Lemma 2.8 the transformed mean curvature $K(h)$ has a quasilinear structure with respect to the height function h and for $h = 0$ the principal part is given by the Laplace-Beltrami operator Δ_Σ. Here $\Delta_\Sigma : X_1 \to X_0$ has maximal L^p-regularity on finite intervals because of Theorem 6.4.3 in Prüss and Simonett [**22**]. To give a simple proof for the local well-posedness, we proceed as in the case of quasilinear parabolic equations, compare e.g. Köhne, Prüss and Wilke [**17**], Theorem 2.1 (for $\mu = 1$ there). We especially cannot apply this result directly since G_T is a non-local operator. That means $G_T(h, v_0)$ at point $x_0 \in \Sigma$ and time $t_0 \in [0, T]$ in general depends on values $h(x, t)$ with $(x, t) \in \Sigma \times [0, T]$ outside a neighbourhood of (x_0, t_0). Nevertheless we can essentially adapt the proof of [**17**] (for $\mu = 1$) since by Corollary 3.5 we have that G_T is sufficiently regular and compatible with restrictions in time on $[0, \tilde{T}]$ for $0 < \tilde{T} \leq T$. Intuitively the velocity and pressure at time T depend on the evolution of the entire interface from the beginning up to time T.

The result for the local well-posedness of the transformed Navier-Stokes/mean curvature flow system (2.4)-(2.11) reads as follows:

THEOREM 4.1. *Let $p > n + 2$ and $2 < q < 3$ with $1 + \frac{n+2}{p} > \frac{n+2}{q}$. For $R > 0$ there exist $T_0 = T_0(R), \varepsilon_0 = \varepsilon_0(R) > 0$ such that for all $0 < T \leq T_0$ and $h_0 \in X_\gamma$ with $\|h_0\|_{X_\gamma} \leq \varepsilon_0$ and $v_0 \in Y_{\gamma,h_0} \cap L^q_\sigma(\Omega)$ with $\|v_0\|_{Y_{\gamma,h_0}} \leq R$ there is a unique solution $(u, \tilde{q}, h)(h_0, v_0) \in Y_T \times V_T$ of (2.4)-(2.11) and*

$$\|(u, \tilde{q}, h)(h_0^1, v_0^1) - (u, \tilde{q}, h)(h_0^2, v_0^2)\|_{Y_T \times V_T} \leq C_R(\|h_0^1 - h_0^2\|_{X_\gamma} + \|\tilde{v}_0^1 - \tilde{v}_0^2\|_{Y_\gamma})$$

holds for $h_0^j \in X_\gamma$ with $\|h_0^j\|_{X_\gamma} \leq \varepsilon_0$ and $v_0^j \in Y_{\gamma,h_j(0)} \cap L^q_\sigma(\Omega)$ with $\|v_0^j\|_{Y_{\gamma,h_j(0)}} \leq R$ as well as $\tilde{v}_0^j := v_0^j \circ \tilde{\Theta}_{h_0^j}$ for $j = 1, 2$.

PROOF. With P, Q as in Lemma 2.8 and U_0 as in Remark 2.5 we denote $\tilde{B}(h) := a_2(h)\, P(h)$ and $\tilde{F}(h) := a_2(h)\, Q(h)$ for all $h \in U_0$. Because of Theorem 6.4.3 in Prüss and Simonett [**22**] there is a $w > 0$ such that $\Delta_\Sigma - wI : X_1 \to X_0$ has maximal L^p-regularity on $\mathbb{R}_+$. Therefore we consider $B := \tilde{B} - wI$ and $F := \tilde{F} + wI$. Since $a_2(0) = 1$ we have $B(0) = \Delta_\Sigma - wI$. Moreover, Remark 2.7 yields $B \in C^1(U_0; \mathcal{L}(X_1, X_0))$ and $F \in C^1(U_0; X_0)$. The embedding $X_\gamma \hookrightarrow L^\infty(\Sigma)$ implies $\overline{B_{X_\gamma}(0, \delta_0)} \subseteq U_0$ for $0 < \delta_0 < R_0$ small. So if δ_0 is sufficiently small, by the mean value theorem, there is an $L > 0$ such that

$$\begin{aligned} \|B(h_1)v - B(h_2)v\|_{X_0} &\leq L\|h_1 - h_2\|_{X_\gamma}\|v\|_{X_1}, \\ \|F(h_1) - F(h_2)\|_{X_0} &\leq L\|h_1 - h_2\|_{X_\gamma} \end{aligned} \tag{4.3}$$

holds for all $h_1, h_2 \in \overline{B_{X_\gamma}(0, \delta_0)}$ and $v \in X_1$.

Now fix $R > 0$ and let $T_0 = T_0(R), \varepsilon = \varepsilon(R) > 0$ be as in Theorem 3.1. Then for $0 < T \leq T_0$ and $h \in V_T$ as well as $h_0 \in X_\gamma$ with $\|h_0\|_{X_\gamma} \leq \varepsilon$ and $v_0 \in Y_{\gamma,h_0} \cap L^q_\sigma(\Omega)$ with $\|v_0\|_{Y_{\gamma,h_0}} \leq R$ the system (2.4)-(2.11) or (4.1)-(4.2), respectively, are equivalent to

$$\partial_t h - B(h)h = F(h) + G_T(h, v_0) \quad \text{in } (0, T), \tag{4.4}$$

$$h(0) = h_0. \tag{4.5}$$

Here $B(h)$ and $F(h)$ are defined in the sense of $V_T \subseteq C^0([0,T];U_0)$ by Remark 2.5. In particular, it holds $B(h) \in C^0([0,T];\mathcal{L}(X_1,X_0))$.

The idea for solving (4.4)-(4.5) is as follows: we fix $B(h)$ at $h=0$ and apply Banach's fixed-point theorem to $H(.;h_0,v_0)$ where $H(h;h_0,v_0) := f$ is the unique solution of

$$\partial_t f - B(0)f = B(h)h - B(0)h + F(h) + G_T(h,v_0) \quad \text{in } (0,T), \tag{4.6}$$

$$f(0) = h_0. \tag{4.7}$$

The unique solvability of (4.6)-(4.7) will follow from Prüss and Simonett [**22**], Theorem 3.5.5 if we extend the right hand side by zero to $\mathbb{R}_+$. To this end we need suitable sets for the height function. For $0<\delta<\min\{\varepsilon,\delta_0\}, 0<r<R_0-\delta_0$ and $0<T\leq T_0$ we set

$$\mathbb{B}_{r,T,h_0} := \{h \in \mathbb{E}_1(T) : h(0)=h_0, \|h\|_{\mathbb{E}_1(T)} \leq r\} \quad \text{for } h_0 \in \overline{B_{X_\gamma}(0,\delta)}.$$

For $r,\delta>0$ small we show $\mathbb{B}_{r,T,h_0} \subseteq V_T \cap C^0([0,T];\overline{B_{X_\gamma}(0,\delta_0)})$. Therefore let $h \in \mathbb{B}_{r,T,h_0}$.

Because of $\|h\|_{\mathbb{E}_1(T)} + \|h_0\|_{X_\gamma} \leq r+\delta < R_0$ and our choice of δ_0 it suffices to show $\|h(t)\|_{X_\gamma} \leq \delta_0$ for all $t \in [0,T]$ if $r,\delta>0$ are small independently of h. By Prüss and Simonett [**22**], Theorem 3.5.5 there is a unique solution $h_0^* \in \mathbb{E}_1(\mathbb{R}_+)$ of

$$\partial_t h_0^* - B(0)h_0^* = 0, \quad h_0^*(0) = h_0. \tag{4.8}$$

Since $h(0)=h_0^*(0)$, we get from Theorem A.3 with a null addition

$$\|h\|_{C^0([0,T];X_\gamma)} \leq C(\|h\|_{\mathbb{E}_1(T)} + \|h_0\|_{X_\gamma}) \leq c(r+\delta). \tag{4.9}$$

For $r,\delta>0$ small we obtain $\|h(t)\|_{X_\gamma} \leq \delta_0$ for all $t\in[0,T]$ and thus the above inclusion.

Now we can define $H(.;h_0,v_0) \equiv H_{r,T}(.;h_0,v_0) : \mathbb{B}_{r,T,h_0} \to \mathbb{E}_1(T) : h \mapsto f$ where f is the unique solution to (4.6)-(4.7). Then $h \in \mathbb{B}_{r,T,h_0}$ is a solution of (4.4)-(4.5) if and only if h is a fixed-point of $H(.;h_0,v_0)$. To apply Banach's theorem we need the following:

For $r,\delta,T>0$ small, all $g_j \in \overline{B_{X_\gamma}(0,\delta)}$ and $h_j \in \mathbb{B}_{r,T,g_j}$ as well as v_0^j in $Y_{\gamma,g_j} \cap L^q_\sigma(\Omega)$ with $\|v_0^j\|_{Y_{\gamma,g_j}} \leq R$ and $\tilde{v}_0^j := v_0^j \circ \tilde{\Theta}_{g_j}$ for $j=1,2$ we will show the following inequality:

$$\|H(h_1;g_1,v_0^1) - H(h_2;g_2,v_0^2)\|_{\mathbb{E}_1(T)} \leq \frac{1}{2}\|h_1-h_2\|_{\mathbb{E}_1(T)} + c_R(\|g_1-g_2\|_{X_\gamma} + \|\tilde{v}_0^1 - \tilde{v}_0^2\|_{Y_\gamma}).$$

Moreover, for all $r>0$ there are $\delta(r),T(r)>0$ such that for all $0<\delta\leq\delta(r)$ and $0<T\leq T(r)$ the estimate $\|H(0;h_0,v_0)\|_{\mathbb{E}_1(T)} \leq \frac{r}{2}$ is valid, where $h_0 \in \overline{B_{X_\gamma}(0,\delta)}$.

Now let us prove this. From Prüss and Simonett [**22**], Theorem 3.5.5 we know that there exists a $C>0$ independent of $T>0$ such that

$$\begin{aligned}\|H(h_1;g_1,v_0^1) - H(h_2;g_2,v_0^2)\|_{\mathbb{E}_1(T)} &\leq C(\|g_1-g_2\|_{X_\gamma} + \|F(h_1)-F(h_2)\|_{\mathbb{E}_0(T)} + \\ &+\|(B(h_1)-B(0))(h_1-h_2) + (B(h_1)-B(h_2))h_2 + G_T(h_1,v_0^1) - G_T(h_2,v_0^2)\|_{\mathbb{E}_0(T)}).\end{aligned}$$

Here because of (4.3) it holds $\|F(h_1)-F(h_2)\|_{\mathbb{E}_0(T)} \leq LT^{\frac{1}{p}}\|h_1-h_2\|_{C^0([0,T];X_\gamma)}$ and with g_j^* as in (4.8) for $g_j, j=1,2$ instead of h_0 we conclude from Theorem A.3

$$\|h_1-h_2\|_{C^0([0,T];X_\gamma)} \leq c(\|h_1-h_2\|_{\mathbb{E}_1(T)} + \|g_1-g_2\|_{X_\gamma}). \tag{4.10}$$

Furthermore (4.3) together with (4.9) and (4.10) implies

$$\begin{aligned}&\|(B(h_1)-B(0))(h_1-h_2)+(B(h_1)-B(h_2))h_2\|_{\mathbb{E}_0(T)}\leq\\ &\leq L(\|h_1\|_{C^0([0,T];X_\gamma)}\|h_1-h_2\|_{\mathbb{E}_1(T)}+\|h_1-h_2\|_{C^0([0,T];X_\gamma)}\|h_2\|_{\mathbb{E}_1(T)})\leq\\ &\leq LC(r+\delta)\|h_1-h_2\|_{\mathbb{E}_1(T)}+Lcr(\|h_1-h_2\|_{\mathbb{E}_1(T)}+\|g_1-g_2\|_{X_\gamma}).\end{aligned}$$

Additionally, for some $s>p$ Corollary 3.5 and Hölder's inequality yield

$$\|G_T(h_1,v_0^1)-G_T(h_2,v_0^2)\|_{\mathbb{E}_0(T)}\leq C_R T^{\frac{1}{p}-\frac{1}{s}}(\|h_1-h_2\|_{\mathbb{E}_1(T)}+\|g_1-g_2\|_{X_\gamma}+\|\tilde{v}_0^1-\tilde{v}_0^2\|_{Y_\gamma}).$$

For $r,\delta,T>0$ small we obtain the claimed difference estimate. It remains to treat $\|H(0;h_0,v_0)\|_{\mathbb{E}_1(T)}$. Theorem 3.5.5 in Prüss and Simonett [**22**] and Corollary 3.5 imply

$$\|H(0;h_0,v_0)\|_{\mathbb{E}_1(T)}\leq C(\delta+\|F(0)\|_{\mathbb{E}_0(T)}+T^{\frac{1}{p}-\frac{1}{s}}C_R).$$

So if $\delta(r),T(r)>0$ are appropriate, we get $\|H(0;h_0,v_0)\|_{\mathbb{E}_1(T)}\leq\frac{r}{2}$ for $0<\delta\leq\delta(r)$ and $0<T\leq T(r)$. Altogether we have proven everything claimed above.

After this preparation we can choose $r,\delta,T_0>0$ small such that everything before holds for r,δ,T_0 and for $2r,\delta,T_0$. Then this also follows for $\tilde{r},\delta,T$ where $\tilde{r}\in\{r,2r\}$ and $0<T\leq T_0$. Now we choose $h_0=g_1=g_2\in\overline{B_{X_\gamma}(0,\delta)}$ and $v_0=v_0^1=v_0^2$ above and obtain that $H_{\tilde{r},T}(.;h_0,v_0)$ is a strict contraction on $\mathbb{B}_{\tilde{r},T,h_0}$ where the latter is a complete metric space as closed subset of a Banach space. Banach's theorem now yields a unique fixed-point $h^*_{\tilde{r},T}(h_0,v_0)\in\mathbb{B}_{\tilde{r},T,h_0}$ of $H_{\tilde{r},T}(.;h_0,v_0)$ in $\mathbb{B}_{\tilde{r},T,h_0}$ which is a solution of (4.4)-(4.5) and

$$(u^*,\tilde{q}^*,h^*)(h_0,v_0):=(F_T(h^*_{r,T}(h_0,v_0),v_0),h^*_{r,T}(h_0,v_0))\in Y_T\times V_T$$

solves the transformed Navier-Stokes/mean curvature flow system (2.4)-(2.11). The Lipschitz-dependence of $(u^*,\tilde{q}^*,h^*)$ on $(h_0,\tilde{v}_0)$ where $\tilde{v}_0:=v_0\circ\tilde{\Theta}_{h_0}$ is obtained as follows: For h^* we use $H_{r,T}(h^*(h_0,v_0);h_0,v_0)=h^*(h_0,v_0)$ in the difference estimate proven before. Then for $(u^*,\tilde{q}^*)$ we utilize Theorem 3.1.

It remains to show uniqueness in $Y_T\times V_T$ for fixed $h_0\in\overline{B_{X_\gamma}(0,\delta)}$ and v_0 as in the theorem. In the following we write $h^*_{\tilde{r},T}$ and $(u^*,\tilde{q}^*,h^*)$ instead of $h^*_{\tilde{r},T}(h_0,v_0)$ and $(u^*,\tilde{q}^*,h^*)(h_0,v_0)$, respectively. In particular it holds $h^*=h^*_{r,T}$. We need that the fixed-point $h^*_{\tilde{r},T}$ is the same for $\tilde{r}\in\{r,2r\}$ and remains the same up to restriction in time if we shrink T. Therefore let $0<\tilde{T}\leq T$ and $\tilde{r}\in\{r,2r\}$. For $h\in\mathbb{B}_{\tilde{r},T,h_0}$ we have $h|_{[0,\tilde{T}]}\in\mathbb{B}_{\tilde{r},\tilde{T},h_0}$ and because of Corollary 3.5 it holds $G_T(h,v_0)|_{[0,\tilde{T}]}=G_{\tilde{T}}(h|_{[0,\tilde{T}]},v_0)$. Hence we obtain

$$H_{\tilde{r},T}(h;h_0,v_0)|_{[0,\tilde{T}]}=H_{\tilde{r},\tilde{T}}(h|_{[0,\tilde{T}]};h_0,v_0).$$

Since $H_{r,\tilde{T}}(h;h_0,v_0)=H_{2r,\tilde{T}}(h;h_0,v_0)$ for all $h\in\mathbb{B}_{r,\tilde{T},h_0}\subseteq\mathbb{B}_{2r,\tilde{T},h_0}$ the uniqueness of the respective fixed-point yields

$$h^*|_{[0,\tilde{T}]}=h^*_{r,\tilde{T}}=h^*_{2r,\tilde{T}}\qquad\text{for all }0<\tilde{T}\leq T.$$

Now the uniqueness in $Y_T\times V_T$ can be shown by a typical contradiction argument. □

Appendix A. Banach-space-valued functions

A.1. Spaces of (Hölder-)Continuous Functions. Let X be a Banach space over $\mathbb{K} = \mathbb{R}$ or $\mathbb{C}$. For a closed interval $I \subseteq \mathbb{R}$ we denote by $C^0(I;X)$, $C_b^0(I;X)$, $BUC(I;X)$, $C^{0,\alpha}(I;X)$ for $\alpha \in (0,1]$ the set of continuous; continuous and bounded; bounded and uniformly continuous; bounded and Hölder-continuous functions $f : I \to X$, respectively. In the latter case we write $[f]_{C^{0,\alpha}(I;X)}$ for the semi-norm and we set $C^\alpha(I;X) := C^{0,\alpha}(I;X)$ if $\alpha \in (0,1)$. Moreover, let $C^{0,0}(I;X) := C^0(I;X)$. If $I \subseteq \mathbb{R}$ is an open interval, then $C^\infty(\overline{I};X)$ is the set of all smooth $f : I \to X$ such that f and all derivatives can be extended continuously to $\overline{I}$ and $C_0^\infty(I;X)$ denotes the set of all $f \in C^\infty(\overline{I};X)$ with compact support in I. Due to lack of space we omit two lemmas treating nonlinear terms in certain Hölder-spaces, cf. preprint [**2**].

A.2. Lebesgue-, Sobolev- and Besov-Spaces. Let $I \subseteq \mathbb{R}$ be measurable, X be a Banach space and $1 \leq p \leq \infty$. Then $L^p(I;X)$ are the usual Bochner spaces and $W_p^k(I;X)$ for I open and $k \in \mathbb{N}$ are the X-valued Sobolev spaces. Moreover, we need vector-valued variants of some fractional order Besov- and Sobolev-spaces: Therefore let $s \in (0,1), 1 \leq p \leq \infty$ and I be an interval in $\mathbb{R}$. Then we define $B_{p,\infty}^s(I;X) := \{f \in L^p(I;X) : \|f\|_{B_{p,\infty}^s(I;X)} < \infty\}$ where the norm is given by $\|f\|_{B_{p,\infty}^s(I;X)} := \|f\|_{L^p(I;X)} + [f]_{B_{p,\infty}^s(I;X)}$ with

$$[f]_{B_{p,\infty}^s(I;X)} := \sup_{0<h\leq 1} \left\| \frac{f(.+h)-f}{h^s} \right\|_{L^p(I_h;X)} \qquad \text{and } I_h := \{x \in I : x+h \in I\}.$$

For $s \in (0,1)$ and $1 \leq p < \infty$ we set $W_p^s(I;X) := \{f \in L^p(I;X) : \|f\|_{W_p^s(I;X)} < \infty\}$ where the norm is given by $\|f\|_{W_p^s(I;X)}^p := \|f\|_{L^p(I;X)}^p + [f]_{W_p^s(I;X)}^p$ and

$$[f]_{W_p^s(I;X)}^p := \int_I \int_I \left(\frac{\|f(x)-f(y)\|}{|x-y|^s} \right)^p \frac{dx\,dy}{|x-y|}.$$

All the above spaces are Banach spaces since $L^p(I;X)$ and $L^1(I \times I)$ are complete. The above definitions equal those in Simon [**25**] up to equivalent norms. Moreover, for $s > 0$ and $1 < p < \infty$ the Bessel-potential spaces are denoted by $H_p^s(I;X)$.

Additionally, we need spaces with mixed regularity. Let $\Omega \subseteq \mathbb{R}^n$ open, $r, s \in \mathbb{N}$, $1 < p < \infty$ and $T > 0$. Then $W_p^{r,s}(\Omega_T) := L^p(0,T;W_p^r(\Omega)) \cap W_p^s(0,T;L^p(\Omega))$. If $T = \infty$ we write $\Omega \times \mathbb{R}_+$ instead of Ω_T. Further anisotropic spaces are defined analogously. For properties of such spaces cf. the appendix in Grubb [**14**] and Denk, Hieber and Prüss [**7**].

We need product estimates for Hölder spaces with suitable Sobolev- and Besov-spaces:

LEMMA A.1. *Let X, Y, Z be Banach spaces over $\mathbb{K} = \mathbb{R}$ or $\mathbb{C}$ and $B : X \times Y \to Z$ be a product. Moreover, let $I \subseteq \mathbb{R}$ be a finite interval, $s \in (0,1)$ and $1 \leq p \leq \infty$. Then $C^s(\overline{I};X) \times B_{p,\infty}^s(I;Y) \to B_{p,\infty}^s(I;Z) : (f,g) \mapsto B(f,g)$ is a product and in the product estimate one can choose the same constant as for B.*

If $s \in (0,1), 1 \leq p < \infty$ and $\varepsilon > 0$ is such that $s + \varepsilon < 1$ holds, then B induces a product on $C^{s+\varepsilon}(\overline{I};X) \times W_p^s(I;Y) \to W_p^s(I;Z)$ and the constant in the estimate can be chosen independently of $|I|$ if $|I| \leq T$ for a fixed $T > 0$.

PROOF. This follows by inserting definitions and using null additions, cf. preprint [**2**]. □

The following theorem yields embeddings for suitable intersection spaces:

THEOREM A.2. *Let X_0, X_1, X be Banach spaces over $\mathbb{K} = \mathbb{R}$ or $\mathbb{C}$ with $X_1 \hookrightarrow X \hookrightarrow X_0$ and $I \subseteq \mathbb{R}$ be an open interval. Assume that for a $\theta \in (0,1)$ the following interpolation inequality holds: $\|x\|_X \leq C_0 \|x\|_{X_0}^{1-\theta} \|x\|_{X_1}^{\theta}$ for all $x \in X_0 \cap X_1$. Then*

(1) *If $1 \leq p, p_0, p_1 \leq \infty$ with $\frac{1}{p} = \frac{1-\theta}{p_0} + \frac{\theta}{p_1}$, then $L^{p_0}(I; X_0) \cap L^{p_1}(I; X_1) \hookrightarrow L^p(I; X)$.*

(2) *If $1 \leq p < \infty$, then $L^p(I; X_1) \cap W^1_p(I; X_0) \hookrightarrow B^{1-\theta}_{p,\infty}(I; X)$.*

In both cases analogous interpolation inequalities hold with C_0 and $3C_0$, respectively.

PROOF. The proof essentially follows from Hölder's inequality and the mean value theorem in the second case, cf. preprint [**2**]. □

The following theorem characterizes the trace of functions in $\mathbb{E}_1(T)$ at $t = 0$.

THEOREM A.3. *Let X_0, X_1 Banach spaces over $\mathbb{K} = \mathbb{R}$ or $\mathbb{C}$ with $X_1 \hookrightarrow X_0$ and let $1 < p < \infty$. Then for $\mathbb{E}_1(\mathbb{R}_+) := L^p(\mathbb{R}_+; X_1) \cap W^1_p(0, \infty; X_0)$ it holds that*

$$X_\gamma := (X_0, X_1)_{1-\frac{1}{p},p} = \{u(0) : u \in \mathbb{E}_1(\mathbb{R}_+)\}$$

and an equivalent norm on X_γ is given by the induced quotient norm. Moreover we have $\mathbb{E}_1(\mathbb{R}_+) \hookrightarrow C^0_b(\mathbb{R}_+; X_\gamma)$ and $\mathbb{E}_1(T) \hookrightarrow C^0([0,T]; X_\gamma)$ for $0 < T < \infty$ where the embedding constant is bounded independently of $T > 0$ if we add $\|u(0)\|_{X_\gamma}$ in the $\mathbb{E}_1(T)$-norm. In particular this also holds for $u(0) = 0$.

PROOF. Up to $\mathbb{E}_1(\mathbb{R}_+) \hookrightarrow C^0_b(\mathbb{R}_+; X_\gamma)$ this directly follows from the trace method, cf. Lunardi [**21**], Section 1.2. Using this one can show the remaining part by extending functions with reflection, cf. preprint [**2**]. □

By interpolation we get a whole scale of embeddings from Theorem A.3:

THEOREM A.4. *Let X_0, X_1 Banach spaces over $\mathbb{K} = \mathbb{R}$ or $\mathbb{C}$ with $X_1 \hookrightarrow X_0$. Moreover, let $1 < p < \infty$ and $T > 0$. Then for $0 < \theta < 1 - \frac{1}{p}$ and $1 \leq q \leq \infty$ it holds*

$$\mathbb{E}_1(T) \hookrightarrow C^{1-\theta-\frac{1}{p}}([0,T]; (X_0, X_1)_{\theta,q})$$

and the embedding constant is bounded independently of $T > 0$ if we add $\|u(0)\|_{X_\gamma}$ in the $\mathbb{E}_1(T)$-norm, where $\mathbb{E}_1(T)$ and X_γ are as in Theorem A.3.

PROOF. It is well-known that $W^1_p(0,T; X_0) \hookrightarrow C^{0,1-\frac{1}{p}}([0,T]; X_0)$ and the semi-norm is estimated by $\|u'\|_{L^p(0,T;X_0)}$ for all $u \in W^1_p(0,T; X_0)$. Now one interpolates with the embedding from Theorem A.3 and uses the reiteration theorem in Lunardi [**21**], Corollary 1.24 and the interpolation inequality in [**21**], Corollary 1.7. □

As another consequence of Theorem A.3 we obtain

LEMMA A.5. *Let $\Omega \subseteq \mathbb{R}^n, n \geq 2$ be a bounded C^∞-domain and $\Sigma \subseteq \Omega$ be a connected, compact and smooth hypersurface that separates Ω in two disjoint, connected domains $\Omega^\pm$ with $\Sigma = \partial\Omega^+$. Moreover, let $0 < T < \infty, 2 \leq p < \infty$ and $\Omega_0 := \Omega^+ \cup \Omega^-$. Then*

$$\mathbb{E}_1(T) := L^p(0,T; X_1) \cap W^1_p(0,T; L^p(\Omega)) \hookrightarrow C^0([0,T]; Y_\gamma)$$

with $X_1 := W^2_p(\Omega_0) \cap W^1_{p,0}(\Omega)$ and $Y_\gamma := \{v \in L^p(\Omega) : v|_{\Omega^\pm} \in W_p^{2-\frac{2}{p}}(\Omega^\pm)\} \cap W^1_{p,0}(\Omega)$. The embedding constant is bounded independently of $T > 0$ if we add $\|u(0)\|_{Y_\gamma}$ in the norm.

PROOF. We define $X_\gamma := (L^p(\Omega), X_1)_{1-\frac{1}{p},p}$ and $\tilde{X}_\gamma := (L^p(\Omega), W_p^2(\Omega_0))_{1-\frac{1}{p},p}$. Then Theorem A.3 implies $\mathbb{E}_1(T) \hookrightarrow C^0([0,T]; X_\gamma)$ and that the space $\tilde{\mathbb{E}}_1(T) := L^p(0,T; W_p^2(\Omega_0)) \cap W_p^1(0,T; L^p(\Omega))$ embeds to $C^0([0,T]; \tilde{X}_\gamma)$. The embedding constants are bounded in a similar way as in the theorem. E.g. from the K-method, cf. Lunardi [**21**], Section 1.1 we obtain

$$\tilde{X}_\gamma = \{v \in L^p(\Omega) : v|_{\Omega^\pm} \in W_p^{2-\frac{2}{p}}(\Omega^\pm)\}$$

with equivalent norms. Hence it remains to show that for the continuous representative of $u \in \mathbb{E}_1(T)$ additionally $u(t) \in W_{p,0}^1(\Omega)$ for all $t \in [0,T]$ holds. Therefore we use the characterization in Theorem A.3 to conclude $X_\gamma \hookrightarrow \tilde{X}_\gamma$. Here X_1 is dense in X_γ by Lunardi [**21**], Proposition 1.17. For $v \in X_1$ it holds $[\![v]\!] = \operatorname{tr}_\Sigma(v|_{\Omega^-}) - \operatorname{tr}_\Sigma(v|_{\Omega^+}) = 0$ and $\operatorname{tr}_{\partial\Omega} v = 0$. By density this carries over to all $v \in X_\gamma$ because of $X_\gamma \hookrightarrow \tilde{X}_\gamma \hookrightarrow W_p^1(\Omega_0)$ for $2 \leq p < \infty$ and the continuity of the respective boundary trace operators. Using the definition of weak derivative and partial integration as well as the trace characterization of $W_{p,0}^1(\Omega)$ we get $X_\gamma \subseteq \tilde{X}_\gamma \cap W_{p,0}^1(\Omega) = Y_\gamma$. □

Acknowledgments

The second author is grateful to Mathias Wilke for many helpful discussions and his lecture series on "maximal regularity".

References

[1] Helmut Abels, *Nonstationary Stokes system with variable viscosity in bounded and unbounded domains*, Discrete Contin. Dyn. Syst. Ser. S **3** (2010), no. 2, 141–157. MR2610556

[2] H. Abels and M. Moser. Well-posedness of a Navier-Stokes/Mean Curvature Flow system. preprint. arXiv:1705.10832.

[3] Helmut Abels and Mathias Wilke, *Well-posedness and qualitative behaviour of solutions for a two-phase Navier-Stokes-Mullins-Sekerka system*, Interfaces Free Bound. **15** (2013), no. 1, 39–75. MR3062573

[4] Robert A. Adams and John J. F. Fournier, *Sobolev spaces*, 2nd ed., Pure and Applied Mathematics (Amsterdam), vol. 140, Elsevier/Academic Press, Amsterdam, 2003. MR2424078

[5] Herbert Amann, *Quasilinear parabolic problems via maximal regularity*, Adv. Differential Equations **10** (2005), no. 10, 1081–1110. MR2162362

[6] J. Bergh and J. Löfström. *Interpolation Spaces*. Springer, Berlin - Heidelberg - New York, 1976.

[7] Robert Denk, Matthias Hieber, and Jan Prüss, *Optimal L^p-L^q-estimates for parabolic boundary value problems with inhomogeneous data*, Math. Z. **257** (2007), no. 1, 193–224. MR2318575

[8] I. V. Denisova, *A priori estimates for the solution of the linear nonstationary problem connected with the motion of a drop in a liquid medium* (Russian), Trudy Mat. Inst. Steklov. **188** (1990), 3–21, 191. Translated in Proc. Steklov Inst. Math. **1991**, no. 3, 1–24; Boundary value problems of mathematical physics, 14 (Russian). MR1100535

[9] I. V. Denisova, *Problem of the motion of two viscous incompressible fluids separated by a closed free interface*, Acta Appl. Math. **37** (1994), no. 1-2, 31–40. Mathematical problems for Navier-Stokes equations (Centro, 1993). MR1308743

[10] I. V. Denisova and V. A. Solonnikov, *Global solvability of the problem of the motion of two incompressible capillary fluids in a container* (Russian, with English and Russian summaries), Zap. Nauchn. Sem. S.-Peterburg. Otdel. Mat. Inst. Steklov. (POMI) **397** (2011), no. Kraevye Zadachi Matematicheskoĭ Fiziki i Smezhnye Voprosy Teorii Funktsiĭ. 42, 20–52, 172; English transl., J. Math. Sci. (N.Y.) **185** (2012), no. 5, 668–686. MR2870107

[11] Joachim Escher and Gieri Simonett, *Classical solutions for Hele-Shaw models with surface tension*, Adv. Differential Equations **2** (1997), no. 4, 619–642. MR1441859

[12] Joachim Escher and Gieri Simonett, *A center manifold analysis for the Mullins-Sekerka model*, J. Differential Equations **143** (1998), no. 2, 267–292. MR1607952

[13] Matthias Geißert, Horst Heck, and Matthias Hieber, *On the equation* $\operatorname{div} u = g$ *and Bogovskiĭ's operator in Sobolev spaces of negative order*, Partial differential equations and functional analysis, Oper. Theory Adv. Appl., vol. 168, Birkhäuser, Basel, 2006, pp. 113–121. MR2240056

[14] Gerd Grubb, *Parameter-elliptic and parabolic pseudodifferential boundary problems in global* L_p *Sobolev spaces*, Math. Z. **218** (1995), no. 1, 43–90. MR1312578

[15] S. Hildebrandt. *Analysis 2*. Springer, Berlin - Heidelberg, 2003.

[16] C. Kneisel. Über das Stefan-Problem mit Oberflächenspannung und thermischer Unterkühlung. Hannover, Univ., Diss., 2007. Access: May 18, 2018. http://edok01.tib.uni-hannover.de/edoks/e01dh07/539840092.pdf

[17] Matthias Köhne, Jan Prüss, and Mathias Wilke, *On quasilinear parabolic evolution equations in weighted* L_p*-spaces*, J. Evol. Equ. **10** (2010), no. 2, 443–463. MR2643804

[18] Matthias Köhne, Jan Prüss, and Mathias Wilke, *Qualitative behaviour of solutions for the two-phase Navier-Stokes equations with surface tension*, Math. Ann. **356** (2013), no. 2, 737–792. MR3048614

[19] L. D. Landau and E. M. Lifshitz, *Fluid mechanics*, Translated from the Russian by J. B. Sykes and W. H. Reid. Course of Theoretical Physics, Vol. 6, Pergamon Press, London-Paris-Frankfurt; Addison-Wesley Publishing Co., Inc., Reading, Mass., 1959. MR0108121

[20] Chun Liu, Norifumi Sato, and Yoshihiro Tonegawa, *Two-phase flow problem coupled with mean curvature flow*, Interfaces Free Bound. **14** (2012), no. 2, 185–203. MR2956324

[21] Alessandra Lunardi, *Interpolation theory*, 2nd ed., Appunti. Scuola Normale Superiore di Pisa (Nuova Serie). [Lecture Notes. Scuola Normale Superiore di Pisa (New Series)], Edizioni della Normale, Pisa, 2009. MR2523200

[22] Jan Prüss and Gieri Simonett, *Moving interfaces and quasilinear parabolic evolution equations*, Monographs in Mathematics, vol. 105, Birkhäuser/Springer, [Cham], 2016. MR3524106

[23] Yoshihiro Shibata and Senjo Shimizu, *Maximal* L_p*-*L_q *regularity for the two-phase Stokes equations; model problems*, J. Differential Equations **251** (2011), no. 2, 373–419. MR2800157

[24] Senjo Shimizu, *Maximal regularity and viscous incompressible flows with free interface*, Parabolic and Navier-Stokes equations. Part 2, Banach Center Publ., vol. 81, Polish Acad. Sci. Inst. Math., Warsaw, 2008, pp. 471–480. MR2548875

[25] Jacques Simon, *Sobolev, Besov and Nikol′skiĭ fractional spaces: imbeddings and comparisons for vector valued spaces on an interval*, Ann. Mat. Pura Appl. (4) **157** (1990), 117–148. MR1108473

[26] V. A. Solonnikov, L_p*-theory of the problem of motion of two incompressible capillary fluids in a container*, J. Math. Sci. (N.Y.) **198** (2014), no. 6, Problems in mathematical analysis. No. 75 (Russian), 761–827. MR3391324

[27] Hans Triebel, *Interpolation theory, function spaces, differential operators*, North-Holland Mathematical Library, vol. 18, North-Holland Publishing Co., Amsterdam-New York, 1978. MR503903

[28] Hans Triebel, *Theory of function spaces*, Monographs in Mathematics, vol. 78, Birkhäuser Verlag, Basel, 1983. MR781540

Fakultät für Mathematik, Universität Regensburg, Universitätsstrasse 31, D-93053 Regensburg, Germany

Email address: helmut.Abels@mathematik.uni-regensburg.de

Fakultät für Mathematik, Universität Regensburg, Universitätsstrasse 31, D-93053 Regensburg, Germany

Email address: maximilian1.moser@mathematik.uni-regensburg.de

Contemporary Mathematics
Volume **710**, 2018
http://dx.doi.org/10.1090/conm/710/14362

PDE analysis of a class of thermodynamically compatible viscoelastic rate-type fluids with stress-diffusion

Miroslav Bulíček, Josef Málek, Vít Průša, and Endre Süli

ABSTRACT. We establish the long-time existence of large-data weak solutions to a system of nonlinear partial differential equations. The system of interest governs the motion of non-Newtonian fluids described by a simplified viscoelastic rate-type model with a stress-diffusion term. The simplified model shares many qualitative features with more complex viscoelastic rate-type models that are frequently used in the modeling of fluids with complicated microstructure. As such, the simplified model provides important preliminary insight into the mathematical properties of these more complex and practically relevant models of non-Newtonian fluids. The simplified model that is analyzed from the mathematical perspective is shown to be thermodynamically consistent, and we extensively comment on the interplay between the thermodynamical background of the model and the mathematical analysis of the corresponding initial-boundary-value problem.

1. Introduction

The main goal of this study is to establish the long-time existence of large-data weak solutions to the following simplified set of governing equations encountered in the mechanics of incompressible non-Newtonian fluids. This simple model shares many standard properties with more complex viscoelastic rate-type fluid models with stress-diffusion, used in applications.

For any given Lipschitz domain $\Omega \subset \mathbb{R}^d$, $d \in \{2,3\}$, $\vec{v}_0 : \Omega \to \mathbb{R}^d$, $b_0 : \Omega \to \mathbb{R}$ and for any $T > 0$, we set $Q := (0,T) \times \Omega$ and we seek the functions $(\vec{v}, p, b) : Q \to \mathbb{R}^d \times \mathbb{R} \times \mathbb{R}$ satisfying, in Q, the system of PDEs:

$$\operatorname{div} \vec{v} = 0, \tag{1.1}$$

$$\varrho(\partial_t \vec{v} + \operatorname{div}(\vec{v} \otimes \vec{v})) - \operatorname{div} \mathbb{T} = 0, \tag{1.2}$$

$$\mathbb{T} = -p\,\mathbb{I} + 2\nu\mathbb{D} - \sigma(\nabla b \otimes \nabla b), \tag{1.3}$$

$$\nu_1 \partial_t b + \nu_1 \operatorname{div}(b\vec{v}) + \mu(b^2 - b) - 2\sigma b^2 \Delta b = 0, \tag{1.4}$$

together with the following boundary conditions, where $\vec{n}$ denotes the unit outward normal vector on $\partial\Omega$:

$$\vec{v} = \vec{0} \quad \text{and} \quad \nabla b \cdot \vec{n} = 0 \qquad \text{on } (0,T) \times \partial\Omega, \tag{1.5}$$

2010 *Mathematics Subject Classification.* Primary 35Q35, 76A05, 76A10.
The authors acknowledge the support of the ERC-CZ project LL1202, financed by MŠMT.

and the initial conditions:

$$\vec{v}(0,\cdot) = \vec{v}_0(\cdot) \quad \text{and} \quad b(0,\cdot) = b_0(\cdot) \qquad \text{in } \Omega. \tag{1.6}$$

Here, $\vec{v}$ is the velocity, $\otimes$ stands for the dyadic product $((\vec{a} \otimes \vec{b})_{ij} = a_i b_j)$, $\mathbb{T}$ is the Cauchy stress, p is the spherical stress (the modified pressure), b is a scalar quantity characterizing the volumetric elastic changes exhibited by the fluid; $\mathbb{D}$ denotes the symmetric part of the velocity gradient, i.e., $\mathbb{D} = (\nabla\vec{v} + (\nabla\vec{v})^{\mathrm{T}})/2$; ν is the viscosity, ϱ is the density, and ν_1, σ and μ are other material constants, all of which are positive.

Provided further that (the definitions of all function spaces appearing in the paper are given in Section 4)

$$\begin{aligned} &\vec{v}_0 \in L^2_{0,\mathrm{div}}(\Omega)^d, \\ &b_0 \in W^{1,2}(\Omega), \quad b_0 > 0 \ \text{a.e. in } \Omega, \quad b_0 \in L^\infty(\Omega) \quad \text{and} \quad \frac{1}{b_0} \in L^\infty(\Omega), \end{aligned} \tag{1.7}$$

we will prove, see Section 5, the following statement:

For any Lipschitz domain $\Omega \subset \mathbb{R}^d$, $T > 0$, $\nu > 0$, $\nu_1 > 0$, $\mu > 0$, $\sigma > 0$, $\varrho > 0$ and for any $(\vec{v}_0, b_0)$ satisfying (1.7), *there exists a couple $(\vec{v}, b)$ solving the problem* (1.1)–(1.6) *in a weak sense.*

The precise definition of *weak solution* is given in Section 4.

Note that if $\sigma = 0$ in the above equations the problem splits into two separate problems: a transport equation with damping for b and the standard Navier–Stokes equations for $(\vec{v}, p)$. Since the work of Leray [**30**] on the incompressible Navier–Stokes equations, see also [**8, 22**], the question of long-time existence of large-data weak solutions has also been explored for more general classes of viscous fluids; further references will be given in the next section. It is then natural to attempt to advance this program by exploring how large the class of fluids might be for which long-time and large-data existence of weak solutions can be established. This task is particularly interesting if one considers fluid models that include terms that are associated with elastic properties of the material. This paper aims to contribute to the study of this question.

Note also that if $\sigma > 0$ then not only is there a diffusion term present in the equation for b, but also the Korteweg stress appears in the expression for the Cauchy stress $\mathbb{T}$ featuring in the equation for the balance of linear momentum. This structure of the governing equations results from a careful derivation of the model based on the thermodynamical approach established for viscoelastic rate-type fluids in [**42**] and further refined in [**39**] (see also [**36**]), and extended to rate-type fluids with stress-diffusion in [**37**]. This thermodynamical approach automatically guarantees that the resulting model is consistent with the laws of thermodynamics. From the point of view of PDE analysis, the approach readily provides the relevant *a priori* estimates upon which the analysis is based, and which would be otherwise completely nontrivial to discover solely from the PDE system (1.1)–(1.4).

Thus, the second goal of this paper is to highlight how the analysis of PDEs for complex systems such as (1.1)–(1.4) is related to the thermodynamical approach used in the derivation of the model and to show that the problem (1.1)–(1.6) therefore merits investigation.

The paper is structured as follows. In the next section, we place the model under consideration into a hierarchy of phenomenological models. In Section 3, we sketch its derivation. As was noted above, this thermodynamical approach automatically guarantees the consistency with the second law of thermodynamics and provides directly the *a priori* estimates available for the system. In Section 4, we re-derive these *a priori* estimates, but now from a purely PDE analytical point of view, and then, after introducing the appropriate function spaces, we precisely state the main result of the paper. The proof of the theorem is contained in Section 5. The existence of solutions to the finite-dimensional approximating system, upon which the proof of the main result is based, is given in the Appendix.

2. Incompressible non-Newtonian fluids: a brief overview

For incompressible fluids, the Cauchy stress tensor $\mathbb{T}$ decomposes as

$$\mathbb{T} = -\phi\mathbb{I} + \mathbb{S}, \tag{2.1}$$

where ϕ is a scalar unknown quantity and $\mathbb{S}$ is related to the symmetric part of the velocity gradient $\mathbb{D}$ via the so-called constitutive relation. Three classes of constitutive relations are frequently used; they are of the following forms:

$$\mathbb{G}(\mathbb{S}, \mathbb{D}) = \mathbb{O}, \tag{2.2}$$

$$\mathbb{G}(\overset{*}{\mathbb{S}}, \mathbb{S}, \overset{*}{\mathbb{D}}, \mathbb{D}) = \mathbb{O}, \tag{2.3}$$

$$\mathbb{G}(\overset{*}{\mathbb{S}}, \mathbb{S}, \overset{*}{\mathbb{D}}, \mathbb{D}) - \Delta\mathbb{S} = \mathbb{O}. \tag{2.4}$$

Here $\mathbb{G}$ stands for an arbitrary continuous tensorial function and $\overset{*}{\mathbb{A}}$ signifies an objective time derivative.

We will give examples and discuss briefly the usefulness of these classes. Then we shall look at the associated initial-boundary-value problems for internal flows from the perspective of the long-time existence of large-data weak solutions.

Besides the (linear, i.e. Newtonian) Navier–Stokes fluid for which the constitutive relation is given by $\mathbb{S} = 2\nu\mathbb{D}$ with $\nu > 0$, the class (2.2) includes for example the Bingham fluid, described by $2\nu\mathbb{D} = \frac{(|\mathbb{S}|-\tau_*)^+}{|\mathbb{S}|}\mathbb{S}$, as well as various generalizations of activated or non-activated (stress) power-law fluids that can be characterized by

$$2\nu(|\mathbb{S}|^2, |\mathbb{D}|^2)(|\mathbb{D}| - d_*)^+\mathbb{D} = 2\alpha(|\mathbb{S}|^2, |\mathbb{D}|^2)(|\mathbb{S}| - \tau_*)^+\mathbb{S},$$

which involves two non-negative activation parameters d_* and τ_*, but only one is assumed to be positive in a given model. The possibility of (2.2) to incorporate nonlinear relations between $\mathbb{S}$ and $\mathbb{D}$ and to include sudden changes from one type of response to another one (within the considered class) makes the class (2.2) suitable for modeling even mixing type phenomena, and the class (2.2) is therefore very popular in many areas (see [**5, 38**] and [**36**] for details and further references).

On the other hand, fluids described by the class (2.2) are not capable of capturing fundamental phenomena such as stress relaxation or nonlinear creep observed in most fluids with complex microstructure. If these phenomena are of interest, then the class (2.3) is a suitable choice.

As was already mentioned above in connection with (2.3), $\overset{*}{\mathbb{A}}$ denotes an objective time derivative. The set of objective derivatives includes the upper-convected Oldroyd derivative defined by

$$\overset{\triangledown}{\mathbb{A}} := \frac{d\mathbb{A}}{dt} - \mathbb{L}\mathbb{A} - \mathbb{A}\mathbb{L}^{\mathrm{T}} \quad \text{where} \quad \mathbb{L} := \nabla\vec{v},$$

the Jaumann–Zaremba (corotational) derivative specified by

$$\overset{\circ}{\mathbb{A}} := \frac{d\mathbb{A}}{dt} - \mathbb{W}\mathbb{A} - \mathbb{A}\mathbb{W}^{T} \quad \text{where} \quad \mathbb{W} := (\mathbb{L} - \mathbb{L}^{T})/2,$$

or the Gordon–Schowalter derivative defined by

$$\overset{\square}{\mathbb{A}} := \overset{\circ}{\mathbb{A}} - a(\mathbb{D}\mathbb{A} - \mathbb{A}\mathbb{D}) \quad \text{where } a \in [-1, 1].$$

Assuming that $\mathbb{T} = -\phi\mathbb{I} + 2\nu\mathbb{D} + \mathbb{A}$, we see that the appropriate choice of material parameter ν and the objective derivative lead to the popular Maxwell, Oldroyd-B and Johnson–Segalman models respectively:

$$\tau\overset{\triangledown}{\mathbb{A}} + \mathbb{A} = 2\nu_1\mathbb{D} \qquad \text{with} \quad \nu = 0,\ \tau = \frac{\nu_1}{E},\ \text{with } \nu_1,\ E > 0,$$

$$\tau\overset{\triangledown}{\mathbb{A}} + \mathbb{A} = 2\nu_1\mathbb{D} \qquad \text{with} \quad \nu = 0,\ \tau = \frac{\nu_1}{E},\ \text{with } \nu_1,\ E > 0,$$

$$\tau\overset{\square}{\mathbb{A}} + \mathbb{S} = 2a\mathbb{D} \qquad \text{with} \quad \nu > 0 \text{ and } a \in [-1, 1].$$

Although these models are widely used, there are several subtle issues regarding their physical underpinnings. These include the ambiguity with respect to the choice of the objective derivative, the possibility of the derivation of the model at a purely macroscopic level, the consistency of the models with the second law of thermodynamics, the extension of the models to the compressible setting, and the inclusion of thermal effects into the models.

To address some of these issues Rajagopal and Srinivasa [**42**] provided a simple, yet general method for the derivation of thermodynamically consistent rate-type fluid models. The method is based on the concept of *natural configuration* and on the knowledge of constitutive relations for *two scalar quantities*: the Helmholtz free energy (characterizing how the material stores energy) and the rate of entropy production (characterizing how the material dissipates energy). Using the method, it is possible to derive new thermodynamically compatible classes of *non-linear* viscoelastic rate-type fluid models, and to investigate the conditions under which these models reduce to the standard Oldroyd-B, Maxwell and Burgers models. More recently, the approach has been refined and extended to compressible fluids and thermal processes (see [**23, 36, 39**]).

Rate-type fluid models with stress-diffusion, c.f. (2.4), are popular in the modelling of shear and vorticity banding phenomena; see, for example, the reviews [**12, 14, 17**]. The *ad hoc* addition of the regularizing term $-\sigma\Delta\mathbb{S}$ to the rate-type equation (2.3) raises doubts about the consistency of the resulting model with the second law of thermodynamics and calls for the specification of appropriate boundary conditions for $\mathbb{S}$. These issues are addressed in the recent study [**37**]. Here, in Section 3, we sketch the derivation under several simplifying assumptions.

As to the question of existence of solutions, for any time interval and any reasonable set of data, a satisfactory theory, within the context of weak solutions, has been established during the last fifty years for a general class of fluids given by (2.2) with a continuous monotone function $\mathbb{G}$ (see [**2–8, 13, 19, 25, 27, 28, 30, 31, 34, 35, 40, 43**]).

Regarding the long-time and large-data existence theory for rate-type fluids in three dimensions much less is known (see [**24, 32, 41**] or [**29**]).

One expects that the addition of the term $-\Delta\mathbb{S}$ to a rate-type equation will automatically improve its mathematical properties and a well-founded theory for a class of fluids of the type (2.4) will therefore emerge. However, most of the results that are in place concern two-dimensional (or steady) flows, see [**1**, **9**, **10**, **15**, **16**, **33**] or require stronger regularization, see [**26**].

Unlike the above two-dimensional results, the general analytical approach developed and presented in Sections 4 and 5 concerns d-dimensional flows. Although the tensor $\mathbb{S}$ is ultimately "replaced" here by a scalar quantity b, the analysis is performed in such a way that it avoids properties which are unlikely to be true for a general tensor $\mathbb{S}$.

3. Derivation of the model

We denote by the superscript $\dot{}$ the material derivative, i.e., $\dot{z} = \frac{\partial z}{\partial t} + (\vec{v}\cdot\nabla)z$ for a scalar function z; if z stands for a longer expression, we write $\dot{\overline{z}}$ instead of $\dot{z}$ in order to avoid confusion. For vector- or tensor-valued quantities, the same notation is used and the above formula is applied to each component.

The fundamental balance equations of mass, linear momentum and energy as well as the formulation of the second law of thermodynamics are the following:

$$\dot{\varrho} = -\varrho \operatorname{div}\vec{v}, \tag{3.1}$$

$$\varrho\dot{\vec{v}} = \operatorname{div}\mathbb{T}, \qquad \mathbb{T} = \mathbb{T}^{\mathrm{T}}, \tag{3.2}$$

$$\varrho\dot{e} = \mathbb{T} : \mathbb{D} - \operatorname{div}\vec{j}_e, \tag{3.3}$$

$$\varrho\dot{\eta} = \varrho\zeta - \operatorname{div}\vec{j}_\eta \quad \text{with } \zeta \geq 0, \tag{3.4}$$

where ϱ is the density, $\vec{v}$ is the velocity, e is the specific internal energy, η is the specific entropy, $\mathbb{D} = (\nabla\vec{v} + (\nabla\vec{v})^{\mathrm{T}})/2$ is the symmetric part of the velocity gradient, $\mathbb{T}$ is the Cauchy stress tensor, $\vec{j}_e$ and $\vec{j}_\eta$ are the energy and entropy fluxes, and ζ stands for the specific rate of entropy production.

Introducing the Helmholtz free energy ψ by

$$\psi := e - \theta\eta,$$

where θ stands for the (positive) temperature, the equations (3.3) and (3.4) lead to

$$\mathbb{T} : \mathbb{D} - \varrho\dot{\psi} - \operatorname{div}(\vec{j}_e - \theta\vec{j}_\eta) - \varrho\eta\dot{\theta} - \nabla\theta\cdot\vec{j}_\eta = \varrho\theta\zeta \quad \text{with } \zeta \geq 0.$$

In what follows, we restrict ourselves to isothermal processes (referring the reader to [**37**] for the development of models where both mechanical and thermal processes are included). Consequently, $\theta > 0$ is constant and the last identity reduces to

$$\mathbb{T} : \mathbb{D} - \varrho\dot{\psi} - \operatorname{div}(\vec{j}_e - \theta\vec{j}_\eta) = \xi \quad \text{with } \xi \geq 0, \tag{3.5}$$

where $\xi := \varrho\theta\zeta$ denotes the rate of dissipation. Note that (3.5) simplifies further to $\mathbb{T} : \mathbb{D} - \varrho\dot{\psi} = \xi$ provided that $\vec{j}_\eta = \frac{\vec{j}_e}{\theta}$, which we do not require here, however.

The approach that we will exploit is based on the concept of natural configuration that splits the total response described by the deformation tensor $\mathbb{F}$ between the current and initial configuration into the purely elastic (reversible, non-dissipative) part $\mathbb{F}_{\kappa_p(t)}$ that operates between the natural and current configuration and the dissipative part $\mathbb{G}$ that maps from the reference to the natural configuration, i.e. $\mathbb{F} = \mathbb{F}_{\kappa_p(t)}\mathbb{G}$. In analogy with the relations $\mathbb{L} = \dot{\mathbb{F}}\mathbb{F}^{-1}$ and $\mathbb{D} = (\mathbb{L} + \mathbb{L}^{\mathrm{T}})/2$, we set $\mathbb{L}_{\kappa_p(t)} := \dot{\mathbb{G}}\mathbb{G}^{-1}$ and define $\mathbb{D}_{\kappa_p(t)} := (\mathbb{L}_{\kappa_p(t)} + (\mathbb{L}_{\kappa_p(t)})^{\mathrm{T}})/2$. We also set $\mathbb{B}_{\kappa_p(t)} := \mathbb{F}_{\kappa_p(t)}\mathbb{F}^{\mathrm{T}}_{\kappa_p(t)}$.

In order to derive the constitutive relations for the Cauchy stress tensor and the energy and entropy fluxes we start with postulating the constitutive relation for the Helmholtz free energy in the form:

$$\psi = \psi_0(\varrho) + \frac{\mu}{2\varrho}(\operatorname{tr}\mathbb{B}_{\kappa_p(t)} - 3 - \ln\det\mathbb{B}_{\kappa_p(t)}) + \frac{\sigma}{2\varrho}|\nabla \operatorname{tr}\mathbb{B}_{\kappa_p(t)}|^2 =: \psi_0(\varrho) + \frac{\pi}{\varrho}, \tag{3.6}$$

with $\mu > 0$, $\sigma > 0$ constant, and

$$\pi := \frac{\mu}{2}(\operatorname{tr}\mathbb{B}_{\kappa_p(t)} - 3 - \ln\det\mathbb{B}_{\kappa_p(t)}) + \frac{\sigma}{2}|\nabla \operatorname{tr}\mathbb{B}_{\kappa_p(t)}|^2.$$

The following identities (see [**39**] for their derivation) hold:

$$\overline{\mathbb{B}_{\kappa_p(t)}}^{\cdot} = \mathbb{L}\mathbb{B}_{\kappa_p(t)} + \mathbb{B}_{\kappa_p(t)}\mathbb{L}^{\mathrm{T}} - 2\,\mathbb{F}_{\kappa_p(t)}\mathbb{D}_{\kappa_p(t)}\mathbb{F}^{\mathrm{T}}_{\kappa_p(t)}, \tag{3.7}$$

$$\overline{\operatorname{tr}\mathbb{B}_{\kappa_p(t)}}^{\cdot} = 2\,\mathbb{B}_{\kappa_p(t)} : \mathbb{D} - 2\,\mathbb{C}_{\kappa_p(t)} : \mathbb{D}_{\kappa_p(t)}, \tag{3.8}$$

$$\overline{\ln\det\mathbb{B}_{\kappa_p(t)}}^{\cdot} = 2\,\mathbb{I} : \mathbb{D} - 2\,\mathbb{I} : \mathbb{D}_{\kappa_p(t)}, \tag{3.9}$$

$$\frac{1}{2}\overline{|\nabla\operatorname{tr}\mathbb{B}_{\kappa_p(t)}|^2}^{\cdot} = \nabla\operatorname{tr}\mathbb{B}_{\kappa_p(t)} \cdot \left(\nabla\overline{\operatorname{tr}\mathbb{B}_{\kappa_p(t)}}^{\cdot} - (\nabla\vec{v})\,\nabla\operatorname{tr}\mathbb{B}_{\kappa_p(t)}\right). \tag{3.10}$$

Setting

$$b := \operatorname{tr}\mathbb{B}_{\kappa_p(t)}$$

and using the formula

$$\nabla b \cdot \nabla\dot{b} = \operatorname{div}(\dot{b}\,\nabla b) - \Delta b\,\dot{b},$$

the above identities (3.6), (3.8) and (3.10) take the form

$$\psi = \psi_0(\varrho) + \frac{\pi}{\varrho} \quad \text{with} \quad \pi = \frac{\mu}{2}(b - 3 - \ln\det\mathbb{B}_{\kappa_p(t)}) + \frac{\sigma}{2}|\nabla b|^2, \tag{3.11}$$

$$\dot{b} = 2\,\mathbb{B}_{\kappa_p(t)} : \mathbb{D} - 2\,\mathbb{C}_{\kappa_p(t)} : \mathbb{D}_{\kappa_p(t)}, \tag{3.12}$$

$$\frac{\overline{|\nabla b|^2}^{\cdot}}{2} = -\left((\nabla b \otimes \nabla b) + 2\Delta b\,\mathbb{B}_{\kappa_p(t)}\right) : \mathbb{D} + \operatorname{div}(\dot{b}\,\nabla b) + 2\Delta b\,\mathbb{C}_{\kappa_p(t)} : \mathbb{D}_{\kappa_p(t)}. \tag{3.13}$$

Consequently, using also (3.1) and (3.9), we compute

$$\begin{aligned}
\varrho\dot{\psi} &= \varrho\psi_0'(\varrho)\dot{\varrho} + \dot{\pi} - \frac{\pi\dot{\varrho}}{\varrho} \\
&= \left(\mu(\mathbb{B}_{\kappa_p(t)} - \mathbb{I}) - \sigma((\nabla b \otimes \nabla b) + 2\Delta b\,\mathbb{B}_{\kappa_p(t)})\right) : \mathbb{D} \\
&\quad - \left(\mu(\mathbb{C}_{\kappa_p(t)} - \mathbb{I}) - 2\sigma\Delta b\,\mathbb{C}_{\kappa_p(t)}\right) : \mathbb{D}_{\kappa_p(t)} + \operatorname{div}(\sigma\dot{b}\,\nabla b) - (\varrho^2\psi_0'(\varrho) - \pi)\operatorname{div}\vec{v}.
\end{aligned}$$

Inserting the result into (3.5) we obtain

$$\begin{aligned}
\xi &= \left(\mathbb{T} - \mu(\mathbb{B}_{\kappa_p(t)} - \mathbb{I}) + \sigma((\nabla b \otimes \nabla b) + 2\Delta b\,\mathbb{B}_{\kappa_p(t)}) + (\varrho^2\psi_0'(\varrho) - \pi)\mathbb{I}\right) : \mathbb{D} \\
&\quad + \left(\mu(\mathbb{C}_{\kappa_p(t)} - \mathbb{I}) - 2\sigma\Delta b\,\mathbb{C}_{\kappa_p(t)}\right) : \mathbb{D}_{\kappa_p(t)} - \operatorname{div}(\vec{j}_e - \theta\vec{j}_\eta + \sigma\dot{b}\,\nabla b) \quad \text{with} \quad \xi \geq 0.
\end{aligned}$$

Finally, setting[1]

$$\begin{aligned}
\vec{j}_\eta &= \frac{\vec{j}_e + \sigma \dot{b}\, \nabla b}{\theta}, \\
p_{\mathrm{th}}^{\mathrm{NS}} &:= \varrho^2\, \psi_0'(\varrho), \\
\mathbb{T}_{\mathrm{el}} &:= -(p_{\mathrm{th}}^{\mathrm{NS}} - \pi)\mathbb{I} + \mu(\mathbb{B}_{\kappa_p(t)} - \mathbb{I}) - \sigma((\nabla b \otimes \nabla b) + 2\Delta b\, \mathbb{B}_{\kappa_p(t)}),
\end{aligned}$$

we arrive at

$$\xi = (\mathbb{T} - \mathbb{T}_{\mathrm{el}}) : \mathbb{D} + \big(\mu(\mathbb{C}_{\kappa_p(t)} - \mathbb{I}) - 2\sigma\Delta b\, \mathbb{C}_{\kappa_p(t)}\big) : \mathbb{D}_{\kappa_p(t)} \quad \text{with} \quad \xi \geq 0, \tag{3.14}$$

which provides key information in the derivation of an entire hierarchy of models, ranging from (compressible/incompressible) Euler and Navier–Stokes fluids, through Maxwell and Oldroyd-B fluids, to diffusive Maxwell and diffusive Oldroyd-B fluids. In addition, the procedure also clarifies how these fluids store and dissipate energy and consequently what quantities are *a priori* controlled by the initial data (which then indicates what *a priori* information/estimates one can use in the analysis of the model).

To illustrate the procedure and to derive a simplified (toy) model that is then analyzed in the second part of this paper, we rewrite (3.14) in the following way. Using the subscript $_\delta$ to signify the deviatoric (traceless) part of a tensor, i.e. $\mathbb{A}_\delta := \mathbb{A} - \frac{1}{3}\mathrm{tr}\,\mathbb{A}\,\mathbb{I}$, noticing that

$$\mathbb{A} : \mathbb{B} = \mathbb{A}_\delta : \mathbb{B}_\delta + \mathrm{tr}\,\mathbb{A}\frac{\mathrm{tr}\,\mathbb{B}}{3}$$

and denoting the *mean normal stress* by m, i.e.,

$$m := \frac{1}{3}\mathrm{tr}\,\mathbb{T},$$

we rewrite (3.14) as

$$\begin{aligned}
\xi = (\mathbb{T}_\delta - [\mathbb{T}_{\mathrm{el}}]_\delta) : \mathbb{D}_\delta &+ \left(m - \frac{\mathrm{tr}\,\mathbb{T}_{\mathrm{el}}}{3}\right) \mathrm{div}\, \vec{v} \\
&+ \big(\mu[\mathbb{C}_{\kappa_p(t)}]_\delta - 2\sigma\Delta b\, [\mathbb{C}_{\kappa_p(t)}]_\delta\big) : [\mathbb{D}_{\kappa_p(t)}]_\delta \\
&+ (\mu(b-3) - 2\sigma b\, \Delta b)\, \frac{\mathrm{tr}\,\mathbb{D}_{\kappa_p(t)}}{3} \quad \text{with} \quad \xi \geq 0,
\end{aligned} \tag{3.15}$$

where

$$\begin{aligned}
[\mathbb{T}_{\mathrm{el}}]_\delta &= \mu[\mathbb{B}_{\kappa_p(t)}]_\delta - \sigma((\nabla b \otimes \nabla b)_\delta + 2\Delta b\, [\mathbb{B}_{\kappa_p(t)}]_\delta), \\
\frac{\mathrm{tr}\,\mathbb{T}_{\mathrm{el}}}{3} &= -p_{\mathrm{th}}^{\mathrm{NS}} + \frac{1}{3}[\mu(b-3) - \sigma|\nabla b|^2 - 2\sigma b\, \Delta b] + \pi \\
&\qquad \text{with } \pi = \frac{\mu}{2}(b - 3 - \ln\det \mathbb{B}_{\kappa_p(t)}) + \frac{\sigma}{2}|\nabla b|^2.
\end{aligned}$$

Referring to [**36**] for details, we wish to emphasize that the structure of (3.15) is rich enough to incorporate the compressible and incompressible Euler and Navier–Stokes, Maxwell, Oldroyd-B, Giesekus, diffusive Maxwell, diffusive Oldroyd-B, and diffusive Giesekus models.

---

[1] Note that the first line states how the entropy flux $\vec{j}_\eta$ is related to the energy flux $\vec{j}_e$, while in the second and third lines we have introduced new notation to simplify the formulae.

Here, we make three simplifications: the fluid is assumed to be incompressible, the density is taken to be uniform and the elastic part of the deformation is supposed to be purely spherical[2]. This means that

$$\operatorname{div} \vec{v} = 0, \quad \varrho \text{ is a positive constant} \quad \text{and} \quad [\mathbb{C}_{\kappa_p(t)}]_\delta = \mathbb{O}. \tag{3.16}$$

Then,

$$\mathbb{C}_{\kappa_p(t)} = [\mathbb{C}_{\kappa_p(t)}]_\delta + \frac{\operatorname{tr} \mathbb{C}_{\kappa_p(t)}}{3}\mathbb{I} = \mathbb{O} + \frac{\operatorname{tr} \mathbb{B}_{\kappa_p(t)}}{3}\mathbb{I} = \frac{b}{3}\mathbb{I},$$

where $\mathbb{O}$ is the zero tensor; note that $[\mathbb{B}_{\kappa_p(t)}]_\delta$ also vanishes. As a consequence of these simplifications, (3.12) and (3.15) reduce to

$$\dot{b} = -\frac{2}{3}\, b \operatorname{tr} \mathbb{D}_{\kappa_p(t)}, \tag{3.17}$$

and

$$\xi = (\mathbb{T}_\delta - [\mathbb{T}_{\mathrm{el}}]_\delta) : \mathbb{D} + (\mu(b-3) - 2\sigma b\, \Delta b)\, \frac{\operatorname{tr} \mathbb{D}_{\kappa_p(t)}}{3} \quad \text{with} \quad \xi \geq 0, \tag{3.18}$$

where

$$[\mathbb{T}_{\mathrm{el}}]_\delta = -\sigma(\nabla b \otimes \nabla b)_\delta.$$

Requiring that

$$\begin{aligned} \mathbb{T}_\delta - [\mathbb{T}_{\mathrm{el}}]_\delta &= 2\nu\, \mathbb{D} \qquad \text{with } \nu > 0, \\ \mu(b-3) - 2\sigma b\, \Delta b &= 2\nu_1 \frac{\operatorname{tr} \mathbb{D}_{\kappa_p(t)}}{3} \qquad \text{with } \nu_1 > 0, \end{aligned} \tag{3.19}$$

we obtain

$$\xi = 2\nu |\mathbb{D}|^2 + 2\nu_1 \frac{|\operatorname{tr} \mathbb{D}_{\kappa_p(t)}|^2}{9}. \tag{3.20}$$

Referring then to (3.17) we deduce that (3.19) leads to

$$\begin{aligned} \mathbb{T} = m\mathbb{I} + 2\nu\mathbb{D} - \sigma(\nabla b \otimes \nabla b)_\delta &= \phi\mathbb{I} + 2\nu\mathbb{D} - \sigma(\nabla b \otimes \nabla b), \\ \nu_1 \frac{\dot{b}}{b} + \mu(b-3) - 2\sigma b\, \Delta b &= 0, \end{aligned} \tag{3.21}$$

and (3.20) implies that

$$\xi = 2\nu |\mathbb{D}|^2 + \frac{\nu_1}{2}\left|\frac{\dot{b}}{b}\right|^2. \tag{3.22}$$

Note that m and consequently $\phi := m - \frac{\sigma}{3}|\nabla b|^2$ cannot be determined constitutively because of the incompressibility constraint. Consequently, it is an additional unknown quantity. In what follows we relabel ϕ by $-p$, as this symbol is used in most studies concerning incompressible fluids. To conclude, recalling also the definition of the material time derivative, the system of governing equations takes the form

$$\operatorname{div} \vec{v} = 0, \tag{3.23}$$

$$\varrho\, (\partial_t \vec{v} + \operatorname{div}(\vec{v} \otimes \vec{v})) - \operatorname{div} \mathbb{T} = 0, \tag{3.24}$$

$$\mathbb{T} = -p\,\mathbb{I} + 2\nu\mathbb{D} - \sigma(\nabla b \otimes \nabla b), \tag{3.25}$$

$$\nu_1 \partial_t b + \nu_1 \operatorname{div}(b\vec{v}) + \mu(b^2 - 3b) - 2\sigma b^2 \Delta b = 0. \tag{3.26}$$

[2]This is reminiscent of the response leading to the Euler fluid.

We wish to emphasize that despite the simplification $[\mathbb{C}_{\kappa_p(t)}]_\delta = \mathbb{O}$ (see (3.16)), the reduced model (3.23)–(3.26) retains all of the material parameters of the original problem in the framework. It is also worth emphasizing that not only does the diffusion term appear in the equation for the mean normal elastic stress b, but also the Korteweg stress is present in the constitutive relation for the Cauchy stress.

Setting

$$\tilde{b} := \frac{b}{3} \quad \text{and} \quad \tilde{\sigma} := 9\sigma$$

we can rewrite (3.23)–(3.26) as follows:

$$\operatorname{div} \vec{v} = 0, \tag{3.27}$$

$$\varrho(\partial_t \vec{v} + \operatorname{div}(\vec{v} \otimes \vec{v})) - \operatorname{div} \mathbb{T} = 0, \tag{3.28}$$

$$\mathbb{T} = -p\,\mathbb{I} + 2\nu\mathbb{D} - \tilde{\sigma}(\nabla\tilde{b} \otimes \nabla\tilde{b}), \tag{3.29}$$

$$\nu_1\partial_t\tilde{b} + \nu_1 \operatorname{div}(\tilde{b}\vec{v}) + 3\mu(\tilde{b}^2 - \tilde{b}) - 2\tilde{\sigma}\tilde{b}^2\Delta\tilde{b} = 0, \tag{3.30}$$

which, upon relabeling $b := \tilde{b}$, $\sigma := \tilde{\sigma}$ and $\mu := 3\mu$ coincides with the system (1.1)–(1.4) that is analyzed in the next sections.

We conclude this section by showing that the choice of the constitutive relations for ψ and ξ directly yields the relevant *a priori* estimates in a straightforward way. Note first that with the simplifications (3.16) and relabeling (9σ by σ and 3μ by μ) the constitutive relation (3.11) reduces to

$$\varrho\psi = \varrho\psi_0(\varrho) + \frac{\mu}{2}(b - 1 - \ln b) + \frac{\sigma}{2}|\nabla b|^2. \tag{3.31}$$

Next, by taking the scalar product of (3.28) with $\vec{v}$ we obtain

$$\partial_t\left(\varrho\frac{|\vec{v}|^2}{2}\right) + \operatorname{div}\left(\varrho\frac{|\vec{v}|^2}{2}\vec{v}\right) - \operatorname{div}(\mathbb{T}\vec{v}) + \mathbb{T} : \mathbb{D} = 0. \tag{3.32}$$

Using the equation (3.5) then leads to

$$\partial_t\left(\varrho\frac{|\vec{v}|^2}{2}\right) + \varrho\dot{\psi} + \xi + \operatorname{div}\left(\varrho\frac{|\vec{v}|^2}{2}\vec{v} - \mathbb{T}\vec{v} + \vec{j}_e - \theta\vec{j}_\eta\right) = 0. \tag{3.33}$$

Inserting (3.20) and (3.31) in (3.33), and recalling that $\dot{\psi} = \partial_t\psi + (\vec{v}\cdot\nabla)\psi$ and $\varrho\psi_0(\varrho) - \frac{\mu}{2}$ is constant, we arrive at

$$\begin{aligned} &\frac{1}{2}\partial_t\left(\varrho|\vec{v}|^2 + \sigma|\nabla b|^2 + \mu(b - \ln b)\right) + 2\nu|\mathbb{D}|^2 + \frac{\nu_1}{2}\left|\frac{\dot{b}}{b}\right|^2 \\ &\quad + \operatorname{div}\left(\frac{1}{2}\left(\varrho|\vec{v}|^2 + \mu(b - \ln b) + \sigma|\nabla b|^2\right)\vec{v} - \mathbb{T}\,\vec{v} + \sigma\dot{b}\nabla b\right) = 0. \end{aligned} \tag{3.34}$$

The next section starts with the derivation of (3.34) from a PDE-analytic point of view.

4. Function spaces and the statement of the main result

In this section, we provide the precise statement of the main theorem regarding the long-time existence of large-data weak solutions to the problem (1.1)–(1.7).

To motivate the choice of function spaces, we first briefly derive the basic energy estimate. For these formal calculations, we assume that all quantities are well defined; in particular, we assume that b is strictly positive. Taking the scalar product of (1.2) with $\vec{v}$ and using the incompressibility constraint (1.1), we obtain the following identity (for the evolution of the kinetic energy):

$$\frac{1}{2}\partial_t(\varrho|\vec{v}|^2) + \frac{1}{2}\operatorname{div}(\varrho\vec{v}|\vec{v}|^2) - \operatorname{div}(\mathbb{T}\,\vec{v}) + \mathbb{T} : \mathbb{D} = 0.$$

Employing further (1.3) and again (1.1) we obtain

$$\frac{1}{2}\partial_t(\varrho|\vec{v}|^2) + 2\nu|\mathbb{D}|^2 = \sigma(\nabla b \otimes \nabla b) : \mathbb{D} - \frac{1}{2}\operatorname{div}(\varrho\vec{v}|\vec{v}|^2) + \operatorname{div}(\mathbb{T}\,\vec{v}). \tag{4.1}$$

Next, dividing (1.4) by b^2 and multiplying the result by $(\partial_t b + \operatorname{div}(b\vec{v}))$, we obtain

$$\begin{aligned}\frac{\nu_1}{b^2}\left|\partial_t b + \operatorname{div}(b\vec{v})\right|^2 &= (-\mu(1-b^{-1}) + 2\sigma\Delta b)(\partial_t b + \operatorname{div}(b\vec{v}))\\ &= -\mu\left(\partial_t(b - \ln b) + \operatorname{div}(\vec{v}(b-\ln b))\right) - \sigma\partial_t|\nabla b|^2 + 2\sigma\operatorname{div}(\partial_t b\nabla b)\\ &\quad + 2\sigma\operatorname{div}(\nabla b(\operatorname{div}(b\vec{v}))) - 2\sigma\nabla b\cdot\nabla(\operatorname{div}(b\vec{v})).\end{aligned} \tag{4.2}$$

Using the fact that $\operatorname{div}\vec{v} = 0$ several times, the last term on the right-hand side can be rewritten as

$$\begin{aligned}\nabla b\cdot\nabla(\operatorname{div}(b\vec{v})) &= \nabla b\cdot\nabla(\vec{v}\cdot\nabla b)\\ &= (\nabla b\otimes\nabla b) : \mathbb{D} + \frac{1}{2}\operatorname{div}(\vec{v}|\nabla b|^2).\end{aligned}$$

Using this expression, (4.2) takes the form

$$\begin{aligned}&\partial_t\left(\sigma|\nabla b|^2 + \mu(b - \ln b)\right) + \frac{\nu_1}{b^2}\left|\partial_t b + \operatorname{div}(b\vec{v})\right|^2 = -2\sigma(\nabla b\otimes\nabla b) : \mathbb{D}\\ &\quad + \operatorname{div}\left(2\sigma(\partial_t b + \operatorname{div}(b\vec{v}))\nabla b - (\mu(b-\ln b) + \sigma|\nabla b|^2)\vec{v}\right).\end{aligned} \tag{4.3}$$

Finally, dividing (4.3) by 2 and adding the result to (4.1), we deduce the energy identity

$$\begin{aligned}&\frac{1}{2}\partial_t\left(\varrho|\vec{v}|^2 + \sigma|\nabla b|^2 + \mu(b-\ln b)\right) + 2\nu|\mathbb{D}|^2 + \frac{\nu_1}{2b^2}\left|\partial_t b + \operatorname{div}(b\vec{v})\right|^2\\ &= \operatorname{div}\left(\mathbb{T}\,\vec{v} + \sigma(\partial_t b + \operatorname{div}(b\vec{v}))\nabla b - \frac{1}{2}\left(\varrho|\vec{v}|^2 + \mu(b-\ln b) + \sigma|\nabla b|^2\right)\vec{v}\right).\end{aligned} \tag{4.4}$$

Hence, integration of the result over $\Omega\times(0,T)$, using integration by parts and the boundary conditions (1.5), we deduce that

$$\begin{aligned}&\sup_{t\in(0,T)}\int_\Omega\left(\varrho|\vec{v}|^2 + \sigma|\nabla b|^2 + \mu(b-\ln b)\right)\,\mathrm{d}x + \int_Q 4\nu|\mathbb{D}|^2 + \frac{\nu_1}{b^2}\left|\partial_t b + \operatorname{div}(b\vec{v})\right|^2\,\mathrm{d}x\,\mathrm{d}t\\ &\quad\le \int_\Omega\left(\varrho|\vec{v}_0|^2 + \sigma|\nabla b_0|^2 + \mu(b_0 - \ln b_0)\right)\,\mathrm{d}x.\end{aligned}$$

This natural energy estimate thus motivates the minimal assumptions on the data $\vec{v}_0$ and b_0 and simultaneously indicates the correct choices of function spaces for the problem. In addition, as it will be seen later, the unknown b also satisfies a version of minimum/maximum principle and therefore we shall also require that b_0 is bounded and uniformly positive on Ω.

Before we formulate the main theorem of the paper, we briefly recall the notations for the relevant function spaces. The standard Lebesgue and Sobolev spaces are denoted by $L^p(\Omega)$ and $W^{k,p}(\Omega)$, respectively, and are equipped with the usual norms denoted by $\|\cdot\|_p$ and $\|\cdot\|_{k,p}$, respectively. Since we also need to deal with vector- and tensor-valued functions, we will emphasize this character by writing

$L^p(\Omega)^d$ or $L^p(\Omega)^{d\times d}$ if necessary. Moreover, since the velocity is a solenoidal function, we define

$$L^2_{0,\mathrm{div}} := \overline{\{\vec{v}\in \mathcal{C}^\infty_0(\Omega;\mathbb{R}^d):\ \operatorname{div}\vec{v}=0\}}^{\|\cdot\|_2},$$
$$W^{1,2}_{0,\mathrm{div}} := \overline{\{\vec{v}\in \mathcal{C}^\infty_0(\Omega;\mathbb{R}^d):\ \operatorname{div}\vec{v}=0\}}^{\|\cdot\|_{1,2}}.$$

Having defined the proper function spaces, we notice that the assumptions on the data, see (1.7), are a natural consequence of (4.4). We can now formulate our main theorem.

THEOREM 4.1. *Let $\Omega\subset\mathbb{R}^d$ be a Lipschitz domain and $T>0$. Assume that ϱ, ν, ν_1, μ and σ are positive constants and that $(\vec{v}_0,b_0)$ satisfies (1.7). Then, there exists a couple $(\vec{v},b)$ such that*

$$\vec{v}\in L^\infty(0,T;L^2_{0,\mathrm{div}})\cap L^2(0,T;W^{1,2}_{0,\mathrm{div}}), \tag{4.5}$$
$$b\in L^\infty(0,T;W^{1,2}(\Omega)),\quad b,b^{-1}\in L^\infty(Q), \tag{4.6}$$
$$\partial_t\vec{v}\in L^1(0,T;(W^{1,2}_{0,\mathrm{div}}\cap W^{d+1,2}(\Omega)^d)^*), \tag{4.7}$$
$$\partial_t b\in L^1(Q) \tag{4.8}$$

and

$$\begin{aligned}(\partial_t b+\operatorname{div}(b\vec{v}))&\in L^2(Q),\\ \Delta b&\in L^2(Q),\end{aligned} \tag{4.9}$$

and this couple solves (1.1)–(1.6) in the following sense: for almost all $t\in(0,T)$ and all $\vec{w}\in W^{1,2}_{0,\mathrm{div}}\cap W^{d+1,2}(\Omega)^d$ we have (recall that $\mathbb{D}$ denotes the symmetric part of $\nabla\vec{v}$)

$$\langle\partial_t(\varrho\vec{v}),\vec{w}\rangle+\int_\Omega(2\nu\mathbb{D}-\sigma\nabla b\otimes\nabla b-\varrho\vec{v}\otimes\vec{v}):\nabla\vec{w}\,\mathrm{d}x=0; \tag{4.10}$$

for almost all $t\in(0,T)$ and all $w\in W^{1,2}(\Omega)$ we have

$$\int_\Omega\left(\frac{\nu_1(\partial_t b+\operatorname{div}(b\vec{v}))}{b^2}+\mu\left(1-\frac{1}{b}\right)\right)w+2\sigma\nabla b\cdot\nabla w\,\mathrm{d}x=0, \tag{4.11}$$

and the initial data are attained in the following sense:

$$\lim_{t\to0_+}(\|\vec{v}(t)-\vec{v}_0\|_2+\|b(t)-b_0\|_{1,2})=0. \tag{4.12}$$

Moreover, the following energy inequality holds for all $t\in(0,T)$:

$$\begin{aligned}&\frac12\int_\Omega\left(\varrho|\vec{v}(t)|^2+\sigma|\nabla b(t)|^2+\mu(b(t)-\ln b(t))\right)\mathrm{d}x\\ &\quad+\int_0^t\int_\Omega 2\nu|\mathbb{D}|^2+\frac{\nu_1}{2b^2}|\partial_t b+\operatorname{div}(b\vec{v})|^2\,\mathrm{d}x\,\mathrm{d}\tau\\ &\quad\le\frac12\int_\Omega\left(\varrho|\vec{v}_0|^2+\sigma|\nabla b_0|^2+\mu(b_0-\ln b_0)\right)\mathrm{d}x.\end{aligned} \tag{4.13}$$

Notice here that (1.1) is automatically satisfied since $\vec{v}\in L^\infty(0,T;L^2_{0,\mathrm{div}})$; the identity (4.10) represents the weak formulation of (1.2), (1.3) and $(1.5)_1$, where we have omitted the presence of the pressure p by choosing divergence-free test functions. Finally, (4.11) is a weak formulation of (1.4) and $(1.5)_2$, which is obtained by division of (1.4) by b^2, multiplying the result by a test function w, integrating over Ω and in the elliptic term using integration by parts together with $(1.5)_2$.

Observe also that because of the assumed regularity (4.6) and (4.9), we could simply say that (1.4) holds almost everywhere in Q, but we would lose the information $(1.5)_2$, which is encoded in (4.11). Nevertheless, (1.4) can be directly obtained from (4.11) by setting $w := b^2\varphi$ with an arbitrary $\varphi \in \mathcal{C}_0^\infty(\Omega)$ and noting that (here we use the fact that $\Delta b \in L^2(\Omega)$ and integration by parts)

$$\int_\Omega \nabla b \cdot \nabla(b^2\varphi)\,\mathrm{d}x = \int_\Omega 2b|\nabla b|^2\varphi + b^2\nabla b\cdot\nabla\varphi\,\mathrm{d}x = \int_\Omega 2b|\nabla b|^2\varphi - \operatorname{div}(b^2\nabla b)\varphi\,\mathrm{d}x$$
$$= -\int_\Omega b^2\Delta b\varphi\,\mathrm{d}x.$$

5. Proof of Theorem 4.1

For the sake of simplicity we set the parameters ν_1 and μ equal to 2 and ϱ, ν and σ equal to 1, and we also divide (1.4) by 2 in order to avoid the presence of all constants. Note that such a simplification does not change the assertion of the theorem, as the proof can be straightforwardly extended to any other values of these parameters thanks to their positivity. We will also frequently use the symbol C to denote a generic positive constant, whose value may change from line to line, and can depend only on the data. In instances when C also depends on the indices of the approximating problem appearing in the proof, this will be clearly indicated.

5.1. Galerkin approximation. First, we introduce the following cut-off function

$$T_n(s) := \min\left\{n, \max\{n^{-1}, s\}\right\} \text{ for } s \in \mathbb{R}.$$

Next, we find a basis $\{\vec{w}_n\}_{n=1}^\infty$ of $W^{1,2}_{0,\mathrm{div}} \cap W^{d+1,2}(\Omega;\mathbb{R}^d)$, which is orthogonal in $L^2_{0,\mathrm{div}}$ and a basis $\{w_n\}_{n=1}^\infty$ of $W^{1,2}(\Omega)$, which is orthonormal in $L^2(\Omega)$ and orthogonal in $W^{1,2}(\Omega)$. The existence of the latter follows from a version of the Hilbert–Schmidt theorem (cf. Lemma 5.1 in [**18**]), according to which w_n, $n \in \mathbb{N}$, can be taken to be the eigenfunctions of the operator $w \mapsto -\Delta w + w$ subject to a homogeneous Neumann boundary condition on $\partial\Omega$. By Theorem 1.2 in [**20**] then $w_n \in W^{1,r}(\Omega)$ for all $r \in (\frac{2d}{d+1} - \epsilon, \frac{2d}{d-1} + \epsilon)$, $d \in \{2,3\}$. Hence, by Morrey's embedding theorem $w_n \in C^{0,\alpha}(\overline{\Omega})$, whereby $\Delta w_n \in C^{0,\alpha}(\overline{\Omega})$, for some $\alpha \in (0, \frac{1}{2})$. Let, for arbitrary $n \in \mathbb{N}$ and $\ell \in \mathbb{N}$, the symbols $\vec{P}^n$ and P^ℓ denote the orthogonal (in L^2) projections onto $\operatorname{span}\{\vec{w}_i\}_{i=1}^n$ and $\operatorname{span}\{w_i\}_{i=1}^\ell$, respectively. Observe also that $\|P^\ell(w)\|_{1,2} \le \|w\|_{1,2}$ for all $w \in W^{1,2}(\Omega)$ and all $\ell \in \mathbb{N}$. Finally, we consider

$$\vec{v}^{n,\ell}(t,x) := \sum_{i=1}^n \alpha_i^{n,\ell}(t)\vec{w}_i(x), \qquad b^{n,\ell}(t,x) := \sum_{i=1}^\ell \beta_i^{n,\ell}(t) w_i(x),$$

satisfying

$$\vec{v}^{n,\ell}(0,x) = \vec{v}_0^n(x) := \vec{P}^n(\vec{v}_0)(x), \qquad b^{n,\ell}(0,x) = b_0^\ell(x) := P^\ell(b_0)(x),$$

and solving, over $(0,T)$, the following system[3] of ordinary differential equations:

$$\int_\Omega \partial_t \vec{v}^{n,\ell} \cdot \vec{w}_i - \vec{v}^{n,\ell} \otimes \vec{v}^{n,\ell} : \nabla \vec{w}_i + \mathbb{S}^{n,\ell} : \nabla \vec{w}_i \,\mathrm{d}x = 0, \qquad i = 1, \dots, n, \tag{5.1}$$

$$\begin{aligned} \int_\Omega \frac{\partial_t b^{n,\ell} w_j}{(T_n(b^{n,\ell}))^2} + \frac{\nabla b^{n,\ell} \cdot \vec{v}^{n,\ell}}{(T_n(b^{n,\ell}))^2} w_j + (1 - (T_n(b^{n,\ell}))^{-1}) w_j \,\mathrm{d}x \\ + \int_\Omega \nabla b^{n,\ell} \cdot \nabla w_j \,\mathrm{d}x = 0, \qquad j = 1, \dots, \ell, \end{aligned} \tag{5.2}$$

where $\mathbb{S}^{n,\ell}$ is defined a.e. in Q by

$$\mathbb{S}^{n,\ell} := 2\mathbb{D}^{n,\ell} - (\nabla b^{n,\ell} \otimes \nabla b^{n,\ell}), \quad \mathbb{D}^{n,\ell} := \frac{1}{2}(\nabla \vec{v}^{n,\ell} + (\nabla \vec{v}^{n,\ell})^{\mathrm{T}}). \tag{5.3}$$

To prove the existence of a solution to the above problem is not a difficult task but since it requires more than invoking standard Carathéodory theory, we provide a sketch of the proof in the Appendix for the sake of completeness.

5.2. The limit $\ell \to \infty$. We start this part by establishing the estimates that will be independent of ℓ but can still be dependent on n. Nevertheless, these estimates will suffice for the purpose of letting $\ell \to \infty$.

First, we define the following primitive function:

$$\Theta_n(s) := \int_0^s \frac{s}{(T_n(s))^2} \,\mathrm{d}s.$$

It follows directly from the properties of the function T_n that

$$n^{-2} s^2 \le 2\Theta_n(s) \le n^2 s^2. \tag{5.4}$$

Then we multiply the j-th equation in (5.2) by $\beta_j^{n,\ell}(t)$ and sum the resulting identities over $j = 1, \dots, \ell$ to obtain the following equality:

$$\int_\Omega \frac{\partial_t b^{n,\ell} b^{n,\ell}}{(T_n(b^{n,\ell}))^2} + \frac{b^{n,\ell} \nabla b^{n,\ell} \cdot \vec{v}^{n,\ell}}{(T_n(b^{n,\ell}))^2} + (1 - (T_n(b^{n,\ell}))^{-1}) b^{n,\ell} + |\nabla b^{n,\ell}|^2 \,\mathrm{d}x = 0. \tag{5.5}$$

Using the definition of Θ_n and the facts that $\operatorname{div} \vec{v}^{n,\ell} = 0$ and $\vec{v}^{n,\ell}$ vanishes on $\partial\Omega$, the second term vanishes and the above identity reduces to the following inequality (using also the properties of T_n and Θ_n):

$$\frac{\mathrm{d}}{\mathrm{d}t} \int_\Omega \Theta_n(b^{n,\ell}(t)) \,\mathrm{d}x + \|\nabla b^{n,\ell}(t)\|_2^2 \le C(n) \left(1 + \int_\Omega \Theta_n(b^{n,\ell}) \,\mathrm{d}x\right).$$

Thus, using Gronwall's lemma, we have the following ℓ-independent bound:

$$\sup_{t \in (0,T)} \|b^{n,\ell}(t)\|_2^2 + \int_0^T \|b^{n,\ell}(t)\|_{1,2}^2 \,\mathrm{d}t \le C(n)(1 + \|b_0\|_2^2). \tag{5.6}$$

Next, we focus on ℓ-independent bounds on $\vec{v}^{n,\ell}$. Since n is fixed, we first recall that (in an n-dimensional space)

$$\|\vec{v}^{n,\ell}(t)\|_{1,\infty} \le C(n) \|\vec{v}^{n,\ell}(t)\|_2. \tag{5.7}$$

[3]Notice that (5.2) is an approximation of (1.4) after division by b^2, i.e., an approximation of (4.11).

Furthermore, multiplying the i-th equation in (5.1) by $\alpha_i^{n,\ell}$ and summing the result over $i = 1, \dots, n$, using the definition of $\mathbb{S}^{n,\ell}$, we get that (using again the fact that $\operatorname{div} \vec{v}^{n,\ell} = 0$ and so the convective term vanishes)

$$
\begin{aligned}
\frac{1}{2}\frac{\mathrm{d}}{\mathrm{d}t}\|\vec{v}^{n,\ell}(t)\|_2^2 + 2\|\mathbb{D}^{n,\ell}(t)\|_2^2 &= \int_\Omega \nabla b^{n,\ell}(t) \otimes \nabla b^{n,\ell}(t) : \mathbb{D}^{n,\ell}(t)\,\mathrm{d}x \\
&\le \|\nabla b^{n,\ell}(t)\|_2^2 \|\vec{v}^{n,\ell}(t)\|_{1,\infty} \\
&\le C(n)\|\nabla b^{n,\ell}(t)\|_2^2 \|\vec{v}^{n,\ell}(t)\|_2 \\
&\le C(n)\|\nabla b^{n,\ell}(t)\|_2^2 + C(n)\|\nabla b^{n,\ell}(t)\|_2^2 \|\vec{v}^{n,\ell}(t)\|_2^2,
\end{aligned}
\tag{5.8}
$$

where we have also used (5.7). Therefore, thanks to the estimate (5.6), Gronwall's lemma implies that

$$
\begin{aligned}
\sup_{t\in(0,T)} \|\vec{v}^{n,\ell}(t)\|_{1,\infty}^2 &\le C(n) \sup_{t\in(0,T)} |\alpha^{n,\ell}(t)|^2 \\
&\le C(n) \sup_{t\in(0,T)} \|\vec{v}^{n,\ell}(t)\|_2^2 \le C(n, \|\vec{v}_0\|_2, \|b_0\|_2).
\end{aligned}
\tag{5.9}
$$

Finally, we derive uniform bounds on time derivatives. To do so, we multiply the j-th equation in (5.2) by $\partial_t \beta_j^{n,\ell}$ to deduce the following identity:

$$
\begin{aligned}
&\frac{1}{2}\frac{\mathrm{d}}{\mathrm{d}t}\int_\Omega |\nabla b^{n,\ell}|^2\,\mathrm{d}x + \int_\Omega \frac{|\partial_t b^{n,\ell}|^2}{(T_n(b^{n,\ell}))^2}\,\mathrm{d}x \\
&\quad = \int_\Omega ((T_n(b^{n,\ell}))^{-1} - 1)\,\partial_t b^{n,\ell} - \frac{\partial_t b^{n,\ell} \nabla b^{n,\ell} \cdot \vec{v}^{n,\ell}}{(T_n(b^{n,\ell}))^2}\,\mathrm{d}x.
\end{aligned}
\tag{5.10}
$$

Hence, by employing the estimate (5.9), Young's inequality and the properties of T_n, we get

$$
\frac{\mathrm{d}}{\mathrm{d}t}\|\nabla b^{n,\ell}(t)\|_2^2 + \|\partial_t b^{n,\ell}(t)\|_2^2 \le C(n)(1 + \|\nabla b^{n,\ell}(t)\|_2^2),
$$

so that by applying Gronwall's lemma again we have

$$
\sup_{t\in(0,T)} \|b^{n,\ell}(t)\|_{1,2}^2 + \int_0^T \|\partial_t b^{n,\ell}(t)\|_2^2\,\mathrm{d}t \le C(n, \|b_0\|_{1,2}, \|\vec{v}_0\|_2). \tag{5.11}
$$

Finally, using the last inequality and also (5.9), we see from (5.1) that

$$
\sup_{t\in(0,T)} \left|\frac{\mathrm{d}}{\mathrm{d}t}\alpha^{n,\ell}(t)\right| \le C(n). \tag{5.12}
$$

Having established bounds that are uniform with respect to ℓ, we can now focus on taking the limit $\ell \to \infty$ in (5.1), (5.2). Indeed, using (5.9) and (5.12), there exists a subsequence that we do not relabel such that

$$
\alpha^{n,\ell} \rightharpoonup^* \alpha^n \qquad \text{weakly}^* \text{ in } W^{1,\infty}(0,T;\mathbb{R}^n), \tag{5.13}
$$

$$
\alpha^{n,\ell} \to \alpha^n \qquad \text{strongly in } \mathcal{C}([0,T];\mathbb{R}^n). \tag{5.14}
$$

Thus, it follows from the definition of $\vec{v}^{n,\ell}$ that

$$
\vec{v}^{n,\ell} \rightharpoonup^* \vec{v}^n \qquad \text{weakly}^* \text{ in } W^{1,\infty}(0,T;W^{1,\infty}(\Omega)^d), \tag{5.15}
$$

$$
\vec{v}^{n,\ell} \to \vec{v}^n \qquad \text{strongly in } \mathcal{C}([0,T];W^{1,\infty}(\Omega)^d), \tag{5.16}
$$

where

$$\vec{v}^n(t,x) = \sum_{i=1}^{n} \alpha_i^n(t)\vec{w}_i(x) \qquad \text{and} \qquad \vec{v}^n(0,x) = \vec{P}^n(\vec{v}_0)(x).$$

Similarly, using (5.11) and compact embedding, we find a subsequence (that is again labeled in the same way) such that

$$b^{n,\ell} \rightharpoonup^* b^n \qquad \text{weakly}^* \text{ in } L^\infty(0,T;W^{1,2}(\Omega)), \tag{5.17}$$

$$b^{n,\ell} \rightharpoonup b^n \qquad \text{weakly in } W^{1,2}(0,T;L^2(\Omega)), \tag{5.18}$$

$$b^{n,\ell} \to b^n \qquad \text{strongly in } L^2(0,T;L^2(\Omega)) \text{ and a.e. in } Q. \tag{5.19}$$

In addition, it follows directly from the construction that $b^n(0) = b_0$.

The convergence results (5.16) and (5.17)–(5.19) allow us to let $\ell \to \infty$ in (5.2) directly, so we conclude that for almost all $t \in (0,T)$ and all $w \in W^{1,2}(\Omega)$ we have

$$\int_\Omega \frac{\partial_t b^n w}{(T_n(b^n))^2} + \frac{\nabla b^n \cdot \vec{v}^n}{(T_n(b^n))^2} w + (1-(T_n(b^n))^{-1})w + \nabla b^n \cdot \nabla w \,\mathrm{d}x = 0. \tag{5.20}$$

In particular, setting $w := b^n(t)$ in the above identity and integrating the result over $(0,T)$ we obtain

$$\int_Q |\nabla b^n|^2 \,\mathrm{d}x\,\mathrm{d}t = -\int_Q \frac{\partial_t b^n b^n}{(T_n(b^n))^2} + \frac{\nabla b^n \cdot \vec{v}^n}{(T_n(b^n))^2} b^n + (1-(T_n(b^n))^{-1})b^n \,\mathrm{d}x\,\mathrm{d}t. \tag{5.21}$$

Furthermore, integration of (5.5) over $(0,T)$ and the use of (5.16) and (5.17)–(5.19) lead to

$$\begin{aligned}
&\lim_{\ell\to\infty} \int_Q |\nabla b^{n,\ell}|^2 \,\mathrm{d}x\,\mathrm{d}t \\
&= -\lim_{\ell\to\infty} \int_Q \frac{\partial_t b^{n,\ell} b^{n,\ell}}{(T_n(b^{n,\ell}))^2} + \frac{b^{n,\ell}\nabla b^{n,\ell} \cdot \vec{v}^{n,\ell}}{(T_n(b^{n,\ell}))^2} + (1-(T_n(b^{n,\ell}))^{-1})b^{n,\ell} \,\mathrm{d}x\,\mathrm{d}t \\
&= -\int_Q \frac{\partial_t b^n b^n}{(T_n(b^n))^2} + \frac{\nabla b^n \cdot \vec{v}^n}{(T_n(b^n))^2} b^n + (1-(T_n(b^n))^{-1})b^n \,\mathrm{d}x\,\mathrm{d}t \\
&= \int_Q |\nabla b^n|^2 \,\mathrm{d}x\,\mathrm{d}t,
\end{aligned} \tag{5.22}$$

where we have used (5.21) for the last equality. Consequently, combining (5.17) and (5.22) we deduce that

$$b^{n,\ell} \to b^n \qquad \text{strongly in } L^2(0,T;W^{1,2}(\Omega)). \tag{5.23}$$

Finally, this strong convergence result together with (5.13) and (5.14) allows us to let $\ell \to \infty$ in (5.1) and (5.3) to get, for a.e. $t \in (0,T)$,

$$\int_\Omega \partial_t \vec{v}^n \cdot \vec{w}_i - \vec{v}^n \otimes \vec{v}^n \cdot \nabla \vec{w}_i + \mathbb{S}^n \cdot \nabla \vec{w}_i \,\mathrm{d}x = 0, \qquad i = 1,\dots,n, \tag{5.24}$$

where $\{\vec{w}_i\}_{i=1}^n$ is the set of Galerkin basis functions defined at the start of the section; furthermore, $\mathbb{S}^n$ is defined by

$$\mathbb{S}^n := 2\mathbb{D}^n - (\nabla b^n \otimes \nabla b^n) \quad \text{and} \quad \mathbb{D}^n := \frac{1}{2}\big(\nabla \vec{v}^n + (\nabla \vec{v}^n)^{\mathrm{T}}\big). \tag{5.25}$$

We note here that the precise passage to the limit in (5.1) as $\ell \to \infty$, leading to (5.24), requires multiplication of (5.1) by any $\varphi \in C_0^\infty(0,T)$ and integration of the resulting equality over $t \in (0,T)$, followed by passage to the limit therein as $\ell \to \infty$, which then results in (5.24) multiplied by $\varphi \in C_0^\infty(0,T)$ and integrated over $t \in (0,T)$. Thanks to the fact that each term under the integral sign in the limiting equality is locally integrable with respect to $t \in (0,T)$ and the equality holds for all $\varphi \in C_0^\infty(0,T)$, the fundamental theorem of the calculus of variations (Du Bois Reymond's lemma) yields the equality (5.24) for a.e. $t \in (0,T)$, as stated.

5.3. Estimates independent of n. We first focus on minimum and maximum principles for b^n. To this end, we define

$$b_{\max} := \max\{1, \|b_0\|_\infty\}, \qquad \frac{1}{b_{\min}} := \max\left\{1, \left\|\frac{1}{b_0}\right\|_\infty\right\},$$

and we also recall the notations $s_+ := \max\{0,s\}$ and $s_- := \min\{0,s\}$. From now on, we shall assume that n is sufficiently large; more precisely, we shall suppose that $n > b_{\max}$ and $n > (b_{\min})^{-1}$.

Then, we set $w := (b^n - b_{\max})_+$ in (5.20) and using the facts that $b_{\max} \geq 1$ and $n > 1$, we have

$$\begin{aligned}
&\int_\Omega \frac{\partial_t b^n (b^n - b_{\max})_+}{(T_n(b^n))^2} + \frac{\nabla b^n \cdot \vec{v}^n}{(T_n(b^n))^2}(b^n - b_{\max})_+ \,\mathrm{d}x \\
&= -\int_\Omega (1 - (T_n(b^n))^{-1})(b^n - b_{\max})_+ + \nabla b^n \cdot \nabla (b^n - b_{\max})_+ \,\mathrm{d}x \\
&= -\int_{b^n > b_{\max}} \frac{T_n(b_n) - 1}{T_n(b^n)}(b^n - b_{\max})_+ + |\nabla b^n|^2 \,\mathrm{d}x \leq 0.
\end{aligned} \tag{5.26}$$

Hence, defining

$$\Gamma_+(t) := \int_{b_{\max}}^t \frac{(s - b_{\max})_+}{(T_n(s))^2} \,\mathrm{d}s \geq 0,$$

we obtain with the help of (5.26) that

$$\begin{aligned}
0 &\geq \int_\Omega \frac{\partial_t b^n (b^n - b_{\max})_+}{(T_n(b^n))^2} + \frac{\nabla b^n \cdot \vec{v}^n}{(T_n(b^n))^2}(b^n - b_{\max})_+ \,\mathrm{d}x \\
&= \int_\Omega \partial_t \Gamma_+(b^n) + \nabla \Gamma_+(b^n) \cdot \vec{v}^n \,\mathrm{d}x = \frac{\mathrm{d}}{\mathrm{d}t}\int_\Omega \Gamma_+(b^n)\,\mathrm{d}x,
\end{aligned} \tag{5.27}$$

where for the last equality we have used integration by parts in the second term and the fact that $\vec{v}^n \in W^{1,2}_{0,\mathrm{div}}$. Since $b_0 \leq b_{\max}$ a.e. in Ω, the last inequality implies (using the positivity of $\Gamma_+(s)$ whenever $s > b_{\max}$) that

$$b^n \leq b_{\max} \text{ a.e. in } Q. \tag{5.28}$$

Similarly, we deduce the minimum principle for b^n. We set $w := (b^n - b_{\min})_-$ in (5.20) and see that (note that $b_{\min} \leq 1$ and $n > 1$)

$$\begin{aligned}
&\int_\Omega \frac{\partial_t b^n (b^n - b_{\min})_-}{(T_n(b^n))^2} + \frac{\nabla b^n \cdot \vec{v}^n}{(T_n(b^n))^2}(b^n - b_{\min})_- \,\mathrm{d}x \\
&= -\int_\Omega (1 - (T_n(b^n))^{-1})(b^n - b_{\min})_- + \nabla b^n \cdot \nabla (b^n - b_{\min})_- \,\mathrm{d}x \\
&= -\int_{b^n < b_{\min}} \frac{T_n(b_n) - 1}{T_n(b^n)}(b^n - b_{\min})_- + |\nabla b^n|^2 \,\mathrm{d}x \leq 0.
\end{aligned} \tag{5.29}$$

Hence, defining

$$\Gamma_-(t) := -\int_t^{b_{\min}} \frac{(s - b_{\min})_-}{(T_n(s))^2} \,\mathrm{d}s \geq 0,$$

we obtain with the help of (5.29) that

$$0 \geq \int_\Omega \partial_t \Gamma_-(b^n) + \nabla\Gamma_-(b^n)\cdot \vec{v}^n \,\mathrm{d}x = \frac{\mathrm{d}}{\mathrm{d}t}\int_\Omega \Gamma_-(b^n)\,\mathrm{d}x. \tag{5.30}$$

Thus, the last inequality implies (using the positivity of $\Gamma_-(s)$ whenever $s < b_{\min}$) that

$$b^n \geq b_{\min} \text{ a.e. in } Q. \tag{5.31}$$

We continue with further uniform estimates. Having (5.28) and (5.31), we see that for sufficiently large n there holds $T_n(b^n) = b^n$, and (5.20) then reduces to

$$\int_\Omega \frac{\partial_t b^n w}{(b^n)^2} + \frac{\nabla b^n \cdot \vec{v}^n}{(b^n)^2} w + (1-(b^n)^{-1})w + \nabla b^n \cdot \nabla w \,\mathrm{d}x = 0 \tag{5.32}$$

for all $w \in W^{1,2}(\Omega)$. However, using (5.17) and (5.18), we see from the above identity that, for each (large) n, $\Delta b^n \in L^2(0,T;L^2(\Omega))$, and consequently, using interior elliptic regularity we have

$$b^n \in L^2(0,T;W^{2,2}(\tilde{\Omega})) \qquad \text{for all } \tilde{\Omega} \subset\subset \Omega. \tag{5.33}$$

In addition, using (5.32), almost everywhere in Q we have

$$\frac{\partial_t b^n}{(b^n)^2} + \frac{\nabla b^n \cdot \vec{v}^n}{(b^n)^2} + (1-(b^n)^{-1}) - \Delta b^n = 0. \tag{5.34}$$

Next, we would like to mimic the procedure that led to the identity (4.4). This in particular means that we would like to multiply (5.34) by $\partial_t b^n + \nabla b^n \cdot \vec{v}^n$, integrate over Ω, and then use integration by parts in the elliptic term. However, for such a procedure, we do not have sufficient regularity of b^n and therefore we need to proceed more carefully. Thus, instead of testing (5.34) by $\partial_t b^n$, we pass to the limit in (5.10). Indeed, integration of (5.10) over $(0,t)$ leads to

$$\begin{aligned}
&\frac{1}{2}\|\nabla b^{n,\ell}(t)\|_2^2 + \int_0^t\int_\Omega \frac{|\partial_t b^{n,\ell}|^2}{(T_n(b^{n,\ell}))^2}\,\mathrm{d}x\,\mathrm{d}s \\
&= \int_0^t\int_\Omega ((T_n(b^{n,\ell}))^{-1}-1)\partial_t b^{n,\ell} - \frac{\partial_t b^{n,\ell}\nabla b^{n,\ell}\cdot \vec{v}^{n,\ell}}{(T_n(b^{n,\ell}))^2}\,\mathrm{d}x\,\mathrm{d}s + \frac{1}{2}\|\nabla P^\ell(b_0)\|_2^2.
\end{aligned}$$

Then, with the help of (5.16), (5.17)–(5.19), (5.23), the fact that $T_n(b^n) = b^n$ and the weak lower semicontinuity of norms, we obtain that for all[4] $t \in (0,T)$ one has the following inequality:

$$\begin{aligned}
&\frac{1}{2}\|\nabla b^n(t)\|_2^2 + \int_0^t\int_\Omega \frac{|\partial_t b^n|^2}{(b^n)^2}\,\mathrm{d}x\,\mathrm{d}s \\
&\leq \int_0^t\int_\Omega ((b^n)^{-1}-1)\partial_t b^n - \frac{\partial_t b^n \nabla b^n\cdot\vec{v}^n}{(b^n)^2}\,\mathrm{d}x\,\mathrm{d}s + \frac{1}{2}\|b_0\|_{1,2}^2 \\
&= -\int_0^t\int_\Omega \frac{\partial_t b^n\nabla b^n\cdot\vec{v}^n}{(b^n)^2}\,\mathrm{d}x\,\mathrm{d}s + \int_\Omega \ln b^n(t) - \ln b_0 - b^n(t) + b_0\,\mathrm{d}x + \frac{1}{2}\|b_0\|_{1,2}^2.
\end{aligned} \tag{5.35}$$

We note in particular that the second term on the left-hand side of (5.35) arises by letting $\ell \to \infty$ in the second term on the left-hand side of the displayed equality preceding (5.35), by applying the weak lower semicontinuity result stated

[4] In fact we first obtain (5.35) only for almost all $t \in (0,T)$. However, thanks to the continuity of b^n in the weak topology, we can extend the result onto the whole time interval $(0,T)$.

in Theorem 3.23 on p.96 of [**11**], thanks to the strong convergence $b^{n,\ell} \to b^n$ in $L^2(0,T;L^2(\Omega))$ as $\ell \to \infty$ (cf. (5.19)), the weak convergence result $\partial_t b^{n,\ell} \rightharpoonup \partial_t b^n$ in $L^2(0,T;L^2(\Omega))$ as $\ell \to \infty$ (cf. (5.18)), and the convexity of the function $\xi \in \mathbb{R} \mapsto |\xi|^2$ featuring in the numerator of the integrand in the term concerned.

Finally, we multiply (5.34) by $\nabla b^n \cdot \vec{v}^n$ and integrate the result over Ω to get (notice here that thanks to the *a priori* estimates, in particular thanks to the fact that $\Delta b^n \in L^2(\Omega)$ for almost all $t \in (0,T)$, such a procedure is rigorous):

$$\int_\Omega \frac{|\nabla b^n \cdot \vec{v}^n|^2}{(b^n)^2} + \frac{\partial_t b^n \nabla b^n \cdot \vec{v}^n}{(b^n)^2}\,\mathrm{d}x = \int_\Omega ((b^n)^{-1}-1)\nabla b^n \cdot \vec{v}^n + \Delta b^n \nabla b^n \cdot \vec{v}^n\,\mathrm{d}x. \tag{5.36}$$

First, we see that since $\operatorname{div}\vec{v}^n = 0$ and $\vec{v}^n$ has zero trace the first integral on the right-hand side can be evaluated with the help of integration by parts as follows:

$$\int_\Omega ((b^n)^{-1}-1)\nabla b^n \cdot \vec{v}^n\,\mathrm{d}x = \int_\Omega \nabla(\ln b^n - b^n)\cdot \vec{v}^n\,\mathrm{d}x = 0. \tag{5.37}$$

Next, we focus on the last term in (5.36). We would like to integrate by parts, but since we do not know that b^n has second derivatives integrable up to the boundary, we have to proceed more carefully. However, since the integral is well defined, we know that

$$\int_\Omega \Delta b^n\,\nabla b^n \cdot \vec{v}^n\,\mathrm{d}x = \lim_{\varepsilon\to 0_+}\int_\Omega \Delta b^n\,\nabla b^n \cdot \vec{v}^n \tau_\varepsilon\,\mathrm{d}x,$$

where τ_ε is smooth and satisfies $|\tau_\varepsilon| \le 1$ and

$$\tau_\varepsilon(x) = \begin{cases} 0 & \text{if dist }(x,\partial\Omega) \le \varepsilon, \\ 1 & \text{if dist }(x,\partial\Omega) \ge 2\varepsilon. \end{cases}$$

Furthermore, since Ω is Lipschitz and $\vec{v}^n \in W^{1,\infty}_{0,\mathrm{div}}$, we can construct such a τ_ε so that also

$$|\vec{v}^n||\nabla\tau_\varepsilon| \le C(\vec{v}^n,\Omega), \tag{5.38}$$

with C independent of ε. Then, using integration by parts recalling again that Ω is Lipschitz, using the fact that τ_ε is compactly supported, the fact that $\operatorname{div}\vec{v}^n = 0$, the estimate (5.38) and the regularity (5.33), we get

$$\begin{aligned}
&\int_\Omega \Delta b^n \nabla b^n \cdot \vec{v}^n \tau_\varepsilon\,\mathrm{d}x = \sum_{i,j=1}^d \int_\Omega \partial_{x_i x_i} b^n \partial_{x_j} b^n \vec{v}^n_j \tau_\varepsilon\,\mathrm{d}x \\
&= -\sum_{i,j=1}^d \int_\Omega \partial_{x_i} b^n \partial_{x_j x_i} b^n \vec{v}^n_j \tau_\varepsilon + \partial_{x_i} b^n \partial_{x_j} b^n \partial_{x_i}\vec{v}^n_j \tau_\varepsilon + \partial_{x_i} b^n \partial_{x_j} b^n \vec{v}^n_j \partial_{x_i}\tau_\varepsilon\,\mathrm{d}x \\
&= -\int_\Omega \frac{1}{2}\nabla|\nabla b^n|^2 \cdot \vec{v}^n\tau_\varepsilon + (\nabla b^n \otimes \nabla b^n)\cdot\nabla\vec{v}^n\tau_\varepsilon + (\nabla b^n \otimes \nabla b^n)\cdot(\vec{v}^n \otimes \nabla\tau_\varepsilon)\,\mathrm{d}x \\
&= -\int_\Omega -\frac{1}{2}|\nabla b^n|^2\vec{v}^n \cdot \nabla\tau_\varepsilon + (\nabla b^n \otimes \nabla b^n)\cdot\nabla\vec{v}^n\tau_\varepsilon + (\nabla b^n \otimes \nabla b^n)\cdot(\vec{v}^n \otimes \nabla\tau_\varepsilon)\,\mathrm{d}x \\
&= -\int_\Omega (\nabla b^n \otimes \nabla b^n)\cdot\nabla\vec{v}^n\tau_\varepsilon\,\mathrm{d}x \\
&\quad + \int_{0<\tau_\varepsilon<1} \frac{1}{2}|\nabla b^n|^2\vec{v}^n \cdot \nabla\tau_\varepsilon - (\nabla b^n \otimes \nabla b^n)\cdot(\vec{v}^n \otimes \nabla\tau_\varepsilon)\,\mathrm{d}x.
\end{aligned}$$

Thus, thanks to the fact that $b^n \in W^{1,2}(\Omega)$ for almost all time and because of (5.38), we get

$$\int_\Omega \Delta b^n \, \nabla b^n \cdot \vec{v}^n \, \mathrm{d}x = \lim_{\varepsilon \to 0+} \int_\Omega \Delta b^n \, \nabla b^n \cdot \vec{v}^n \tau_\varepsilon \, \mathrm{d}x = -\int_\Omega (\nabla b^n \otimes \nabla b^n) \cdot \nabla \vec{v}^n \, \mathrm{d}x.$$

Hence, returning to (5.36) and using also (5.37), we obtain

$$\int_\Omega \frac{|\nabla b^n \cdot \vec{v}^n|^2}{(b^n)^2} + \frac{\partial_t b^n \, \nabla b^n \cdot \vec{v}^n}{(b^n)^2} \, \mathrm{d}x = -\int_\Omega (\nabla b^n \otimes \nabla b^n) \cdot \nabla \vec{v}^n \, \mathrm{d}x. \tag{5.39}$$

Finally, we integrate (5.39) with respect to time over $(0,t)$ and add the result to (5.35) to deduce, after simple algebraic manipulations, that

(5.40)

$$\begin{aligned}
&\frac{\|\nabla b^n(t)\|_2^2}{2} + \int_0^t \int_\Omega \frac{|\partial_t b^n + \nabla b^n \cdot \vec{v}^n|^2}{(b^n)^2} \, \mathrm{d}x \, \mathrm{d}s \\
&\quad \le -\int_0^t \int_\Omega (\nabla b^n \otimes \nabla b^n) \cdot \mathbb{D}^n \, \mathrm{d}x \, \mathrm{d}s + \int_\Omega \ln b^n(t) - \ln b_0 - b^n(t) + b_0 \, \mathrm{d}x + \frac{\|b_0\|_{1,2}^2}{2}.
\end{aligned}$$

Now, our goal is to combine this estimate with the balance of the kinetic energy. Therefore, we multiply the i-th equation in (5.24) by $\alpha_i^n(t)$ and sum over $i = 1, \ldots, n$ to get

$$\int_\Omega \partial_t \vec{v}^n \cdot \vec{v}^n - \vec{v}^n \otimes \vec{v}^n \cdot \nabla \vec{v}^n + \mathbb{S}^n \cdot \nabla \vec{v}^n \, \mathrm{d}x = 0.$$

Then, for the second term we use integration by parts, so it vanishes thanks to $\operatorname{div} \vec{v}^n = 0$ and for the last term we use (5.25) to obtain

$$\frac{1}{2} \frac{\mathrm{d}}{\mathrm{d}t} \|\vec{v}^n\|_2^2 + 2\|\mathbb{D}^n\|_2^2 - \int_\Omega (\nabla b^n \otimes \nabla b^n) \cdot \mathbb{D}^n \, \mathrm{d}x \, \mathrm{d}s = 0. \tag{5.41}$$

Finally, we integrate (5.41) with respect to time over $(0,t)$ and add the result to (5.40) and obtain the final inequality

$$\begin{aligned}
&\frac{1}{2}\left(\|\vec{v}^n(t)\|_2^2 + \|\nabla b^n(t)\|_2^2\right) + \int_0^t \int_\Omega \frac{|\partial_t b^n + \nabla b^n \cdot \vec{v}^n|^2}{(b^n)^2} + 2|\mathbb{D}^n|^2 \, \mathrm{d}x \, \mathrm{d}s \\
&\quad \le \int_\Omega \ln b^n(t) - \ln b_0 - b^n(t) + b_0 \, \mathrm{d}x + \frac{1}{2}\left(\|b_0\|_{1,2}^2 + \|\vec{P}^n(\vec{v}_0)\|_2^2\right).
\end{aligned} \tag{5.42}$$

To conclude this subsection, we summarize the estimates we have obtained. It directly follows from (5.31), (5.28) and (5.40) and Korn's inequality that

$$\begin{aligned}
0 < b_{\min} \le b^n &\le b_{\max} < \infty, \\
\sup_{t \in (0,T)} \left(\|\vec{v}^n(t)\|_2 + \|\nabla b^n(t)\|_2\right) &\le C, \\
\int_0^T \|\vec{v}^n\|_{1,2}^2 + \|\partial_t b^n + \nabla b^n \cdot \vec{v}^n\|_2^2 \, \mathrm{d}t &\le C.
\end{aligned} \tag{5.43}$$

Consequently, using the interpolation inequality

$$\|\vec{v}^n\|_{\frac{2(d+2)}{d}}^{\frac{2(d+2)}{d}} \le C \|\vec{v}^n\|_2^{\frac{4}{d}} \|\vec{v}^n\|_{1,2}^2$$

we also get from (5.43) that

$$\int_0^T \|\vec{v}^n\|_{\frac{2(d+2)}{d}}^{\frac{2(d+2)}{d}} \, \mathrm{d}t \le C. \tag{5.44}$$

Finally, we focus on bounds on time derivatives. First, using (5.24) and the *a priori* estimates (5.43) we obtain the bound (here we also use the Sobolev embedding $W^{d+1,2}(\Omega) \hookrightarrow W^{1,\infty}(\Omega)$)

$$\int_0^T \|\partial_t \vec{v}^n\|^{\frac{d+2}{d}}_{(W^{1,2}_{0,\mathrm{div}} \cap W^{d+1,2}(\Omega)^d)^*} \,\mathrm{d}t \le C. \tag{5.45}$$

Similarly, with the help of (5.43), (5.44) and Hölder's inequality, we deduce that

$$\begin{aligned}\int_0^T \|\partial_t b^n\|^{\frac{d+2}{d+1}}_{\frac{d+2}{d+1}} \,\mathrm{d}t &\le C \int_0^T \|\partial_t b^n + \nabla b^n \cdot \vec{v}^n\|^{\frac{d+2}{d+1}}_{\frac{d+2}{d+1}} + \||\nabla b^n||\vec{v}^n|\|^{\frac{d+2}{d+1}}_{\frac{d+2}{d+1}} \,\mathrm{d}t \\ &\le C + C \int_0^T \|\partial_t b^n + \nabla b^n \cdot \vec{v}^n\|_2^2 + \|\nabla b^n\|_2^2 + \|\vec{v}^n\|^{\frac{2(d+2)}{d}}_{\frac{2(d+2)}{d}} \,\mathrm{d}t \le C.\end{aligned} \tag{5.46}$$

5.4. The limit $n \to \infty$. In this final subsection, we let $n \to \infty$ and show that the sequence of solutions to the approximating problem indexed by n tends to a solution of the original problem. Having the uniform estimates (5.28), (5.31), (5.43) and (5.44)–(5.46), there exist $\vec{v}$ and b and subsequences that we do not relabel, such that

$$\vec{v}^n \rightharpoonup^* \vec{v} \qquad \text{weakly}^* \text{ in } L^\infty(0,T; L^2_{0,\mathrm{div}}), \tag{5.47}$$

$$\vec{v}^n \rightharpoonup \vec{v} \qquad \text{weakly in } L^2(0,T; W^{1,2}_{0,\mathrm{div}}) \cap L^{\frac{2(d+2)}{d}}(0,T; L^{\frac{2(d+2)}{d}}(\Omega)^d), \tag{5.48}$$

$$\partial_t \vec{v}^n \rightharpoonup \partial_t \vec{v} \qquad \text{weakly in } L^{\frac{d+2}{d}}(0,T; (W^{1,2}_{0,\mathrm{div}} \cap W^{d+1,2}(\Omega)^d)^*), \tag{5.49}$$

$$b^n \rightharpoonup^* b \qquad \text{weakly}^* \text{ in } L^\infty(0,T; W^{1,2}(\Omega)), \tag{5.50}$$

$$b^n \rightharpoonup^* b \qquad \text{weakly}^* \text{ in } L^\infty(0,T; L^\infty(\Omega)), \tag{5.51}$$

$$\partial_t b^n \rightharpoonup \partial_t b \qquad \text{weakly in } L^{\frac{d+2}{d+1}}(0,T; L^{\frac{d+2}{d+1}}(\Omega)). \tag{5.52}$$

Using the Aubin–Lions lemma we then deduce that

$$\vec{v}^n \to \vec{v} \qquad \text{strongly in } L^r(0,T; L^r(\Omega)^d) \tag{5.53}$$

for all $r \in [1, 2(d+2)/d)$. In addition,

$$\partial_t b^n + \nabla b^n \cdot \vec{v}^n \rightharpoonup \partial_t b + \nabla b \cdot \vec{v} \qquad \text{weakly in } L^2(0,T; L^2(\Omega)), \tag{5.54}$$

$$b^n \to b \qquad \text{strongly in } \mathcal{C}([0,T]; L^q(\Omega)) \tag{5.55}$$

for all $q \in [1, \infty)$. In particular, (5.55) follows from (5.50) and (5.52), the compact embedding of $L^\infty(0,T; W^{1,2}(\Omega)) \cap W^{1,\frac{d+2}{d+1}}(0,T; L^{\frac{d+2}{d+1}}(\Omega))$ into $\mathcal{C}([0,T]; L^q(\Omega))$ for all $q \in [1,\infty)$ when $d = 2$ and all $q \in [1,6)$ when $d = 3$, guaranteed by the Aubin–Lions lemma; and, in the case of $d = 3$, we have also used (5.51) in order to extend the range of q from $[1,6)$ to $[1,\infty)$ using the boundedness of the sequence $(b^n)_{n\ge 1}$ in $L^\infty(0,T; L^\infty(\Omega))$, guaranteed by (5.51), in conjunction with the strong convergence of this sequence in $\mathcal{C}([0,T]; L^q(\Omega))$ for $q \in [1,6)$.

In addition, it follows from (5.31), (5.28) and (5.55) that almost everywhere in Q we have

$$0 < b_{\min} \le b(t,x) \le b_{\max} < \infty. \tag{5.56}$$

Furthermore, we can easily let $n \to \infty$ in (5.32) to obtain, for almost all $t \in (0,T)$, that

$$\int_\Omega \frac{\partial_t b\, w}{b^2} + \frac{\nabla b \cdot \vec{v}}{b^2} w + (1 - b^{-1}) w + \nabla b \cdot \nabla w \,\mathrm{d}x = 0 \qquad \text{for all } w \in W^{1,2}(\Omega), \tag{5.57}$$

which is nothing else but (4.11) with our special choices of ν_1, μ and σ. To obtain also the strong convergence of ∇b^n, we set $w := b$ in (5.57) and $w := b^n$ in (5.32) and with the help of (5.54) and (5.55), we deduce that

$$\begin{aligned}\lim_{n\to\infty}\int_0^T \|\nabla b^n\|_2^2 \, \mathrm{d}t &= -\lim_{n\to\infty}\int_Q \frac{\partial_t b^n + \nabla b^n \cdot \vec{v}^n}{b^n} + (b^n - 1)\, \mathrm{d}x\, \mathrm{d}t \\ &= -\int_Q \frac{\partial_t b + \nabla b \cdot \vec{v}}{b} + (b-1)\, \mathrm{d}x\, \mathrm{d}t = \int_0^T \|\nabla b\|_2^2 \, \mathrm{d}t.\end{aligned}$$

Thus, this identity, combined also with (5.50), leads to

$$b^n \to b \qquad \text{strongly in } L^2(0,T;W^{1,2}(\Omega)). \tag{5.58}$$

Consequently, by (5.47)–(5.49), (5.53) and (5.58), we can let $n \to \infty$ in (5.24), (5.25) to conclude that for almost all $t \in (0,T)$ and all $\vec{w} \in W^{1,2}_{0,\mathrm{div}} \cap W^{d+1,2}(\Omega)^d$ we have

$$\langle \partial_t \vec{v}, \vec{w}\rangle + \int_\Omega \mathbb{S} \cdot \nabla \vec{w} - (\vec{v} \otimes \vec{v}) : \nabla \vec{w} \, \mathrm{d}x = 0, \tag{5.59}$$

with $\mathbb{S}$ given as

$$\mathbb{S} = 2\mathbb{D} - (\nabla b \otimes \nabla b), \quad \mathbb{D} = \frac{1}{2}(\nabla \vec{v} + (\nabla \vec{v})^{\mathrm{T}}). \tag{5.60}$$

Hence, (4.10) is also satisfied. Furthermore, letting $n \to \infty$ in (5.40), using lower semicontinuity of norms and the convergence results (5.47), (5.48), (5.50), (5.54) and (5.55), we also obtain the energy inequality (4.13). Thus, to complete the proof of Theorem 4.1 it only remains to prove the attainment of the initial conditions.

First, a standard argument yields that

$$\begin{aligned} \vec{v}(t) &\rightharpoonup \vec{v}_0 \quad \text{weakly in } L^2_{0,\mathrm{div}}, \\ b(t) &\rightharpoonup b_0 \quad \text{weakly in } L^2(\Omega). \end{aligned} \tag{5.61}$$

Next, thanks to (5.50) and (5.52), we can strengthen the above convergence results as follows:

$$\begin{aligned} \nabla b(t) &\rightharpoonup \nabla b_0 \quad \text{weakly in } L^2(\Omega)^d, \\ b(t) &\to b_0 \quad \text{strongly in } L^2(\Omega). \end{aligned} \tag{5.62}$$

In addition, using (5.55) and weak lower semicontinuity of the norm function and neglecting the terms with the correct sign, we can let $n \to \infty$ in (5.40) to obtain for all $t \in (0,T)$ that

$$\frac{\|\vec{v}(t)\|_2^2 + \|\nabla b(t)\|_2^2}{2} \le \int_\Omega \ln\left(\frac{b(t)}{b_0}\right) - b(t) + b_0 \, \mathrm{d}x + \frac{\|\nabla b_0\|_2^2 + \|\vec{v}_0\|_2^2}{2}. \tag{5.63}$$

Hence, it follows (by noting the strong convergence results (5.62) and (5.56)) that

$$\limsup_{t\to 0} \|\vec{v}(t)\|_2^2 + \|\nabla b(t)\|_2^2 \le \|\vec{v}_0\|_2^2 + \|\nabla b_0\|_2^2,$$

which, when combined with (5.61) and (5.62), then leads to (4.12). The proof of the theorem is thereby complete.

Appendix A. Existence of Galerkin approximation

Here, we sketch the proof of the existence of the Galerkin approximation, i.e., a solution to (5.1)–(5.3). First, we recall the problem we would like to investigate, but we omit writing all superscripts related to the indices of the approximating sequences. Hence, we are interested in finding a solution to

$$\int_\Omega \partial_t \vec{v}\cdot\vec{w}_i - \vec{v}\otimes\vec{v} : \nabla\vec{w}_i + \mathbb{S} : \nabla\vec{w}_i \,\mathrm{d}x = 0, \qquad i = 1,\dots,n, \tag{A.1}$$

$$\begin{aligned}\int_\Omega \frac{\partial_t b\, w_j}{(T_n(b))^2} + \frac{\nabla b\cdot\vec{v}}{(T_n(b))^2} w_j + (1-(T_n(b))^{-1})w_j \,\mathrm{d}x&\\ + \int_\Omega \nabla b\cdot\nabla w_j\,\mathrm{d}x = 0, \qquad j = 1,\dots,\ell,&\end{aligned} \tag{A.2}$$

where $\mathbb{S}$ is defined a.e. in Q by

$$\mathbb{S} := 2\mathbb{D} - (\nabla b\otimes\nabla b), \quad \mathbb{D} := \frac{1}{2}(\nabla\vec{v} + (\nabla\vec{v})^{\mathrm{T}}), \tag{A.3}$$

where $(\vec{v}, b)$ are given by

$$\vec{v}(t,x) := \sum_{i=1}^{n}\alpha_i(t)\vec{w}_i(x), \qquad b(t,x) := \sum_{i=1}^{\ell}\beta_i(t)w_i(x)$$

with the initial conditions

$$\vec{v}(0,x) = \vec{P}^n(\vec{v}_0)(x), \qquad b(0,x) = P^\ell(b_0)(x).$$

Because of the presence of the nonlinearity in the term involving $\partial_t b$ we cannot simply invoke standard existence results from the theory of ordinary differential equations; we therefore proceed slightly differently. We denote

$$f(s) := \int_0^s \frac{1}{(T_n(t))^2}\,\mathrm{d}t, \qquad F(s) := \int_0^s f(t)\,\mathrm{d}t$$

and observe that F is uniformly convex, satisfying, with some $C_1, C_2 > 0$,

$$C_1 s^2 \le F(s) \le C_2 s^2, \qquad s \ge 0. \tag{A.4}$$

In addition, we can rewrite (A.2) as

$$\begin{aligned}\int_\Omega \partial_t f(b)\, w_j + \nabla f(b)\cdot\vec{v} w_j + (1-(T_n(b))^{-1})w_j\,\mathrm{d}x&\\ + \int_\Omega \nabla b\cdot\nabla w_j\,\mathrm{d}x = 0, \qquad j = 1,\dots,\ell.&\end{aligned} \tag{A.5}$$

We shall approximate (A.5) with the help of Rothe's method, i.e., we consider the following system of $N\times\ell$ equations:

$$\begin{aligned}&\int_\Omega \frac{f(b^{k+1}) - f(b^k)}{\tau} w_j + \nabla f(b^{k+1})\cdot\vec{v}^k w_j + (1-(T_n(b^{k+1}))^{-1})w_j\,\mathrm{d}x\\ &+ \int_\Omega \nabla b^{k+1}\cdot\nabla w_j\,\mathrm{d}x = 0, \qquad j = 1,\dots,\ell, \qquad k = 0,\dots,N-1,\end{aligned} \tag{A.6}$$

where $\tau := T/N$ and

$$b^0(x) := P^\ell b_0(x), \quad \vec{v}^k(x) := \frac{1}{\tau}\int_{t_k}^{t_{k+1}} \vec{v}(t,x)\,\mathrm{d}t, \quad t_k := kT/N,$$

and

$$b^k(x) := \sum_{i=1}^{\ell} \beta_i^k w_i(x).$$

Finally, we define

$$b(t,x) := \frac{t_{k+1}-t}{\tau} b^k(x) + \frac{t-t_k}{\tau} b^{k+1}(x) \qquad \text{for } t \in [t_k, t_{k+1}]. \tag{A.7}$$

Notice that for given $\vec{v}$, the problem (A.6) has a solution.[5] Moreover, this solution and consequently also b depends continuously on $\vec{v}$ and therefore, we can now use Carathéodory's theory to deduce the existence of a solution to the problem (A.1), (A.3) coupled to (A.6), (A.7). Finally, we will establish *a priori* estimates in which one can let $N \to \infty$ to deduce the solvability of the original problem (A.1)–(A.3)

Multiplying (A.6) by β_j^{k+1}, summing with respect to j and using $\operatorname{div} \vec{v} = 0$, we obtain

(A.8)

$$\int_\Omega (f(b^{k+1}) - f(b^k))\, b^{k+1} + \tau(1 - (T_n(b^{k+1}))^{-1})\, b^{k+1} \,\mathrm{d}x + \tau \|\nabla b^{k+1}\|_2^2 = 0$$

for all $k = 0, \dots, N-1$. Next, we denote by F^* the convex conjugate of the uniformly convex function F, i.e.,

$$F^*(s) := \sup_t (st - F(t)), \qquad \text{with } \tilde{C}_1 s^2 \le F^*(s) \le \tilde{C}_2 s^2,$$

and with the help of Young's inequality we have

$$\begin{aligned}(f(b^{k+1}) - f(b^k))b^{k+1} &= F(b^{k+1}) + F^*(f(b^{k+1})) - f(b^k)b^{k+1} \\ &\ge F^*(f(b^{k+1})) - F^*(f(b^k)).\end{aligned}$$

We note further that, since $1/n \le T_n(s) \le n$ for all $s \in \mathbb{R}$, we have $f(s) \ge s/n^2$ for all $s \ge 0$; note furthermore that if $b^{k+1}(x) \le 0$ for some $x \in \Omega$ then $(1 - (T_n(b^{k+1}(x)))^{-1})\, b^{k+1}(x) \ge 0$. Hence, by using Young's inequality and the

[5] The proof of existence of a solution $b^{k+1} \in W_\ell := \operatorname{span}\{w_i\}_{i=1}^{\ell}$ proceeds, by using a simple consequence of Brouwer's fixed point theorem, as follows. For a given, fixed, $\vec{v}^k$ as above and a given, fixed, b^k, with $k \in \{0, \dots, N-1\}$, we define the (continuous) mapping $\Phi_k : W_\ell \to W_\ell$ by

$$(\Phi_k(b), w) := \int_\Omega \frac{f(b) - f(b^k)}{\tau} w + \nabla f(b) \cdot \vec{v}^k w + (1 - (T_n(b))^{-1}) w \,\mathrm{d}x + \int_\Omega \nabla b \cdot \nabla w \,\mathrm{d}x,$$

for all $w \in W_\ell$. An argument analogous to the one leading to (A.8) results in

$$\tau(\Phi_k(b), b) = \int_\Omega (f(b) - f(b^k))\, b + \tau(1 - (T_n(b))^{-1})\, b \,\mathrm{d}x + \tau\|\nabla b\|_2^2 \qquad \forall\, b \in W_\ell.$$

As $(1 - (T_n(s))^{-1})\, s \ge -1$ for all $s \in \mathbb{R}$, it follows that $\tau(\Phi_k(b), b) \ge (f(b) - f(b^k), b) - \tau|\Omega|$. By noting that $T_n(s) \le n$ for all $s \in \mathbb{R}$ and recalling the definition of f, we have

$$\tau(\Phi_k(b), b) \ge (1/n^2)\|b\|_2^2 - (f(b^k), b) - \tau|\Omega| \ge (1/2n^2)\|b\|_2^2 - (n^2/2)\|f(b^k)\|_2^2 - \tau|\Omega|.$$

Hence, $(\Phi_k(b), b) \ge 0$ for all $b \in W_\ell$ such that $\|b\|^2 \ge n^4\|f(b^k)\|_2^2 + 2\tau n^2|\Omega|$, and the existence of a $b = b^{k+1} \in W_\ell$ such that $\Phi_k(b) = 0$ then follows from Corollary 1.1 on p.279 in **[21]**.

properties of the function F^*, we have that

$$\begin{aligned}\int_\Omega \tau(1-(T_n(b^{k+1}))^{-1})\, b^{k+1}\,\mathrm{d}x &\geq \int_\Omega \tau(1-(T_n(b^{k+1}))^{-1})\, b^{k+1}\chi_{b^{k+1}\geq 0}\,\mathrm{d}x \\ &\geq -\frac{\tau n^4}{2\tilde{C}_1}\int_\Omega (1-(T_n(b^{k+1}))^{-1})^2\chi_{b^{k+1}\geq 0}\,\mathrm{d}x - \frac{\tilde{C}_1\tau}{2n^4}\int_\Omega (b^{k+1})^2\chi_{b^{k+1}\geq 0}\,\mathrm{d}x \\ &\geq -\frac{\tau n^4}{2\tilde{C}_1}(n-1)^2|\Omega| - \frac{\tilde{C}_1\tau}{2}\int_\Omega (f(b^{k+1}))^2\chi_{b^{k+1}\geq 0}\,\mathrm{d}x \\ &\geq -\frac{\tau n^4}{2\tilde{C}_1}(n-1)^2 - \frac{\tau}{2}\int_\Omega F^*(f(b^{k+1}))\,\chi_{b^{k+1}\geq 0}\,\mathrm{d}x \\ &\geq -\frac{\tau n^4}{2\tilde{C}_1}(n-1)^2 - \frac{\tau}{2}\int_\Omega F^*(f(b^{k+1}))\,\mathrm{d}x.\end{aligned}$$

Thus, using these inequalities in (A.8) and summing the result over $k = 1, \dots, m$ with arbitrary $m \leq N-1$, we get, with $C = Tn^4(n-1)^2/(2\tilde{C}_1)$, that

$$\begin{aligned}\int_\Omega F^*(f(b^{m+1})) - F^*(f(b^0))\,\mathrm{d}x + \sum_{k=0}^{m}\tau\|\nabla b^{k+1}\|_2^2 \\ \leq C + \frac{1}{2}\sum_{k=0}^{m}\tau\int_\Omega F^*(f(b^{k+1}))\,\mathrm{d}x.\end{aligned} \tag{A.9}$$

Hence, using the discrete version of the Gronwall's lemma and also the definition of b, we obtain

$$\sup_{t\in(0,T)}\|b(t)\|_2^2 + \int_0^T \|\nabla b(t)\|_2^2\,\mathrm{d}t \leq C. \tag{A.10}$$

With this estimate in hand, we can now also obtain an estimate for $\vec{v}$ (see the proof of the main theorem):

$$\sup_{t\in(0,T)}\|\vec{v}(t)\|_2^2 \leq C(n,\ell) \implies \sup_{t\in(0,T)}\|\nabla\vec{v}(t)\|_\infty^2 \leq C(n,\ell). \tag{A.11}$$

Finally, we multiply (A.6) by $(\beta_j^{k+1} - \beta_j^k)$ and sum with respect to j to obtain

$$\begin{aligned}\int_\Omega \frac{(f(b^{k+1})-f(b^k))(b^{k+1}-b^k)}{\tau} + (1-(T_n(b^{k+1}))^{-1})(b^{k+1}-b^k)\,\mathrm{d}x \\ + \int_\Omega \nabla b^{k+1}\cdot\nabla(b^{k+1}-b^k)\,\mathrm{d}x = 0.\end{aligned} \tag{A.12}$$

Thus, summing with respect to $k = 0, \dots, m$ and using standard inequalities, we have

$$\sum_{k=0}^{m}\tau\left\|\frac{b^{k+1}-b^k}{\tau}\right\|_2^2 + \|\nabla b^m\|_2^2 \leq C(1+\|\nabla b^0\|_2^2). \tag{A.13}$$

Consequently, using the definition of b we get

$$\int_0^T \|\partial_t b\|_2^2\,\mathrm{d}t + \sup_{t\in(0,T)}\|\nabla b(t)\|_2^2 \leq C. \tag{A.14}$$

Therefore, with the help of (A.10), (A.11) and (A.14), it is quite easy to let $N \to \infty$ to establish the existence of a solution to the original problem.

References

[1] J. W. Barrett and S. Boyaval, *Existence and approximation of a (regularized) Oldroyd-B model*, Math. Models Methods Appl. Sci. **21** (2011), no. 9, 1783–1837. MR2843021

[2] H. Bellout, Frederick Bloom, and J. Nečas, *Young measure-valued solutions for non-Newtonian incompressible fluids*, Comm. Partial Differential Equations **19** (1994), no. 11-12, 1763–1803. MR1301173

[3] D. Breit, L. Diening, and S. Schwarzacher, *Solenoidal Lipschitz truncation for parabolic PDEs*, Math. Models Methods Appl. Sci. **23** (2013), no. 14, 2671–2700. MR3119635

[4] M. Bulíček, F. Ettwein, P. Kaplický, and D. Pražák, *On uniqueness and time regularity of flows of power-law like non-Newtonian fluids*, Math. Methods Appl. Sci. **33** (2010), no. 16, 1995–2010. MR2744616

[5] M. Bulíček, P. Gwiazda, J. Málek, and A. Świerczewska-Gwiazda, *On unsteady flows of implicitly constituted incompressible fluids*, SIAM J. Math. Anal. **44** (2012), no. 4, 2756–2801. MR3023393

[6] M. Bulíček and J. Málek, *On unsteady internal flows of Bingham fluids subject to threshold slip on the impermeable boundary*, Recent developments of mathematical fluid mechanics, Adv. Math. Fluid Mech., Birkhäuser/Springer, Basel, 2016, pp. 135–156. MR3524183

[7] M. Bulíček, J. Málek, and K. R. Rajagopal, *Navier's slip and evolutionary Navier-Stokes-like systems with pressure and shear-rate dependent viscosity*, Indiana Univ. Math. J. **56** (2007), no. 1, 51–85. MR2305930

[8] L. Caffarelli, R. Kohn, and L. Nirenberg, *Partial regularity of suitable weak solutions of the Navier-Stokes equations*, Comm. Pure Appl. Math. **35** (1982), no. 6, 771–831. MR673830

[9] L. Chupin and S. Martin, *Stationary Oldroyd model with diffusive stress: mathematical analysis of the model and vanishing diffusion process*, J. Non-Newton. Fluid Mech. **218** (2015), 27–39. MR3447498

[10] P. Constantin and M. Kliegl, *Note on global regularity for two-dimensional Oldroyd-B fluids with diffusive stress*, Arch. Ration. Mech. Anal. **206** (2012), no. 3, 725–740. MR2989441

[11] B. Dacorogna, *Direct Methods in the Calculus of Variations*, vol. 78, Springer Science & Business Media, 2007.

[12] J. K. G. Dhont and W. J. Briels, *Gradient and vorticity banding*, Rheol. Acta **47** (2008), no. 3, 257–281 (English).

[13] L. Diening, M. Růžička, and J. Wolf, *Existence of weak solutions for unsteady motions of generalized Newtonian fluids*, Ann. Sc. Norm. Super. Pisa Cl. Sci. (5) **9** (2010), no. 1, 1–46. MR2668872

[14] T. Divoux, M. A. Fardin, S. Manneville, and S. Lerouge, *Shear banding of complex fluids*, Annual review of fluid mechanics. Vol. 48, Annu. Rev. Fluid Mech., vol. 48, Annual Reviews, Palo Alto, CA, 2016, pp. 81–103. MR3726380

[15] A. W. El-Kareh and L. G. Leal, *Existence of solutions for all Deborah numbers for a non-Newtonian model modified to include diffusion*, J. Non-Newton. Fluid Mech. **33** (1989), no. 3, 257–287.

[16] T. M. Elgindi and F. Rousset, *Global regularity for some Oldroyd-B type models*, Comm. Pure Appl. Math. **68** (2015), no. 11, 2005–2021. MR3403757

[17] M. A. Fardin, T. J. Ober, C. Gay, G. Gregoire, G. H. McKinley, and S. Lerouge, *Potential "ways of thinking" about the shear-banding phenomenon*, Soft Matter **8** (2012), 910–922.

[18] L. E. Figueroa and E. Süli, *Greedy approximation of high-dimensional Ornstein-Uhlenbeck operators*, Found. Comput. Math. **12** (2012), no. 5, 573–623. MR2970851

[19] J. Frehse, J. Málek, and M. Steinhauer, *On existence results for fluids with shear dependent viscosity—unsteady flows*, Partial differential equations (Praha, 1998), Chapman & Hall/CRC Res. Notes Math., vol. 406, Chapman & Hall/CRC, Boca Raton, FL, 2000, pp. 121–129. MR1713880

[20] J. Geng, *$W^{1,p}$ estimates for elliptic problems with Neumann boundary conditions in Lipschitz domains*, Adv. Math. **229** (2012), no. 4, 2427–2448. MR2880228

[21] V. Girault and P.-A. Raviart, *Finite element methods for Navier-Stokes equations*, Springer Series in Computational Mathematics, vol. 5, Springer-Verlag, Berlin, 1986. Theory and algorithms. MR851383

[22] E. Hopf, *Über die Anfangswertaufgabe für die hydrodynamischen Grundgleichungen* (German), Math. Nachr. **4** (1951), 213–231. MR0050423

[23] J. Hron, V. Miloš, V. Průša, O. Souček, and K. Tůma, *On thermodynamics of viscoelastic rate type fluids with temperature dependent material coefficients*, Int. J. Non-Linear Mech. **95** (2017), no. 4, 193–208.DOI 10.1016/j.ijnonlinmec.2017.06.011

[24] D. Hu and T. Lelièvre, *New entropy estimates for Oldroyd-B and related models*, Commun. Math. Sci. **5** (2007), no. 4, 909–916. MR2375053

[25] A. A. Kiselev and O. A. Ladyženskaya, *On the existence and uniqueness of the solution of the nonstationary problem for a viscous, incompressible fluid* (Russian), Izv. Akad. Nauk SSSR. Ser. Mat. **21** (1957), 655–680. MR0100448

[26] O. Kreml, M. Pokorný, and P. Šalom, *On the global existence for a regularized model of viscoelastic non-Newtonian fluid*, Colloq. Math. **139** (2015), no. 2, 149–163. MR3337214

[27] O. A. Ladyženskaja, *New equations for the description of the motions of viscous incompressible fluids, and global solvability for their boundary value problems* (Russian), Trudy Mat. Inst. Steklov. **102** (1967), 85–104. MR0226907

[28] O. A. Ladyzhenskaya, *The mathematical theory of viscous incompressible flow*, Second English edition, revised and enlarged. Translated from the Russian by Richard A. Silverman and John Chu. Mathematics and its Applications, Vol. 2, Gordon and Breach, Science Publishers, New York-London-Paris, 1969. MR0254401

[29] C. Le Bris and T. Lelièvre, *Micro-macro models for viscoelastic fluids: modelling, mathematics and numerics*, Sci. China Math. **55** (2012), no. 2, 353–384. MR2886543

[30] J. Leray, *Sur le mouvement d'un liquide visqueux emplissant l'espace* (French), Acta Math. **63** (1934), no. 1, 193–248. MR1555394

[31] J.-L. Lions, *Quelques méthodes de résolution des problèmes aux limites non linéaires* (French), Dunod; Gauthier-Villars, Paris, 1969. MR0259693

[32] P. L. Lions and N. Masmoudi, *Global solutions for some Oldroyd models of non-Newtonian flows*, Chinese Ann. Math. Ser. B **21** (2000), no. 2, 131–146. MR1763488

[33] M. Lukáčová-Medvid'ová, H. Mizerová, and Š. Nečasová, *Global existence and uniqueness result for the diffusive Peterlin viscoelastic model*, Nonlinear Anal. **120** (2015), 154–170. MR3348052

[34] J. Málek, J. Nečas, and M. Růžička, *On the non-Newtonian incompressible fluids*, Math. Models Methods Appl. Sci. **3** (1993), no. 1, 35–63. MR1203271

[35] J. Málek, J. Nečas, and M. Růžička, *On weak solutions to a class of non-Newtonian incompressible fluids in bounded three-dimensional domains: the case $p \geq 2$*, Adv. Differential Equations **6** (2001), no. 3, 257–302. MR1799487

[36] J. Málek and V. Průša, *Derivation of equations for continuum mechanics and thermodynamics of fluids*, Handbook of Mathematical Analysis in Mechanics of Viscous Fluids (Y. Giga and A. Novotný, eds.), Springer, 2017, pp. 1–70.

[37] J. Málek, V. Průša, T. Skřivan, and E. Süli, *Thermodynamics of viscoelastic rate-type fluids with stress diffusion*, Phys. Fluids **30** (1995), no. 6, 789–812, DOI 10.1063/1.5018172.

[38] J. Málek, K. R. Rajagopal, and M. Růžička, *Existence and regularity of solutions and the stability of the rest state for fluids with shear dependent viscosity*, Math. Models Methods Appl. Sci. **5** (1995), no. 6, 789–812. MR1348587

[39] J. Málek, K. R. Rajagopal, and K. Tůma, *On a variant of the Maxwell and Oldroyd-B models within the context of a thermodynamic basis*, Int. J. Non-Linear Mech. **76** (2015), 42–47.

[40] E. Maringová and J. Žabenský, *On a Navier–Stokes–Fourier-like system capturing transitions between viscous and inviscid fluid regimes and between no-slip and perfect-slip boundary conditions*, Nonlinear Anal. Real World Appl. **41** (2018), no. 6, 152–178, DOI 10.1016/j.nonrwa.2017.10.008.

[41] N. Masmoudi, *Global existence of weak solutions to macroscopic models of polymeric flows* (English, with English and French summaries), J. Math. Pures Appl. (9) **96** (2011), no. 5, 502–520. MR2843224

[42] K. R. Rajagopal and A. R. Srinivasa, *A thermodynamic frame work for rate type fluid models*, Journal of Non-Newtonian Fluid Mechanics **88** (2000), no. 3, 207–227.

[43] J. Wolf, *Existence of weak solutions to the equations of non-stationary motion of non-Newtonian fluids with shear rate dependent viscosity*, J. Math. Fluid Mech. **9** (2007), no. 1, 104–138. MR2305828

Mathematical Institute, Charles University, Faculty of Mathematics and Physics, Sokolovská 83, 186 75 Prague 8, Czech Republic
Email address: mbul8060@karlin.mff.cuni.cz

Mathematical Institute, Charles University, Faculty of Mathematics and Physics, Sokolovská 83, 186 75 Prague 8, Czech Republic
Email address: malek@karlin.mff.cuni.cz

Mathematical Institute, Charles University, Faculty of Mathematics and Physics, Sokolovská 83, 186 75 Prague 8, Czech Republic
Email address: prusv@karlin.mff.cuni.cz

Mathematical Institute, University of Oxford, Woodstock Road, Oxford OX2 6GG, United Kingdom
Email address: endre.suli@maths.ox.ac.uk

Contemporary Mathematics
Volume **710**, 2018
http://dx.doi.org/10.1090/conm/710/14363

Local stability of energy estimates for the Navier–Stokes equations.

Diego Chamorro, Pierre Gilles Lemarié-Rieusset, and Kawther Mayoufi

ABSTRACT. We study the regularity of the weak limit of a sequence of dissipative solutions to the Navier–Stokes equations when no assumption is made on the behavior of the pressures.

1. Local weak solutions

In this paper, we are interested in local properties (regularity, local energy estimates) of weak solutions of Navier–Stokes equations.

DEFINITION 1.1 (Local weak solutions). Let Ω be a domain in $\mathbb{R} \times \mathbb{R}^3$ and $\vec{f} \in L^2_{\text{loc}}(\Omega)$ a divergence-free time-dependent vector field. A vector field $\vec{u}$ will be said to be a *local weak solution* of the Navier–Stokes equations on Ω (associated to the force $\vec{f}$) if, for each cylinder $Q = I \times O$ (where I is an open interval in $\mathbb{R}$ and O an open subset of $\mathbb{R}^3$) such that $\bar{Q}$ is a compact subset of Ω, we have $\vec{u} \in L^\infty_t L^2_x(Q) \cap L^2_t H^1_x(Q)$, $\vec{u}$ is divergence-free and, for every smooth compactly supported divergence-free vector field $\vec{\phi} \in \mathcal{D}(Q)$ we have

$$\iint_Q \vec{u} \cdot (\partial_t \vec{\phi} + \Delta \vec{\phi}) + \vec{u} \cdot (\vec{u} \cdot \vec{\nabla} \vec{\phi}) + \vec{f} \cdot \vec{\phi} \, dt \, dx = 0. \tag{1.1}$$

More precisely, we shall address the behavior of a weak limit of regular solutions.

DEFINITION 1.2 (Regular local solutions). Let Ω be a domain in $\mathbb{R} \times \mathbb{R}^3$ and $\vec{f} \in L^2_{\text{loc}}(\Omega)$ a divergence-free time-dependent vector field and $\vec{u}$ a local weak solution of the Navier–Stokes equations on Ω (associated to the force $\vec{f}$).

A) $\vec{u}$ is a *regular local solution* if, for each cylinder $Q \subset\subset \Omega$, we have $\vec{u} \in L^\infty_{t,x}(Q)$,
B) The set $R(\vec{u})$ of *regular points* of $\vec{u}$ is the largest open subset of Ω on which $\vec{u}$ is a regular solution. The set $\Sigma(\vec{u})$ of *singular points* is the complement of $R(\vec{u}) : \Sigma(\vec{u}) = \Omega \setminus R(\vec{u})$.

Our result is then the following one **[M]** :

2010 *Mathematics Subject Classification.* Primary 35Q30, 76D05.
Key words and phrases. Navier–Stokes equations; partial regularity; Caffarelli, Kohn and Nirenberg theory; local energy inequalities.

THEOREM 1.3 (Singular points of a weak limit.). *Let Ω be a domain in $\mathbb{R}\times\mathbb{R}^3$. Assume that we have sequences $\vec{f}_n$ of divergence-free time-dependent vector fields and $\vec{u}_n$ of local weak solutions of the Navier–Stokes equations on Ω (associated to the forces $\vec{f}_n$) such that, for each cylinder $Q \subset\subset \Omega$, we have*

- *$\vec{f}_n \in L^2_t H^1_x(Q)$ and $\vec{f}_n$ converges weakly in $L^2_t H^1_x$ to a limit $\vec{f}$,*
- *the sequence $\vec{u}_n$ is bounded in $L^\infty_t L^2_x(Q) \cap L^2_t H^1_x(Q)$ and converges weakly in $L^2_t H^1_x(Q)$ to a limit $\vec{u}$,*
- *for every n, $\vec{u}_n$ is bounded on Q.*

Then the limit $\vec{u}$ is a local weak solution on Ω of the Navier–Stokes equations associated to the force $\vec{f}$, and its set $\Sigma(\vec{u})$ has parabolic one-dimensional Hausdorff measure equal to 0.

As we shall see, the main tool of the proof is an extension of the Caffarelli–Kohn–Nirenberg theory [**Ca**] to the case where we have no control on the pressure (i.e. the case of generalized suitable solutions [**W**] or dissipative solutions [**Ch**]).

2. Pressure

Equations (1.1) can classically be rewritten as an equation involving a pressure term. See for instance [**W**]. In the following, we shall only need the pressure inside spherical cylinders $Q = I \times B$ (where I is an open interval in $\mathbb{R}$ and B an open ball of $\mathbb{R}^3$). In that case, it is very easy to define a pressure p such that

$$\partial_t \vec{u} = \Delta\vec{u} - \vec{u}\cdot\vec{\nabla}\vec{u} - \vec{\nabla}p + \vec{f} \quad \text{in } \mathcal{D}'(Q). \tag{2.1}$$

Indeed, let Q, $Q^\#$, and Q^* be three relatively compact cylinders in Ω with $\overline{Q} \subset Q^\#$ and $\overline{Q^\#} \subset Q^*$ and ψ a cut off smooth function supported in Q^* and equal to 1 on a neighboorhood of $\overline{Q^\#}$. The function

$$p_0 = -\frac{1}{\Delta}\left(\sum_{i=1}^3\sum_{j=1}^3 \partial_i\partial_j(\psi u_i u_j)\right)$$

(where we write, for a tempered distribution T, $\frac{1}{\Delta}T$ the unique tempered distribution S (if it exists) such that $\Delta S = T$ and $\lim_{t\to+\infty} e^{t\Delta}S = 0$) belongs to $L^2_t L^{3/2}_x$ and, on $Q^\#$, the distribution

$$\vec{T} = \partial_t\vec{u} - \Delta\vec{u} + \vec{u}\cdot\vec{\nabla}\vec{u} + \vec{\nabla}p_0 - \vec{f}$$

satisfies

$$\text{curl } \vec{T} = 0 \text{ and div } \vec{T} = 0.$$

Moreover, $\vec{T}_0 = \vec{T} - \partial_t\vec{u}$ belongs to $L^2_t H^{-2}_x(Q^\#)$. Picking $t_0 \in I$, we define $\vec{S} = \vec{u} + \int_{t_0}^t \vec{T}_0(s,.)\,ds$. We have $\vec{S} \in L^\infty_t H^{-2}_x(Q^\#)$. Moreover, we have ∂_t curl $\vec{S} = 0$ and ∂_t div $\vec{S} = 0$. Thus, if $\alpha \in \mathcal{D}(I)$ with $\int \alpha\,dt = 1$, we find that

$$\vec{S}_0 = \vec{S} - \int_I \alpha(s)\vec{S}(s,.)\,ds$$

satisfies

$$\partial_t\vec{S}_0 = \vec{T}, \quad \text{curl } \vec{S}_0 = 0 \text{ and div } \vec{S}_0 = 0.$$

In particular,

$$\Delta\vec{S}_0 = \vec{\nabla}(\text{ div } \vec{S}_0) - \vec{\nabla}\wedge(\text{ curl } \vec{S}_0) = 0.$$

Thus, we get that $\vec{S}_0$ is smooth in the space variable; in particular $\vec{S}_0 \in L^\infty_t W^{1,\infty}_x(Q)$. If $x_0 \in B$ and if we define

$$\varpi(t,x) = \int_0^1 \vec{S}_0(t,(1-\theta)x_0 + \theta x)\cdot(x-x_0)\,d\theta,$$

we find that $\varpi \in L^\infty_{t,x}(Q)$ and $\vec{\nabla}\varpi = \vec{S}_0$. Defining $p = p_0 - \partial_t\varpi$, we find the equality (2.1).

Of course, the pressure may be singular in time (as $\partial_t\varpi$ is only the derivative of a bounded function). We shall comment further on this in Sections 3 and 5.

3. Energy balance

This section is devoted to the study of $\partial_t|\vec{u}|^2$, as it is the main tool to estimate the partial regularity of $\vec{u}$. If $\vec{u}$ and the pressure p were regular, we could write from equality (2.1)

$$\partial_t|\vec{u}|^2 = 2\vec{u}\cdot\partial_t\vec{u} = 2\vec{u}\cdot\Delta\vec{u} - 2\vec{u}\cdot(\vec{u}\cdot\vec{\nabla}\vec{u} + \vec{\nabla}p) + 2\vec{u}\cdot\vec{f}$$

and rewrite

$$2\vec{u}\cdot\Delta\vec{u} = \Delta(|\vec{u}|^2) - 2|\vec{\nabla}\otimes\vec{u}|^2$$

and, since $\operatorname{div}\vec{u} = 0$,

$$2\vec{u}\cdot(\vec{u}\cdot\vec{\nabla}\vec{u} + \vec{\nabla}p) = \operatorname{div}((|\vec{u}|^2 + 2p)\vec{u}).$$

This would give the following local energy balance in Q

$$\partial_t|\vec{u}|^2 = \Delta(|\vec{u}|^2) - 2|\vec{\nabla}\otimes\vec{u}|^2 - \operatorname{div}((|\vec{u}|^2 + 2p)\vec{u}) + 2\vec{u}\cdot\vec{f}. \tag{3.1}$$

However, local weak solutions (and their associates pressures) are not regular enough to allow those computations : the problem lies in the fact that the terms $\vec{u}\cdot(\vec{u}\cdot\nabla\vec{u})$ and $\vec{u}\cdot\vec{\nabla}p$ are not well defined in $\mathcal{D}'$. If the pressure is regular enough (for instance, $p \in L^{3/2}_{t,x}(Q)$) then one first smoothens $\vec{u}$ with a mollifier $\varphi_\epsilon = \frac{1}{\epsilon^3}\varphi(\frac{x}{\epsilon})$, defining $\vec{u}_\epsilon = \varphi_\epsilon * \vec{u}$. One then finds

$$\partial_t|\vec{u}_\epsilon|^2 = \Delta(|\vec{u}_\epsilon|^2) - 2|\vec{\nabla}\otimes\vec{u}_\epsilon|^2 - 2\vec{u}_\epsilon\cdot\varphi_\epsilon*(\vec{u}.\vec{\nabla}\vec{u}) - 2\operatorname{div}((p*\varphi_\epsilon)\vec{u}_\epsilon) + 2\vec{u}_\epsilon\cdot(\varphi_\epsilon*\vec{f}).$$

The limit $\epsilon \to 0$ gives then

$$\partial_t|\vec{u}|^2 = \Delta(|\vec{u}|^2) - 2|\vec{\nabla}\otimes\vec{u}|^2 - 2\lim_{\epsilon\to 0}\vec{u}_\epsilon\cdot\varphi_\epsilon*(\vec{u}\cdot\vec{\nabla}\vec{u}) - 2\operatorname{div}(p\vec{u}) + 2\vec{u}\cdot\vec{f}.$$

In order to compare this expression with (3.1), we define

$$M_\epsilon(\vec{u}) = -\operatorname{div}(|\vec{u}|^2\vec{u}) + 2\vec{u}_\epsilon\cdot\varphi_\epsilon*(\vec{u}\cdot\vec{\nabla}\vec{u})$$

and write

$$\partial_t|\vec{u}|^2 = \Delta(|\vec{u}|^2) - 2|\vec{\nabla}\otimes\vec{u}|^2 - \operatorname{div}((|\vec{u}|^2 + 2p)\vec{u}) + 2\vec{u}\cdot\vec{f} - \lim_{\epsilon\to 0}M_\epsilon(\vec{u}).$$

However, our assumptions on weak solutions don't allow us to make all those computations, as the pressure we can define on Q has no regularity with respect to the time variable, so that $p\vec{u}$ is not well defined in $\mathcal{D}'$. Thus, one must smoothens as well $\vec{u}$ with respect to the time variable, with a mollifier $\psi_\eta(t) = \frac{1}{\eta}\psi(\frac{t}{\eta})$. Defining $\vec{u}_{\epsilon,\eta} = \psi_\eta *_t \varphi_\epsilon *_x \vec{u} = \xi_{\eta,\epsilon} *_{t,x} \vec{u}$, one finds

$$\begin{aligned}\partial_t|\vec{u}_{\epsilon,\eta}|^2 &= \Delta(|\vec{u}_{\epsilon,\eta}|^2) - 2|\vec{\nabla}\otimes\vec{u}_{\epsilon,\eta}|^2 - 2\vec{u}_{\epsilon,\eta}\cdot\xi_{\eta,\epsilon}*(\vec{u}\cdot\vec{\nabla}\vec{u})\\ &\quad -2\operatorname{div}((p*\xi_{\eta,\epsilon})\vec{u}_{\epsilon,\eta}) + 2\vec{u}_{\epsilon,\eta}\cdot(\xi_{\eta,\epsilon}*\vec{f}).\end{aligned}$$

The limit $\eta \to 0$ gives then

$$\partial_t |\vec u_\epsilon|^2 = \Delta(|\vec u_\epsilon|^2) - 2|\vec\nabla \otimes \vec u_\epsilon|^2 - 2\vec u_\epsilon \cdot \varphi_\epsilon * (\vec u \cdot \vec\nabla \vec u) - 2 \lim_{\eta\to 0} \operatorname{div}\left((p * \xi_{\eta,\epsilon})\vec u_{\epsilon,\eta}\right) + 2\vec u_\epsilon \cdot (\varphi_\epsilon * \vec f).$$

The limit $\epsilon \to 0$ gives finally

$$\begin{aligned}\partial_t |\vec u|^2 &= \Delta(|\vec u|^2) - 2|\vec\nabla \otimes \vec u|^2 \\ &\quad - 2 \lim_{\epsilon\to 0}\left(\vec u_\epsilon \cdot \varphi_\epsilon * (\vec u \cdot \vec\nabla \vec u) + \lim_{\eta\to 0} \operatorname{div}\left((p * \xi_{\eta,\epsilon})\vec u_{\epsilon,\eta}\right)\right) + 2\vec u \cdot \vec f.\end{aligned} \tag{3.2}$$

In order to circumvent the problems of lack of regularity for the pressure, we introduce the notion of harmonic correction :

DEFINITION 3.1 (Harmonic corrections). Let Ω be a domain in $\mathbb{R} \times \mathbb{R}^3$, $\vec f \in L^2_{\mathrm{loc}}(\Omega)$ a divergence-free time-dependent vector field and $\vec u$ a local weak solution of the Navier–Stokes equations on Ω (associated to the force $\vec f$). A *harmonic correction* $\vec H$ on a cylinder $Q \subset\subset \Omega$ is a vector field such that

- $\operatorname{div} \vec H = 0$ and $\Delta \vec H = 0$,
- $\vec H \in L^\infty_{t,x}(Q)$ and $\partial_i \vec H \in L^\infty_{t,x}(Q)$ for $i = 1, 2, 3$,
- there exists $\vec F \in L^2_{t,x}(Q)$ and $P \in L^{3/2}_{t,x}(Q)$ such that the vector field $\vec U = \vec u + \vec H$ satisfies
$$\partial_t \vec U = \Delta \vec U - \vec U \cdot \vec\nabla \vec U - \vec\nabla P + \vec F.$$

In the literature, one can find at least two such harmonic corrections for local weak solutions :

LEMMA 3.2. *Let Ω be a domain in $\mathbb{R} \times \mathbb{R}^3$, $\vec f \in L^2_{\mathrm{loc}}(\Omega)$ a divergence-free time-dependent vector field and $\vec u$ a local weak solution of the Navier–Stokes equations on Ω (associated to the force $\vec f$). Let Q be a spherical cylinder in Ω. Then:*

A) *the decomposition of the pressure p as $p = p_0 - \partial_t \varpi$ described in Section 1 provides a harmonic correction $\vec H = -\vec\nabla \varpi$ of $\vec u$ on Q,*

B) *Let $\psi(t,x) = \alpha(t)\beta(x)$ be a smooth cut-off function supported by a cylinder $Q^* \subset\subset \Omega$ and equal to 1 on a neighborhood of Q. Then $\vec U = -\frac{1}{\Delta}\vec\nabla \wedge (\psi \vec\nabla \vec u)$ is such that $\vec H = \vec U - \vec u$ is a harmonic correction of $\vec u$ on Q.*

PROOF. The case of $\vec H = -\vec\nabla \varpi$ has been discussed by Wolf [**W**]. For $\vec U = \vec u - \vec\nabla \varpi$, we have $\vec\nabla \wedge \vec U = \vec\nabla \wedge \vec u$ and $\Delta \vec U = \Delta \vec u$, so that

$$\begin{aligned}\partial_t \vec U - \Delta \vec U + \vec U \cdot \vec\nabla \vec U &= \partial_t \vec u - \partial_t \vec\nabla \varpi - \Delta \vec u + (\vec\nabla \wedge \vec u) \wedge (\vec u - \vec\nabla \varpi) + \vec\nabla(\frac{|\vec U|^2}{2}) \\ &= \vec\nabla(\frac{|\vec U|^2}{2} - \frac{|\vec u|^2}{2} - p_0) + \vec f - (\vec\nabla \wedge \vec u) \wedge \vec\nabla \varpi\end{aligned}$$

We may then decompose $(\vec\nabla \wedge \vec u) \wedge \vec\nabla \varpi \in L^2_t L^2_x(Q)$ into $\vec f_1 + \vec\nabla p_1$ with $\vec f_1 \in L^2_t L^2_x$ and $\operatorname{div} \vec f_1 = 0$ and $p_1 \in L^2_t L^6_x(Q)$ (for instance, by extending $(\vec\nabla \wedge \vec u) \wedge \vec\nabla \varpi$ by 0 outside Q and then using the Leray projection operator). We thus find

$$P = p_0 + \frac{|\vec u|^2}{2} - \frac{|\vec U|^2}{2} + p_1 \text{ and } \vec F = \vec f - \vec f_1.$$

The case of $\vec U = -\frac{1}{\Delta}\vec\nabla \wedge (\psi \vec\nabla \vec u)$ has been discussed by Chamorro, Lemarié-Rieusset and Mayoufi in [**Ch, Le**]. It is worth noticing that the pressure P they obtain belongs to $L^2_t L^q_x(Q)$ for every $q < 3/2$.

Note that, in both cases, even if $\vec{f}$ is assumed to be more regular, we cannot get a better regularity for $\vec{F}$ than $L^2_t L^2_x$, because of the contribution of $(\vec{\nabla} \wedge \vec{u}) \wedge \vec{H}$ to the force. □

An important result of Chamorro, Lemarié-Rieusset and Mayoufi is the following one [**Ch**, **Le**] :

THEOREM 3.3 (Energy balance). *Let Ω be a domain in $\mathbb{R} \times \mathbb{R}^3$, $\vec{f} \in L^2_{\mathrm{loc}}(\Omega)$ a divergence-free time-dependent vector field and $\vec{u}$ a local weak solution of the Navier–Stokes equations on Ω (associated to the force $\vec{f}$). Let Q be a spherical cylinder in Ω and p the pressure associated to $\vec{u}$ on Q. Then:*

A) *The quantities*

$$M(\vec{u}) = \lim_{\epsilon \to 0} \Big(- \text{ div } (|\vec{u}|^2 \vec{u}) + 2\vec{u}_\epsilon \cdot \varphi_\epsilon * (\vec{u} \cdot \vec{\nabla} \vec{u}) \Big)$$

and

$$<< \text{div } (p\vec{u}) >> = \lim_{\epsilon \to 0} \lim_{\eta \to 0} \text{div } ((p * \xi_{\eta,\epsilon}) \vec{u}_{\epsilon,\eta})$$

are well defined in $\mathcal{D}'(Q)$.

B) *We have the energy balance on Q :*

$$\partial_t |\vec{u}|^2 = \Delta(|\vec{u}|^2) - 2|\vec{\nabla} \otimes \vec{u}|^2 - \text{ div } (|\vec{u}|^2 \vec{u}) - 2 << \text{div } (p\vec{u}) >> + 2\vec{u} \cdot \vec{f} - M(\vec{u}).$$

C) *$M(\vec{u})$ can be computed as a defect of regularity. More precisely, we have, for*

$$A_{k,\epsilon}(\vec{u}) = \frac{(u_k(t,x-y) - u_k(t,x))(\vec{u}(t,x-y) - \vec{u}(t,x))}{\epsilon} \cdot \int \varphi_\epsilon(z)(\vec{u}(t,x-z) - \vec{u}(t,x))\, dz$$

and

$$B_{k,\epsilon}(\vec{u}) = \frac{(u_k(t,x-y) - u_k(t,x))|\vec{u}(t,x-y) - \vec{u}(t,x)|^2}{\epsilon},$$

the identity

$$M_\epsilon(\vec{u}) = \sum_{k=1}^{3} \int \frac{1}{\epsilon^3} \partial_k \varphi(\frac{y}{\epsilon})(2A_{k,\epsilon}(\vec{u}) - B_{k,\epsilon}(\vec{u}))\, dy - C_\epsilon(\vec{u}) \tag{3.3}$$

where $\lim_{\epsilon \to 0} C_\epsilon(\vec{u}) = 0$ in $\mathcal{D}'(Q)$.

D) *If $\vec{U} = \vec{u} + \vec{H}$ where $\vec{H}$ is a harmonic correction of $\vec{u}$, then $M(\vec{U}) = M(\vec{u})$.*

PROOF. The key tool is identity (3.3) which has been described by Duchon and Robert [**D**] for any divergence-free vector field $\vec{u}$ in $L^\infty_t L^2_x(Q) \cap L^2_t H^1_x(Q)$. Let us remark that if w_1 and w_2 belong to $L^\infty_t L^2_x(Q) \cap L^2_t H^1_x(Q)$ and w_3 to $L^\infty_t \mathrm{Lip}_x(Q)$ then we have obviously

$$\lim_{\epsilon \to 0} \int \frac{1}{\epsilon^3} \partial_k \varphi(\frac{y}{\epsilon}) \frac{(w_1(x-y) - w_1(x))(w_2(x-y) - w_2(x))(w_3(x-y) - w_3(x))}{\epsilon}\, dy = 0.$$

Thus, if $\vec{H}$ is a harmonic correction of $\vec{u}$, we have $\lim_{\epsilon \to 0} M_\epsilon(\vec{u} + \vec{H}) - M_\epsilon(\vec{u}) = 0$. Since the limits $\lim_{\epsilon \to 0} M_\epsilon(\vec{u} + \vec{H})$ and $\lim_{\epsilon \to 0}(M_\epsilon(\vec{u}) + 2 \lim_{\eta \to 0} \text{div } ((p * \xi_{\eta,\epsilon}) \vec{u}_{\epsilon,\eta}))$ are well defined in $\mathcal{D}'(Q)$, we find that $M(\vec{u})$ and $<<$ div $(p\vec{u})$ $>>$ are well defined and that $M(\vec{u}) = M(\vec{u} + \vec{H})$. □

Of course, if $\vec{u}$ is regular enough, we have $M(\vec{u}) = 0$. Due to formula (3.3), Duchon and Robert [**D**] could see that when $\vec{u}$ belongs locally to $L^3_t(B^{1/3}_{3,q})_x$ with $q < +\infty$, then $M(\vec{u}) = 0$. This is the case when the classical criterion $\vec{u} \in L^4_{t,x}(\Omega)$ is fulfilled, since $L^4_t L^4_x \cap L^2_t H^1_x \subset L^3_t(B^{1/3}_{3,3})_x$. In particular, the support of the distribution $M(\vec{u})$ is a subset of the set $\Sigma(\vec{u})$ of singular points.

4. Dissipativity and partial regularity

The best result we know about (partial) regularity of weak solutions has been given in 1982 by Caffarelli, Kohn and Nirenberg [**Ca**, **La**]. Their result is based on the notion of suitable solutions (due to Scheffer [**Sc**]):

Definition 4.1 (Suitable solutions). Let $\vec{u}$ be a local weak solution of the Navier–Stokes solutions on a domain $\Omega \subset \mathbb{R} \times \mathbb{R}^3$. Then $\vec{u}$ is suitable if if satisfies the following two conditions :

- the pressure p is locally in $L^{3/2}_{t,x}$,
- $M(\vec{u}) \geq 0$ (i.e. $M(\vec{u})$ is a non-negative locally finite Borel measure).

Let us define now the parabolic metric $\rho((t,x),(s,y)) = \max(\sqrt{|t-s|}, |x-y|)$ and the parabolic cylinders $Q_r(t,x) = \{(s,y) : \rho((t,x),(s,y)) < r\}$.

Theorem 4.2 (Caffarelli, Kohn and Nirenberg's regularity theorem). *Let Ω be a domain in $\mathbb{R} \times \mathbb{R}^3$, $\vec{f} \in L^2_{\text{loc}}(\Omega)$ a divergence-free time-dependent vector field and $\vec{u}$ a local weak solution of the Navier–Stokes equations on Ω (associated to the force $\vec{f}$). Assume that moreover*

- *$\vec{u}$ is suitable,*
- *the force $\vec{f}$ is regular : $\vec{f}$ belongs locally to $L^2_t H^1_x$,*

Then:

- *if $(t,x) \notin \Sigma(\vec{u})$, there exists a neighborhood of (t,x) on which $\vec{u}$ is Hölderian (with respect to the parabolic metric ρ) and we have*
$$\lim_{r \to 0} \frac{1}{r} \iint_{Q_r(t,x)} |\vec{\nabla} \otimes \vec{u}|^2 \, ds \, dy = 0.$$
- *if $(t,x) \in \Sigma(\vec{u})$, then*
$$\limsup_{r \to 0} \frac{1}{r} \iint_{Q_r(t,x)} |\vec{\nabla} \otimes \vec{u}|^2 \, ds \, dy > \epsilon^*,$$
where ϵ^* is a positive constant (which doesn't depend on $\vec{u}$, $\vec{f}$ nor Ω).

The size of $\Sigma(\vec{u})$ is then easily controlled with the following lemma :

Lemma 4.3 (Parabolic Hausdorff dimension.). *Let u belongs locally to $L^2_t H^1_x$ and let Σ be the set defined by*
$$(t,x) \in \Sigma \Leftrightarrow \limsup_{r \to 0} \frac{1}{r} \iint_{Q_r(t,x)} |\vec{\nabla} u|^2 \, ds \, dy > 0.$$
Then Σ has parabolic one-dimensional Hausdorff measure equal to 0.

Chamorro, Lemarié–Rieusset and Mayoufi [**Ch**] have considered the case where no integrability assumptions were made on the pressure p. This implies to change the definition of suitable solutions. Following [**D**], they introduced the notion of dissipative solutions :

DEFINITION 4.4 (Dissipative solutions). Let $\vec{u}$ be a local weak solutions of the Navier–Stokes solutions on a domain $\Omega \subset \mathbb{R}\times\mathbb{R}^3$. Then $\vec{u}$ is dissipative if $M(\vec{u}) \geq 0$.

A similar notion has been given by Wolf [**W**]. Indeed, if $\vec{u}$ is dissipative and if we use the harmonic correction $\vec{H} = -\vec{\nabla}\varpi$ described in Lemma 3.2, we find, for $\vec{U} = \vec{u} + \vec{H}$:

$$\begin{aligned} M(\vec{U}) &= -\partial_t|\vec{U}|^2 + \Delta(|\vec{U}|^2) - 2|\vec{\nabla}\otimes\vec{U}|^2 - \text{ div } (|\vec{U}|^2\vec{U}) - 2 \text{ div } (P\vec{U}) + 2\vec{U}\cdot\vec{F} \\ &= -\partial_t|\vec{U}|^2 + \Delta(|\vec{U}|^2) - 2|\vec{\nabla}\otimes\vec{U}|^2 - \text{ div } ((|\vec{U}|^2 + 2p_0)\vec{U}) \\ &\quad +2\vec{U}\cdot\vec{f} - 2\vec{U}\cdot\vec{f_1} - 2 \text{ div } (p_1\vec{U}) \\ &= -\partial_t|\vec{U}|^2 + \Delta(|\vec{U}|^2) - 2|\vec{\nabla}\otimes\vec{U}|^2 - \text{ div } ((|\vec{U}|^2 + 2p_0)\vec{U}) \\ &\quad +2\vec{U}\cdot\vec{f} + 2\vec{U}\cdot(\vec{\nabla}\varpi\wedge(\vec{\nabla}\wedge\vec{U})). \end{aligned}$$

Writing $M(\vec{U}) \geq 0$ is exactly expressing that $\vec{u}$ is a generalized suitable solution, as defined by Wolf.

Another tool used by Chamorro, Lemarié–Rieusset and Mayoufi is the notion of parabolic Morey space :

DEFINITION 4.5 (Parabolic Morrey spaces). A function θ belongs to the parabolic Morrey space $\mathcal{M}^{s,\tau}(\Omega)$ if

$$\sup_{x_0,t_0,r} \frac{1}{r^{5(1-\frac{s}{\tau})}} \iint_\Omega 1_{|t-t_0|<r^2} 1_{|x-x_0|<r} |\theta(t,x)|^s \, dt \, dx < +\infty.$$

Parabolic Morrey spaces have been used by Kukavica [**K**] in a variant of Caffarelli, Kohn and Nirenberg's theorem [**Ca**], and by O'Leary [**O**, **Le**] in a variant of Serrin's regularity theorem [**Se**] :

THEOREM 4.6 (Kukavica's theorem). *There exists a positive constant ϵ^* such that the following holds : If $\vec{U}$ is a solution of the Navier–Stokes equations on a domain Ω_1, associated to a force $\vec{F}$ and a pressure P and if x_0, t_0, $\vec{U}$, P and $\vec{F}$ satisfy the following assumptions*

- *$\vec{U}$ belongs to $L^\infty_t L^2_x \cap L^2_t H^1_x$,*
- *$P \in L^{3/2}_{t,x}(\Omega)$,*
- *div $\vec{F} = 0$ and $\vec{F} \in L^2_{t,x}(\Omega_1)$,*
- *$\vec{U}$ is suitable,*
- *$(t_0, x_0) \in \Omega_1$ and*

$$\limsup_{r\to 0} \frac{1}{r} \iint_{(t_0-r^2,t_0+r^2)\times B(x_0,r)} |\vec{\nabla}\otimes\vec{U}|^2 \, ds \, dx < \epsilon^*,$$

then there exists $\tau > 5$ and a neighborhood Ω_2 of (t_0, x_0) such that $\vec{U} \in \mathcal{M}^{3,\tau}(\Omega_2)$.

THEOREM 4.7 (O'Leary's theorem). *If $\vec{u}$ is a solution of the Navier–Stokes equations on a domain Ω_2, associated to a force $\vec{f}$ and if $\vec{u}$ and $\vec{f}$ satisfy the following assumptions*

- *$\vec{u}$ belongs to $L^\infty_t L^2_x \cap L^2_t H^1_x$,*
- *div $\vec{f} = 0$ and $\vec{f} \in L^2_t H^k_x(\Omega_2)$ for some $k \in \mathbb{N}$,*
- *$\vec{u} \in \mathcal{M}^{s,\tau}(\Omega_2)$ with $\tau > 5$ and $2 < s \leq \tau$,*

then, for every subdomain Ω_3 which is relatively compact in Ω_2, we have

$$\vec{u} \in L^\infty_t H^{k+1}_x \cap L^2_t H^{k+2}_x(\Omega_3).$$

Using those theorems, Chamorro, Lemarié–Rieusset and Mayoufi [**Ch**] could prove the following theorem (which is essentially the result proved previously by Wolf [**W**]) :

THEOREM 4.8 (Wolf's theorem). *Let Ω be a domain in $\mathbb{R}\times\mathbb{R}^3$, $\vec f \in L^2_{\rm loc}(\Omega)$ a divergence-free time-dependent vector field and $\vec u$ a local weak solution of the Navier–Stokes equations on Ω (associated to the force $\vec f$). Assume that moreover*

- *$\vec u$ is dissipative,*
- *the force $\vec f$ is regular : $\vec f$ belongs locally to $L^2_t H^1_x$,*

Then:

- *if $(t,x)\notin\Sigma(\vec u)$, then*
$$\lim_{r\to 0}\frac{1}{r}\iint_{Q_r(t,x)}|\vec\nabla\otimes\vec u|^2\,ds\,dy = 0.$$
- *if $(t,x)\in\Sigma(\vec u)$, then*
$$\limsup_{r\to 0}\frac{1}{r}\iint_{Q_r(t,x)}|\vec\nabla\otimes\vec u|^2\,ds\,dy \geq \epsilon^*$$
where ϵ^* is a positive constant (which doesn't depend on $\vec u$, $\vec f$ nor Ω).

PROOF. We sketch the proof given in [**Ch, Le**]. Let ϵ^* be the constant in Kukavica's theorem. Let $(t_0,x_0)\in\Omega$ with
$$\limsup_{r\to 0}\frac{1}{r}\iint_{Q_r(t_0,x_0)}|\vec\nabla\otimes\vec u|^2\,ds\,dy < \epsilon^*.$$
We introduce a harmonic correction $\vec H$ on a cylindric neighborhood of (x_0,t_0) and consider the vector field $\vec U=\vec u+\vec H$. If $\vec u$ is dissipative, then $\vec U$ is suitable, associated to a force $\vec F\in L^2_tL^2_x(Q)$ and a pressure $P\in L^{3/2}_tL^{3/2}_x(Q)$. Moreover,
$$\limsup_{r\to 0}\frac{1}{r}\iint_{Q_r(t_0,x_0)}|\vec\nabla\otimes\vec U|^2\,ds\,dy = \limsup_{r\to 0}\frac{1}{r}\iint_{Q_r(t_0,x_0)}|\vec\nabla\otimes\vec u|^2\,ds\,dy < \epsilon^*.$$
Thus, by Kukavica's theorem, there exists $\tau>5$ and a neighborhood $\Omega_2\subset Q$ of (t_0,x_0) such that $\vec U\in\mathcal{M}^{3,\tau}(\Omega_2)$. As $\vec u=\vec U-\vec H$, we see that we have as well $\vec u\in\mathcal{M}^{3,\tau}(\Omega_2)$. As $\vec f\in L^2_tH^1_x$, we may use O'Leary's theorem and find that, on a cylindric neighborhood Ω_3 of (t_0,x_0), we have $\vec u\in L^\infty_tH^2_x(\Omega_3)\subset L^\infty_{t,x}(\Omega_3)$. Thus, $(t_0,x_0)\notin\Sigma(\vec u)$. □

5. Weak convergence of local weak solutions.

In this final section, we prove Theorem 1.3. Recall that we consider a sequence $(\vec f_n)_{n\in\mathbb{N}}$ of divergence-free time-dependent vector fields on a domain $\Omega\subset\mathbb{R}\times\mathbb{R}^3$ and a sequence $(\vec u_n)_{n\in\mathbb{N}}$ of local weak solutions of the Navier–Stokes equations on Ω (associated to the forces $\vec f_n$) such that, for each cylinder $Q\subset\subset\Omega$, we have

- $\vec f_n\in L^2_tH^1_x(Q)$ and $\vec f_n$ converges weakly in $L^2_tH^1_x$ to a limit $\vec f$,
- the sequence $\vec u_n$ is bounded in $L^\infty_tL^2_x(Q)\cap L^2_tH^1_x(Q)$ and converges weakly in $L^2_tH^1_x(Q)$ to a limit $\vec u$,
- for every n, $\vec u_n$ is bounded on Q (the bound depending on n).

Recall that we have described in Lemma 3.3 the distribution $M(\vec{u}_n)$ associated to the energy balance for the local weak solution $\vec{u}_n$. With the regularity assumption $\vec{u}_n \in L^\infty_{t,x}(Q) \subset L^4_t L^4_x(Q)$, we know that $M(\vec{u}_n) = 0$. We have shown how to associate a pressure p_n on Q to the solution $\vec{u}_n$ and that $M(\vec{u}_n)$ could be computed as

$$\begin{aligned} M(\vec{u}_n) &= -\partial_t|\vec{u}_n|^2 + \Delta(|\vec{u}_n|^2) - 2|\vec{\nabla} \otimes \vec{u}_n|^2 - \text{ div } (|\vec{u}_n|^2\vec{u}_n) \\ &\quad -2 << \text{div } (p_n\vec{u}_n) >> +2\vec{u}_n \cdot \vec{f}_n. \end{aligned}$$

Our aim is then to prove that the limit $\vec{u}$ is a solution to the Navier–Stokes aquations associated to the limit $\vec{f}$ and that this solution is dissipative :

$$M(\vec{u}) \geq 0.$$

We cannot give a direct proof, as it is possible that no term in the definition of $M(\vec{u}_n)$ converge to the corresponding term in $M(\vec{u})$: p is not the limit in $\mathcal{D}'$ of p_n and $|\vec{u}|^2$ is not the limit in $\mathcal{D}'$ of $|\vec{u}_n|^2$... It is easy to find an example of such a bad behavior by studying Serrin's example of smooth in space and singular in time solution to the Navier–Stokes equations [**Se**] :

EXAMPLE 5.1 (Serrin's example). Let ψ be defined on a neighborhood of $B(x_0, r_0)$ and be harmonic, $\Delta\psi = 0$, and let $\vec{f} = 0$ and

$$\vec{u} = \alpha(t)\vec{\nabla}\psi(x),$$

where $\alpha \in L^\infty((a,b))$. Then $\vec{u}$ is a local weak solution of the Navier–Stokes equations on $(a,b) \times B(x_0, r_0)$:

$$\partial_t\vec{u} = \Delta\vec{u} - \vec{u} \cdot \vec{\nabla}\vec{u} - \vec{\nabla}(-\dot{\alpha}\psi - \frac{|\vec{u}|^2}{2}) + \vec{f}.$$

Clearly, if α is not regular, the pressure p has no integrability in the time variable (because of the presence of the singular term $\dot{\alpha}(t)$)and $\vec{u}$ has no regularity in the time variable. Thus, $\vec{u}$ is dissipative (as a matter of fact, $M(\vec{u}) = 0$) but not suitable, as it violates both assumptions and conclusions of the Caffarelli, Kohn and Nirenberg theorem.

Let us adapt this example to our problem. We define

$$\vec{u}_n(t,x) = \cos(nt) \begin{pmatrix} x_1 \\ -x_2 \\ 0 \end{pmatrix}$$

- $\vec{u}_n$ is a solution on $\mathbb{R} \times \mathbb{R}^3$ of

$$\begin{cases} \partial_t\vec{u}_n = \Delta\vec{u}_n - (\vec{u}_n \cdot \vec{\nabla})\vec{u}_n - \vec{\nabla}p_n \\ \text{div}\vec{u}_n = 0 \end{cases}$$

- In this example, we have for a bounded domain Ω_0

$$\vec{u}_n \rightharpoonup 0$$

in $L^2_t H^1_x(\Omega_0)$ and

$$(\vec{u}_n \cdot \vec{\nabla})\vec{u}_n \rightharpoonup \frac{1}{2}\begin{pmatrix} x_1 \\ x_2 \\ 0 \end{pmatrix} \neq 0,$$

in $\mathcal{D}'(\Omega_0)$.

In order to circumvent this problem of non-convergence, we shall use two tools : equations on vorticities $\vec{\omega}_n = \vec{\nabla} \wedge \vec{u}_n$ and on harmonic corrections $\vec{U}_n = \vec{u}_n + \vec{H}_n = -\frac{1}{\Delta}\vec{\nabla} \wedge \Big(\psi(\vec{\nabla} \wedge \vec{u}_n)\Big)$.

Step 1 : Vorticities.

On a cylinder $Q \subset\subset \Omega$, we may write the Navier–Stokes equations on the divergence-free vector field $\vec{u}_n$ in many ways. The first one is given by equation (1.1) : for every smooth compactly supported divergence-free vector field $\vec{\phi} \in \mathcal{D}(Q)$ we have

$$\iint_Q \vec{u}_n \cdot (\partial_t \vec{\phi} + \Delta \vec{\phi}) + \vec{u}_n \cdot (\vec{u}_n \cdot \vec{\nabla}\vec{\phi}) + \vec{f}_n \cdot \vec{\phi}\, dt\, dx = 0.$$

We may rewrite this equation as:

$$\partial_t \vec{u}_n = \Delta \vec{u}_n - \vec{u}_n \cdot \vec{\nabla}\vec{u}_n + \vec{f}_n \text{ in } (\mathcal{D}_\sigma(Q))'$$

where $\mathcal{D}_\sigma(Q)$ is the space of smooth compactly supported divergence-free vector fields on Q.

The second one is given by equations (2.1): for a distribution $p_n \in \mathcal{D}'(Q)$, we have

$$\partial_t \vec{u}_n = \Delta \vec{u}_n - \vec{u}_n \cdot \vec{\nabla}\vec{u}_n - \vec{\nabla} p_n + \vec{f}_n \quad \text{ in } \mathcal{D}'(Q).$$

The next one is based on the identity

$$\vec{u}_n \cdot \vec{\nabla}\vec{u}_n = \vec{\omega}_n \wedge \vec{u}_n + \vec{\nabla}(\frac{|\vec{u}_n|^2}{2})$$

from which we get

$$\partial_t \vec{u}_n = \Delta \vec{u}_n - \vec{\omega}_n \wedge \vec{u}_n + \vec{f}_n \text{ in } (\mathcal{D}_\sigma(Q))'.$$

We have seen that, in some cases, we don't have the convergence of $(\vec{u}_n \cdot \vec{\nabla})\vec{u}_n$ to $\vec{u} \cdot \vec{\nabla}\vec{u}$ in $\mathcal{D}'(\Omega_0)$. But we shall prove the following lemma :

LEMMA 5.2 (Convergence of the non-linear term). *We have the following convergence results :*

$$\vec{\omega}_n \wedge \vec{u}_n \rightharpoonup \vec{\omega} \wedge \vec{u} \text{ in } \mathcal{D}'(Q)$$

so that

$$(\vec{u}_n \cdot \vec{\nabla})\vec{u}_n \rightharpoonup \vec{u} \cdot \vec{\nabla}\vec{u} \text{ in } (\mathcal{D}_\sigma(Q))'.$$

Thus, this lemma will prove the first half of Theorem 1.3: the limit $\vec{u}$ is a local weak solution on Ω of the Navier–Stokes equations associated to the force $\vec{f}$. The proof of the lemma is based on the following variant of the classical Rellich lemma [**Le**, **M**] :

LEMMA 5.3 (Rellich's lemma). *Let $-\infty < \sigma_1 < \sigma_2 < +\infty$. Let Ω be a domain in $\mathbb{R} \times \mathbb{R}^3$. If a sequence of distribution T_n is weakly convergent to a distribution T in $(L^2_t H^{\sigma_2}_x)_{\text{loc}}$ and if the sequence $(\partial_t T_n)$ is bounded in $(L^2_t H^{\sigma_1}_x)_{\text{loc}}$, then T_n is strongly convergent in $(L^2_t H^{\sigma}_x)_{\text{loc}}$ for every $\sigma < \sigma_2$.*

We apply Rellich's lemma to $\vec{\omega}_n$. We have

$$\partial_t \vec{\omega}_n = \Delta \vec{\omega}_n - \text{ div } (\vec{u}_n \otimes \vec{\omega}_n - \vec{\omega}_n \otimes \vec{u}_n) - \vec{\nabla} \wedge \vec{f}_n,$$

so that the sequence $(\partial_t \vec{\omega}_n)$ is bounded in $(L^2_t H^{\sigma_1}_x)_{\text{loc}}$ for all $\sigma_1 < -5/2$. Moreover, $\vec{\omega}_n$ is weakly convergent to $\vec{\omega}$ in $(L^2_t L^2_x)_{\text{loc}}$. Thus, $\vec{\omega}_n$ is strongly convergent in $(L^2_t H^{-1}_x)_{\text{loc}}$. As $\vec{u}_n$ is weakly convergent to $\vec{u}$ in $(L^2_t H^1_x)_{\text{loc}}$, we find that $\vec{\omega}_n \wedge \vec{u}_n$ is weakly convergent to $\vec{\omega} \wedge \vec{u}$ in $\mathcal{D}'(\Omega)$.

Step 2 : Harmonic corrections.

We now end the proof of Theorem 1.3 by checking the dissipativity of the limit $\vec{u}$. We restate the theorem as a result of stability for dissipativity :

THEOREM 5.4 (Dissipative limits). *Let Ω be a domain in $\mathbb{R}\times\mathbb{R}^3$. Assume that we have sequences $\vec{f}_n$ of divergence-free time-dependent vector fields and $\vec{u}_n$ of local weak solutions of the Navier–Stokes equations on Ω (associated to the forces $\vec{f}_n$) such that, for each cylinder $Q \subset\subset \Omega$, we have*

- *$\vec{f}_n \in L^2_t H^1_x(Q)$ and $\vec{f}_n$ converges weakly in $L^2_t H^1_x$ to a limit $\vec{f}$,*
- *the sequence $\vec{u}_n$ is bounded in $L^\infty_t L^2_x(Q) \cap L^2_t H^1_x(Q)$ and converges weakly in $L^2_t H^1_x(Q)$ to a limit $\vec{u}$,*
- *for every n, $\vec{u}_n$ is dissipative.*

Then the limit $\vec{u}$ is a dissipative local weak solution on Ω of the Navier–Stokes equations associated to the force $\vec{f}$.

PROOF. We already know that $\vec{u}$ is a local weak solution on Ω of the Navier–Stokes equations associated to the force $\vec{f}$. We have to prove its dissipativity.

Let $Q \subset\subset \Omega$ be a cylinder and $\psi \in \mathcal{D}(\Omega)$ be a cut-off function which is equal to 1 on a neighborhood of Q. In order to prove that $\vec{u}$ is dissipative, we shall prove that the harmonic correction $\vec{U} = \vec{H} + \vec{u} = -\frac{1}{\Delta}\vec{\nabla} \wedge \left(\psi(\vec{\nabla} \wedge \vec{u})\right)$ is suitable.

We define as well $\vec{U}_n = -\frac{1}{\Delta}\vec{\nabla} \wedge \left(\psi(\vec{\nabla} \wedge \vec{u}_n)\right)$. The weak convergence of $\vec{u}_n$ in $(L^2_t H^1_x)_{\mathrm{loc}}(\Omega)$ implies the weak convergence of $\vec{U}_n$ to $\vec{U}$ in $L^2_t H^1_x(Q)$. Moreover, the uniform boundedness of the sequence $(\vec{u}_n)_{n\in\mathbb{N}}$ in $(L^2_t H^1_x \cap L^\infty_t L^2_x)_{\mathrm{loc}}(\Omega)$ and of the sequence $(\vec{f}_n)_{n\in\mathbb{N}}$ in $(L^2_t H^1_x)_{\mathrm{loc}}(\Omega)$ implies that the sequences of pressure P_n and of forces $\vec{F}_n$ associated to $\vec{U}_n$ are uniformly bounded (respectively in $L^{3/2}_t L^{3/2}_x(Q) \cap L^2_t L^{6/5}_x(Q)$ and in $L^2_t L^2_x(Q)$). Thus, $\partial_t \vec{U}_n$ is bounded in $L^2_t H^{-2}_x(Q)$ and Rellich's lemma gives us that $\vec{U}_n$ is strongly convergent to $\vec{U}$ in $(L^2_t L^2_x)_{\mathrm{loc}}(Q)$ (and, since $\vec{U}_n$ is bounded in $L^{10/3}_t L^{10/3}_x(Q)$, we have strong convergence in $(L^3_t L^3_x)_{\mathrm{loc}}(Q)$ as well).

Taking subsequences, we may assume that the bounded sequences P_n (in $L^{3/2}_t L^{3/2}_x(Q)$), $\vec{F}_n$ (in $L^2_t L^2_x(Q)$) and $|\vec{\nabla} U_n|^2$ (in $L^1_t L^1_x(Q)$) converge weakly in $\mathcal{D}'$ to limits $P_\infty \in L^{3/2}_t L^{3/2}_x(Q)$, $\vec{F}_\infty \in L^2_t L^2_x(Q)$ and ν_∞ (a non-negative finite measure on Q). In particular, we have enough convergence to see that every term in the right-hand side of equality

$$M(\vec{u}_n) = -\partial_t|\vec{U}_n|^2 + \Delta(|\vec{U}_n|^2) - 2|\vec{\nabla}\otimes\vec{U}_n|^2 - \operatorname{div}(|\vec{U}_n|^2\vec{U}_n) - 2\operatorname{div}(P_n\vec{U}_n) + 2\vec{U}_n\cdot\vec{U}_n$$

has a limit, so that $\nu_1 = \lim_{n\to+\infty} M(\vec{U}_n)$ exists and

$$\nu_1 = -\partial_t|\vec{U}|^2 + \Delta(|\vec{U}|^2) - 2\nu_\infty - \operatorname{div}(|\vec{U}|^2\vec{U}) - 2\operatorname{div}(P_\infty\vec{U}) + 2\vec{U}\cdot\vec{F}_\infty.$$

As $M(\vec{U}_n) \geq 0$, we find that $\nu_1 \geq 0$. Moreover, by the Banach–Steinhaus theorem, we find that $\nu_2 = \nu_\infty - |\vec{\nabla}\otimes\vec{U}|^2 \geq 0$. As $M(\vec{U}) = \nu_1 + 2\nu_2$, we have $M(\vec{U}) \geq 0$. Hence, $\vec{U}$ is suitable and $\vec{u}$ is dissipative. □

Now, the proof of Theorem 1.3 is just a combination of all the previous results. First, under the assumptions of Theorem 1.3, we see that the solutions $\vec{u}_n$ are dissipative (with $M(\vec{u}_n) = 0$). Thus, we may apply Theorem 5.4 and find that the limit $\vec{u}$ is a dissipative local weak solution. Theorem 4.8 then describes the set of singular points of $\vec{u}$ and Lemma 4.3 describes the size of this set. The proof is then complete.

References

[Ca] L. Caffarelli, R. Kohn, and L. Nirenberg, *Partial regularity of suitable weak solutions of the Navier-Stokes equations*, Comm. Pure Appl. Math. **35** (1982), no. 6, 771–831. MR673830

[Ch] D. Chamorro, PG. Lemarié-Rieusset & K. Mayoufi. *The role of the pressure in the partial regularity theory for weak solutions of the Navier–Stokes equations.* https://arxiv.org/abs/1602.06137, (2016).

[D] Jean Duchon and Raoul Robert, *Inertial energy dissipation for weak solutions of incompressible Euler and Navier-Stokes equations*, Nonlinearity **13** (2000), no. 1, 249–255. MR1734632

[K] Igor Kukavica, *On partial regularity for the Navier-Stokes equations*, Discrete Contin. Dyn. Syst. **21** (2008), no. 3, 717–728. MR2399434

[La] O. A. Ladyzhenskaya and G. A. Seregin, *On partial regularity of suitable weak solutions to the three-dimensional Navier-Stokes equations*, J. Math. Fluid Mech. **1** (1999), no. 4, 356–387. MR1738171

[Le] Pierre Gilles Lemarié-Rieusset, *The Navier-Stokes problem in the 21st century*, CRC Press, Boca Raton, FL, 2016. MR3469428

[M] K. Mayoufi. *Les inégalités d'énergie locales dans la théorie des équations de Navier–Stokes*, Ph. D. Univ. Évry, (2017).

[O] Mike O'Leary, *Conditions for the local boundedness of solutions of the Navier-Stokes system in three dimensions*, Comm. Partial Differential Equations **28** (2003), no. 3-4, 617–636. MR1976466

[Sc] Vladimir Scheffer, *Hausdorff measure and the Navier-Stokes equations*, Comm. Math. Phys. **55** (1977), no. 2, 97–112. MR0510154

[Se] James Serrin, *On the interior regularity of weak solutions of the Navier-Stokes equations*, Arch. Rational Mech. Anal. **9** (1962), 187–195. MR0136885

[W] Jörg Wolf, *A direct proof of the Caffarelli-Kohn-Nirenberg theorem*, Parabolic and Navier-Stokes equations. Part 2, Banach Center Publ., vol. 81, Polish Acad. Sci. Inst. Math., Warsaw, 2008, pp. 533–552. MR2548880

LaMME, Univ Evry, CNRS, Université Paris-Saclay, 91025, Evry, France
Current address: IBGBI, 23 bd de France, 91037 Évry cedex, France
Email address: diego.chamorro@univ-evry.fr

LaMME, Univ Evry, CNRS, Université Paris-Saclay, 91025, Evry, France
Current address: IBGBI, 23 bd de France, 91037 Évry cedex, France
Email address: pierregilles.lemarierieusset@univ-evry.fr

LaMME, Univ Evry, CNRS, Université Paris-Saclay, 91025, Evry, France
Current address: IBGBI, 23 bd de France, 91037 Évry cedex, France
Email address: kawther.mayoufi@univ-evry.fr

Contemporary Mathematics
Volume **710**, 2018
http://dx.doi.org/10.1090/conm/710/14364

Blow up criteria for the compressible Navier–Stokes equations

Hi Jun Choe and Minsuk Yang

ABSTRACT. We study the strong solution to the 3-D compressible Navier–Stokes equations. We propose a new blow up criterion for barotropic gases in terms of the integral norm of density ρ and the divergence of the velocity $\mathbf{u}$ without any restriction on the physical viscosity constants. Our blow up criteria can be seen as a partial realization of the underlying principle that higher integrability implies boundedness and then eventual regularity. We also present a similar blow-up criterion for heat conducting gases.

1. Introduction

The motion of the compressible barotropic gases is governed by the viscous compressible Navier–Stokes equations, which consist of the equation of the conservation of mass

$$\partial_t \rho + \operatorname{div}(\rho \mathbf{u}) = 0 \tag{1.1}$$

and the equation of the conservation of momentum

$$\partial_t(\rho \mathbf{u}) + \operatorname{div}(\rho \mathbf{u} \otimes \mathbf{u}) = L\mathbf{u} - \nabla p \tag{1.2}$$

in $(0,T) \times \Omega$ where the state variables $\mathbf{u}$ and ρ denote the velocity field and the density of the fluid, respectively. The Lamé operator L is given by

$$L\mathbf{u} = \mu \Delta \mathbf{u} + (\lambda + \mu)\nabla \operatorname{div} \mathbf{u} \tag{1.3}$$

where the viscosity constants satisfy the physical conditions

$$\mu > 0 \quad \text{and} \quad 3\lambda + 2\mu > 0. \tag{1.4}$$

The pressure p is given by the barotropic constitutive relation

$$p = \alpha \rho^\gamma \tag{1.5}$$

for some constants $\gamma > 1$ and $\alpha > 0$. In this paper, the space domain is assumed to be a torus $\Omega = \mathbb{T}^3 = \mathbb{R}^3/\mathbb{Z}^3$. The periodic boundary conditions are assumed and the initial conditions

$$\rho|_{t=0} = \rho_0, \qquad \mathbf{u}|_{t=0} = \mathbf{u}_0$$

2010 *Mathematics Subject Classification.* Primary 76N10.

H. Choe has been supported by the National Research Foundation of Korea (NRF) grant funded by the Korea government(MSIP) (No. 2015R1A5A1009350).

M. Yang has been supported by the National Research Foundation of Korea (NRF) grant funded by the Korea government(MSIP) (No. 2016R1C1B2015731).

are assumed to guarantee the existence of strong solutions. The constant α in (1.5) are inconsequential in our argument, so we simply assume $\alpha = 1$.

We also consider the motion of the compressible heat conducting gases. The viscous compressible heat conducting Navier–Stokes equations satisfy

$$\partial_t(\rho\theta) + \operatorname{div}(\rho\theta\mathbf{u}) + \rho\theta \operatorname{div}\mathbf{u} - \kappa\Delta\theta = \frac{\mu}{2}|\nabla\mathbf{u} + \nabla\mathbf{u}^t|^2 + \lambda|\operatorname{div}\mathbf{u}|^2 \tag{1.6}$$

with the conservation of mass (1.2) and the conservation of momentum (1.3) in $(0,T)\times\Omega$ where the constant $\kappa > 0$ denotes the heat conductivity. The pressure p is given by the constitutive relation

$$p = \alpha\rho\theta. \tag{1.7}$$

for some constant $\alpha > 0$. The space domain is also assumed to be a torus $\Omega = \mathbb{T}^3 = \mathbb{R}^3/\mathbb{Z}^3$. The periodic boundary conditions are assumed and the initial conditions

$$\rho|_{t=0} = \rho_0, \qquad \mathbf{u}|_{t=0} = \mathbf{u}_0, \qquad \theta|_{t=0} = \theta_0$$

are assumed to guarantee the existence of strong solutions. Since the constants α and κ plays no role in our argument, we simply assume $\alpha = \kappa = 1$.

When the initial data are in suitable energy classes, global existence of weak solutions for the compressible barotropic gases was proved for $\gamma \geq 9/5$ by Lions [**11**]. Later, the result was extended to the case $\gamma > 3/2$ by Feireisl–Novotný–Petzeltová [**7**]. Although there is a large volume of literature on the existence results according to various conditions on the initial data (e.g., Solonnikov [**13**], Matsumura–Nishida [**12**], Vaigant–Kazhikhov [**15**], and Hoff [**8**]), we only mention the works closely related to this paper.

In a series of papers [**1**–**4**] Cho, Choe, Jin, and Kim proved short time existence of strong solutions allowing vacuum region. When vacuum region initially exists, Xin [**17**] proved that the smooth solution blows up in finite time. Huang–Li–Xin [**9**] established Beal-Kato-Majda type blow up criterion in terms of velocity. Li–Li–Xin [**10**] proved the strong solution blows up as vacuum states vanish. Recently, Sun–Wang–Zhang [**14**] presented a sophisticated characterization of the maximal existence time of strong solution for barotropic gases in terms of the essential superemum norm of density when the viscosity constants λ and μ satisfy $\lambda < 7\mu$. Wen–Zhu [**16**] improved and extended the result in Sun–Wang–Zhang [**14**] by making weaker assumptions on such viscosity constants. We refer the reader to the paper [**16**] for summary and references of other important blow up criteria and extensions.

Our aim in this paper is to present new blow up criteria, Theorem 2.3 and Theorem 2.5, without any restriction on the physical constants λ and μ. We use the integral norm of density ρ and the divergence of the velocity $\mathbf{u}$ which is notably significant if we consider the continuity equation and the weak compactness for approximations. It seems that the condition on the divergence of the velocity $\mathbf{u}$ commensurates with such viscosity restrictions. We also present similar blow up criteria for heat conducting gases. Our blow up criteria can be seen as a partial realization of the underlying principle that higher integrability implies boundedness and then eventual regularity.

2. Preliminaries and main results

We shall use the standard notations $L^q(\Omega)$ and $W^{k,q}(\Omega)$ for the Lebesgue spaces and the Sobolev spaces, respectively. We denote $H^k(\Omega) = W^{k,2}(\Omega)$ and

$$D^{k,q}(\Omega) = \left\{ \mathbf{u} \in L^1_{loc}(\Omega) : \left\| \nabla^k \mathbf{u} \right\|_{L^q(\Omega)} < \infty \right\}.$$

We define the space $D_0^{k,q}(\Omega)$ to be the closure of $C_c^\infty(\Omega)$ in $D^{k,q}(\Omega)$.

The following results on the local wellposedness in time and blow up criteria for the strong solution for the initial data with vacuum are established by Cho, Choe, and Kim in a series of papers [**1**], [**2**], and [**3**].

We recall the following theorem for the existence of the strong solution corresponding to barotropic gases.

THEOREM 2.1. *Let Ω be a bounded smooth domain or $\mathbb{R}^3$. Suppose $\mathbf{u}_0 \in D_0^1(\Omega) \cap D^2(\Omega)$ and*

$$\rho_0 \in W^{1,q}(\Omega) \cap H^1(\Omega) \cap L^1(\Omega)$$

for some $q \in (3, 6]$. If ρ_0 is nonnegative and the initial data satisfy the compatibility condition

$$L\mathbf{u}_0 + \nabla p(\rho_0) = \sqrt{\rho_0} g$$

for some vector field $g \in L^2(\Omega)$, then there exist a time $T \in (0, \infty]$ and a unique solution

$$(\rho, \mathbf{u}) \in C([0,T] : H^1 \cap W^{1,q}(\Omega)) \times C([0,T] : D^2(\omega)) \cap L^2(0,T; D^{2,q}(\Omega)).$$

Moreover, if the maximal existence time $T^\star$ of the solution is finite, then

$$\lim \sup_{t \to T^\star} ||\rho||_{W^{1,q}}(t) + ||\mathbf{u}||_{D^1(\Omega)} = \infty.$$

We recall the following theorem for the existence of the strong solution corresponding to heat conducting gases.

THEOREM 2.2. *Let Ω be a bounded smooth domain or $\mathbb{R}^3$. Suppose $\mathbf{u}_0, \theta_0 \in D_0^1(\Omega) \cap D^2(\Omega)$ and*

$$\rho_0 \in W^{1,q}(\Omega) \cap H^1(\Omega) \cap L^1(\Omega)$$

for some $q \in (3, 6]$. If ρ_0 is nonnegative and the initial data satisfy the compatibility conditions

$$\begin{aligned} L\mathbf{u}_0 + \nabla p(\rho_0) &= \sqrt{\rho_0} g_1 \\ \Delta\theta_0 + \frac{\mu}{2} |\nabla \mathbf{u}_0 + (\nabla \mathbf{u}_0)^t|^2 + \lambda (\operatorname{div} \mathbf{u}_0)^2 &= \sqrt{\rho_0} g_2 \end{aligned}$$

for a vector field $g_1, g_2 \in L^2(\Omega)$, then there exist a time $T \in (0, \infty]$ and a unique solution

$$\begin{aligned} (\rho, \mathbf{u}, \theta) &\in C([0,T) : H^1 \cap W^{1,q}) \times C([0,T) : D^2 \cap D_0^1) \times L^2([0,T) : D^{2,q}) \\ (\rho_t, \mathbf{u}_t, \theta_t) &\in C([0,T) : L^2 \cap L^q) \times L^2([0,T) : D_0^1) \times L^2([0,T) : D_0^1) \\ (\rho^{1/2}\mathbf{u}_t, \rho^{1/2}\theta_t) &\in L^\infty([0,T) : L^2) \times L^\infty([0,T) : L^2). \end{aligned}$$

Throughout this paper we consider the strong solution in the above regularity function classes. The results in Theorem 2.1 and 2.2 continue to hold for $\Omega = \mathbb{T}^3$ with periodic boundary conditions. First, we present the following theorem as our blow up criterion for compressible gases.

THEOREM 2.3. *Suppose* $(\rho, \mathbf{u})$ *is the unique strong solution in Theorem* 2.1. *If the maximal existence time* $T^\star$ *is finite, then there exists a number* $\beta \in (1, \infty)$ *depending only on* γ *such that*

$$\limsup_{t \to T^\star} (\|\rho\|_{L^\beta}(t) + \|\operatorname{div} \mathbf{u}\|_{L^3}(t)) = \infty.$$

REMARK 2.4. For example, if $\gamma = 3/2$, then the integral norm of density with the integrable exponent $\beta \geq 16$ blows up. Actually, we propose a range of the exponent β in the proof. Naturally, it is an important question to know the minimal integrability of the exponent β.

Our blow up criterion for compressible heat conducting gases is the following theorem.

THEOREM 2.5. *Suppose* $(\rho, \mathbf{u}, \theta)$ *is the unique strong solution in Theorem* 2.2. *If the maximal existence time* T^* *is finite, then there exists a number* $\delta \in (1, \infty)$ *depending only on fluid dynamic parameters such that*

$$\limsup_{t \to T^*} (\|\rho\|_{L^\delta}(t) + \|\theta\|_{L^\infty}(t) + \|\Delta\theta\|_{L^2}(t) + \|\operatorname{div} \mathbf{u}\|_{L^3}(t)) = \infty.$$

We end this section by providing a few notations.

- We denote by $\widetilde{f}$ the derivative
$$\widetilde{f} = (\partial_t + u \cdot \nabla) f.$$
Previously, the notation $\dot{f}$ was commonly used in papers, but we use $\widetilde{f}$ for easier visual discrimination.
- We denote the $L^q(\Omega)$ norm simply by
$$\|f\|_q = \|f\|_{L^q(\Omega)}.$$
- We denote $X \lesssim Y$ if there is a generic positive constant C such that $|X| \leq C|Y|$.

3. Reduction of the problem

If we have proved $\sup_{0 \leq t \leq T} \|\rho\|_\infty < \infty$ without assuming $7\mu > \lambda$, then the argument in Section 5 of [**14**] immediately yields

$$\rho \in L^\infty(0, T; W^{1,q}), \qquad q \in (3, 6].$$

This implies T is not the maximal existence time of the strong solution. Therefore, the goal of this section is to realize that the integrability of ρ and $\nabla \mathbf{u}$ yields the boundedness of ρ.

It is well known that the effective viscous flux plays an important role in the existence theory of weak solutions. The following simple observation is profoundly useful to our argument. Let $\mathcal{P}$ be the Helmholz–Leray projection to solenoidal spaces. Taking the operator $\mathcal{P}$ to the momentum equation $\rho\widetilde{\mathbf{u}} = L\mathbf{u} - \nabla p$ yields

$$\mathcal{P}(\rho\widetilde{\mathbf{u}}) = \mu\Delta\mathcal{P}\mathbf{u}. \tag{3.1}$$

Let us define

$$G := (\lambda + 2\mu) \operatorname{div} \mathbf{u} - p.$$

Then

$$\Delta G = \operatorname{div}(\rho\widetilde{\mathbf{u}}) = \operatorname{div}(L\mathbf{u} - \nabla p) = \Delta\left((\lambda + 2\mu) \operatorname{div} \mathbf{u} - p\right).$$

Observing that $\lambda + 2\mu > 0$ and

$$\operatorname{div} L\mathbf{u} = (\lambda + 2\mu)\Delta \operatorname{div} \mathbf{u},$$

we may take the divergence operator to the momentum equation to obtain

$$\operatorname{div}(\rho\tilde{\mathbf{u}}) = \Delta G \tag{3.2}$$

in the sense of distributions.

The following lemma shows a sufficient condition for boundedness of density.

LEMMA 3.1. *Suppose*

$$\sup_{0\le t\le T}\left(\|\nabla\mathbf{u}\|_A + \|\rho\|_\beta + \left\|\rho^{1/C}\mathbf{u}\right\|_C\right) < \infty \tag{3.3}$$

for some positive numbers A, B, C, β *satisfying* $A < B < C$, $1/A + 1/B < 1/3$, *and*

$$(C-1)B/(C-B) \le \beta. \tag{3.4}$$

Then

$$\sup_{0\le t\le T}\|\rho\|_\infty < \infty.$$

PROOF. We define the Lagrangean flow X of $\mathbf{u}$ by $X(s,s,x) = x$ and

$$\frac{\partial}{\partial t}X(t,s,x) = \mathbf{u}(t, X(t,s,x)).$$

From the continuity equation we have

$$\tilde{\rho} = (\partial_t + \mathbf{u}\cdot\nabla)\rho = -\rho \operatorname{div}\mathbf{u}.$$

Hence we can write

$$\begin{aligned}\rho(t, X(t,0,x)) &= \rho_0(x)\exp\left(\int_0^t \frac{d}{ds}\ln\rho(s, X(s,0,x))ds\right)\\ &= \rho_0(x)\exp\left(-\int_0^t (\operatorname{div}\mathbf{u})(s, X(s,0,x))ds\right).\end{aligned}$$

Since $(\lambda+2\mu)\operatorname{div}\mathbf{u} = G + p$ and the pressure p is non-negative, we can deduce

$$\rho(t, X(t,0,x)) \le \rho_0(x)\exp\left(-(\lambda+2\mu)^{-1}\int_0^t G(s, X(s,0,x))ds\right).$$

To conclude the lemma, it suffices to prove that the following integral is bounded

$$K(x) := -\int_0^t G(s, X(s,0,x))ds.$$

A direct computation reveals

$$\rho\tilde{\mathbf{u}} = \widetilde{\rho\mathbf{u}} - \tilde{\rho}\mathbf{u} = \widetilde{\rho\mathbf{u}} + \rho\operatorname{div}\mathbf{u}\mathbf{u}.$$

From the relation (3.2) we have

$$\begin{aligned}K(x) = &-\Delta^{-1}\operatorname{div}\int_0^t \widetilde{\rho\mathbf{u}}(s, X(s,0,x))ds\\ &-\int_0^t \Delta^{-1}\operatorname{div}(\rho\operatorname{div}\mathbf{u}\mathbf{u})(s, X(s,0,x))ds\end{aligned}$$

where we notice that $\Delta^{-1}\operatorname{div}(\widetilde{\rho\mathbf{u}})$ and $\Delta^{-1}\operatorname{div}(\rho\operatorname{div}\mathbf{uu})$ are well-defined. Integrating $\widetilde{\rho\mathbf{u}}$ on the flow can be calculated by

$$\begin{aligned}\int_0^t \widetilde{\rho\mathbf{u}}(s,X(s,0,x))ds &= \int_0^t \frac{d}{ds}(\rho\mathbf{u})(s,X(s,0,x))ds \\ &= (\rho\mathbf{u})(t,X(t,0,x)) - (\rho\mathbf{u})(0,x).\end{aligned}$$

Thus, we have

$$\begin{aligned}K(x) \le \Delta^{-1}\operatorname{div}(\rho_0\mathbf{u}_0)(x) &- \Delta^{-1}\operatorname{div}(\rho\mathbf{u})(t,X(t,0,x)) \\ &- \int_0^t \Delta^{-1}\operatorname{div}(\rho\operatorname{div}\mathbf{uu})(s,X(s,0,x))ds.\end{aligned}$$

Since $3 < AB/(A+B)$, the Sobolev-type estimate yields

$$\left\|\Delta^{-1}\operatorname{div}(\rho\operatorname{div}\mathbf{uu})\right\|_\infty \lesssim \|\rho\operatorname{div}\mathbf{uu}\|_{AB/(A+B)} \le \|\nabla\mathbf{u}\|_A\|\rho\mathbf{u}\|_B.$$

Thus, using the condition (3.3), we have

$$\begin{aligned}\|K\|_\infty &\lesssim \sup_{0\le t\le T}\left\|\Delta^{-1}\operatorname{div}(\rho\mathbf{u})\right\|_\infty + \int_0^T \left\|\Delta^{-1}\operatorname{div}(\rho\operatorname{div}\mathbf{uu})\right\|_\infty \\ &\lesssim \left(1 + T\sup_{0\le t\le T}\|\nabla\mathbf{u}\|_A\right)\sup_{0\le t\le T}\|\rho\mathbf{u}\|_B \\ &\lesssim \sup_{0\le t\le T}\|\rho\mathbf{u}\|_B\end{aligned}$$

and for all $0\le t\le T$

$$\begin{aligned}\|\rho\mathbf{u}\|_B &\le \left\|\rho^{(C-1)/C}\right\|_{CB/(C-B)}\left\|\rho^{1/C}\mathbf{u}\right\|_C \\ &= \|\rho\|_{(C-1)B/(C-B)}^{(C-1)/C}\left\|\rho^{1/C}\mathbf{u}\right\|_C < \infty.\end{aligned}$$

This proves the lemma. □

4. Proof of Theorem 2.3

We assume that the strong solution $(\rho,\mathbf{u})$ stated in Theorem 2.1 satisfies the following integrability condition for some positive number β

$$\sup_{0\le t\le T}\left(\|\rho\|_\beta + \|\operatorname{div}\mathbf{u}\|_3\right) < \infty. \tag{4.1}$$

The number β will be specified in each lemma.

If $(\rho,\mathbf{u})$ is a strong solution in $[0,T)$, then it is apparent from the continuity equation that for all $t\in[0,T)$

$$\int_\Omega \rho(t) = \int_\Omega \rho_0.$$

Additionally, multiplying the momentum equation by $\mathbf{u}$, we see that for all $t\in[0,T)$

$$\begin{aligned}&\frac{1}{2}\int_\Omega \rho|\mathbf{u}|^2(t) + \frac{1}{\gamma-1}\int_\Omega \rho^\gamma(t) + \mu\int_0^t\int_\Omega |\nabla\mathbf{u}|^2 + (\mu+\lambda)\int_0^t\int_\Omega |\operatorname{div}\mathbf{u}|^2 \\ &= \frac{1}{2}\int_\Omega \rho_0|\mathbf{u}_0|^2 + \frac{1}{\gamma-1}\int_\Omega \rho_0^\gamma.\end{aligned}$$

We shall show that the condition (4.1) with a suitable β yields $\sup_{0\le t\le T}\left\|\rho^{1/C}\mathbf{u}\right\|_C<\infty$. In order to prove it, we derive the following estimate by testing with $C|\mathbf{u}|^{C-2}\mathbf{u}$ in the momentum equation.

LEMMA 4.1. *Let* $2<C<\infty$. *For all* $t\in[0,T]$

$$\frac{d}{dt}\int_\Omega \rho|\mathbf{u}|^C + C\mu\int_\Omega |\mathbf{u}|^{C-2}|\nabla\mathbf{u}|^2 \lesssim \left(\int_\Omega |\operatorname{div}\mathbf{u}|^3\right)^{C/3} + \int_\Omega p^2|\mathbf{u}|^{C-2}.$$

PROOF. A direct calculation gives

$$\int_\Omega \rho\partial_t|\mathbf{u}|^C = C\int_\Omega \rho\partial_t\mathbf{u}\cdot|\mathbf{u}|^{C-2}\mathbf{u}.$$

Using the continuity equation and integrating by parts we have

$$\int_\Omega \partial_t\rho|\mathbf{u}|^C = -\int_\Omega \operatorname{div}(\rho\mathbf{u})|\mathbf{u}|^C = C\int_\Omega \rho(\mathbf{u}\cdot\nabla)\mathbf{u}\cdot|\mathbf{u}|^{C-2}\mathbf{u}.$$

Summing the two identity yields

$$\frac{d}{dt}\int_\Omega \rho|\mathbf{u}|^C = C\int_\Omega \rho\tilde{\mathbf{u}}\cdot|\mathbf{u}|^{C-2}\mathbf{u}.$$

Integrating by parts yields

$$\begin{aligned}-\int_\Omega L\mathbf{u}\cdot|\mathbf{u}|^{C-2}\mathbf{u} &= \mu\int_\Omega |\mathbf{u}|^{C-2}|\nabla\mathbf{u}|^2 + (C-2)\mu\int_\Omega |\mathbf{u}|^{C-2}|\nabla|\mathbf{u}||^2\\ &\quad + (\lambda+\mu)\int_\Omega |\mathbf{u}|^{C-2}|\operatorname{div}\mathbf{u}|^2 + (\lambda+\mu)\int_\Omega (\operatorname{div}\mathbf{u})\mathbf{u}\cdot\nabla|\mathbf{u}|^{C-2}.\end{aligned}$$

Hence, we have

$$\begin{aligned}&\frac{1}{C}\frac{d}{dt}\int_\Omega \rho|\mathbf{u}|^C + \mu\int_\Omega |\mathbf{u}|^{C-2}|\nabla\mathbf{u}|^2\\ &\quad + (C-2)\mu\int_\Omega |\mathbf{u}|^{C-2}|\nabla|\mathbf{u}||^2 + (\lambda+\mu)\int_\Omega |\mathbf{u}|^{C-2}|\operatorname{div}\mathbf{u}|^2\\ &= -(\lambda+\mu)\int_\Omega (\operatorname{div}\mathbf{u})\mathbf{u}\cdot\nabla|\mathbf{u}|^{C-2} - \int_\Omega \nabla p\cdot|\mathbf{u}|^{C-2}\mathbf{u}\\ &=: I_1+I_2.\end{aligned}\tag{4.2}$$

By the Sobolev inequality

$$\begin{aligned}I_1 &\le (C-2)(\lambda+\mu)\int_\Omega |\operatorname{div}\mathbf{u}||\mathbf{u}|^{C-2}|\nabla|\mathbf{u}||\\ &\lesssim (C-2)\left(\int_\Omega |\operatorname{div}\mathbf{u}|^3\right)^{1/3}\left(\int_\Omega |\mathbf{u}|^{3C}\right)^{(C-2)/6C}\left(\int_\Omega |\mathbf{u}|^{C-2}|\nabla|\mathbf{u}||^2\right)^{1/2}\\ &\lesssim (C-2)C^{(C-2)/C}\left(\int_\Omega |\operatorname{div}\mathbf{u}|^3\right)^{1/3}\left(\int_\Omega |\mathbf{u}|^{C-2}|\nabla|\mathbf{u}||^2\right)^{(C-1)/C}.\end{aligned}$$

By the Young inequality we have for $\varepsilon>0$

$$I_1 \le \varepsilon\int_\Omega |\mathbf{u}|^{C-2}|\nabla|\mathbf{u}||^2 + \alpha_\varepsilon(C-2)^C C^{(C-2)}\left(\int_\Omega |\operatorname{div}\mathbf{u}|^3\right)^{C/3}$$

and

$$I_2 = \int_\Omega p|\mathbf{u}|^{C-2} \operatorname{div} \mathbf{u} + p\mathbf{u} \cdot \nabla |\mathbf{u}|^{C-2}$$
$$\leq \varepsilon \int_\Omega (|\mathbf{u}|^{C-2} |\operatorname{div} \mathbf{u}|^2 + |\mathbf{u}|^{C-2} |\nabla |\mathbf{u}||^2) + \alpha_\varepsilon \int_\Omega p^2 |\mathbf{u}|^{C-2}$$

for some constant α_ε. Combining the identity (4.2) and the estimates for I_1 and I_2 with a small fixed number ε, we conclude the result. We note that all the integrals in the proof converge. □

LEMMA 4.2. *Let* $2 < C < \infty$. *If the condition* (4.1) *holds for some* β *satisfying*

$$\frac{2C\gamma - C + 2}{2} \leq \beta,$$

then

$$\sup_{0 \leq t \leq T} \left\| \rho^{1/C} \mathbf{u} \right\|_C < \infty.$$

PROOF. From Lemma 4.1 we note that for all $t \in [0, T]$

$$\frac{d}{dt} \int_\Omega \rho |\mathbf{u}|^C \lesssim 1 + \int_\Omega p^2 |\mathbf{u}|^{C-2}.$$

Since $p = \rho^\gamma$, using Hölder's inequality we have

$$\int_\Omega p^2 |\mathbf{u}|^{C-2} \leq \left(\int_\Omega \rho^{(2C\gamma - C + 2)/2} \right)^{2/C} \left(\int_\Omega \rho |\mathbf{u}|^C \right)^{(C-2)/C}.$$

Young's inequality yields

$$\frac{d}{dt} \int_\Omega \rho |\mathbf{u}|^C \lesssim \int_\Omega \rho |\mathbf{u}|^C + 1.$$

The result follows from the Gronwall lemma. □

LEMMA 4.3. *Let* $8 \leq C < \infty$. *If the condition* (4.1) *holds for some* β *satisfying*

$$\max \left\{ \frac{2C\gamma - C + 2}{2}, \frac{3C - 6}{C - 6} \right\} \leq \beta,$$

then

$$\int_0^T \int_\Omega \rho |\tilde{\mathbf{u}}|^2 + \sup_{0 \leq t \leq T} \int_\Omega |\nabla \mathbf{u}|^2 < \infty.$$

PROOF. We divide the proof into the following five steps.

Step 1) Testing the equation (1.2) with $\partial_t \mathbf{u}$ and integrating by parts yields

$$0 = \int_\Omega (\rho \tilde{\mathbf{u}} - L\mathbf{u} + \nabla p) \cdot \partial_t \mathbf{u}$$
$$= \int_\Omega \rho \tilde{\mathbf{u}} \cdot (\tilde{\mathbf{u}} - (\mathbf{u} \cdot \nabla)\mathbf{u}) - \int_\Omega L\mathbf{u} \cdot \partial_t \mathbf{u} - \int_\Omega p \partial_t \operatorname{div} \mathbf{u}$$
$$= \int_\Omega \rho |\tilde{\mathbf{u}}|^2 + \frac{1}{2} \frac{d}{dt} \int_\Omega (\mu |\nabla \mathbf{u}|^2 + (\lambda + \mu) |\operatorname{div} \mathbf{u}|^2 - 2p \operatorname{div} \mathbf{u})$$
$$+ \int_\Omega \partial_t p \operatorname{div} \mathbf{u} - \int_\Omega \rho \tilde{\mathbf{u}} \cdot (\mathbf{u} \cdot \nabla)\mathbf{u}.$$

Since $(\lambda+2\mu)\operatorname{div}\mathbf{u} = p + G$ and $\lambda + 2\mu > 0$, we have

$$\begin{aligned}(\lambda+2\mu)&\int_\Omega \partial_t p \operatorname{div}\mathbf{u}\\ &= \frac{1}{2}\frac{d}{dt}\int_\Omega p^2 + \int_\Omega \partial_t p G\\ &= \frac{1}{2}\frac{d}{dt}\int_\Omega p^2 + \int_\Omega (\partial_t p + \operatorname{div}(p\mathbf{u}))G + \int_\Omega p\mathbf{u}\cdot\nabla G.\end{aligned}$$

Thus,

$$\begin{aligned}&\int_\Omega \rho|\widetilde{\mathbf{u}}|^2 + \frac{1}{2}\frac{d}{dt}\int_\Omega \left(\mu|\nabla\mathbf{u}|^2 + (\lambda+\mu)|\operatorname{div}\mathbf{u}|^2 - 2p\operatorname{div}\mathbf{u} + \frac{1}{\lambda+2\mu}p^2\right)\\ &= -\frac{1}{\lambda+2\mu}\int_\Omega (\partial_t p + \operatorname{div}(p\mathbf{u}))G - \frac{1}{\lambda+2\mu}\int_\Omega p\mathbf{u}\cdot\nabla G + \int_\Omega \rho\widetilde{\mathbf{u}}\cdot(\mathbf{u}\cdot\nabla)\mathbf{u}\\ &=: I_1 + I_2 + I_3.\end{aligned} \tag{4.3}$$

Step 2) Since $p = \rho^\gamma$ and $\partial_t\rho = -\operatorname{div}(\rho\mathbf{u})$, we have

$$\partial_t p + \operatorname{div}(p\mathbf{u}) = -(\gamma-1)p\operatorname{div}\mathbf{u}$$

and hence

$$I_1 \lesssim \int_\Omega |p\operatorname{div}\mathbf{u}G|.$$

Because of the periodic boundary condition, we have $\int_\Omega \operatorname{div}\mathbf{u} = 0$ and hence by the Jensen inequality $\langle G\rangle = -\langle p\rangle \lesssim \|p\|_{3/2}$. Using the Sobolev-type inequality and the relation $\operatorname{div}(\rho\widetilde{\mathbf{u}}) = \Delta G$ we deduce that

$$\begin{aligned}I_1 &\le \int_\Omega |p\operatorname{div}\mathbf{u}||G - \langle G\rangle| + |\langle G\rangle|\int_\Omega |p\operatorname{div}\mathbf{u}|\\ &\lesssim \|p\|_{(3C-6)/C}\|\operatorname{div}\mathbf{u}\|_3\|\nabla G\|_{(3C-6)/(2C-6)} + \|p\|_{3/2}^2\|\operatorname{div}\mathbf{u}\|_3\\ &\lesssim \|p\|_{(3C-6)/C}\|\rho\widetilde{\mathbf{u}}\|_{(3C-6)/(2C-6)} + \|p\|_{3/2}^2.\end{aligned}$$

An elementary calculation shows that for $8 \le C < \infty$

$$\frac{\gamma(3C-6)}{C} \le \frac{2C\gamma - C + 2}{2} \le \beta$$

and hence $\|p\|_{3/2} \lesssim \|p\|_{(3C-6)/C} < \infty$. Hölder's inequality yields

$$\begin{aligned}\|\rho\widetilde{\mathbf{u}}\|_{(3C-6)/(2C-6)} &\le \left\|\rho^{1/2}\right\|_{(6C-12)/(C-6)}\left\|\rho^{1/2}\widetilde{\mathbf{u}}\right\|_2\\ &= \|\rho\|_{(3C-6)/(C-6)}^{1/2}\left\|\rho^{1/2}\widetilde{\mathbf{u}}\right\|_2\\ &\lesssim \left\|\rho^{1/2}\widetilde{\mathbf{u}}\right\|_2.\end{aligned}$$

Therefore

$$I_1 \lesssim \left\|\rho^{1/2}\widetilde{\mathbf{u}}\right\|_2 + 1. \tag{4.4}$$

Step 3) Using Lemma 5.2 and the relation $\operatorname{div}(\rho\widetilde{\mathbf{u}}) = \Delta G$, we obtain

$$
\begin{aligned}
I_2 &= -\frac{1}{\lambda+2\mu}\int_\Omega p\mathbf{u}\cdot\nabla G \\
&\lesssim \left\|\rho^{1/C}\mathbf{u}\right\|_C \left\|\rho^{(C\gamma-1)/C}\right\|_{(3C^2-6C)/(C^2-3C+6)} \|\nabla G\|_{(3C-6)/(2C-6)} \\
&\lesssim \|\rho\|^{(C\gamma-1)/C}_{(3C^2\gamma-6C\gamma-3C+6)/(C^2-3C+6)} \|\rho\widetilde{\mathbf{u}}\|_{(3C-6)/(2C-6)}.
\end{aligned}
$$

An elementary calculation shows that for $8 \le C < \infty$

$$
\frac{3C^2\gamma - 6C\gamma - 3C + 6}{C^2 - 3C + 6} \le \frac{2C\gamma - C + 2}{2} \le \beta
$$

and hence $\|\rho\|_{(3C^2\gamma-6C\gamma-3C+6)/(C^2-3C+6)} < \infty$. Similarly, Hölder's inequality provides

$$
I_2 \lesssim \|\rho\widetilde{\mathbf{u}}\|_{(3C-6)/(2C-6)} \lesssim \left\|\rho^{1/2}\widetilde{\mathbf{u}}\right\|_2. \tag{4.5}
$$

Step 4) Hölder's inequality yields

$$
\begin{aligned}
I_3 &= \int_\Omega \rho\widetilde{\mathbf{u}}\cdot(\mathbf{u}\cdot\nabla)\mathbf{u} \\
&\lesssim \left\|\rho^{1/2-1/C}\right\|_{6C/(C-6)} \left\|\rho^{1/C}\mathbf{u}\right\|_C \left\|\rho^{1/2}\widetilde{\mathbf{u}}\right\|_2 \|\nabla\mathbf{u}\|_3 \\
&\lesssim \|\rho\|^{(C-2)/(2C)}_{(3C-6)/(C-6)} \left\|\rho^{1/2}\widetilde{\mathbf{u}}\right\|_2 \|\nabla\mathbf{u}\|_3 \\
&\lesssim \left\|\rho^{1/2}\widetilde{\mathbf{u}}\right\|_2 \|\nabla\mathbf{u}\|_3.
\end{aligned}
$$

Using the Helmholz–Leray decomposition of $\mathbf{u}$ and an interpolation, we determine that for each $\varepsilon > 0$ there is a positive constant α_ε such that

$$
\begin{aligned}
\|\nabla\mathbf{u}\|_3 &\lesssim \|\operatorname{div}\mathbf{u}\|_3 + \|\nabla\mathcal{P}\mathbf{u}\|_3 \\
&\lesssim \varepsilon\left\|\nabla^2\mathcal{P}\mathbf{u}\right\|_{5/3} + \alpha_\varepsilon \\
&\lesssim \varepsilon\|\rho\widetilde{\mathbf{u}}\|_{5/3} + \alpha_\varepsilon \\
&\le \varepsilon\left\|\rho^{1/2}\right\|_{10}\left\|\rho^{1/2}\widetilde{\mathbf{u}}\right\|_2 + \alpha_\varepsilon.
\end{aligned}
$$

An elementary calculation shows that for $8 \le C < \infty$

$$
5 \le \frac{2C\gamma - C + 2}{2} \le \beta
$$

and hence $\left\|\rho^{1/2}\right\|_{10} = \|\rho\|_5^{1/2} < \infty$. Therefore, we conclude that for each $\varepsilon > 0$ there is a positive constant α_ε such that

$$
I_3 \lesssim \left\|\rho^{1/2}\widetilde{\mathbf{u}}\right\|_2 \|\nabla\mathbf{u}\|_3 \lesssim \varepsilon\left\|\rho^{1/2}\widetilde{\mathbf{u}}\right\|_2^2 + \alpha_\varepsilon. \tag{4.6}
$$

Step 5) Combining the identity (4.3) and the estimates for I_1, I_2, and I_3 with a fixed small $\varepsilon > 0$, we obtain that for some positive constant α

$$\begin{aligned}&\int_\Omega \rho|\widetilde{\mathbf{u}}|^2 + \frac{d}{dt}\int_\Omega \left(\mu|\nabla \mathbf{u}|^2 + (\lambda+\mu)|\operatorname{div}\mathbf{u}|^2 - 2p\operatorname{div}\mathbf{u} + \frac{p^2}{\lambda+2\mu}\right)\\ &\le \frac{1}{4}\left\|\rho^{1/2}\widetilde{\mathbf{u}}\right\|_2^2 + \alpha(\left\|\rho^{1/2}\widetilde{\mathbf{u}}\right\|_2 + 1)\\ &\le \frac{1}{2}\left\|\rho^{1/2}\widetilde{\mathbf{u}}\right\|_2^2 + \alpha^2 + \alpha.\end{aligned}$$

Integrate in time over $[0,t]$ we obtain

$$\begin{aligned}&\frac{1}{2}\int_0^t\int_\Omega \rho|\widetilde{\mathbf{u}}|^2 + \int_\Omega \left(\mu|\nabla \mathbf{u}|^2 + (\lambda+\mu)|\operatorname{div}\mathbf{u}|^2 + \frac{p^2}{\lambda+2\mu}\right)(t)\\ &\le 2\sup_{0\le t\le T}\int_\Omega |p\operatorname{div}\mathbf{u}| + (\alpha^2+\alpha)T + \alpha_0\end{aligned}$$

where the constant α_0 depends only on the initial data and the physical constants μ and λ. Since $\int_\Omega |p\operatorname{div}\mathbf{u}| \le \|p\|_{3/2}\|\operatorname{div}\mathbf{u}\|_3 < \infty$, we conclude the result by taking supremum for $t \in [0,T]$.

This completes the proof of Lemma 4.3. □

We derive a uniform energy estimate for the material derivative by modifying the arguement made by Sun–Wang–Zhang [**14**].

LEMMA 4.4. *Let* $8 \le C < \infty$. *If the condition* (4.1) *holds for some* β *satisfying*

$$\max\left\{\frac{2C\gamma - C + 2}{2}, \frac{3C-6}{C-6}\right\} \le \beta,$$

then

$$\sup_{0\le t\le T}\int_\Omega \rho|\widetilde{\mathbf{u}}|^2 + \int_0^T\int_\Omega |\nabla\widetilde{\mathbf{u}}|^2 < \infty.$$

PROOF. We present the proof in five steps.

Step 1) Taking the material derivative to $L\mathbf{u}$ yields

$$L(\partial_t\mathbf{u}) + \operatorname{div}(L\mathbf{u}\otimes\mathbf{u}) - \operatorname{div}\mathbf{u}L\mathbf{u}.$$

While taking the material derivative to $\rho\widetilde{\mathbf{u}} + \nabla p$ yields

$$\begin{aligned}&\partial_t(\rho\widetilde{\mathbf{u}}) + (\mathbf{u}\cdot\nabla)(\rho\widetilde{\mathbf{u}}) + \partial_t\nabla p + (\mathbf{u}\cdot\nabla)\nabla p\\ &= -\operatorname{div}(\rho\mathbf{u})\widetilde{\mathbf{u}} + \rho\partial_t\widetilde{\mathbf{u}} + (\nabla\rho\cdot\mathbf{u})\widetilde{\mathbf{u}} + \rho(\mathbf{u}\cdot\nabla)\widetilde{\mathbf{u}}\\ &\quad + \nabla\partial_t p + \operatorname{div}(\nabla p\otimes\mathbf{u}) - (\nabla p)\operatorname{div}\mathbf{u}\\ &= \rho\partial_t\widetilde{\mathbf{u}} + \rho(\mathbf{u}\cdot\nabla)\widetilde{\mathbf{u}} + \nabla\partial_t p + \operatorname{div}(\nabla p\otimes\mathbf{u}) - \operatorname{div}\mathbf{u}L\mathbf{u}.\end{aligned}$$

Subsequently, from the momentum equation, we obtain

$$[\rho\partial_t\widetilde{\mathbf{u}} + \rho(\mathbf{u}\cdot\nabla)\widetilde{\mathbf{u}}] - [L(\partial_t\mathbf{u}) + \operatorname{div}(L\mathbf{u}\otimes\mathbf{u})] = -[\nabla\partial_t p + \operatorname{div}(\nabla p\otimes\mathbf{u})].$$

Multiplying $\widetilde{\mathbf{u}}$ and integrating on Ω yields

$$\begin{aligned}&\frac{1}{2}\frac{d}{dt}\int_\Omega \rho|\widetilde{\mathbf{u}}|^2 - \int_\Omega [L(\partial_t\mathbf{u}) + \operatorname{div}(L\mathbf{u}\otimes\mathbf{u})]\cdot\widetilde{\mathbf{u}}\\ &= \int_\Omega \partial_t p\operatorname{div}\widetilde{\mathbf{u}} + (\mathbf{u}\cdot\nabla)\widetilde{\mathbf{u}}\cdot\nabla p.\end{aligned} \tag{4.7}$$

Step 2) Now, we estimate the nonlinear terms. Integrating by parts leads to

$$\begin{aligned}&\int_\Omega |\nabla \widetilde{\mathbf{u}}|^2 + \int_\Omega [\Delta \partial_t \mathbf{u} + \operatorname{div}(\Delta u \otimes \mathbf{u})] \cdot \widetilde{\mathbf{u}} \\ &= \int_\Omega [-\Delta((\mathbf{u}\cdot\nabla)\mathbf{u}) + \operatorname{div}(\Delta u \otimes \mathbf{u})] \cdot \widetilde{\mathbf{u}} \\ &\lesssim \int_\Omega |\nabla \mathbf{u}|^2 |\nabla \widetilde{\mathbf{u}}|.\end{aligned}$$

Similarly, we have

$$\begin{aligned}&\int_\Omega |\operatorname{div} \widetilde{\mathbf{u}}|^2 + \int_\Omega [\nabla \operatorname{div} \partial_t \mathbf{u} + \operatorname{div}(\nabla \operatorname{div} \mathbf{u} \otimes \mathbf{u})] \cdot \widetilde{\mathbf{u}} \\ &= \int_\Omega [-\nabla \operatorname{div}((\mathbf{u}\cdot\nabla)\mathbf{u}) + \operatorname{div}(\nabla \operatorname{div} \mathbf{u} \otimes \mathbf{u})] \cdot \widetilde{\mathbf{u}} \\ &\lesssim \int_\Omega |\nabla \mathbf{u}|^2 |\nabla \widetilde{\mathbf{u}}|.\end{aligned}$$

Therefore, using the Cauchy inequality we deduce that for $\varepsilon > 0$

$$\begin{aligned}&\mu \int_\Omega |\nabla \widetilde{\mathbf{u}}|^2 + (\lambda+\mu) \int_\Omega |\operatorname{div} \widetilde{\mathbf{u}}|^2 + \int_\Omega [L(\partial_t \mathbf{u}) + \operatorname{div}(L\mathbf{u} \otimes \mathbf{u})] \cdot \widetilde{\mathbf{u}} \\ &\lesssim \int_\Omega |\nabla \mathbf{u}|^2 |\nabla \widetilde{\mathbf{u}}| \lesssim \varepsilon \int_\Omega |\nabla \widetilde{\mathbf{u}}|^2 + \alpha_\varepsilon \int_\Omega |\nabla \mathbf{u}|^4\end{aligned} \tag{4.8}$$

for some positive number α_ε. For the computations in Step 1 and 2, we follow the rationale in Sun–Wang–Zhang [**14**].

Step 3) For barotropic gases, through direct computation, we can easily verify the relation

$$\partial_t p + \operatorname{div}(p\mathbf{u}) = -(\gamma-1) p \operatorname{div} \mathbf{u},$$

which is simply the renormalized continuity equation. Hence, integrating by parts yields

$$\begin{aligned}&\int_\Omega \partial_t p \operatorname{div} \widetilde{\mathbf{u}} + (\mathbf{u}\cdot\nabla)\widetilde{\mathbf{u}} \cdot \nabla p \\ &= \int_\Omega \partial_t p \operatorname{div} \widetilde{\mathbf{u}} - p\mathbf{u} \cdot \nabla \operatorname{div} \widetilde{\mathbf{u}} - p(\nabla \mathbf{u})^T : \nabla \widetilde{\mathbf{u}} \\ &= \int_\Omega (\partial_t p + \operatorname{div}(p\mathbf{u})) \operatorname{div} \widetilde{\mathbf{u}} - p(\nabla \mathbf{u})^T : \nabla \widetilde{\mathbf{u}} \\ &\le (\|\partial_t p + \operatorname{div}(p\mathbf{u})\|_2 + \|p\nabla \mathbf{u}\|_2)\|\nabla \widetilde{\mathbf{u}}\|_2 \\ &\lesssim \|p\|_4 \|\nabla \mathbf{u}\|_4 \|\nabla \widetilde{\mathbf{u}}\|_2.\end{aligned}$$

An elementary calculation shows that for $8 \le C < \infty$

$$4\gamma < \frac{2C\gamma - C + 2}{2}$$

and hence $\|p\|_4 \le \|\rho\|_{4\gamma}^\gamma < \infty$. By the Cauchy inequality we observe that for $\varepsilon > 0$

$$\int_\Omega \partial_t p \operatorname{div} \widetilde{\mathbf{u}} + (\mathbf{u}\cdot\nabla)\widetilde{\mathbf{u}} \cdot \nabla p \lesssim \varepsilon \int_\Omega |\nabla \widetilde{\mathbf{u}}|^2 + \alpha_\varepsilon \left(\int_\Omega |\nabla \mathbf{u}|^4\right)^{1/2} \tag{4.9}$$

for some positive number α_ε.

Step 4) Combining (4.7), (4.8), and (4.9) with a fixed small number ε, we obtain

$$\frac{d}{dt}\int_\Omega \rho|\widetilde{\mathbf{u}}|^2 + \int_\Omega |\nabla\widetilde{\mathbf{u}}|^2 \lesssim \int_\Omega |\nabla\mathbf{u}|^4 + 1.$$

Using the Helmholz–Leray decomposition of $\mathbf{u}$ and an interpolation we have

$$\begin{aligned}\|\nabla\mathbf{u}\|_4 &\lesssim \|\operatorname{div}\mathbf{u}\|_4 + \|\nabla\mathcal{P}u\|_4\\ &\lesssim \|G-\langle G\rangle\|_4 + \|p-\langle p\rangle\|_4 + \|\nabla\mathcal{P}u\|_4\\ &\lesssim \|\nabla G\|_{12/7} + \left\|\nabla^2\mathcal{P}u\right\|_{12/7} + \|p\|_4\\ &\lesssim \|\rho\widetilde{\mathbf{u}}\|_{12/7} + 1\\ &\lesssim \left\|\rho^{1/2}\right\|_{12}\left\|\rho^{1/2}\widetilde{\mathbf{u}}\right\|_2 + 1\\ &\lesssim \left\|\rho^{1/2}\widetilde{\mathbf{u}}\right\|_2 + 1.\end{aligned}$$

If we denote

$$\begin{aligned}Y(t) &= \int_\Omega \rho|\widetilde{\mathbf{u}}|^2(t)\\ Z(t) &= \int_\Omega |\nabla\widetilde{\mathbf{u}}|^2(t),\end{aligned}$$

then we have, for some positive constant α, the inequality

$$\frac{d}{dt}Y(t) + Z(t) \le \alpha Y(t)^2 + \alpha. \tag{4.10}$$

Step 5) We observe that

$$\begin{aligned}&\frac{d}{dt}\left(Y(t)\exp\left(-\alpha\int_0^t Y(s)ds\right)\right)\\ &= \left(\frac{d}{dt}Y(t) - \alpha Y(t)^2\right)\exp\left(-\alpha\int_0^t Y(s)ds\right)\\ &\le \alpha\exp\left(-\alpha\int_0^t Y(s)ds\right).\end{aligned}$$

Integrating over $[0,\tau]$ yields

$$Y(\tau) \le Y(0)\exp\left(\alpha\int_0^\tau Y(s)ds\right) + \alpha\int_0^\tau \exp\left(\alpha\int_t^\tau Y(s)ds\right)dt.$$

From Lemma 4.3, we deduce

$$\eta := \int_0^T Y(t)dt = \int_0^T\int_\Omega \rho|\widetilde{\mathbf{u}}|^2 < \infty.$$

Thus, for all $\tau \in [0,T]$

$$Y(\tau) \le Y(0)e^{\alpha\eta} + \alpha e^{\alpha\eta}T.$$

Since $\tau \in [0,T]$ is arbitrary, we obtain that

$$\sup_{0\le\tau\le T} Y(\tau) < \infty.$$

Moreover, we can reapply the inequality (4.10) to conclude that

$$\int_0^T Z(t)dt = \int_0^T \int_\Omega |\nabla \widetilde{\mathbf{u}}|^2 \leq \alpha T \left(1 + \sup_{0\leq t\leq T} Y(t)^2\right).$$

This completes the proof. □

The following lemma show that higher integrability of ρ and $\rho^{1/2}\widetilde{\mathbf{u}}$ yields higher integrability of $\nabla \mathbf{u}$.

LEMMA 4.5. *Let $3 < A < 6$. Suppose*

$$\sup_{0\leq t\leq T} \left\|\rho^{1/2}\widetilde{\mathbf{u}}\right\|_2 < \infty$$

and $\sup_{0\leq t\leq T} \|\rho\|_\beta < \infty$ for some β satisfying

$$\max\left\{A\gamma, \frac{3A}{6-A}\right\} \leq \beta.$$

Then

$$\sup_{0\leq t\leq T} \|\nabla \mathbf{u}\|_A < \infty.$$

PROOF. First we recall that $(\lambda + 2\mu)\operatorname{div}\mathbf{u} = p + G$ and $\lambda + 2\mu > 0$. Due to the periodic boundary condition, we have $\int_\Omega \operatorname{div}\mathbf{u} = 0$ and hence

$$\langle G\rangle + \langle p\rangle = 0$$

where we use the notation $\langle f\rangle = ⨍_\Omega f$. We next recall the relations $\operatorname{div}(\rho\widetilde{\mathbf{u}}) = \Delta G$ and $\mathcal{P}(\rho\widetilde{\mathbf{u}}) = \mu\Delta\mathcal{P}\mathbf{u}$. Using the Helmholz–Leray decomposition of $\mathbf{u}$ and the elliptic estimate for G and $\mathcal{P}\mathbf{u}$, we obtain that

$$\begin{aligned}
\|\nabla \mathbf{u}\|_A &\lesssim \|\operatorname{div}\mathbf{u}\|_A + \|\nabla\mathcal{P}\mathbf{u}\|_A \\
&\lesssim \|G - \langle G\rangle\|_A + \|p - \langle p\rangle\|_A + \|\nabla\mathcal{P}\mathbf{u}\|_A \\
&\lesssim \|\nabla G\|_{3A/(3+A)} + \left\|\nabla^2\mathcal{P}\mathbf{u}\right\|_{3A/(3+A)} + \|p\|_A \\
&\lesssim \|\rho\widetilde{\mathbf{u}}\|_{3A/(3+A)} + \|p\|_A.
\end{aligned}$$

Since we know $\|p\|_A = \|\rho\|_{A\gamma}^\gamma$ and

$$\|\rho\widetilde{\mathbf{u}}\|_{3A/(3+A)} \leq \left\|\rho^{1/2}\right\|_{6A/(6-A)}\left\|\rho^{1/2}\widetilde{\mathbf{u}}\right\|_2 \leq \|\rho\|_{3A/(6-A)}^{1/2}\left\|\rho^{1/2}\widetilde{\mathbf{u}}\right\|_2$$

via Hölder inequality, we conclude that

$$\|\nabla \mathbf{u}\|_A \lesssim \left\|\rho^{1/2}\widetilde{\mathbf{u}}\right\|_2 + 1.$$

Integrating in time arrives at the result. □

Now, we are ready to prove Theorem 2.3.

PROOF OF THEOREM 2.3. Let $3 < A < 6 < B \leq 8 \leq C < \infty$ with $1/A + 1/B < 1/3$. Suppose the condition (4.1) holds for

$$\max\left\{A\gamma, \frac{3A}{6-A}, \frac{2C\gamma - C + 2}{2}, \frac{3C-6}{C-6}, \frac{(C-1)B}{C-B}\right\} \leq \beta.$$

Combining Lemma 4.2, Lemma 4.3, Lemma 4.4, and Lemma 4.5 reveals

$$\sup_{0\leq t\leq T}\left(\|\nabla \mathbf{u}\|_A + \|\rho\|_\beta + \left\|\rho^{1/C}\mathbf{u}\right\|_C\right) < \infty.$$

Therefore from Lemma 3.1 we conclude $\sup_{0\le t\le T}\|\rho\|_\infty<\infty$. □

REMARK 4.6. If $\gamma=3/2$, then we may take $A=5$, $B=8$, $C=15$, and $\beta=16$ so that all the conditions of the lemmas are fulfilled.

5. Proof of Theorem 2.5

In this section, we assume that $(\rho,\mathbf{u},\theta)$ is the strong solution with the regularity stated in Theorem 2.2 and satisfies for some positive number δ the following integrability condition

$$\sup_{0\le t\le T}\left(\|\rho\|_\delta+\|\theta\|_\infty+\|\Delta\theta\|_2+\|\operatorname{div}\mathbf{u}\|_3\right)<\infty. \tag{5.1}$$

The constant δ will be specified in each lemmas.

LEMMA 5.1. *If the condition* (5.1) *holds for some $\delta\ge 3/2$, then*

$$\sup_{0\le t\le T}\int_\Omega\rho(|\mathbf{u}|^2+|\theta|^2)+\int_0^T\int_\Omega(|\nabla\mathbf{u}|^2+|\nabla\theta|^2)<\infty.$$

PROOF. Testing the equation (1.2) with $\mathbf{u}$ yields

$$\frac{1}{2}\frac{d}{dt}\int_\Omega\rho|\mathbf{u}|^2+\mu\int_\Omega|\nabla\mathbf{u}|^2+(\mu+\lambda)\int_\Omega|\operatorname{div}\mathbf{u}|^2=\int_\Omega p\operatorname{div}\mathbf{u}.$$

By Hölder's inequality

$$\int_\Omega p\operatorname{div}\mathbf{u}\le\|\rho\|_{3/2}\|\theta\|_\infty\|\operatorname{div}\mathbf{u}\|_3<\infty.$$

Similarly, testing the equation (1.6) with θ yields

$$\begin{aligned}&\frac{1}{2}\frac{d}{dt}\int_\Omega\rho|\theta|^2+\int_\Omega|\nabla\theta|^2\\&=-\int_\Omega\rho\theta^2\operatorname{div}\mathbf{u}+\frac{\mu}{2}\int_\Omega|\nabla\mathbf{u}+\nabla\mathbf{u}^t|^2\theta+\lambda\int_\Omega|\operatorname{div}\mathbf{u}|^2\theta.\end{aligned}$$

The right hand side is bounded by

$$\|\rho\|_{3/2}\|\theta\|_\infty^2\|\operatorname{div}\mathbf{u}\|_3+\|\theta\|_\infty\|\nabla\mathbf{u}\|_2^2.$$

Integrating in time yields the result. □

LEMMA 5.2. *Let $2<C<\infty$. If the condition* (5.1) *holds for some δ satisfying $(C+2)/2\le\delta$, then*

$$\sup_{0\le t\le T}\left\|\rho^{1/C}\mathbf{u}\right\|_C<\infty.$$

PROOF. From Lemma 4.1

$$\frac{d}{dt}\int_\Omega\rho|\mathbf{u}|^C\lesssim 1+\int_\Omega p^2|\mathbf{u}|^{C-2}.$$

By Hölder's inequality

$$\int_\Omega p^2|\mathbf{u}|^{C-2}=\int_\Omega\rho^2\theta^2|\mathbf{u}|^{C-2}\le\|\theta\|_\infty^2\left(\int_\Omega\rho^{(C+2)/2}\right)^{2/C}\left(\int_\Omega\rho|\mathbf{u}|^C\right)^{(C-2)/C}$$

and by Young's inequality

$$\frac{d}{dt}\int_\Omega\rho|\mathbf{u}|^C\lesssim 1+\int_\Omega\rho|\mathbf{u}|^C.$$

The result follows from the Gronwall lemma. □

LEMMA 5.3. *Let* $10 \le C < \infty$. *If the condition* (5.1) *holds for some* δ *satisfying* $(C+2)/2 \le \delta$, *then*

$$\int_0^T \int_\Omega \rho|\widetilde{\mathbf{u}}|^2 + \sup_{0\le t\le T} \int_\Omega |\nabla \mathbf{u}|^2 < \infty.$$

PROOF. We present the proof in three steps.

Step 1) Recall the following identity, which was derived in the first step proving Lemma 4.3,

$$\begin{aligned}
&\int_\Omega \rho|\widetilde{\mathbf{u}}|^2 + \frac{1}{2}\frac{d}{dt}\int_\Omega \left(\mu|\nabla \mathbf{u}|^2 + (\lambda+\mu)|\operatorname{div}\mathbf{u}|^2 - 2p\operatorname{div}\mathbf{u} + \frac{1}{\lambda+2\mu}p^2\right) \\
&= -\frac{1}{\lambda+2\mu}\int_\Omega (\partial_t p + \operatorname{div}(p\mathbf{u}))G - \frac{1}{\lambda+2\mu}\int_\Omega p\mathbf{u}\cdot\nabla G + \int_\Omega \rho\widetilde{\mathbf{u}}\cdot(\mathbf{u}\cdot\nabla)\mathbf{u} \qquad (5.2)\\
&=: I_1 + I_2 + I_3.
\end{aligned}$$

Step 2) The equation (1.6) can be rewritten as

$$\partial_t p + \operatorname{div}(p\mathbf{u}) = -p\operatorname{div}\mathbf{u} + \Delta\theta + \frac{\mu}{2}|\nabla\mathbf{u} + \nabla\mathbf{u}^t|^2 + \lambda|\operatorname{div}\mathbf{u}|^2.$$

Since $\|p\|_3 \le \|\rho\|_3\|\theta\|_\infty < \infty$, the condition (5.1) yields

$$\begin{aligned}
I_1 &\lesssim \int_\Omega (|p\operatorname{div}\mathbf{u}| + |\Delta\theta| + |\nabla\mathbf{u}|^2)|G| \\
&\lesssim \left(\|p\operatorname{div}\mathbf{u}\|_{3/2} + \|\Delta\theta\|_{3/2} + \||\nabla\mathbf{u}|^2\|_{3/2}\right)\|G\|_3 \\
&\lesssim \left(\|p\|_3\|\operatorname{div}\mathbf{u}\|_3 + \|\Delta\theta\|_2 + \|\nabla\mathbf{u}\|_3^2\right)(\|\operatorname{div}\mathbf{u}\|_3 + \|p\|_3) \\
&\lesssim 1 + \|\nabla\mathbf{u}\|_3^2.
\end{aligned}$$

Using Lemma 5.2 and the relation $\operatorname{div}(\rho\widetilde{\mathbf{u}}) = \Delta G$, we obtain

$$\begin{aligned}
I_2 &\lesssim \int_\Omega |p\mathbf{u}||\nabla G| \\
&\le \|\theta\|_\infty \left\|\rho^{1/C}\mathbf{u}\right\|_C \left\|\rho^{(C-1)/C}\right\|_{(3C^2-2C)/(C^2-3C+2)} \|\nabla G\|_{(3C-2)/(2C-2)} \\
&\lesssim \|\rho\|_{(3C-2)/(C-2)}^{(C-1)/C}\|\rho\widetilde{\mathbf{u}}\|_{(3C-2)/(2C-2)}.
\end{aligned}$$

An elementary calculation shows that for $10 \le C < \infty$,

$$(3C-2)/(C-2) \le (C+2)/2 \le \delta$$

and hence by Hölder's inequality

$$I_2 \lesssim \|\rho\widetilde{\mathbf{u}}\|_{(3C-2)/(2C-2)} \lesssim \left\|\rho^{1/2}\right\|_{(6C-4)/(C-2)}\left\|\rho^{1/2}\widetilde{\mathbf{u}}\right\|_2 \lesssim \left\|\rho^{1/2}\widetilde{\mathbf{u}}\right\|_2.$$

Another elementary calculation shows also that for $10 \le C < \infty$,

$$(3C-6)/(C-6) \le (C+2)/2 \le \delta$$

and hence by Hölder's inequality

$$
\begin{aligned}
I_3 &\lesssim \left\|\rho^{1/2-1/C}\right\|_{6C/(C-6)}\left\|\rho^{1/C}\mathbf{u}\right\|_C\left\|\rho^{1/2}\widetilde{\mathbf{u}}\right\|_2\|\nabla\mathbf{u}\|_3 \\
&\lesssim \|\rho\|_{(3C-6)/(C-6)}^{(C-2)/(2C)}\left\|\rho^{1/2}\widetilde{\mathbf{u}}\right\|_2\|\nabla\mathbf{u}\|_3 \\
&\lesssim \left\|\rho^{1/2}\widetilde{\mathbf{u}}\right\|_2\|\nabla\mathbf{u}\|_3.
\end{aligned}
$$

Step 3) Combining (5.2) and the estimates for I_1, I_2, and I_3, we obtain that

$$
\begin{aligned}
&\int_\Omega \rho|\widetilde{\mathbf{u}}|^2 + \frac{1}{2}\frac{d}{dt}\int_\Omega\left(\mu|\nabla\mathbf{u}|^2 + (\lambda+\mu)|\operatorname{div}\mathbf{u}|^2 - 2p\operatorname{div}\mathbf{u} + \frac{1}{\lambda+2\mu}p^2\right) \\
&\lesssim 1 + \|\nabla\mathbf{u}\|_3^2 + \left\|\rho^{1/2}\widetilde{\mathbf{u}}\right\|_2 + \left\|\rho^{1/2}\widetilde{\mathbf{u}}\right\|_2\|\nabla\mathbf{u}\|_3.
\end{aligned}
\tag{5.3}
$$

Using the Helmholz–Leray decomposition of $\mathbf{u}$ and an interpolation we have

$$
\|\nabla\mathbf{u}\|_3 \lesssim \|\operatorname{div}\mathbf{u}\|_3 + \|\nabla\mathcal{P}\mathbf{u}\|_3 \lesssim \varepsilon\left\|\nabla^2\mathcal{P}\mathbf{u}\right\|_{8/5} + \alpha_\varepsilon
$$

for some positive constant α_ε. Using the relation $\mathcal{P}(\rho\widetilde{\mathbf{u}}) = \mu\Delta\mathcal{P}\mathbf{u}$ in (3.1) we have

$$
\left\|\nabla^2\mathcal{P}\mathbf{u}\right\|_{8/5} \lesssim \|\rho\widetilde{\mathbf{u}}\|_{8/5} \le \left\|\rho^{1/2}\right\|_8\left\|\rho^{1/2}\widetilde{\mathbf{u}}\right\|_2 \lesssim \left\|\rho^{1/2}\widetilde{\mathbf{u}}\right\|_2.
$$

Hence

$$
\|\nabla\mathbf{u}\|_3 \lesssim \varepsilon\left\|\rho^{1/2}\widetilde{\mathbf{u}}\right\|_2 + \alpha_\varepsilon.
$$

We fix a small number ε so that, by Young's inequality, the estimate (5.3) becomes

$$
\begin{aligned}
&\int_\Omega \rho|\widetilde{\mathbf{u}}|^2 + \frac{d}{dt}\int_\Omega\left(\mu|\nabla\mathbf{u}|^2 + (\lambda+\mu)|\operatorname{div}\mathbf{u}|^2 - 2p\operatorname{div}\mathbf{u} + \frac{1}{\lambda+2\mu}p^2\right) \\
&\le \frac{1}{2}\int_\Omega \rho|\widetilde{\mathbf{u}}|^2 + \alpha
\end{aligned}
$$

for some constant α. Since $\int_\Omega |p\operatorname{div}\mathbf{u}| \le \|p\|_{3/2}\|\operatorname{div}\mathbf{u}\|_3 < \infty$, we arrive at the result by integrating in time over $[0,t]$ and then by taking the supremum for $t \in [0,T]$.

This completes the proof of Lemma 5.3. □

LEMMA 5.4. *Let* $10 \le C < \infty$. *If the condition* (5.1) *holds for some* δ *satisfying* $(C+2)/2 \le \delta$, *then*

$$
\sup_{0\le t\le T}\int_\Omega \rho|\widetilde{\mathbf{u}}|^2 + \int_0^T\int_\Omega |\nabla\widetilde{\mathbf{u}}|^2 < \infty.
$$

PROOF. Since the proof is essentially the same as the proof of Lemma 4.4, we only describe the descrepancies. Taking the material derivative to the momentum equation we obtain

$$
\frac{1}{2}\frac{d}{dt}\int_\Omega \rho|\widetilde{\mathbf{u}}|^2 - \int_\Omega [L(\partial_t\mathbf{u}) + \operatorname{div}(L\mathbf{u}\otimes\mathbf{u})]\cdot\widetilde{\mathbf{u}} = \int_\Omega \partial_t p\operatorname{div}\widetilde{\mathbf{u}} + (\mathbf{u}\cdot\nabla)\widetilde{\mathbf{u}}\cdot\nabla p.
\tag{5.4}
$$

Integrating by parts yields

$$\begin{aligned}(5.5)\qquad &\mu\int_\Omega |\nabla\widetilde{\mathbf{u}}|^2 + (\lambda+\mu)\int_\Omega |\operatorname{div}\widetilde{\mathbf{u}}|^2 + \int_\Omega [L(\partial_t\mathbf{u}) + \operatorname{div}(L\mathbf{u}\otimes\mathbf{u})]\cdot\widetilde{\mathbf{u}} \\ &\lesssim \int_\Omega |\nabla\mathbf{u}|^2|\nabla\widetilde{\mathbf{u}}|\end{aligned}$$

and

$$\begin{aligned}&\int_\Omega \partial_t p \operatorname{div}\widetilde{\mathbf{u}} + (\mathbf{u}\cdot\nabla)\widetilde{\mathbf{u}}\cdot\nabla p \\ &= \int_\Omega \partial_t p\operatorname{div}\widetilde{\mathbf{u}} - p\mathbf{u}\cdot\nabla\operatorname{div}\widetilde{\mathbf{u}} - p(\nabla\mathbf{u})^T : \nabla\widetilde{\mathbf{u}} \\ &= \int_\Omega (\partial_t p + \operatorname{div}(p\mathbf{u}))\operatorname{div}\widetilde{\mathbf{u}} - p(\nabla\mathbf{u})^T : \nabla\widetilde{\mathbf{u}} \\ &\le (\|\partial_t p + \operatorname{div}(p\mathbf{u})\|_2 + \|p\nabla\mathbf{u}\|_2)\|\nabla\widetilde{\mathbf{u}}\|_2.\end{aligned}$$

The equation (1.6) can be rewritten as

$$\partial_t p + \operatorname{div}(p\mathbf{u}) = -p\operatorname{div}\mathbf{u} + \Delta\theta + \frac{\mu}{2}|\nabla\mathbf{u}+\nabla\mathbf{u}^t|^2 + \lambda|\operatorname{div}\mathbf{u}|^2.$$

Since $\|p\|_4 \le \|\rho\|_4\|\theta\|_\infty < \infty$, we have

$$\begin{aligned}&\|\partial_t p + \operatorname{div}(p\mathbf{u})\|_2 + \|p\nabla\mathbf{u}\|_2 \\ &\lesssim \|p\|_4\|\nabla\mathbf{u}\|_4 + \|\Delta\theta\|_2 + \||\nabla\mathbf{u}|^2\|_2 \\ &\lesssim 1 + \|\nabla\mathbf{u}\|_4^2.\end{aligned}$$

By the Cauchy inequality we observe that for $\varepsilon > 0$

$$(5.6)\qquad \int_\Omega \partial_t p\operatorname{div}\widetilde{\mathbf{u}} + (\mathbf{u}\cdot\nabla)\widetilde{\mathbf{u}}\cdot\nabla p \lesssim \varepsilon\int_\Omega|\nabla\widetilde{\mathbf{u}}|^2 + \alpha_\varepsilon\left(1+\int_\Omega|\nabla\mathbf{u}|^4\right)$$

for some positive number α_ε. Combining (5.4), (5.5), and (5.6) with a fixed small number ε, we get

$$\frac{d}{dt}\int_\Omega \rho|\widetilde{\mathbf{u}}|^2 + \int_\Omega|\nabla\widetilde{\mathbf{u}}|^2 \lesssim \int_\Omega|\nabla\mathbf{u}|^4 + 1.$$

From here the arguments provided in steps 4 and 5 of Lemma 4.4 completes the proof. □

LEMMA 5.5. *Let $3 < A < 6$. Suppose*

$$\sup_{0\le t\le T}\left\|\rho^{1/2}\widetilde{\mathbf{u}}\right\|_2 < \infty$$

and $\sup_{0\le t\le T}\|\rho\|_\beta < \infty$ *for some β satisfying*

$$\max\left\{A, \frac{3A}{6-A}\right\} \le \beta.$$

Then

$$\sup_{0\le t\le T}\|\nabla\mathbf{u}\|_A < \infty.$$

PROOF. Using the Helmholz–Leray decomposition of $\mathbf{u}$ and the Sobolev-type estimate we obtain

$$\begin{aligned}\|\nabla \mathbf{u}\|_A &\lesssim \|\operatorname{div} \mathbf{u}\|_A + \|\nabla \mathcal{P}\mathbf{u}\|_A \\ &\lesssim \|G - \langle G\rangle\|_A + \|p - \langle p\rangle\|_A + \|\nabla \mathcal{P}\mathbf{u}\|_A \\ &\lesssim \|\nabla G\|_{3A/(3+A)} + \left\|\nabla^2 \mathcal{P}\mathbf{u}\right\|_{3A/(3+A)} + \|p\|_A \\ &\lesssim \|\rho\tilde{\mathbf{u}}\|_{3A/(3+A)} + \|p\|_A.\end{aligned}$$

Considering $\|p\|_A \leq \|\rho\|_A \|\theta\|_\infty$ and Hölder inequality, we observe

$$\|\rho\tilde{\mathbf{u}}\|_{3A/(3+A)} \leq \left\|\rho^{1/2}\right\|_{6A/(6-A)} \left\|\rho^{1/2}\tilde{\mathbf{u}}\right\|_2 \leq \|\rho\|_{3A/(6-A)}^{1/2} \left\|\rho^{1/2}\tilde{\mathbf{u}}\right\|_2$$

and therefore

$$\|\nabla \mathbf{u}\|_A \lesssim \left\|\rho^{1/2}\tilde{\mathbf{u}}\right\|_2 + 1.$$

Integrating in time yields the result. □

Now, we are ready to prove Theorem 2.5.

PROOF OF THEOREM 2.5. Let $3 < A < 6 < B < C < \infty$ with $1/A + 1/B < 1/3$. Suppose the condition (5.1) holds for

$$\max\left\{\frac{3A}{6-A}, \frac{C+2}{2}, \frac{(C-1)B}{C-B}\right\} \leq \beta.$$

Combining Lemma 5.2, Lemma 5.3, Lemma 5.4, and Lemma 5.5 we obtain that

$$\sup_{0\leq t\leq T}\left(\|\nabla \mathbf{u}\|_A + \|\rho\|_\beta + \left\|\rho^{1/C}\mathbf{u}\right\|_C\right) < \infty.$$

Therefore, from Lemma 3.1 we conclude $\sup_{0\leq t\leq T}\|\rho\|_\infty < \infty$. □

REMARK 5.6. We may take $A = 24/5$, $B = 41/5$, $C = 22$, and $\beta = 13$ so that all the conditions of the lemmas are fulfilled.

References

[1] Yonggeun Cho, Hi Jun Choe, and Hyunseok Kim, *Unique solvability of the initial boundary value problems for compressible viscous fluids* (English, with English and French summaries), J. Math. Pures Appl. (9) **83** (2004), no. 2, 243–275. MR2038120

[2] Yonggeun Cho and Hyunseok Kim, *On classical solutions of the compressible Navier-Stokes equations with nonnegative initial densities*, Manuscripta Math. **120** (2006), no. 1, 91–129. MR2223483

[3] Hi Jun Choe and Hyunseok Kim, *Strong solutions of the Navier-Stokes equations for isentropic compressible fluids*, J. Differential Equations **190** (2003), no. 2, 504–523. MR1970039

[4] Hi Jun Choe and Bum Ja Jin, *Regularity of weak solutions of the compressible Navier-Stokes equations*, J. Korean Math. Soc. **40** (2003), no. 6, 1031–1050. MR2013486

[5] R. Danchin, *Global existence in critical spaces for compressible Navier-Stokes equations*, Invent. Math. **141** (2000), no. 3, 579–614. MR1779621

[6] Benoît Desjardins, *Regularity of weak solutions of the compressible isentropic Navier-Stokes equations*, Comm. Partial Differential Equations **22** (1997), no. 5-6, 977–1008. MR1452175

[7] Eduard Feireisl, Antonín Novotný, and Hana Petzeltová, *On the existence of globally defined weak solutions to the Navier-Stokes equations*, J. Math. Fluid Mech. **3** (2001), no. 4, 358–392. MR1867887

[8] David Hoff, *Global solutions of the Navier-Stokes equations for multidimensional compressible flow with discontinuous initial data*, J. Differential Equations **120** (1995), no. 1, 215–254. MR1339675

[9] X. Hunag, J. Li and Z. Xin, *A blow-up criterion for the compressible Navier-Stokes equations*, Commun. Math. Phys. **301** (2011), 23–35

[10] Hai-Liang Li, Jing Li, and Zhouping Xin, *Vanishing of vacuum states and blow-up phenomena of the compressible Navier-Stokes equations*, Comm. Math. Phys. **281** (2008), no. 2, 401–444. MR2410901

[11] Pierre-Louis Lions, *Mathematical topics in fluid mechanics. Vol. 2*, Oxford Lecture Series in Mathematics and its Applications, vol. 10, The Clarendon Press, Oxford University Press, New York, 1998. Compressible models; Oxford Science Publications. MR1637634

[12] Akitaka Matsumura and Takaaki Nishida, *Initial-boundary value problems for the equations of motion of compressible viscous and heat-conductive fluids*, Comm. Math. Phys. **89** (1983), no. 4, 445–464. MR713680

[13] V. A. Solonnikov, *The solvability of initial-boundary value problem for the equations of motion of s viscous compressible fluid*, J. Sov. Math. **14** (1980), 1120–1133.

[14] Yongzhong Sun, Chao Wang, and Zhifei Zhang, *A Beale-Kato-Majda blow-up criterion for the 3-D compressible Navier-Stokes equations* (English, with English and French summaries), J. Math. Pures Appl. (9) **95** (2011), no. 1, 36–47. MR2746435

[15] V. A. Vaĭgant and A. V. Kazhikhov, *On the existence of global solutions of two-dimensional Navier-Stokes equations of a compressible viscous fluid* (Russian, with Russian summary), Sibirsk. Mat. Zh. **36** (1995), no. 6, 1283–1316, ii; English transl., Siberian Math. J. **36** (1995), no. 6, 1108–1141. MR1375428

[16] Huanyao Wen and Changjiang Zhu, *Blow-up criterions of strong solutions to 3D compressible Navier-Stokes equations with vacuum*, Adv. Math. **248** (2013), 534–572. MR3107521

[17] Zhouping Xin, *Blowup of smooth solutions to the compressible Navier-Stokes equation with compact density*, Comm. Pure Appl. Math. **51** (1998), no. 3, 229–240. MR1488513

Center for Mathematical Analysis and Computations, Yonsei University, 50 Yonseiro, Seodaemungu, Seoul, Republic of Korea
Email address: `choe@yonsei.ac.kr`

Center for Mathematical Analysis and Computations, Yonsei University, 50 Yonseiro, Seodaemungu, Seoul, Republic of Korea
Email address: `yangm@kias.re.kr`

Contemporary Mathematics
Volume **710**, 2018
http://dx.doi.org/10.1090/conm/710/14365

Asymptotic stability of equilibrium to 3-D MHD system

Wen Deng and Ping Zhang

Abstract. In this article, we are going to present the main result obtained in [**8**] concerning the optimal energy decay rates for the solutions of 3-D incompressible Magnetic hydrodynamical system with the initial magnetic field and velocity being a small perturbation of the equilibrium state $(e_3, 0)$.

1. Introduction

In this article, we investigate the large time behavior of the global smooth solutions to the following three-dimensional incompressible magnetic hydrodynamical (or MHD in short) system with initial data being sufficiently close to the equilibrium state $(e_3, 0)$:

$$(1.1) \qquad \begin{cases} \partial_t b + u\cdot\nabla b = b\cdot\nabla u, \qquad (t,x)\in\mathbb{R}^+\times\mathbb{R}^3, \\ \partial_t u + u\cdot\nabla u - \Delta u + \nabla p = b\cdot\nabla b, \\ \operatorname{div} u = \operatorname{div} b = 0, \\ (b,u)|_{t=0} = (b_0, u_0) \quad \text{with} \quad b_0 = e_3 + \varepsilon\phi, \end{cases}$$

where $b = (b^1, b^2, b^3)$ denotes the magnetic field, $u = (u^1, u^2, u^3)$ and p stand for the velocity and scalar pressure of the fluid respectively. This MHD system (1.1) with zero diffusivity in the magnetic field equation can be applied to model plasmas when the plasmas are strongly collisional, or the resistivity due to these collisions are extremely small. One may check the references [**5, 7**] for more explanations to this system.

Concerning the well-posedness issue of the system (1.1), Chemin et al [**4**] proved the local well-posedness of (1.1) with initial data in the critical Besov spaces. Lin and the second author [**15**] proved the global well-posedness to a modified three-dimensional MHD system with initial data sufficiently close to the equilibrium state (see [**16**] for a simplified proof). Lin, Xu and the second author [**14**] established the global well-posedness of (1.1) in 2-D provided that the initial data is near the equilibrium state $(e_d, 0)$ and the initial magnetic field, b_0, satisfies sort of admissible

2010 *Mathematics Subject Classification.* Primary 35Q30, 76D03.
Key words and phrases. MHD system, Nash-Moser iteration scheme, Littlewood-Paley theory.
The second author was partially supported by NSF of China under Grants 11371347 and 11688101 and innovation grant from National Center for Mathematics and Interdisciplinary Sciences.

condition, namely

$$\int_{\mathbb{R}} (b_0 - e_3)(Z(t,\alpha))\, dt = 0 \quad \text{for all} \quad \alpha \in \mathbb{R}^d \times \{0\} \tag{1.2}$$

with $Z(t,\alpha)$ being determined by

$$\begin{cases} \dfrac{dZ(t,\alpha)}{dt} = b_0(Z(t,\alpha)), \\ Z(t,\alpha)|_{t=0} = \alpha. \end{cases}$$

Similar result in three space dimension was proved by Xu and the second author in [**18**].

In the 2-D case, the restriction (1.2) was removed by Ren, Wu, Xiang and Zhang in [**17**] by carefully exploiting the divergence structure of the velocity field. Moreover, the authors proved that

$$\|\partial_{x_2}^k b(t)\|_{L^2} + \|\partial_{x_2}^k u(t)\|_{L^2} \leq C\langle t\rangle^{-\frac{1+2k}{4}+\epsilon} \quad \text{for any } \epsilon \in]0,1/2[, \ k = 0,1,2, \tag{1.3}$$

where $\langle t\rangle \overset{\text{def}}{=} (1+t^2)^{\frac{1}{2}}$. A more elementary existence proof was also given by Zhang in [**19**]. Very recently, Abidi and the second author removed the restriction (1.2) in [**1**] for the 3-D MHD system. Moreover, if the initial magnetic field equals to e_3 and with other technical assumptions, this solution decays to zero according to

$$\|u(t)\|_{H^2} + \|b(t) - e_3\|_{H^2} \leq C\langle t\rangle^{-\frac{1}{4}}. \tag{1.4}$$

Note that (1.4) corresponding to the critical case of (1.3), that is, $\epsilon = 0$ in (1.3).

This idea of considering the global well-posedness of MHD system with initial data close to the equilibrium sate $(e_d, 0)$ goes back to the work by Bardos, Sulem and Sulem [**2**] for the global well-posedness of ideal incompressible MHD system. In general, it is not known whether or not classical solutions of (1.1) can develop finite time singularities even in two dimension.

The main result of [**8**] states as follows:

THEOREM 1.1. *Let $e_3 = (0,0,1)$, $b_0 = e_3 + \varepsilon\phi$ with $\phi = (\phi_1,\phi_2,\phi_3) \in C_c^\infty$ and $\operatorname{div}\phi = 0$, let $u_0 \in W^{N_0,1} \cap H^{N_0}$ for some integer N_0 sufficiently large. Then there exist sufficiently small positive constants ε_0, c_0 such that if*

$$\|u_0\|_{W^{N_0,1}} + \|u_0\|_{H^{N_0}} \leq c_0 \quad \textit{and} \quad \varepsilon \leq \varepsilon_0, \tag{1.5}$$

(1.1) has a unique global solution (b,u) so that for any $T > 0$, $b \in C^2([0,T]\times\mathbb{R}^3)$, $u \in C^2([0,T]\times\mathbb{R}^3)$. Moreover, for some $\kappa > 0$, there hold

$$\begin{aligned} &\|u(t)\|_{W^{2,\infty}} \leq C_\kappa\langle t\rangle^{-\frac{5}{4}+\kappa}, \qquad \|b(t) - e_3\|_{W^{2,\infty}} \leq C_\kappa\langle t\rangle^{-\frac{3}{4}+\kappa} \quad \textit{and} \\ &\|u(t)\|_{H^2} + \|b(t) - e_3\|_{H^2} \leq C\langle t\rangle^{-\frac{1}{2}}, \quad \|\nabla u(t)\|_{L^2} \leq C\langle t\rangle^{-1}. \end{aligned} \tag{1.6}$$

Let us remark that the above theorem recovers the global well-posedness result of the system (1.1) in [**1**]. Moreover, the bigger the integer N_0, the smaller the positive constant κ. The main idea of the proof here works in both two space dimension and in three space dimension. The L^∞ decay rates of the solution in (1.6) are completely new. The L^2 decay rates of the solution are optimal in the sense that these decay rates coincide with those of the linearized system (see Propositions 3.2 and 3.11 below), which greatly improves the rate given by (1.4). We can also work on the decay rates for the higher order derivatives of the solutions. But we choose not to pursue on this direction here.

2. Structure and strategies of the proof

2.1. Lagrangian formulation of (1.1). As observed in the previous references ([**14, 18**]), the linearized system of (1.1) around the equilibrium state $(e_3, 0)$ reads

$$\begin{cases} Y_{tt} - \Delta Y_t - \partial_3^2 Y = \boldsymbol{f} \quad \text{in } \mathbb{R}^+ \times \mathbb{R}^3, \\ Y|_{t=0} = Y^{(0)}, \quad Y_t|_{t=0} = Y^{(1)}. \end{cases} \tag{2.1}$$

It is easy to calculate that this system has two different eigenvalues

$$\lambda_1(\xi) = -\frac{|\xi|^2}{2} + \sqrt{\frac{|\xi|^4}{4} - \xi_3^2} \quad \text{and} \quad \lambda_2(\xi) = -\frac{|\xi|^2}{2} - \sqrt{\frac{|\xi|^4}{4} - \xi_3^2}. \tag{2.2}$$

The Fourier modes corresponding to $\lambda_2(\xi)$ decays like $e^{-t|\xi|^2}$. Whereas the decay property of the Fourier modes corresponding to $\lambda_1(\xi)$ varies with directions of ξ as

$$\lambda_1(\xi) = -\frac{2\xi_3^2}{|\xi|^2\big(1 + \sqrt{1 - \frac{4\xi_3^2}{|\xi|^4}}\big)} \to -1 \quad \text{as} \quad |\xi| \to \infty$$

only in the ξ_3 direction. This simple analysis shows that the dissipative properties of the system (2.1) may be more complicated than that for the linearized system of isentropic compressible Navier-Stokes system (see [**6**] for instance). Moreover, it is well-known that it is in general impossible to propagate the anisotropic regularities for the transport equation. This motivates us to use the Lagrangian formulation of the system (1.1).

Let us now recall the Lagrangian formulation of (1.1) from [**1**]. Let (b, u) be a smooth enough solution of (1.1), we define

$$\begin{aligned} &X(t,y) = y + \int_0^t u(t', X(t',y))dt' \stackrel{\text{def}}{=} y + Y(t,y), \\ &\boldsymbol{u}(t,y) \stackrel{\text{def}}{=} u(t, X(t,y)), \quad \boldsymbol{b}(t,y) \stackrel{\text{def}}{=} b(t, X(t,y)), \quad \boldsymbol{p}(t,y) \stackrel{\text{def}}{=} p(t, X(t,y)), \\ &\mathcal{A} \stackrel{\text{def}}{=} (Id + \nabla_y Y)^{-1} \quad \text{and} \quad \nabla_Y \stackrel{\text{def}}{=} \mathcal{A}^t \nabla_y. \end{aligned} \tag{2.3}$$

Then $(Y, \boldsymbol{b}, \boldsymbol{u}, \boldsymbol{p})$ solves

$$\begin{cases} \boldsymbol{b}(t,y) = \partial_{b_0} X(t,y), \quad \nabla_Y \cdot \boldsymbol{b} = 0, \\ Y_{tt} - \Delta_y Y_t - \partial_{b_0}^2 Y = \partial_{b_0} b_0 + g, \\ Y_{|t=0} = Y^{(0)} = 0, \qquad Y_{t|t=0} = Y^{(1)} = u_0(y), \end{cases} \tag{2.4}$$

where

$$\begin{aligned} &g = \operatorname{div}_y\big[(\mathcal{A}\mathcal{A}^t - Id)\nabla_y Y_t\big] - \mathcal{A}^t \nabla_y \boldsymbol{p}, \quad \partial_{b_0} \stackrel{\text{def}}{=} b_0 \cdot \nabla_y, \quad \text{and} \\ &(\Delta_x p)(t, X(t,y)) = \sum_{i,j=1}^3 \nabla_{Y^i} \nabla_{Y^j} \big(\partial_{b_0} X^i \partial_{b_0} X^j - Y_t^i Y_t^j\big)(t,y). \end{aligned} \tag{2.5}$$

In what follows, we assume that

$$\operatorname{supp}(b_0(x_{\rm h}, \cdot) - e_3) \subset [0, K] \quad \text{and} \quad b_0^3 \neq 0. \tag{2.6}$$

Due to the difficulty of the variable coefficients for the linearized system of (2.4), we shall use Frobenius Theorem type argument to find a new coordinate system $\{z\}$ so that $\partial_{b_0} = \partial_{z_3}$. Toward this, let us define

$$\begin{cases} \dfrac{dy_1}{dy_3} = \dfrac{b_0^1}{b_0^3}(y_1, y_2, y_3), & y_1|_{y_3=0} = w_1, \\ \dfrac{dy_2}{dy_3} = \dfrac{b_0^2}{b_0^3}(y_1, y_2, y_3), & y_2|_{y_3=0} = w_2, \\ y_3 = w_3, \end{cases} \tag{2.7}$$

and

$$z_1 = w_1, \quad z_2 = w_2, \quad z_3 = w_3 + \int_0^{w_3} \Big(\frac{1}{b_0^3(y(w))} - 1 \Big)\, dw_3'. \tag{2.8}$$

Then we have

$$\partial_{b_0} f(y) = \frac{\partial f(y(w(z)))}{\partial z_3}, \quad \nabla_y = \nabla_Z = \mathcal{B}^t(z)\nabla_z \text{ with } \mathcal{B}(z) = \Big(\frac{\partial y(w(z))}{\partial z}\Big)^{-1}. \tag{2.9}$$

It is easy to observe that

$$\begin{aligned} \mathcal{B}(z) = \Big(\frac{\partial y(w(z))}{\partial z}\Big)^{-1} &= \Big(\frac{\partial y(w(z))}{\partial w} \times \frac{\partial w(z)}{\partial z}\Big)^{-1} \\ &= \Big(\frac{\partial w(z)}{\partial z}\Big)^{-1} \Big(\frac{\partial y(w(z))}{\partial w}\Big)^{-1} = \Big(\frac{\partial z}{\partial w}\Big) \Big(\frac{\partial y(w(z))}{\partial w}\Big)^{-1}. \end{aligned} \tag{2.10}$$

Yet it follows from (2.7) that

$$\begin{aligned} \Big(\frac{\partial y(w)}{\partial w}\Big) &= \begin{pmatrix} 1 & 0 & \frac{b_0^1}{b_0^3} \\ 0 & 1 & \frac{b_0^2}{b_0^3} \\ 0 & 0 & 1 \end{pmatrix} \\ &+ \begin{pmatrix} \int_0^{w_3} \frac{\partial}{\partial y_1}\big(\frac{b_0^1}{b_0^3}\big) dy_3' & \int_0^{w_3} \frac{\partial}{\partial y_2}\big(\frac{b_0^1}{b_0^3}\big) dy_3' & 0 \\ \int_0^{w_3} \frac{\partial}{\partial y_1}\big(\frac{b_0^2}{b_0^3}\big) dy_3' & \int_0^{w_3} \frac{\partial}{\partial y_2}\big(\frac{b_0^2}{b_0^3}\big) dy_3' & 0 \\ 0 & 0 & 0 \end{pmatrix} \begin{pmatrix} \frac{\partial y_1}{\partial w_1} & \frac{\partial y_1}{\partial w_2} & \frac{\partial y_1}{\partial w_3} \\ \frac{\partial y_2}{\partial w_1} & \frac{\partial y_2}{\partial w_2} & \frac{\partial y_2}{\partial w_3} \\ \frac{\partial y_3}{\partial w_1} & \frac{\partial y_3}{\partial w_2} & \frac{\partial y_3}{\partial w_3} \end{pmatrix} \\ &\stackrel{\text{def}}{=} A_1(y(w)) + A_2(y(w))\Big(\frac{\partial y(w)}{\partial w}\Big), \end{aligned} \tag{2.11}$$

which gives

$$\Big(\frac{\partial y(w)}{\partial w}\Big) = \big(Id - A_2(y(w))\big)^{-1} A_1(y(w)). \tag{2.12}$$

While it is easy to observe that

$$\Big(\frac{\partial z(w)}{\partial w}\Big) = \begin{pmatrix} 1 & 0 & 0 \\ 0 & 1 & 0 \\ \int_0^{w_3} \frac{\partial}{\partial w_1}\big(\frac{1}{b_0^3(y(w))}\big) dw_3' & \int_0^{w_3} \frac{\partial}{\partial w_2}\big(\frac{1}{b_0^3(y(w))}\big) dy_3' & \frac{1}{b_0^3} \end{pmatrix} \stackrel{\text{def}}{=} A_3(w). \tag{2.13}$$

As a consequence, we obtain

$$\begin{aligned} & y(w) = (y_{\rm h}(w_{\rm h}, w_3), w_3), \;\; w(z) = (z_{\rm h}, w_3(z)), \\ & y(w(z)) = \big(y_{\rm h}(z_{\rm h}, w_3(z)), w_3(z)\big), \\ & \mathcal{B}(z) = A_3(w(z)) A_1^{-1}(y(w(z))\big(Id - A_2(w(z))\big), \end{aligned} \tag{2.14}$$

with the matrices A_1, A_2, A_3 being determined by (2.11) and (2.13) respectively.

For simplicity, let us abuse the notation that $Y(t,z) = Y(t, y(w(z)))$. Then the system (2.4) becomes

(2.15)
$$\begin{cases} Y_{tt} - \Delta_z Y_t - \partial_{z_3}^2 Y = (\nabla_Z \cdot \nabla_Z - \Delta_z)Y_t + \partial_{z_3} b_0(y(w(z))) + g(y(w(z))), \\ Y_{|t=0} = Y_0 = 0, \qquad Y_{t|t=0} = Y_1(z) = u_0(y(w(z))), \end{cases}$$

for g given by (2.4). Since $\partial_{z_3} b_0(y(w(z)))$ in the source term is a time independent function, we now introduce a smooth cut-off function $\eta(z_3)$ with $\eta(z_3) = \begin{cases} 0, & z_3 \geq 2+K, \\ 1, & -1 \leq z_3 \leq 1+K, \\ 0, & z_3 \leq -2, \end{cases}$ and a correction term $\tilde{Y}$ so that $Y = \tilde{Y} + \bar{Y}$ and

(2.16)
$$\widetilde{Y}(z) = \eta(z_3)\Big(\int_{-1}^{z_3} (e_3 - b_0(y(w(z_{\rm h}, z_3')))) \, dz_3' - \int_{-1}^{K+1} (e_3 - b_0(y(w(z_{\rm h}, z_3')))) \, dz_3'\Big),$$

which satisfies

$$\partial_{z_3} \tilde{Y}(z) = e_3 - b_0(y(w(z))), \quad \text{and} \quad \partial_{z_3}\big(\partial_{z_3} \tilde{Y} + b_0(y(w(z)))\big) = 0. \tag{2.17}$$

Then in view of (2.23), (2.24) and (2.30) of [**1**], $\bar{Y}$ solves

$$\begin{cases} \bar{Y}_{tt} - \Delta_z \bar{Y}_t - \partial_{z_3}^2 \bar{Y} = f, \\ \bar{Y}_{|t=0} = \bar{Y}^{(0)} = -\tilde{Y}, \qquad \bar{Y}_{t|t=0} = Y^{(1)}, \end{cases} \tag{2.18}$$

with

$$\begin{aligned} &\mathcal{A} = \Big(Id + \mathcal{B}^t \nabla_z \widetilde{Y} + \mathcal{B}^t \nabla_z \bar{Y}\Big)^{-1}, \quad \text{and} \\ &f = \mathcal{B}^t \nabla_z \cdot \big[(\mathcal{A}\mathcal{A}^t - Id)\mathcal{B}^t \nabla_z \bar{Y}_t\big] + \mathcal{B}^t \nabla_z \cdot (\mathcal{B}^t \nabla_z \bar{Y}_t) - \Delta_z \bar{Y}_t - (\mathcal{B}\mathcal{A})^t \nabla_z \boldsymbol{p}, \\ &\nabla_z \boldsymbol{p} = -\nabla_z \Delta_z^{-1} \mathrm{div}_z\big(\det(\mathcal{B}^{-1})(\mathcal{B}\mathcal{A}\mathcal{A}^t\mathcal{B}^t - Id)\nabla_z \boldsymbol{p}\big) \\ &\qquad - \nabla_z \Delta_z^{-1} \mathrm{div}_z\big((\det(\mathcal{B}^{-1})Id - Id)\nabla_z \boldsymbol{p}\big) \\ &\qquad + \nabla_z \Delta_z^{-1} \mathrm{div}_z\Big(\mathcal{B}\mathcal{A}\mathrm{div}_z\, \big(\det(\mathcal{B}^{-1})\mathcal{B}\mathcal{A}(\partial_3 \bar{Y} \otimes \partial_3 \bar{Y} - \bar{Y}_t \otimes \bar{Y}_t)\big)\Big). \end{aligned} \tag{2.19}$$

2.2. The proof of Theorem 1.1. Before presenting the main result for the system (2.18-2.19), let us first introduce notations of the norms: For $f : \mathbb{R}_y^3 \to \mathbb{R}$, $u : \mathbb{R}^+ \times \mathbb{R}_y^3 \to \mathbb{R}$, and $p \in [1, +\infty]$, $N \in \mathbb{N}$, we denote

$$\|f\|_{W^{N,p}} \stackrel{\text{def}}{=} \sum_{|\alpha| \leq N} \|D_y^\alpha f\|_{L^p} \quad \text{and} \quad \|u\|_{L^p;k,N} \stackrel{\text{def}}{=} \sup_{t>0}(1+t)^k \|u(t)\|_{W^{N,p}}.$$

In particular, when $p = 1$, $p = 2$ and $p = \infty$, we simplify the notations as

$$\begin{aligned} &|||f|||_N \stackrel{\text{def}}{=} \|f\|_{W^{N,1}}, \qquad \|f\|_N \stackrel{\text{def}}{=} \|f\|_{H^N}, \qquad |f|_N \stackrel{\text{def}}{=} \|f\|_{W^{N,\infty}} \\ &\text{and} \qquad \|u\|_{k,N} \stackrel{\text{def}}{=} \|u\|_{L^2;k,N}, \qquad |u|_{k,N} \stackrel{\text{def}}{=} \|u\|_{L^\infty;k,N}. \end{aligned} \tag{2.20}$$

THEOREM 2.1. *There exist an integer L_0 and small constants $\eta, \varepsilon_0 > 0$ such that if*

$$|||(\bar{Y}^{(0)}, Y^{(1)})|||_{L_0} + \|(\bar{Y}^{(0)}, Y^{(1)})\|_{L_0} \leq \eta \quad \text{and} \quad \varepsilon \leq \varepsilon_0. \tag{2.21}$$

Then the system (2.18) has a unique global solution $\bar{Y} \in C^2([0,\infty); C^{N_1-4}(\mathbb{R}^3))$, where $N_1 = [(L_0 - 12)/2]$. Furthermore, for some $\kappa > 0$, there hold

$$|\partial_3 \bar{Y}|_{\frac{3}{4}-\kappa,2} + |\bar{Y}_t|_{\frac{5}{4}-\kappa,2} + |\bar{Y}|_{\frac{1}{4}-\kappa,2} \leq C_\kappa \eta, \tag{2.22}$$

and

(2.23)

$$\begin{aligned}&\||D|^{-1}(\partial_3\bar{Y},\bar{Y}_t)\|_{0,N_1+2}+\|\nabla\bar{Y}\|_{0,N_1+1}+\|(\bar{Y}_t,\partial_3\bar{Y})\|_{\frac{1}{2},N_1+1}+\|\nabla\bar{Y}_t\|_{1,N_1-1}\\&\quad+\|\bar{Y}_t\|_{L^2_t(H^{N_1+2})}+\big\|(\partial_3\bar{Y},\langle t\rangle^{\frac{1}{2}}\nabla\bar{Y}_t)\big\|_{L^2_t(H^{N_1+1})}+\|\bar{Y}_{tt}\|_{\frac{1}{2},N_1-2}\leq C.\end{aligned}$$

Admitting Theorem 2.1 for the time being, let us now turn to the proof of Theorem 1.1.

PROOF OF THEOREM 1.1. Indeed, in view of (2.3), one has

$$\begin{aligned}&Y_1(z)=u_0(y_{\rm h}(z_{\rm h},w_3(z)),w_3(z))\quad\text{and}\quad \boldsymbol{u}(t,y)=Y_t(t,y+Y(t,y)),\\&\boldsymbol{b}(t,y)=b_0(y)+b_0(y)\cdot\nabla_yY(t,y)\quad\text{with}\\&Y(t,(y_{\rm h}(z_{\rm h},w_3(z)),w_3(z)))=\widetilde{Y}(z)+\bar{Y}(t,z),\end{aligned}\tag{2.24}$$

with $\widetilde{Y}(z)$ and $\bar{Y}(t,z)$ being determined by (2.16) and (2.18) respectively.

Whereas in view of (2.11), (2.13) and (2.14), we get, by a similar proof of Lemma 4.3 of [**1**] that for any $N\in\mathbb{N}$,

$$|(\mathcal{B}-Id)|_N\leq C_N\varepsilon.\tag{2.25}$$

So that under the assumptions of (1.5), there holds (2.21). Then Theorem 2.1 ensures that the system (2.18-2.19) has a unique global classical solution $\bar{Y}\in C^2([0,\infty);C^{N_1-4}(\mathbb{R}^3))$, which verifies (2.22) and (2.23). In particular, it follows from (2.16) and (2.22) that

$$|\nabla_zY|_{0,1}\leq|\partial_z\widetilde{Y}|_1+|\nabla\bar{Y}|_{0,1}\leq C(\varepsilon+\eta),$$

which together with (2.24) ensures that $\boldsymbol{u}\in C^2([0,\infty)\times\mathbb{R}^3)$ and $\boldsymbol{b}\in C^2([0,\infty)\times\mathbb{R}^3)$. Furthermore due to

$$\Big|\frac{\partial X}{\partial y}-Id\Big|_{0,1}=\big|{}^t\mathcal{B}\nabla_zY\big|_{0,1}\leq C(\varepsilon+\eta),$$

we deduce from (2.3) that $u\in C^2([0,\infty)\times\mathbb{R}^3)$ and $b\in C^2([0,\infty)\times\mathbb{R}^3)$ which verifies the system (1.1) thanks to the derivation at the beginning of Subsection 2.1.

On the other hand, by virtue of (2.17), we have

$$\boldsymbol{b}(t,y(w(z)))=b_0(y(w(z)))+\partial_3\widetilde{Y}(z)+\partial_3\bar{Y}(t,z)=e_3+\partial_3\bar{Y}(t,z),$$

which together with (2.22), (2.23) and (2.24) implies that there holds (1.6). This completes the proof of Theorem 1.1. □

3. Strategies of the proof to Theorem 2.1

Observing from the calculations in [**1**] that under the assumptions of Theorem 1.1, the matrix $\mathcal{B}$ given by (2.14) is sufficiently close to the identity matrix in the norms of $W^{N_0,1}$ and H^{N_0} as long as ε is sufficiently small. To avoid cumbersome calculation, here we just prove Theorem 2.1 for the system (2.1) with

(3.1)

$$\begin{aligned}&\mathcal{A}=(Id+\nabla_yY)^{-1},\quad f=\nabla_y\cdot\big((\mathcal{A}\mathcal{A}^t-Id)\nabla_yY_t\big)-\mathcal{A}^t\nabla_y\boldsymbol{p},\quad\text{and}\\&\boldsymbol{p}=-\Delta_y^{-1}\text{div}_y\big((\mathcal{A}\mathcal{A}^t-Id)\nabla_y\boldsymbol{p}\big)+\Delta_y^{-1}\text{div}_y\Big(\mathcal{A}\text{div}_y\big(\mathcal{A}(\partial_{y_3}Y\otimes\partial_{y_3}Y-Y_t\otimes Y_t)\big)\Big),\end{aligned}$$

which corresponds to $\mathcal{B} = Id$ in (2.18). The general case follows along the same line.

Let us remark that the system (2.1) is not scaling, rotation and Lorentz invariant, so that Klainerman's vector field method ([**13**]) can not be applied here. Yet the ideas developed by Klainerman in the seminar papers [**10**–**12**] can be well adapted for this system. We now recall the classical result on the global well-posedness to some evolutionary system from [**11**]. Let us consider the following system

$$\left\{\begin{array}{ll} u_t - \mathcal{L}u = F(u, Du) \quad \text{with} \quad Du = (u_t, u_{x_1}, \cdots, u_{x_d}), \\ u_{|t=0} = u, \qquad Pu_0 = 0, \end{array}\right. \tag{3.2}$$

where $\mathcal{L} \stackrel{\text{def}}{=} \sum_{|\alpha|\leq\gamma} a_\alpha D_x^\alpha$ with a_α being $r\times r$ matrices with constant entries. Under the assumptions that

(1) $\mathcal{L}$ satisfies a dissipative condition of the following type: there exists a positive definite $r \times r$ matrix A such that

$$\text{either} \int_{\mathbb{R}^d} \Re(A\mathcal{L}f, f)\,dx \leq 0 \quad \text{or} \quad \int_{\mathbb{R}^d} \Re(A\mathcal{L}f, f)\,dx \leq -\|\nabla f\|_{L^2}^2$$

for any $f \in C_c^\infty$;

(2) Let $\Gamma(t)u_0$ be the solution of

$$\partial_t u - \mathcal{L}u = 0 \quad \text{and} \quad u(0, x) = u_0(x).$$

There is a differential matrix P such that

$$|\Gamma(t)u_0|_0 \leq C\langle t\rangle^{-k_0} |||u_0|||_d$$

for any $u_0 \in W^{d,1} \cap L^\infty$ that satisfies $Pu_0 = 0$;

(3) $AF_{u_t}, AF_{u_{x_i}}, i = 1, \cdots, d$, are symmetric matrices and F_{u_t} is independent of u_t. Moreover

$$|F(u, Du)| \leq C(|u| + |Du|)^{p+1} \quad \text{for } |u| + |Du| \text{ sufficiently small;} \tag{3.3}$$

(4) p is an integer and F is a smooth function so that there holds

$$\frac{1}{p}\left(1 + \frac{1}{p}\right) < k_0; \tag{3.4}$$

Klainerman proved in [**11**] the following celebrated theorem:

THEOREM 3.1 (Theorem 1 of [**11**]). *There exist an integer $N_0 > 0$ and a small constant $\eta > 0$ such that if*

$$|||u_0|||_{N_0} + \|u_0\|_{N_0} \leq \eta,$$

(3.2) has a unique solution $u \in C^1([0, T]; C^\gamma)$ for any $T > 0$. Moreover, the solution behaves, for t large, like

$$|u(t, x)| = O\left(t^{-\frac{1+\varepsilon}{p}}\right) \quad as \quad t \to \infty, \tag{3.5}$$

for some small $\varepsilon > 0$. Also

$$\|u(t)\|_{L^2} = O(1) \quad as \quad t \to \infty. \tag{3.6}$$

Let us remark that due to the appearance of the double Riesz transform in the expression of f in (3.1), the source term f in (2.1) can not satisfy the growth condition (3.3); secondly, even if we can assume the source term f is in quadratic growth of $(Y_t, \partial_3 Y)$, that corresponds to $p = 1$ in (3.3), the growth rate obtained below does not meet the requirement of (3.4). This makes it impossible to apply Theorem 3.1 for the system (2.1). Yet by considering the specific anisotropic structure of the system (2.1), we can still succeed in applying Nash-Moser scheme to establish the global existence as well as the large time behavior of solutions to (2.1) with (3.1).

Now we outline the proof of Theorem 2.1. According to the strategy in [**10**–**12**], the first step is to study the decay properties of the linear system:

$$\begin{cases} \mathcal{Y}_{tt} - \Delta \mathcal{Y}_t - \partial_3^2 \mathcal{Y} = 0, \\ \mathcal{Y}_{|t=0} = \mathcal{Y}_0, \qquad \mathcal{Y}_{t|t=0} = \mathcal{Y}_1. \end{cases} \tag{3.7}$$

PROPOSITION 3.2. *Let $\mathcal{Y}(t)$ be a smooth enough solution of* (3.7). *Given $\delta \in [0,1]$, $N \in \mathbb{N}$, there exist $C_{\delta,N}, C_N > 0$ such that*

$$\begin{aligned} &|\partial_3 \mathcal{Y}|_{1,N} + |\partial_t \mathcal{Y}|_{\frac{3}{2}-\delta,N} + |\mathcal{Y}|_{\frac{1}{2},N} \\ &\qquad \le C_{\delta,N}\big(\||D|^{2\delta}(\mathcal{Y}_0,\mathcal{Y}_1)\|_{L^1} + \||D|^{N+4}(\Delta\mathcal{Y}_0,\mathcal{Y}_1)\|_{L^1}\big); \end{aligned} \tag{3.8}$$

$$\begin{aligned} &\|(\partial_t \mathcal{Y}, \partial_3 \mathcal{Y})\|_{L^\infty_t(H^{N+1})} + \|\Delta \mathcal{Y}\|_{L^\infty_t(H^N)} + \|\nabla \partial_t \mathcal{Y}\|_{L^2_t(H^{N+1})} \\ &\qquad + \|\nabla \partial_3 \mathcal{Y}\|_{L^2_t(H^N)} \le C_N\big(\|(\partial_3 \mathcal{Y}_0, \mathcal{Y}_1)\|_{N+1} + \|\Delta \mathcal{Y}_0\|_N\big); \end{aligned} \tag{3.9}$$

$$\begin{aligned} &\|\langle t\rangle^{\frac{1}{2}}(\partial_t \mathcal{Y}, \partial_3 \mathcal{Y})\|_{L^\infty_t(H^N)} + \|\langle t\rangle^{\frac{1}{2}} \nabla \partial_t \mathcal{Y}\|_{L^2_t(H^N)} \\ &\qquad \le C_N\big(\||D|^{-1}(\partial_3 \mathcal{Y}_0, \mathcal{Y}_1)\|_{N+1} + \|\nabla \mathcal{Y}_0\|_N\big); \end{aligned} \tag{3.10}$$

$$\|\langle t\rangle \Delta \partial_t \mathcal{Y}\|_{L^\infty_t(H^N)} \le C_N \|(\Delta \mathcal{Y}_0, \mathcal{Y}_1)\|_{N+2}. \tag{3.11}$$

We emphasize here the estimates of (3.8) and (3.9) are of anisotropic type, which means that the decay rates of the partial derivatives of the solution to (3.7) are different, which is consistent with the heuristic discussions at the beginning of Section 2. Moreover, the estimate of (3.8) is valid for $\delta = 0$. Similar estimates as (3.10) and (3.11) were not proved in [**10**–**12**]. They are purely due to the special structure of the linearized system (3.7).

With the above proposition, we next turn to the decay estimates for the solutions of the following inhomogeneous equation of (3.7)

$$\begin{cases} Y_{tt} - \Delta Y_t - \partial_3^2 Y = g, \\ Y|_{t=0} = Y_t|_{t=0} = 0. \end{cases} \tag{3.12}$$

PROPOSITION 3.3. *Let $\delta \in [0, 1/4[$ and $\theta \in [1, \infty[$. We assume that $g(t) = 0$ if $t \ge \theta$. Then the solution Y to* (3.12) *verifies for any $N \ge 0$,*

$$|\partial_3 Y|_{1,N} + |\partial_t Y|_{\frac{3}{2}-\delta,N} + |Y|_{\frac{1}{2},N} \le C_{\delta,N} R_{N,\theta}(g), \tag{3.13}$$

where

$$\begin{aligned} R_{N,\theta}(g) &\overset{def}{=} \|\!|g|\!\|_{L^1_t(\delta,N)} + \theta^{\frac{1}{2}} \big\|\langle t\rangle^{\frac{1}{2}} |D|^{-1} g\big\|_{L^2_t(H^{N+3})} \\ &\qquad + \log\langle\theta\rangle \big\||D|^{-1} g\big\|_{\frac{3}{2}-\delta,N+3}, \end{aligned} \tag{3.14}$$

where

$$\begin{aligned} &\|\|g\|\|_{\delta,N} \overset{def}{=} \|\||D|^{2\delta}g\|_{L^1} + \|\||D|^{N+4}g\|_{L^1} \\ \text{and} \quad &\|\|g\|\|_{L^p_t(\delta,N)} \overset{def}{=} \Big(\int_0^t \|\|g(t')\|\|^p_{\delta,N}\,dt'\Big)^{\frac{1}{p}}. \end{aligned} \tag{3.15}$$

Next we calculate the linearized system of (2.1), which reads

$$\begin{cases} X_{tt} - \Delta X_t - \partial_3^2 X = f'(Y;X) + g, \\ X|_{t=0} = X_t|_{t=0} = 0, \end{cases} \tag{3.16}$$

where $f'(Y;X) = f_0'(Y;X) + f_1'(Y;X) + f_2'(Y;X)$, and $f_0'(Y;X)$, $f_1'(Y;X)$ and $f_2'(Y;X)$ are determined respectively by

$$\begin{aligned} f_0'(Y;X) &= \nabla\cdot\Big((\mathcal{A}\mathcal{A}^t - Id)'(Y;X)\nabla Y_t\Big) + \nabla\cdot\big((\mathcal{A}\mathcal{A}^t - Id)\nabla X_t\big) \\ &= \nabla\cdot\Big(\big(\mathcal{A}(-\nabla X)\mathcal{A}\mathcal{A}^t + \mathcal{A}\mathcal{A}^t(-\nabla X)^t\mathcal{A}^t\big)\nabla Y_t\Big) + \nabla\cdot\big((\mathcal{A}\mathcal{A}^t - Id)\nabla X_t\big). \end{aligned} \tag{3.17}$$

For $m = 1, 2$, we have

$$\begin{aligned} f_m'(Y;X) &= (\mathcal{A}^t)'(Y;X)\nabla \boldsymbol{p}_m(Y) + \mathcal{A}^t\nabla \boldsymbol{p}_m'(Y;X) \\ &= -\mathcal{A}^t(\nabla X)^t\mathcal{A}^t(\nabla \boldsymbol{p}_m)(Y) + \mathcal{A}^t\nabla \boldsymbol{p}_m'(Y;X) \end{aligned} \tag{3.18}$$

with $\boldsymbol{p}_m'(Y;X), m = 1, 2$, given by

$$\begin{aligned} \boldsymbol{p}_1'(Y;X) = \Delta^{-1}\mathrm{div}\Big(&-(\mathcal{A}\mathcal{A}^t - Id)\nabla \boldsymbol{p}_1'(Y;X) \\ &- \big(\mathcal{A}(-\nabla X)\mathcal{A}\mathcal{A}^t + \mathcal{A}\mathcal{A}^t(-\nabla X)\mathcal{A}^t\big)\nabla \boldsymbol{p}_1(Y) \\ &+ \mathcal{A}\mathrm{div}\big((\mathcal{A}(-\nabla X)\mathcal{A})(\partial_3 Y\otimes\partial_3 Y)\big) \\ &+ (\mathcal{A}(-\nabla X)\mathcal{A})\mathrm{div}\big(\mathcal{A}(\partial_3 Y\otimes\partial_3 Y)\big) \\ &+ \mathcal{A}\mathrm{div}\big(\mathcal{A}(\partial_3 Y\otimes\partial_3 X + \partial_3 X\otimes\partial_3 Y)\big)\Big), \end{aligned} \tag{3.19}$$

and

$$\begin{aligned} \boldsymbol{p}_2'(Y;X) = \Delta^{-1}\mathrm{div}\Big(&-(\mathcal{A}\mathcal{A}^t - Id)\nabla \boldsymbol{p}_2'(Y;X) \\ &- \big(\mathcal{A}(-\nabla X)\mathcal{A}\mathcal{A}^t + \mathcal{A}\mathcal{A}^t(-\nabla X)\mathcal{A}^t\big)\nabla \boldsymbol{p}_2(Y) \\ &+ \mathcal{A}\mathrm{div}\big((\mathcal{A}(-\nabla X)\mathcal{A})(Y_t\otimes Y_t)\big) \\ &+ (\mathcal{A}(-\nabla X)\mathcal{A})\mathrm{div}\big(\mathcal{A}(Y_t\otimes Y_t)\big) \\ &+ \mathcal{A}\mathrm{div}\big(\mathcal{A}(Y_t\otimes X_t + X_t\otimes Y_t)\big)\Big). \end{aligned} \tag{3.20}$$

Furthermore, the second derivative of $f''(Y;X,W) = \sum_{m=0}^2 f_m''(Y;X,W)$ with

$$\begin{aligned} f_0''(Y;X,W) &= \nabla\cdot\Big((\mathcal{A}\mathcal{A}^t - Id)''(Y;X,W)\nabla Y_t\Big) \\ &\quad + \nabla\cdot\big((\mathcal{A}\mathcal{A}^t - Id)'(Y;X)\nabla W_t\big) + \nabla\cdot\big((\mathcal{A}\mathcal{A}^t - Id)'(Y;W)\nabla X_t\big), \end{aligned} \tag{3.21}$$

and for $f_m(Y) = \mathcal{A}^t\nabla \boldsymbol{p}_m$, $m = 1, 2$, we have

$$\begin{aligned} f_m''(Y;X,W) &= (\mathcal{A}^t)''(Y;X,W)\nabla \boldsymbol{p}_m(Y) + \mathcal{A}^t\nabla\big(\boldsymbol{p}_m''(Y;X,W)\big) \\ &\quad + (\mathcal{A}^t)'(Y;X)\nabla\big(\boldsymbol{p}_m'(Y;W)\big) + (\mathcal{A}^t)'(Y;W)\nabla\big(\boldsymbol{p}_m'(Y;X)\big), \end{aligned} \tag{3.22}$$

where $\boldsymbol{p}_1''(Y;X,W), m=1,2$, are given by

(3.23)
$$\begin{aligned}
&\boldsymbol{p}_1''(Y;X,W) = -\Delta^{-1}\mathrm{div}\Big((\mathcal{A}\mathcal{A}^t - Id)\nabla \boldsymbol{p}_1''(Y;X,W)\\
&+ (\mathcal{A}\mathcal{A}^t - Id)''(Y;X,W)\nabla \boldsymbol{p}_1(Y) + (\mathcal{A}\mathcal{A}^t - Id)'(Y;X)\nabla \boldsymbol{p}_1'(Y;W)\\
&+ (\mathcal{A}\mathcal{A}^t - Id)'(Y;W)\nabla \boldsymbol{p}_1'(Y;X) - \mathcal{A}\mathrm{div}\big[\mathcal{A}(\partial_3 X\otimes\partial_3 W + \partial_3 W\otimes \partial_3 X)\\
&+ \mathcal{A}''(Y;X,W)(\partial_3 Y\otimes\partial_3 Y) + \mathcal{A}'(Y;X)(\partial_3 Y\otimes \partial_3 W + \partial_3 W\otimes\partial_3 Y))\\
&+ \mathcal{A}'(Y;W)(\partial_3 Y\otimes\partial_3 X + \partial_3 X\otimes\partial_3 Y)\big] - \mathcal{A}''(Y;X,W)\mathrm{div}\big(\mathcal{A}(\partial_3 Y\otimes\partial_3 Y)\big)\\
&- \mathcal{A}'(Y;X)\mathrm{div}\big[\mathcal{A}'(Y;W)(\partial_3 Y\otimes\partial_3 Y) + \mathcal{A}(\partial_3 Y\otimes\partial_3 W + \partial_3 W\otimes\partial_3 Y)\big]\\
&- \mathcal{A}'(Y;W)\mathrm{div}\big[\mathcal{A}'(Y;X)(\partial_3 Y\otimes\partial_3 Y) + \mathcal{A}(\partial_3 Y\otimes\partial_3 X + \partial_3 X\otimes\partial_3 Y)\big]\Big),
\end{aligned}$$

and

$$\begin{aligned}
&\boldsymbol{p}_2''(Y;X,W) = -\Delta^{-1}\mathrm{div}\Big((\mathcal{A}\mathcal{A}^t - Id)\nabla \boldsymbol{p}_2''(Y;X,W)\\
&+ (\mathcal{A}\mathcal{A}^t - Id)''(Y;X,W)\nabla \boldsymbol{p}_2(Y) + (\mathcal{A}\mathcal{A}^t - Id)'(Y;X)\nabla \boldsymbol{p}_2'(Y;W)\\
&+ (\mathcal{A}\mathcal{A}^t - Id)'(Y;W)\nabla \boldsymbol{p}_2'(Y;X) - \mathcal{A}\mathrm{div}\big[\mathcal{A}(X_t\otimes W_t + W_t\otimes X_t)\\
&+ \mathcal{A}''(Y;X,W)(Y_t\otimes Y_t) + \mathcal{A}'(Y;X)(Y_t\otimes W_t + W_t\otimes Y_t)\\
&+ \mathcal{A}'(Y;W)(Y_t\otimes X_t + X_t\otimes Y_t)\big] - \mathcal{A}''(Y;X,W)\mathrm{div}\big(\mathcal{A}(Y_t\otimes Y_t)\\
&- \mathcal{A}'(Y;X)\mathrm{div}\big[\mathcal{A}(Y_t\otimes W_t + W_t\otimes Y_t) + \mathcal{A}'(Y;W)(Y_t\otimes Y_t)\big]\\
&- \mathcal{A}'(Y;W)\mathrm{div}\big[\mathcal{A}(Y_t\otimes X_t + X_t\otimes Y_t) + \mathcal{A}'(Y;X)(Y_t\otimes Y_t)\big]\Big).
\end{aligned} \tag{3.24}$$

The main result concerning the estimates for $f'(Y;X)$ reads

PROPOSITION 3.4. *Let the functionals, $f_0'(Y;X), f_1'(Y;X), f_2'(Y;X)$, be given by (3.17) and (3.18) respectively, and the norm $|||\cdot|||_{\delta,N}$ be given by (3.15). Then under the assumptions that $\delta>0$, and*

$$\|\nabla Y\|_{\dot B^{\frac{3}{2}}_{2,1}} \leq \delta_1 \quad \text{and} \quad \|\nabla Y\|_{\dot B^{\frac{5}{2}}_{2,1}} \leq 1, \tag{3.25}$$

for some $\delta_1>0$ sufficiently small, we have

$$\begin{aligned}
|||f_0'(Y;X)|||_{\delta,N} &\lesssim \|\nabla Y\|_0\|\nabla X_t\|_{N+6} + \|\nabla Y\|_{N+6}\|\nabla X_t\|_0\\
&+ \|\nabla Y_t\|_0\|\nabla X\|_{N+6} + \big(\|\nabla Y_t\|_{N+6} + \|\nabla Y\|_{N+6}|\nabla Y_t|_0\big)\|\nabla X\|_0,
\end{aligned} \tag{3.26}$$

and

$$|||f_1'(Y;X)|||_{\delta,N} \lesssim \mathfrak{f}_1(\partial_3 Y,\partial_3 X) \quad \text{and} \tag{3.27}$$

$$|||f_2'(Y;X)|||_{\delta,N} \lesssim \mathfrak{f}_1(Y_t, X_t), \tag{3.28}$$

where the functional $\mathfrak{f}_1(\mathfrak{x},\mathfrak{y})$ is given by

$$\begin{aligned}
\mathfrak{f}_1(\mathfrak{x},\mathfrak{y}) \stackrel{def}{=} \|\mathfrak{x}\|_0\big(\|\mathfrak{y}\|_{N+6} + |\mathfrak{x}|_0\|\nabla X\|_{N+6}\big) + \|\mathfrak{y}\|_1\big(\|\mathfrak{x}\|_{N+6} + \|\nabla Y\|_{N+6}|\mathfrak{x}|_1\big)\\
+\big(\|\mathfrak{x}\|_{N+6} + \|\nabla Y\|_{N+6}\|\mathfrak{x}\|_3\big)|\mathfrak{x}|_1\|\nabla X\|_1.
\end{aligned}$$

PROPOSITION 3.5. *Under the assumption of Proposition 3.4, we have*

$$\begin{aligned}
\||D|^{-1}f_0'(Y;X)\|_{N+1} &\lesssim |\nabla Y|_0\|\nabla X_t\|_{N+1} + |\nabla Y|_{N+1}\|\nabla X_t\|_0\\
&+ |Y_t|_1\|\nabla X\|_{N+1} + \big(|Y_t|_{N+2} + |Y_t|_1|\nabla Y|_{N+1}\big)\|\nabla X\|_0,
\end{aligned} \tag{3.29}$$

and

$$\||D|^{-1}f_1'(Y;X)\|_{N+1} \lesssim \mathfrak{f}_2(\partial_3 Y, \partial_3 X) \quad \textit{and} \tag{3.30}$$

$$\||D|^{-1}f_2'(Y;X)\|_{N+1} \lesssim \mathfrak{f}_2(Y_t, X_t), \tag{3.31}$$

where the functional $\mathfrak{f}_2(\mathfrak{x},\mathfrak{y})$ *is given by*

$$\begin{aligned}\mathfrak{f}_2(\mathfrak{x},\mathfrak{y}) \stackrel{def}{=} &\big(|\mathfrak{x}|_1^{\frac{4}{3}}\|\mathfrak{x}\|_0^{\frac{2}{3}} + |\mathfrak{x}|_1^2\big)\big(\|\nabla X\|_{N+1} + |\nabla Y|_{N+1}\|\nabla X\|_1\big) + |\mathfrak{x}|_0\|\mathfrak{y}\|_{N+1}\\ &+ \big(|\mathfrak{x}|_{N+1} + |\nabla Y|_{N+1}|\mathfrak{x}|_1\big)\|\mathfrak{y}\|_1 + \big(|\mathfrak{x}|_0^{\frac{1}{3}}\|\mathfrak{x}\|_0^{\frac{2}{3}} + |\mathfrak{x}|_0\big)|\mathfrak{x}|_{N+1}\|\nabla X\|_1.\end{aligned}$$

Let us remark that Riesz transform does not map continuously from L^1 to L^1. Nevertheless due to (3.19) and (3.20), we can not avoid estimates of this type. To overcome this difficulty, a natural replacement of $\dot{W}^{s,1}$ will be the Besov space $\dot{B}^s_{1,1}$, which satisfies

$$\|\nabla(-\Delta)^{-\frac{1}{2}}|D|^s(f)\|_{L^1} \lesssim \|f\|_{\dot{B}^s_{1,1}} \quad \forall\, s\in\mathbb{R}\,.$$

We now recall the precise definition of the Besov norms from [**3**] for instance.

DEFINITION 3.6. Let us consider a smooth function φ on $\mathbb{R}$, the support of which is included in $[3/4, 8/3]$ such that

$$\forall \tau>0\,,\ \sum_{j\in\mathbb{Z}}\varphi(2^{-j}\tau) = 1 \quad \text{and} \quad \chi(\tau) \stackrel{\text{def}}{=} 1 - \sum_{j\geq 0}\varphi(2^{-j}\tau) \in \mathcal{D}([0,4/3]).$$

Let us define

$$\Delta_j a = \mathcal{F}^{-1}(\varphi(2^{-j}|\xi|)\widehat{a}), \quad \text{and} \quad S_j a = \mathcal{F}^{-1}(\chi(2^{-j}|\xi|)\widehat{a}).$$

Let (p,r) be in $[1,+\infty]^2$ and s in $\mathbb{R}$. We define the Besov norm by

$$\|a\|_{\dot{B}^s_{p,r}} \stackrel{\text{def}}{=} \big\|\big(2^{js}\|\Delta_j a\|_{L^p}\big)_j\big\|_{\ell^r(\mathbb{Z})}.$$

We remark that in the special case when $p = r = 2$, the Besov spaces $\dot{B}^s_{p,r}$ coincides with the classical homogeneous Sobolev spaces $\dot{H}^s$.

The estimates of the second derivations of f_0, f_1 and f_2 can be listed as follows:

PROPOSITION 3.7. *Let* f_0'', f_1'', f_2'' *be given by* (3.21) *and* (3.22) *respectively. Then under the assumption of* (3.25), *we have*

(3.32)
$$\begin{aligned}&\||D|^{-1}f_0''(Y;X,W)\|_N \lesssim |Y_t|_1\big(|\nabla X|_N\|\nabla W\|_0 + |\nabla X|_0\|\nabla W\|_N\big)\\ &+ \big((|Y_t|_{N+1} + |\nabla Y|_N|Y_t|_1)|\nabla X|_0 + |X_t|_{N+1}\big)\|\nabla W\|_0 + |\nabla X|_N\|\nabla W_t\|_0\\ &+ |\nabla X|_0\|\nabla W_t\|_N + |\nabla Y|_N\big(|\nabla X|_0\|\nabla W_t\|_0 + |X_t|_1\|\nabla W\|_0\big) + |X_t|_1\|\nabla W\|_N,\end{aligned}$$

and

$$\||D|^{-1}f_1''(Y;X,W)\|_N \lesssim \mathfrak{f}_3(\partial_3 Y, \partial_3 X, \partial_3 W) \quad \textit{and} \tag{3.33}$$

$$\||D|^{-1}f_2''(Y;X,W)\|_N \lesssim \mathfrak{f}_3(Y_t, X_t, W_t), \tag{3.34}$$

where the functional $\mathfrak{f}_3(\mathfrak{x},\mathfrak{y},\mathfrak{z})$ *is given by*

$$\begin{aligned}\mathfrak{f}_3(\mathfrak{x},\mathfrak{y},\mathfrak{z}) \overset{def}{=} & (|\mathfrak{y}|_N + |\nabla Y|_N|\mathfrak{y}|_0)\|\mathfrak{z}\|_0 + |\mathfrak{y}|_0\|\mathfrak{z}\|_N \\ & + \big(|\mathfrak{x}|_0 + |\mathfrak{x}|_1^{\frac13}\|\mathfrak{x}\|_1^{\frac23}\big)\big(|\nabla X|_N\|\mathfrak{z}\|_0 + |\nabla X|_0\|\mathfrak{z}\|_N + |\mathfrak{y}|_1\|\nabla W\|_N \\ & + (|\mathfrak{y}|_N + |\mathfrak{x}|_N|\nabla X|_1)\|\nabla W\|_1\big) + \big(|\mathfrak{x}|_N + |\nabla Y|_N|\mathfrak{x}|_1\big)|\nabla X|_1\|\mathfrak{z}\|_1 \\ & + \big(|\mathfrak{x}|_1^{\frac43}\|\mathfrak{x}\|_0^{\frac23} + |\mathfrak{x}|_1^2\big)\big((|\nabla X|_N + |\nabla Y|_N|\nabla X|_1)\|\nabla W\|_1 + |\nabla X|_1\|\nabla W\|_N\big) \\ & + \big(|\mathfrak{x}|_N + |\mathfrak{x}|_N^{\frac13}\|\mathfrak{x}\|_N^{\frac23} + |\nabla Y|_N(|\mathfrak{x}|_1 + |\mathfrak{x}|_0^{\frac13}\|\mathfrak{x}\|_0^{\frac23})\big)|\mathfrak{y}|_1\|\nabla W\|_1.\end{aligned}$$

PROPOSITION 3.8. *Let* f''_m, $m = 0,1,2$ *be given in* (3.21) *and* (3.22) , *the norm* $|||\cdot|||_{\delta,N}$ *be given by* (3.15). *Then under the assumption of* (3.25), *we have*

$$\begin{aligned}|||f''_0(Y;X,W)|||_{\delta,N} \lesssim & |Y_t|_1\big(\|\nabla X\|_{N+6}\|\nabla W\|_0 + \|\nabla X\|_0\|\nabla W\|_{N+6}\big) \\ & + \big(\|\nabla Y_t\|_{N+6} + |Y_t|_1\|\nabla Y\|_{N+6}\big)\big(|\nabla X|_0\|\nabla W\|_0 + \|\nabla X\|_0|\nabla W|_0\big) \\ & + \|\nabla X\|_0\|\nabla W_t\|_{N+6} + \big(\|\nabla X\|_{N+6} + |\nabla X|_0\|\nabla Y\|_{N+6}\big)\|\nabla W_t\|_0 \\ & + \|\nabla W\|_0\|\nabla X_t\|_{N+6} + \big(\|\nabla W\|_{N+6} + |\nabla W|_0\|\nabla Y\|_{N+6}\big)\|\nabla X_t\|_0,\end{aligned} \tag{3.35}$$

and

$$|||f''_1(Y;X,W)|||_{\delta,N} \lesssim \mathfrak{f}_4(\partial_3 Y,\partial_3 X,\partial_3 W) \quad \textit{and} \tag{3.36}$$

$$|||f''_2(Y;X,W)|||_{\delta,N} \lesssim \mathfrak{f}_4(Y_t,X_t,W_t), \tag{3.37}$$

where the functional $\mathfrak{f}_4(\mathfrak{x},\mathfrak{y},\mathfrak{z})$ *is given by*

$$\begin{aligned}\mathfrak{f}_4(\mathfrak{x},\mathfrak{y},\mathfrak{z}) \overset{def}{=} & \big(\|\mathfrak{z}\|_0 + |\mathfrak{x}|_0\|\nabla W\|_0 + \|\mathfrak{x}\|_0|\nabla W|_0\big)\|\mathfrak{y}\|_{N+6} + \big(\|\mathfrak{x}\|_0|\nabla X|_0 + \|\mathfrak{y}\|_0\big)\|\mathfrak{z}\|_{N+6} \\ & + |\mathfrak{y}|_0\|\mathfrak{z}\|_0\|\nabla Y\|_{N+6} + |\mathfrak{x}|_0\big(\|\mathfrak{z}\|_0\|\nabla X\|_{N+6} + \|\nabla X\|_0\|\mathfrak{z}\|_{N+6} + \|\mathfrak{y}\|_0\|\nabla W\|_{N+6}\big) \\ & + \big(\|\mathfrak{x}\|_{N+6} + |\mathfrak{x}|_0\|\nabla Y\|_{N+6}\big)\big(\|\nabla X\|_0|\mathfrak{z}|_0 + |\nabla X|_1\|\mathfrak{z}\|_1\big) \\ & + \big(\|\mathfrak{x}\|_{N+6} + \|\mathfrak{x}\|_3\|\nabla Y\|_{N+6}\big)\big(|\mathfrak{y}|_1\|\nabla W\|_1 + \|\mathfrak{y}\|_1|\nabla W|_0\big) \\ & + |\mathfrak{x}|_1\|\mathfrak{x}\|_3\big(\|\nabla X\|_{N+6}(\|\nabla W\|_0 + |\nabla W|_0) + (\|\nabla X\|_0 + |\nabla X|_0)\|\nabla W\|_{N+6}\big) \\ & + \big(|\mathfrak{x}|_1\|\mathfrak{x}\|_{N+6} + |\mathfrak{x}|_1\|\mathfrak{x}\|_3\|\nabla Y\|_{N+6}\big)\big(|\nabla X|_0\|\nabla W\|_1 + \|\nabla X\|_1|\nabla W|_0\big).\end{aligned}$$

With the above estimates, we can investigate energy estimates for the solutions of the linearized equation (3.16).

THEOREM 3.9. *Let* Y *be a smooth enough vector field and* X *be a smooth solution to the linearized equation* (3.16). *We assume that* Y *satisfies* (3.25) *and*

$$\|Y_t\|_{0,0} \le 1, \quad \textit{and} \quad |Y_t|_{0,1} \le 1. \tag{3.38}$$

Then for any $\varepsilon > 0$, *we have*

$$\begin{aligned}&\mathcal{E}_0(t) \le C_\varepsilon\|\langle t\rangle^{\frac{1+\varepsilon}{2}}|D|^{-1}g\|_{L^2_t(H^1)}E_\varepsilon(Y) \quad \textit{and} \quad \textit{for} \quad N \ge 1 \\ &\mathcal{E}_N(t) \le C_{\varepsilon,N}\Big(\|\langle t\rangle^{\frac{1+\varepsilon}{2}}g\|_{L^2_t(H^N)} + \gamma_{\varepsilon,N+1}(Y)\|\langle t\rangle^{\frac{1+\varepsilon}{2}}|D|^{-1}g\|_{L^2_t(H^1)}\Big)E_\varepsilon(Y),\end{aligned} \tag{3.39}$$

where

(3.40)

$$\mathcal{E}_N(t) \overset{def}{=} |||D|^{-1}(X_t,\partial_3 X)\|_{0,N+2} + \|\nabla X\|_{0,N+1} + \|X_t\|_{L^2_t(H^{N+2})} + \|\partial_3 X\|_{L^2_t(H^{N+1})};$$

$$E_\varepsilon(Y) \overset{def}{=} \exp\Big(C\big(|\partial_3 Y|^{\frac43}_{\frac12+\varepsilon,1}\|\partial_3 Y\|^{\frac23}_{L^2_t(L^2)} + |\partial_3 Y|^2_{\frac12+\varepsilon,1} + |Y_t|_{1+\varepsilon,2}\big)\Big),$$

and

$$(3.41)\quad \begin{aligned}\gamma_{\varepsilon,N+1}(Y) \stackrel{def}{=} & 1+|\partial_3 Y|_{\frac12+\varepsilon,N+1}\big(1+|\partial_3 Y|_{\frac12+\varepsilon,1}\big)+|Y_t|_{1+\varepsilon,N+2}\\ &+|\partial_3 Y|^{\frac13}_{\frac12+\varepsilon,1}\|\partial_3 Y\|^{\frac23}_{L^2_t(L^2)}\times\big(|\partial_3 Y|_{\frac12+\varepsilon,N+1}+|\nabla Y|_{0,N+1}\big)\\ &+|\nabla Y|_{0,N+1}\big(1+|\partial_3 Y|^2_{\frac12+\varepsilon,0}+|\partial_3 Y|_{\frac12+\varepsilon,0}+|Y_t|_{1+\varepsilon,1}\big).\end{aligned}$$

We notice that when we perform the energy estimates for the derivatives of the solutions to (3.16), we are not able to treat the term $\nabla\cdot\big((\mathcal{A}\mathcal{A}^t-Id)\nabla X_t\big)$, which appears in $f_0'(Y;X)$ (see (3.17)), as a source term. Instead, we need to rewrite (3.16) as

$$(3.42)\quad X_{tt}-\nabla\cdot\partial_t\big(\mathcal{A}\mathcal{A}^t\nabla X\big)-\partial_3^2 X=\widetilde{f}'(Y;X)+g$$

where $\widetilde{f}'(Y;X)=\widetilde{f}_0'(Y;X)+f_1'(Y;X)+f_2'(Y;X)$ with $f_m'(Y;X), m=1,2,$ given by (3.18), and $\widetilde{f}_0'(Y;X)$ by

$$(3.43)\quad \widetilde{f}_0'(Y;X)=-\nabla\cdot\big(\mathcal{A}(\nabla X\mathcal{A}+\mathcal{A}^t(\nabla X)^t)\mathcal{A}^t\nabla Y_t\big)-\nabla\cdot\big(\partial_t(\mathcal{A}\mathcal{A}^t)\nabla X\big).$$

With the energy estimates obtained in Theorem 3.9, we can work on the time-weighted energy estimate for the solutions of (3.16).

COROLLARY 3.10. *Under the assumptions of Theorem* 3.9, *we have*

$$(3.44)\quad \mathcal{E}_0+\|(X_t,\partial_3 X)\|_{\frac12,1}+\|\langle t\rangle^{\frac12}\nabla X_t\|_{L^2_t(H^1)}\leq C_\varepsilon\|\langle t\rangle^{\frac{1+\varepsilon}{2}}|D|^{-1}g\|_{L^2_t(H^1)}E_\varepsilon(Y),$$

and for $N\geq 1$,

$$(3.45)\quad \begin{aligned}&\mathcal{E}_N+\|(X_t,\partial_3 X)\|_{\frac12,N+1}+\|\langle t\rangle^{\frac12}\nabla X_t\|_{L^2_t(H^{N+1})}\\ &\leq C_{\varepsilon,N}\Big(\|\langle t\rangle^{\frac{1+\varepsilon}{2}}g\|_{L^2_t(H^N)}+\gamma_{\varepsilon,N+1}(Y)\|\langle t\rangle^{\frac{1+\varepsilon}{2}}|D|^{-1}g\|_{L^2_t(H^1)}\Big)E_\varepsilon(Y).\end{aligned}$$

PROPOSITION 3.11. *Under the assumptions of Theorem* 3.9, *we have for* $N\geq 0$,

$$(3.46)\quad \begin{aligned}&\|\nabla X_t\|_{1,N}\leq C_{\varepsilon,N}\big(\||D|^{-1}g\|_{1+\varepsilon,N+2}+\|\nabla Y\|_{N+2}\||D|^{-1}g\|_{1+\varepsilon,2}\big)\\ &+C_{\varepsilon,N}\Big(\|\langle t\rangle^{\frac{1+\varepsilon}{2}}g\|_{L^2_t(H^{N+1})}+\gamma_{\varepsilon,N+2}(Y)\|\langle t\rangle^{\frac{1+\varepsilon}{2}}|D|^{-1}g\|_{L^2_t(H^1)}\\ &+\|\nabla Y\|_{0,N+2}\big(\|\langle t\rangle^{\frac{1+\varepsilon}{2}}g\|_{L^2_t(H^1)}+\gamma_{\varepsilon,2}(Y)\|\langle t\rangle^{\frac{1+\varepsilon}{2}}|D|^{-1}g\|_{L^2_t(H^1)}\big)\Big)E_\varepsilon(Y).\end{aligned}$$

We emphasize that the decay estimates (3.46) can not be obtained by energy estimate. In fact, we will have to exploit anisotropic Littlewood-Paley analysis and the dissipative properties of the linear system (2.1).

Let us summarize that under the assumptions (3.25), (3.38) and if we assume moreover

$$(3.47)\quad |\partial_3 Y|^{\frac43}_{\frac12+\varepsilon,1}\|\partial_3 Y\|^{\frac23}_{L^2_t(L^2)}+|\partial_3 Y|^2_{\frac12+\varepsilon,1}+|Y_t|_{1+\varepsilon,2}\leq 1,$$

we have the following energy estimates: for $N \geq 0$, (we make the convention $\|u\|_{k,-1} = 0$)

(3.48)

$$\begin{aligned}
&\mathcal{E}_N + \|(X_t, \partial_3 X)\|_{\frac{1}{2},N+1} + \|\langle t\rangle^{\frac{1}{2}}\nabla X_t\|_{L^2_t(H^{N+1})} + \|\nabla X_t\|_{1,N-1}\\
&\leq C_{\varepsilon,N}\Big(\||D|^{-1}g\|_{1+\varepsilon,N+1} + \|\langle t\rangle^{\frac{1+\varepsilon}{2}} g\|_{L^2_t(H^N)}\\
&\qquad + \widetilde{\gamma}_{\varepsilon,N+1}(Y)\big(\||D|^{-1}g\|_{1+\varepsilon,2} + \|\langle t\rangle^{\frac{1+\varepsilon}{2}}|D|^{-1}g\|_{L^2_t(H^1)}\big)\Big) \quad \text{with}\\
&\widetilde{\gamma}_{\varepsilon,N+1}(Y) \leq C\big(1 + |\partial_3 Y|_{\frac{1}{2}+\varepsilon,N+1} + |Y_t|_{1+\varepsilon,N+2} + |\nabla Y|_{0,N+1} + \|\nabla Y\|_{0,N+1}\big).
\end{aligned}$$

With the preparations in the previous sections, we can now exploit Nash-Moser iteration scheme to prove Theorem 2.1. In order to do so, we first recall some basic properties of the smoothing operator from [**10, 11**]. Let $\chi(t) \in C^\infty(\mathbb{R}; [0,1])$ be such that

$$\chi(t) = 1 \text{ for } t \leq \frac{1}{2}, \quad \chi(t) = 0 \text{ for } t \geq 1.$$

Define for $\theta \geq 1$, the (cutoff-in-time) operator

$$S^{(1)}(\theta)Y(t,y) \stackrel{\text{def}}{=} \chi\Big(\frac{t}{\theta}\Big)Y(t,y). \tag{3.49}$$

Then we have

$$|S^{(1)}(\theta)Y|_{k,N} \leq C_{k,s}\theta^{k-s}|Y|_{s,N}, \quad \text{if } \ k \geq s \geq 0,$$

and

$$|(1 - S^{(1)}(\theta))Y|_{s,N} \leq C_{k,s}\theta^{-(k-s)}|Y|_{k,N} \quad \text{if } \ k \geq s \geq 0.$$

For $\theta' \geq 1$, we define the usual mollifying operator $S^{(2)}(\theta')$ in the space variables by

$$S^{(2)}(\theta')Y(t,y) \stackrel{\text{def}}{=} \widehat{\varphi}\Big(\frac{D_y}{\theta'}\Big)Y(t,y) = (\theta')^3\int_{\mathbb{R}^3}\varphi(\theta'(y-z))Y(t,z)dz, \tag{3.50}$$

where $\varphi \in \mathcal{S}(\mathbb{R}^3)$ satisfies

$$\widehat{\varphi}(\xi) = 1 \text{ for } |\xi| \leq \frac{1}{2}, \quad \widehat{\varphi}(\xi) = 0 \text{ for } |\xi| \geq 1,$$

so that

$$\int_{\mathbb{R}^3}\varphi(y)dy = 1, \quad \int_{\mathbb{R}^3} y^\alpha\varphi(y)dy = 0, \ \forall\ |\alpha| > 0.$$

We then have

$$|S^{(2)}(\theta')Y|_{k,N} \leq C_{N,M}(\theta')^{N-M}|Y|_{k,M} \quad \text{if } \ N \geq M \geq 0,$$

as well as

$$|(1 - S^{(2)}(\theta'))Y|_{k,M} \leq C_{N,M}(\theta')^{-(N-M)}|Y|_{k,N} \quad \text{if } N \geq M \geq 0.$$

Define the operator

$$S(\theta,\theta') \stackrel{\text{def}}{=} S^{(1)}(\theta)S^{(2)}(\theta'), \quad \text{for} \quad \theta,\ \theta' \geq 1. \tag{3.51}$$

Then it follows that

$$\begin{aligned}
&|S(\theta,\theta')Y|_{k,N} \leq C\theta^{k-s}(\theta')^{N-M}|Y|_{s,M},\\
&\|\langle t\rangle^k S(\theta,\theta')g\|_{L^p_t(H^N)} \leq C\theta^{k-s}(\theta')^{N-M}\|\langle t\rangle^s g\|_{L^p_t(H^M)},
\end{aligned} \tag{3.52}$$

if $k \geq s \geq 0$, $N \geq M \geq 0$. Moreover, due to

$$1 - S(\theta, \theta') = \big(1 - S^{(1)}(\theta)\big) + S^{(1)}(\theta)\big(1 - S^{(2)}(\theta')\big),$$

one has

$$\begin{aligned} |(1 - S(\theta, \theta'))Y|_{s,M} &\leq C\theta^{-(k-s)}|Y|_{k,M} + C(\theta')^{-(N-M)}|Y|_{s,N}, \\ \|\langle t\rangle^s(1 - S(\theta, \theta'))g\|_{L^p_t(H^M)} &\leq C\theta^{-(k-s)}\|\langle t\rangle^k g\|_{L^p_t(H^M)} \\ &\quad + C(\theta')^{-(N-M)}\|\langle t\rangle^s g\|_{L^p_t(H^N)} \end{aligned} \tag{3.53}$$

provided that $k \geq s \geq 0$, $N \geq M \geq 0$.

Let us denote

$$\Phi(Y) \stackrel{\text{def}}{=} Y_{tt} - \Delta Y_t - \partial_3^2 Y - f(Y),$$

for f given by (3.1). Then we can write (2.1) equivalently as

$$\Phi(Y) = 0, \quad Y(0, y) = Y^{(0)}, \quad Y_t(0, y) = Y^{(1)}. \tag{3.54}$$

We aim to solve (3.54) via Nash-Moser iteration scheme.

Let us define Y_0 via

$$\begin{cases} \partial_{tt}Y_0 - \Delta\partial_t Y_0 - \partial_3^2 Y_0 = 0, \\ Y_0(0, y) = Y^{(0)}, \quad \partial_t Y_0(0, y) = Y^{(1)}. \end{cases} \tag{3.55}$$

Inductively, assume that we already determine Y_p. In order to define Y_{p+1}, we introduce a mollified version of $\Phi'(Y_p)$ as follows

$$L_p X \stackrel{\text{def}}{=} \Phi'(S_p Y_p)X = X_{tt} - \Delta X_t - \partial_3^2 X - f'(S_p Y_p; X), \tag{3.56}$$

where S_p is the smoothing operator defined by

$$S_p = S(\theta_p, \theta'_p), \quad \text{with} \quad \theta_p = 2^p, \quad \theta'_p = \theta_p^{\bar{\varepsilon}} = 2^{\bar{\varepsilon}p}, \quad \text{and} \quad p \geq 0, \tag{3.57}$$

where $S(\theta, \theta')$ is defined in (3.51) and $\bar{\varepsilon} > 0$ is a small constant to be chosen later on. Then it follows from (3.52) and (3.53) that

$$\begin{aligned} |S_p Y|_{k,N} &\leq C\theta_p^{k-s}\theta_p^{\bar{\varepsilon}(N-M)}|Y|_{s,M} \\ \|\langle t\rangle^k S_p g\|_{L^2_t(H^N)} &\leq C\theta_p^{k-s}\theta_p^{\bar{\varepsilon}(N-M)}\|\langle t\rangle^s g\|_{L^2_t(H^M)} \\ |||S_p g|||_{L^1_t(\delta,N)} &\leq C\theta_p^{\bar{\varepsilon}(N-M)}|||g|||_{L^1_t(\delta,M)}, \end{aligned} \tag{S I}$$

and

$$\begin{aligned} |(1 - S_p)Y|_{0,0} &\leq C_{k,N}\big(\theta_p^{-k}|Y|_{k,0} + \theta_p^{-\bar{\varepsilon}N}|Y|_{0,N}\big), \\ \|\langle t\rangle^s(1 - S_p)g\|_{L^2_t(L^2)} &\leq C_{k,N}\big(\theta_p^{-(k-s)}\|\langle t\rangle^k g\|_{L^2_t(L^2)} + \theta_p^{-\bar{\varepsilon}N}\|\langle t\rangle^s g\|_{L^2_t(H^N)}\big), \end{aligned} \tag{S II}$$

for $k \geq s \geq 0$, $N \geq M \geq 0$, where the norm $|||\cdot|||_{L^1_t(\delta,N)}$ is given by (3.15).

Following Hörmander's version of Nash-Moser Scheme ([**9**]) (see also Klainerman's seminar papers [**10, 11**]), we define

$$Y_{p+1} = Y_p + X_p, \quad \text{with} \quad X_p = L_p^{-1} g_p, \tag{3.58}$$

where L_p^{-1} is a right inverse operator of L_p with zero initial data, that is: $X = L_p^{-1} g_p$ solves

$$\begin{cases} L_p X = g_p \quad \text{with } L_p \text{ given by (3.56)}, \\ X(0, y) = 0, \quad X_t(0, y) = 0. \end{cases} \tag{3.59}$$

In order to prove the convergence of the scheme, we define

$$e'_p \stackrel{\text{def}}{=} \big(\Phi'(Y_p) - L_p\big)X_p, \quad e''_p \stackrel{\text{def}}{=} \Phi(Y_{p+1}) - \Phi(Y_p) - \Phi'(Y_p)X_p, \tag{3.60}$$

and $e_p \stackrel{\text{def}}{=} e'_p + e''_p$, from which, we infer

$$\begin{aligned}\Phi(Y_{p+1}) - \Phi(Y_p) &= \Phi'(Y_p)X_p + e''_p = \Phi'(Y_p)L_p^{-1}g_p + e''_p\\ &= \big(\Phi'(Y_p) - L_p\big)L_p^{-1}g_p + g_p + e''_p = e'_p + e''_p + g_p.\end{aligned}$$

As a result, it comes out

$$\Phi(Y_{p+1}) - \Phi(Y_p) = e_p + g_p \quad \text{and} \quad \Phi(Y_{p+1}) - \Phi(Y_0) = \sum_{j=0}^{p}(e_j + g_j). \tag{3.61}$$

To achieve that the above limit as $p \to \infty$ is equal to $-\Phi(Y_0)$, we set

$$\sum_{j=0}^{p} g_j + S_pE_p = -S_p\Phi(Y_0) \quad \text{with} \quad E_p \stackrel{\text{def}}{=} \sum_{j=0}^{p-1} e_j. \tag{3.62}$$

The last relation defines g_p as follows

$$g_0 = -S_0\Phi(Y_0), \; g_p = -(S_p - S_{p-1})E_{p-1} - S_pe_{p-1} - (S_p - S_{p-1})\Phi(Y_0). \tag{3.63}$$

Let us fix the small constants: ε, $\bar\varepsilon$, $\delta > 0$, so that

$$\bar\varepsilon \le \frac{1}{20}, \quad \delta + 5\bar\varepsilon \le \frac{1}{4}, \quad \delta + \varepsilon + 4\bar\varepsilon \le \frac{1}{4}. \tag{3.64}$$

Let us take

$$\gamma = \frac{1}{4} - \bar\varepsilon, \quad \beta = \frac{1}{4} + \bar\varepsilon, \tag{3.65}$$

and $N_0 \in \mathbb{N}$ is chosen such that

$$\bar\varepsilon N_0 \ge \frac{1}{2} = \gamma + \beta. \tag{3.66}$$

We shall inductively prove the following statements:

Proposition 3.12. *Let $\delta_1 > 0$ be determined by Propositions 3.4, 3.5, 3.7, 3.8 and Theorem 3.9. Then for the constants $\beta, \gamma, N_0, \varepsilon, \bar\varepsilon$ and δ given by (3.64-3.66), for any $0 \le N \le N_0$, we have*

(P1, p)

$$\begin{aligned}&\big\||D|^{-1}(\partial_3X_p, \partial_tX_p)\big\|_{0,N+2} + \|\nabla X_p\|_{0,N+1} + \|(\partial_tX_p, \partial_3X_p)\|_{\frac{1}{2},N+1}\\ &+ \|\partial_tX_p\|_{L^2_t(H^{N+2})} + \big\|(\partial_3X_p, \langle t\rangle^{\frac{1}{2}}\nabla\partial_tX_p)\big\|_{L^2_t(H^{N+1})} + \|\nabla\partial_tX_p\|_{1,N-1} \le \eta\theta_p^{-\beta+\bar\varepsilon N};\end{aligned}$$

and

$$\text{(P2, } p\text{)} \qquad \begin{aligned}&|\partial_3X_p|_{k,N} \le \eta\theta_p^{k-\frac{1}{2}-\gamma+\bar\varepsilon N} && \text{if } \tfrac{1}{2} \le k \le 1,\\ &|\partial_tX_p|_{k,N} \le \eta\theta_p^{k-(1-\delta)-\gamma+\bar\varepsilon N} && \text{if } 1-\delta \le k \le \tfrac{3}{2} - \delta,\\ &|X_p|_{k,N} \le \eta\theta_p^{k-\gamma+\bar\varepsilon N} && \text{if } 0 \le k \le \tfrac{1}{2};\end{aligned}$$

and

$$(\mathrm{P3},p)\quad \begin{aligned}&\|\nabla Y_p\|_{L^\infty_t(\dot B^{\frac32}_{2,1})}\leq \delta_1,\ \|\nabla Y_p\|_{L^\infty_t(\dot B^{\frac52}_{2,1})}\leq 1,\ \|\partial_t Y_p\|_{0,0}\leq 1,\ |\partial_t Y_p|_{0,1}\leq 1,\\ &|\partial_3 Y_p|^{\frac43}_{\frac12+\varepsilon,1}\|\partial_3 Y_p\|^{\frac23}_{L^2_t(L^2)}+|\partial_3 Y_p|^2_{\frac12+\varepsilon,1}+|\partial_t Y_p|_{1+\varepsilon,2}\leq 1.\end{aligned}$$

Recall the convention that $\|u\|_{k,-1}=0$. We shall deduce the following propositions from Proposition 3.12.

PROPOSITION 3.13. *Under the assumptions of Proposition 3.12, we have, for $N\geq 0$,*

(I) (i)

$$\begin{aligned}|S_{p+1}\partial_3 Y_{p+1}|_{k,N}&\leq C_{k,N}\eta\theta_{p+1}^{k-\frac12-\gamma+\bar\varepsilon N} && \text{if } k\geq\frac12,\ k-\frac12-\gamma+\bar\varepsilon N\geq\bar\varepsilon,\\ |S_{p+1}\partial_t Y_{p+1}|_{k,N}&\leq C_{k,N}\eta\theta_{p+1}^{k-(1-\delta)-\gamma+\bar\varepsilon N} && \text{if } k\geq 1-\delta,\ k-(1-\delta)-\gamma+\bar\varepsilon N\geq\bar\varepsilon,\\ |S_{p+1}Y_{p+1}|_{k,N}&\leq C_{k,N}\eta\theta_{p+1}^{k-\gamma+\bar\varepsilon N} && \text{if } k\geq 0,\ k-\gamma+\bar\varepsilon N\geq\bar\varepsilon;\end{aligned}$$

(I) (ii)

$$\begin{aligned}\Delta_{p+1}&\overset{def}{=}\big\||D|^{-1}S_{p+1}(\partial_3 Y_{p+1},\partial_t Y_{p+1})\big\|_{0,N+2}+\|S_{p+1}\nabla Y_{p+1}\|_{0,N+1}\\ &+\|S_{p+1}(\partial_t Y_{p+1},\partial_3 Y_{p+1})\|_{\frac12,N+1}+\big\|(S_{p+1}\partial_3 Y_{p+1},\langle t\rangle^{\frac12}S_{p+1}\nabla\partial_t Y_{p+1})\big\|_{L^2_t(H^{N+1})}\\ &+\|S_{p+1}\partial_t Y_{p+1}\|_{L^2_t(H^{N+2})}+\|S_{p+1}\nabla\partial_t Y_{p+1}\|_{1,N-1}\leq C_N\eta\theta_{p+1}^{-\beta+\bar\varepsilon N}\quad \text{if } -\beta+\bar\varepsilon N\geq\bar\varepsilon;\end{aligned}$$

$$(\mathrm{II})\ (\mathrm{i})\quad\begin{aligned}|S_{p+1}\partial_3 Y_{p+1}|_{k,N}&\leq C_{k,N}\eta && \text{if } k\geq\frac12, && k-\frac12-\gamma+\bar\varepsilon N\leq-\bar\varepsilon,\\ |S_{p+1}\partial_t Y_{p+1}|_{k,N}&\leq C_{k,N}\eta && \text{if } k\geq 1-\delta, && k-(1-\delta)-\gamma+\bar\varepsilon N\leq-\bar\varepsilon,\\ |S_{p+1}Y_{p+1}|_{k,N}&\leq C_{k,N}\eta && \text{if } k\geq 0, && k-\gamma+\bar\varepsilon N\leq-\bar\varepsilon;\end{aligned}$$

$$(\mathrm{II})\ (\mathrm{ii})\qquad \Delta_{p+1}\leq C_N\eta\qquad \text{if}\quad -\beta+\bar\varepsilon N\leq-\bar\varepsilon;$$

(III)

$$\begin{aligned}|(1-S_{p+1})\partial_3 Y_{p+1}|_{k,N}&\leq C_{k,N}\eta\theta_{p+1}^{k-\frac12-\gamma+\bar\varepsilon N} && \text{if } \frac12\leq k\leq 1,\ N\leq N_0,\\ |(1-S_{p+1})\partial_t Y_{p+1}|_{k,N}&\leq C_{k,N}\eta\theta_{p+1}^{k-(1-\delta)-\gamma+\bar\varepsilon N} && \text{if } 1-\delta\leq k\leq\frac32-\delta,\ N\leq N_0,\\ |(1-S_{p+1})Y_{p+1}|_{k,N}&\leq C_{k,N}\eta\theta_{p+1}^{k-\gamma+\bar\varepsilon N} && \text{if } 0\leq k\leq\frac12,\ N\leq N_0.\end{aligned}$$

PROPOSITION 3.14. *Let e_p, g_p and $R_{N,\theta}(g)$ be given by (3.60), (3.63) and (3.14) respectively. Let $\alpha\overset{def}{=}\frac12-\delta-\bar\varepsilon>0$. Then there hold*

<u>(1) *Estimates for e_p.*</u>

(IV) (i)

$$\|\langle t\rangle^{k+\frac12}|D|^{-1}e_p\|_{L^2_t(H^{N+1})}\lesssim\eta^2\theta_p^{k+\delta-\gamma-\beta+\bar\varepsilon(N+3)}\ \text{if}\ 0\leq k\leq\alpha,\ 0\leq N\leq N_0-2,$$

$$(\mathrm{IV})\ (\mathrm{ii})\ \ \||D|^{-1}e_p\|_{1+k,N+1}\lesssim\eta^2\theta_p^{k+\delta-\gamma-\beta+\bar\varepsilon(N+2)}\ \text{if}\ \ 0\leq k\leq\frac12-\delta,\ N\leq N_0-2,$$

$$(\mathrm{IV})\ (\mathrm{iii})\quad \|\langle t\rangle^{\frac12}e_p\|_{L^2_t(\delta,N)}\lesssim\eta^2\theta_p^{-\gamma+\bar\varepsilon(N+5)}\qquad\qquad \text{if}\ 0\leq N\leq N_0-6;$$

(2) *Estimates for g_{p+1}.*

(V) (i) $\|\langle t\rangle^{k+\frac{1}{2}}|D|^{-1}g_{p+1}\|_{L^2_t(H^{N+1})} \le C\eta^2\theta_{p+1}^{k+\delta-\gamma-\beta+\bar\varepsilon(N+3)}$ *if* $k\ge 0,\ N\ge 0,$

(V) (ii) $\||D|^{-1}g_{p+1}\|_{1+k,N+1} \lesssim \eta^2\theta_{p+1}^{k+\delta-\gamma-\beta+\bar\varepsilon(N+2)}$ *if* $k\ge 0,\ N\ge 0,$

(V) (iii) $\|g_{p+1}\|_{L^1_t(N)} \le C\eta^2\theta_{p+1}^{-\gamma+\bar\varepsilon(N+6)}$ *if* $-\gamma+\bar\varepsilon(N+5)\ge\bar\varepsilon,$

(V) (iv) $\|g_{p+1}\|_{L^1_t(N)} \le C\eta^2\theta_{p+1}^{\bar\varepsilon}$ *if* $-\gamma+\bar\varepsilon(N+5)\le -\bar\varepsilon;$

(3) *Estimates for $R_{N,\theta_{p+1}}(g_{p+1})$.*

(VI) (i) $R_{N,\theta_{p+1}}(g_{p+1}) \le C\eta^2\theta_{p+1}^{\frac{1}{2}-\gamma+\bar\varepsilon N}$ *if* $-\gamma+\bar\varepsilon(N+5)\ge\bar\varepsilon,$

(VI) (ii) $R_{0,\theta_{p+1}}(g_{p+1}) \le C\eta^2\theta_{p+1}^{\frac{1}{2}-\gamma}.$

The following interpolation lemma will be crucial in the proof of the above propositions, whose proof is exactly the same as that of Lemma 6.1 of [**10**], which we omit the details here.

LEMMA 3.15 (Interpolation lemma). *Let $p\in[1,+\infty]$, $\theta\ge 1$ and $\bar\varepsilon>0$, which satisfy*

$$\beta>\bar\varepsilon,\quad k_0-\beta\ge\bar\varepsilon,\quad -\beta+\bar\varepsilon N_0\ge\bar\varepsilon.$$

Assume that $u\in C^\infty([0,+\infty)\times\mathbb{R}^n)$ satisfies

$$\begin{aligned}&\|u\|_{L^p_t(L^2)}\le C\theta^{-\beta},\\ &\|\langle t\rangle^k u\|_{L^p_t(H^N)}\le C\theta^{k-\beta+\bar\varepsilon N},\end{aligned}\tag{3.67}$$

for $0\le k\le k_0$, $0\le N\le N_0$ such that $k-\beta+\bar\varepsilon N\ge\bar\varepsilon$. Then for all $0\le k\le k_0$, $0\le N\le N_0$,

$$\|\langle t\rangle^k u\|_{L^p_t(H^N)}\le C_{k_0,N_0}\theta^{k-\beta+\bar\varepsilon N}.$$

Finally with the previous propositions, we can prove the convergence of the approximate solutions constructed by (3.58), and this completes the proof of Theorem 2.1. One may check [**8**] for details.

Acknowledgments

P. Zhang would like to thank Professor Fanghua Lin and Professor Jalal Shatah for profitable discussions.

References

[1] Hammadi Abidi and Ping Zhang, *On the global solution of a 3-D MHD system with initial data near equilibrium*, Comm. Pure Appl. Math. **70** (2017), no. 8, 1509–1561. MR3666563

[2] C. Bardos, C. Sulem, and P.-L. Sulem, *Longtime dynamics of a conductive fluid in the presence of a strong magnetic field*, Trans. Amer. Math. Soc. **305** (1988), no. 1, 175–191. MR920153

[3] Hajer Bahouri, Jean-Yves Chemin, and Raphaël Danchin, *Fourier analysis and nonlinear partial differential equations*, Grundlehren der Mathematischen Wissenschaften [Fundamental Principles of Mathematical Sciences], vol. 343, Springer, Heidelberg, 2011. MR2768550

[4] Jean-Yves Chemin, David S. McCormick, James C. Robinson, and Jose L. Rodrigo, *Local existence for the non-resistive MHD equations in Besov spaces*, Adv. Math. **286** (2016), 1–31. MR3415680

[5] T. G. Cowling and D. Phil, *Magnetohydrodynnamics*, The Institute of Physics, 1976.

[6] R. Danchin, *Global existence in critical spaces for compressible Navier-Stokes equations*, Invent. Math. **141** (2000), no. 3, 579–614. MR1779621

[7] P. A. Davidson, *An introduction to magnetohydrodynamics*, Cambridge Texts in Applied Mathematics, Cambridge University Press, Cambridge, 2001. MR1825486

[8] W. Deng and P. Zhang, Large time behavior of solutions to 3-D MHD system with initial data near equilibrium, arXiv:1702.05260.

[9] L. Hörmander, *Implicit Function Theorems*, Lectures at Stanford University, Summer 1977.

[10] Sergiu Klainerman, *Global existence for nonlinear wave equations*, Comm. Pure Appl. Math. **33** (1980), no. 1, 43–101. MR544044

[11] Sergiu Klainerman, *Long-time behavior of solutions to nonlinear evolution equations*, Arch. Rational Mech. Anal. **78** (1982), no. 1, 73–98. MR654553

[12] Sergiu Klainerman, *Long time behaviour of solutions to nonlinear wave equations*, Proceedings of the International Congress of Mathematicians, Vol. 1, 2 (Warsaw, 1983), PWN, Warsaw, 1984, pp. 1209–1215. MR804771

[13] Sergiu Klainerman, *Uniform decay estimates and the Lorentz invariance of the classical wave equation*, Comm. Pure Appl. Math. **38** (1985), no. 3, 321–332. MR784477

[14] Fanghua Lin, Li Xu, and Ping Zhang, *Global small solutions of 2-D incompressible MHD system*, J. Differential Equations **259** (2015), no. 10, 5440–5485. MR3377532

[15] Fanghua Lin and Ping Zhang, *Global small solutions to an MHD-type system: the three-dimensional case*, Comm. Pure Appl. Math. **67** (2014), no. 4, 531–580. MR3168121

[16] Fanghua Lin and Ting Zhang, *Global small solutions to a complex fluid model in three dimensional*, Arch. Ration. Mech. Anal. **216** (2015), no. 3, 905–920. MR3325778

[17] Xiaoxia Ren, Jiahong Wu, Zhaoyin Xiang, and Zhifei Zhang, *Global existence and decay of smooth solution for the 2-D MHD equations without magnetic diffusion*, J. Funct. Anal. **267** (2014), no. 2, 503–541. MR3210038

[18] Li Xu and Ping Zhang, *Global small solutions to three-dimensional incompressible magnetohydrodynamical system*, SIAM J. Math. Anal. **47** (2015), no. 1, 26–65. MR3296601

[19] Ting Zhang, *Global solutions to the 2D viscous, non-resistive MHD system with large background magnetic field*, J. Differential Equations **260** (2016), no. 6, 5450–5480. MR3448784

Academy of Mathematics & Systems Science and Hua Loo-Keng Key Laboratory of Mathematics, Chinese Academy of Sciences, Beijing 100190, CHINA

Email address: dengwen@amss.ac.cn

Academy of Mathematics & Systems Science and Hua Loo-Keng Key Laboratory of Mathematics, Chinese Academy of Sciences, Beijing 100190, CHINA, and School of Mathematical Sciences, University of Chinese Academy of Sciences, Beijing 100049, China.

Email address: zp@amss.ac.cn

Contemporary Mathematics
Volume **710**, 2018
http://dx.doi.org/10.1090/conm/710/14366

On some properties of weak solutions to elliptic equations with divergence-free drifts

Nikolay Filonov and Timofey Shilkin

ABSTRACT. We discuss the local properties of weak solutions to the equation $-\Delta u + b \cdot \nabla u = 0$. The Dirichlet boundary value problem for this equation is studied as well. The corresponding theory is well-known in the case $b \in L_n$, where n is the dimension of the space. Our main interest is focused on the case $b \in L_2$. In this case the structure assumption $\operatorname{div} b = 0$ turns out to be crucial.

1. Introduction

Assume $n \geq 2$, $\Omega \subset \mathbb{R}^n$ is a bounded domain with the boundary $\partial\Omega \subset C^2$, and $b : \Omega \to \mathbb{R}^n$, $f : \Omega \to \mathbb{R}$. In this paper we investigate the properties of weak solutions $u : \Omega \to \mathbb{R}$ to the following scalar equation

$$(1.1) \qquad -\Delta u + b \cdot \nabla u \;=\; f \qquad \text{in} \quad \Omega.$$

This equation describes the diffusion in a stationary incompressible flow. If it is not stated otherwise, we always impose the following conditions

$$b \in L_2(\Omega), \qquad f \in W_2^{-1}(\Omega)$$

(see the list of notation at the end of this section). We use the following

DEFINITION 1.1. Assume $b \in L_2(\Omega)$, $f \in W_2^{-1}(\Omega)$. The function $u \in W_2^1(\Omega)$ is called a weak solution to the equation (1.1) if the following integral identity holds:

$$(1.2) \qquad \int\limits_\Omega \nabla u \cdot (\nabla \eta + b\eta)\, dx \;=\; \langle f, \eta \rangle, \qquad \forall\, \eta \in C_0^\infty(\Omega).$$

Together with the equation (1.1) one can consider the formally conjugate (up to the sign of the drift) equation

$$(1.3) \qquad -\Delta u + \operatorname{div}(bu) \;=\; f \qquad \text{in} \quad \Omega.$$

2010 *Mathematics Subject Classification.* Primary 35B65.
Key words and phrases. elliptic equations, weak solutions, regularity.
Both authors are supported by RFBR grant 17-01-00099-a.
The research of the second author leading to these results has received funding from the People Programme (Marie Curie Actions) of the European Union's Seventh Framework Programme FP7/2007-2013/ under REA grant agreement n°. 319012 and from the Funds for International Co-operation under Polish Ministry of Science and Higher Education grant agreement n°. 2853/7.PR/2013/2. The author also thanks the Technische Universität of Darmstadt for its hospitality.

DEFINITION 1.2. Assume $b \in L_2(\Omega)$, $f \in W_2^{-1}(\Omega)$. The function $u \in W_2^1(\Omega)$ is called a weak solution to the equation (1.3) if

$$(1.4)\qquad \int_\Omega (\nabla u - bu)\cdot \nabla \eta \, dx \ = \ \langle f, \eta\rangle, \qquad \forall\, \eta \in C_0^\infty(\Omega).$$

The advantage of the equation (1.3) is that it allows one to define weak solutions for a drift b belonging to a weaker class than $L_2(\Omega)$. Namely, Definition 1.2 makes sense for $u \in W_2^1(\Omega)$ if

$$(1.5)\qquad b \in L_s(\Omega) \qquad \text{where} \qquad s \ = \ \begin{cases} \frac{2n}{n+2}, & n \geq 3, \\ 1+\varepsilon, \ \varepsilon > 0, & n = 2. \end{cases}$$

Nevertheless, it is clear that for a divergence-free drift $b \in L_2(\Omega)$ the Definitions 1.1 and 1.2 coincide.

Together with the equation (1.1) we discuss boundary value problems with Dirichlet boundary conditions:

$$(1.6)\qquad \begin{cases} -\Delta u + b\cdot \nabla u = f \quad \text{in} \quad \Omega, \\ u|_{\partial\Omega} = \varphi. \end{cases}$$

For weak solutions the boundary condition is understood in the sense of traces. Assume f is "good enough" and $b \in L_2(\Omega)$, $\operatorname{div} b = 0$. Our main observation is that the regularity of solution u inside Ω can depend on the behaviour of its boundary values. If the function φ is bounded, then the solution u is also bounded (see Theorem 3.4 below). If the function φ is unbounded on $\partial\Omega$, then the solution u can become infinite in internal points of Ω (see Example 3.6 below). So, we distinguish between two cases: the case of general boundary data $\varphi \in W_2^{1/2}(\partial\Omega)$, and the case of bounded boundary data

$$(1.7)\qquad \varphi \in L_\infty(\partial\Omega) \cap W_2^{1/2}(\partial\Omega).$$

Discussing the properties of weak solutions to the problem (1.6) we also distinguish between another two cases: in Section 2 we consider sufficiently regular drifts, namely, $b \in L_n(\Omega)$, and in Section 3 we focus on the case of drifts b from $L_2(\Omega)$ satisfying $\operatorname{div} b = 0$. Section 4 is devoted to possible ways of relaxation of the condition $b \in L_n(\Omega)$ in the framework of the regularity theory. In Appendix for reader's convenience we present proofs of some results formulated in previous sections. We believe most of the proofs gathered in Appendix are either known or can be obtained as straightforward modification of the known theory. So, our main results which seems to be new (or at least the authors could not identify them in the literature) are Example 3.6 and Theorem 3.7 which (being combined with already known results) establish an interesting phenomenon discussed in Section 3.2.

Together with the elliptic equation (1.1) it is possible to consider its parabolic analogue

$$(1.8)\qquad \partial_t u - \Delta u + b\cdot\nabla u \ = \ f \qquad \text{in} \quad \Omega\times(0,T),$$

but it should be a subject of a separate survey. We address the interested readers to the related papers [**Z**], [**LZ**], [**NU**], [**Sem**], [**SSSZ**], [**SV**], [**SVZ**] and references there.

In the paper we explore the following notation. For any a, $b \in \mathbb{R}^n$ we denote by $a\cdot b$ its scalar product in $\mathbb{R}^n$. We denote by $L_p(\Omega)$ and $W_p^k(\Omega)$ the usual Lebesgue and Sobolev spaces. The space $\overset{\circ}{W}{}^1_p(\Omega)$ is the closure of $C_0^\infty(\Omega)$ in $W^1_p(\Omega)$ norm. The negative Sobolev space $W_p^{-1}(\Omega)$, $p \in (1, +\infty)$, is the set of all distributions which are bounded functionals on $\overset{\circ}{W}{}^1_{p'}(\Omega)$ with $p' := \frac{p}{p-1}$. For any $f \in W_p^{-1}(\Omega)$ and $w \in \overset{\circ}{W}{}^1_{p'}(\Omega)$ we denote by $\langle f, w\rangle$ the value of the distribution f on the function w. We use the notation $W_2^{1/2}(\partial\Omega)$ for the Slobodetskii–Sobolev space. By $C(\bar\Omega)$ and $C^\alpha(\bar\Omega)$, $\alpha \in (0,1)$ we denote the spaces of continuous and Hölder continuous functions on $\bar\Omega$. The space $C^{1+\alpha}(\bar\Omega)$ consists of functions u whose gradient ∇u is Hölder continuous. The index "loc" in notation of the functional spaces $L_{\infty,loc}(\Omega)$, $C^\alpha_{loc}(\Omega)$, $C^{1+\alpha}_{loc}(\Omega)$ etc implies that the function belongs to the corresponding functional class over every compact set which is contained in Ω. The symbols $\rightharpoonup$ and $\to$ stand for the weak and strong convergence respectively. We denote by $B_R(x_0)$ the ball in $\mathbb{R}^n$ of radius R centered at x_0 and write B_R if $x_0 = 0$. We write also B instead of B_1.

2. Regular drifts

2.1. Local properties. For sufficiently regular drifts we have the local Hölder continuity of a solution.

THEOREM 2.1. *Assume*

$$b \in L_n(\Omega) \quad \text{if} \quad n \ge 3, \qquad \int_\Omega |b|^2 \ln(2 + |b|^2)\, dx \; < \infty \quad \text{if} \quad n = 2. \tag{2.1}$$

Let $u \in W^1_2(\Omega)$ be a weak solution to (1.1) *with f satisfying*

$$f \in L_p(\Omega), \qquad p > \frac{n}{2}\ .$$

Then

$$u \in C^\alpha_{loc}(\Omega) \qquad \text{with} \qquad \begin{cases} \alpha = 2 - \frac{n}{p}, & p < n, \\ \forall\, \alpha < 1, & p \ge n. \end{cases}$$

The local Hölder continuity of weak solutions in Theorem 2.1 with some $\alpha \in (0,1)$ is well-known, see [**St**, Theorem 7.1] or [**NU**, Corollary 2.3] in the case $f \equiv 0$. The Hölder continuity with arbitrary $\alpha \in (0,1)$ was proved in the case $f \equiv 0$, for example, in [**F**]. The extension of this result for non-zero right hand side is routine.

If b possesses more integrability then the first gradient of a weak solution is locally Hölder continuous.

THEOREM 2.2. *Let $b \in L_p(\Omega)$ with $p > n$, and $u \in W^1_2(\Omega)$ be a weak solution to* (1.1) *with $f \in L_p(\Omega)$. Then $u \in C^{1+\alpha}_{loc}(\Omega)$ with $\alpha = 1 - \frac{n}{p}$.*

For the proof see [**LU**, Chapter III, Theorem 15.1].

2.2. Boundary value problem. We consider the second term $\int_\Omega \nabla u \cdot b\, \eta\, dx$ in the equation (1.2) as a bilinear form in $\overset{\circ}{W}{}^1_2(\Omega)$. It defines a linear operator $T : \overset{\circ}{W}{}^1_2(\Omega) \to \overset{\circ}{W}{}^1_2(\Omega)$ by the relation

$$\int_\Omega \nabla(Tu) \cdot \nabla \eta\, dx = \int_\Omega \nabla u \cdot b\, \eta\, dx, \quad \forall\, u, \eta \in \overset{\circ}{W}{}^1_2(\Omega). \tag{2.2}$$

The following result is well-known.

THEOREM 2.3. *Let b satisfy* (2.1). *Then the operator $T : \overset{\circ}{W}{}^1_2(\Omega) \to \overset{\circ}{W}{}^1_2(\Omega)$ defined by* (2.2) *is compact.*

Indeed, if $n \geq 3$ then the estimate

$$(2.3) \qquad \left| \int\limits_\Omega \nabla u \cdot b\, \eta\, dx \right| \leq C_b \|\nabla u\|_{L_2(\Omega)} \|\nabla \eta\|_{L_2(\Omega)} \quad \forall\, u, \eta \in \overset{\circ}{W}{}^1_2(\Omega)$$

follows by the imbedding theorem and the Hölder inequality. In the case $n = 2$ such estimate can be found for example in [**F**, Lemma 4.3]. Next, the operator T can be approximated in the operator norm by compact linear operators T_ε generated by the bilinear forms $\int\limits_\Omega \nabla u \cdot b_\varepsilon\, \eta\, dx$ where $b_\varepsilon \in C^\infty(\bar{\Omega})$.

REMARK 2.4. The condition $b \in L_2(\Omega)$ in the case $n = 2$ is not sufficient. For example, one can take $\Omega = B_{1/3}$,

$$b(x) = \frac{x}{|x|^2\, |\ln |x||^{3/4}}, \qquad u(x) = \eta(x) = |\ln |x||^{3/8} - (\ln 3)^{3/8}.$$

Then $\int\limits_\Omega \nabla u \cdot b\, \eta\, dx = \infty$, and therefore, the corresponding operator T is unbounded.

REMARK 2.5. The issue of boundedness and compactness of the operator T in the case of the whole space, $\Omega = \mathbb{R}^n$, is investigated in full generality in [**MV**], see Theorem 4.1 below. In this section we restrict ourselves by considering assumptions on b only in L_p–scale.

Now, the problem (1.6) with $\varphi \equiv 0$ reduces to the equation $u + Tu = h$ in $\overset{\circ}{W}{}^1_2(\Omega)$ with an appropriate right hand side h. The solvability of the last equation follows from the Fredholm theory. Roughly speaking, "the existence follows from the uniqueness".

The uniqueness in the case $b \in L_n(\Omega)$, $n \geq 3$, and $\operatorname{div} b = 0$ is especially simple. In this situation

$$\int\limits_\Omega b \cdot \nabla u\, u\, dx = 0 \qquad \forall\, u \in \overset{\circ}{W}{}^1_2(\Omega),$$

and the uniqueness for the problem (1.6) follows. In the general case of drifts satisfying (2.1) without the condition $\operatorname{div} b = 0$ the proof of the uniqueness is more sophisticated. It requires the maximum principle which can be found, for example, in [**NU**], see Corollary 2.2 and remarks at the end of Section 2 there.

THEOREM 2.6. *Let b satisfy* (2.1). *Assume $u \in W^1_2(\Omega)$ is a weak solution to the problem* (1.6) *with $f \equiv 0$ and $\varphi \in L_\infty(\partial\Omega) \cap W^{1/2}_2(\partial\Omega)$. Then either $u \equiv const$ in Ω or the following estimate holds:*

$$\operatorname{essinf}_{\partial\Omega} \varphi \; < \; u(x) \; < \; \operatorname{esssup}_{\partial\Omega} \varphi, \qquad \forall\, x \in \Omega.$$

COROLLARY 2.7. *Let b satisfy* (2.1). *Then a weak solution to the problem* (1.6) *is unique in the space $W^1_2(\Omega)$.*

Now, the solvability of the problem (1.6) is straightforward.

THEOREM 2.8. *Let b satisfy* (2.1). *Then for any $f \in W^{-1}_2(\Omega)$ and $\varphi \in W^{1/2}_2(\partial\Omega)$ the problem* (1.6) *has the unique weak solution $u \in W^1_2(\Omega)$, and*

$$\|u\|_{W^1_2(\Omega)} \leq C \left(\|f\|_{W^{-1}_2(\Omega)} + \|\varphi\|_{W^{1/2}_2(\partial\Omega)} \right).$$

PROOF. For $\varphi \equiv 0$ Theorem 2.8 follows from Fredholm's theory. In the general case the problem (1.6) can be reduced to the corresponding problem with homogeneous boundary conditions for the function $v := u - \tilde{\varphi}$, where $\tilde{\varphi}$ is some extension of φ from $\partial\Omega$ to Ω with the control of the norm $\|\tilde{\varphi}\|_{W_2^1(\Omega)} \le c\|\varphi\|_{W_2^{1/2}(\partial\Omega)}$. The function v can be determined as a weak solution to the problem

$$\left\{ \begin{array}{c} -\Delta v + b\cdot\nabla v = f + \Delta\tilde{\varphi} - b\cdot\nabla\tilde{\varphi} \quad \text{in} \quad \Omega, \\ v|_{\partial\Omega} = 0 \end{array} \right. \tag{2.4}$$

Under assumption (2.1) the right hand side belongs to $W_2^{-1}(\Omega)$ due to Theorem 2.3. □

Note that for $n \ge 3$ the problems (1.6) and (2.4) are equivalent only in the case of "regular" drifts $b \in L_n(\Omega)$. If $b \in L_2(\Omega)$ and additionally $\operatorname{div} b = 0$, then $b\cdot\nabla\tilde{\varphi} \in W_{n'}^{-1}(\Omega)$, $n' = \frac{n}{n-1}$, and the straightforward reduction of the problem (1.6) to the problem with homogeneous boundary data is not possible.

Finally, to investigate in Section 3 the problem (1.6) with divergence-free drifts from $L_2(\Omega)$ we need the following maximum estimate.

THEOREM 2.9. *Let b satisfy (2.1). Assume φ satisfies (1.7) and let $u \in W_2^1(\Omega)$ be a weak solution to (1.6) with some $f \in L_p(\Omega)$, $p > n/2$. Then*

1) $u \in L_\infty(\Omega)$ and

$$\|u\|_{L_\infty(\Omega)} \le \|\varphi\|_{L_\infty(\partial\Omega)} + C\,\|f\|_{L_p(\Omega)}. \tag{2.5}$$

2) If $\operatorname{div} b = 0$ then $C = C(n,p,\Omega)$ does not depend on b.

We believe Theorem 2.9 is known though it is difficult for us to identify the precise reference to the statement we need. So, we present its proof in Appendix.

REMARK 2.10. For $n \ge 3$ consider the following example:

$$\Omega = B, \qquad u(x) = \ln|x|, \qquad b(x) = (n-2)\frac{x}{|x|^2}.$$

The statements of Theorem 2.1, Theorem 2.6 and Corollary 2.7 are violated for these functions. On the other hand, $-\Delta u + b\cdot\nabla u = 0$, $u \in \overset{\circ}{W}{}_2^1(\Omega)$ and $b \in L_p(\Omega)$ for any $p < n$. It means that for non-divergence free drifts the condition $b \in L_n(\Omega)$ in (2.1) is sharp.

REMARK 2.11. For $n = 2$ the condition $b \in L_2(\Omega)$ is not sufficient. The statements of Theorem 2.1, Theorem 2.6 and Corollary 2.7 are violated for the functions

$$u(x) = \ln|\ln|x||, \qquad b(x) = -\frac{x}{|x|^2\ln|x|}$$

in a ball $\Omega = B_{1/e}$, nevertheless $b \in L_2(\Omega)$.

Conversely, if in the case $n = 2$ we assume that $b \in L_2(\Omega)$ and $\operatorname{div} b = 0$, then the estimate (2.3) is fulfilled (see [**MV**] or [**F**]), and all statements of this section (Theorems 2.1, 2.3, 2.6, 2.8 and 2.9) hold true, see [**F**] or [**NU**]. So, this case can be considered as the regular one. See also Remark 4.3 below.

3. Non-regular divergence-free drifts

In this section we always assume that $\operatorname{div} b = 0$. It turns out that this assumption plays the crucial role in local boundedness of weak solutions if one considers drifts $b \in L_p(\Omega)$ with $p < n$, $n \geq 3$. Recall that the case $n = 2$, $b \in L_2(\Omega)$ and $\operatorname{div} b = 0$ can be considered as a regular case, see Remark 2.11. Thus, below we restrict ourselves to the case $n \geq 3$.

3.1. Boundary value problem. We have the following approximation result.

THEOREM 3.1. *Assume* $b \in L_2(\Omega)$, $\operatorname{div} b = 0$, $f \in W_2^{-1}(\Omega)$, *and let* $u \in W_2^1(\Omega)$ *be a weak solution to* (1.1). *Assume also* $b_k \in L_n(\Omega)$, $\operatorname{div} b_k = 0$ *is an arbitrary sequence satisfying*

$$b_k \to b \quad in \quad L_2(\Omega),$$

and let $u_k \in W_2^1(\Omega)$ *be the unique weak solution to the problem*

$$\left\{ \begin{array}{c} -\Delta u_k + b_k \cdot \nabla u_k = f, \\ u_k|_{\partial\Omega} = \varphi, \end{array} \right. \tag{3.1}$$

where $\varphi = u|_{\partial\Omega}$. *Then*

$$u_k \to u \quad in \quad L_q(\Omega) \quad for\ any \quad q < \frac{n}{n-2}. \tag{3.2}$$

Moreover, if $\varphi \in L_\infty(\partial\Omega)$ *then*

$$u_k \rightharpoonup u \quad in \quad W_2^1(\Omega). \tag{3.3}$$

The convergence (3.2) is proved (in its parabolic version) for $q = 1$ in [**Z**, Proposition 2.4]. Note that the proof in [**Z**] uses the uniform Gaussian upper bound of the Green functions of the operators $\partial_t u - \Delta u + b_k \cdot \nabla u$ (sf. [**A**]). In Appendix we present an elementary proof of Theorem 3.1 based on the maximum estimate in Theorem 2.9 and duality arguments.

Theorem 3.1 has several consequences. The first of them is the uniqueness of weak solutions, see [**Z**] and [**Zhi**]:

THEOREM 3.2. *Let* $b \in L_2(\Omega)$, $\operatorname{div} b = 0$. *Then a weak solution to the problem* (1.6) *is unique in the class* $W_2^1(\Omega)$.

Indeed, u is a L_q-limit of the approximating sequence u_k, and such limit is unique. The alternative proof of the uniqueness (which is in a sense "direct", i.e. it does not hang upon the approximation result of Theorem 3.1) for $b \in L_2(\Omega)$, $\operatorname{div} b = 0$, can be found in [**Zhi**] (see also some development in [**Su**]). Moreover, in [**Zhi**] it was shown that under conditions $b \in L_2(\Omega)$, $\operatorname{div} b = 0$ and $\varphi \equiv 0$ the weak solution to the problem (1.6) satisfies the energy identity

$$\int\limits_\Omega |\nabla u|^2\, dx \;=\; \langle f, u\rangle.$$

Note also that in [**Zhi**] it was also shown that the uniqueness can break for weak solutions to the equation (1.3) if b satisfy (1.5) (actually a little better than (1.5)) and $\operatorname{div} b = 0$, but $b \notin L_2(\Omega)$.

Another consequence of Theorem 3.1 is the existence of weak solution.

THEOREM 3.3. *Let* $b \in L_2(\Omega)$, $\operatorname{div} b = 0$. *Then for any* $f \in W_2^{-1}(\Omega)$ *and any* φ *satisfying* (1.7) *there exists a weak solution to the problem* (1.6).

Theorem 3.3 is proved in Appendix.

Finally, Theorem 3.1 allows one to establish the global boundedness of weak solutions whenever the boundary data are bounded.

THEOREM 3.4. *Let $b \in L_2(\Omega)$, $\operatorname{div} b = 0$, $f \in L_p(\Omega)$, $p > n/2$, and φ satisfies (1.7). Assume $u \in W_2^1(\Omega)$ is a weak solution to (1.6). Then $u \in L_\infty(\Omega)$ and*

$$\|u\|_{L_\infty(\Omega)} \ \le \ \|\varphi\|_{L_\infty(\partial\Omega)} + C\ \|f\|_{L_p(\Omega)}, \tag{3.4}$$

where the constant $C = C(n, p, \Omega)$ is independent on b.

Theorem 3.4 is proved in Appendix.

3.2. Local properties. Note that any weak solution to (1.1) belonging to the class $W_2^1(\Omega)$ can be viewed as a weak solution to the problem (1.6) with some $\varphi \in W_2^{1/2}(\Omega)$.

THEOREM 3.5. *Assume* $\operatorname{div} b = 0$ *and*

$$b \in L_p(B) \quad \textit{where} \quad p = 2 \quad \textit{if} \quad n = 3 \quad \textit{and} \quad p > \frac{n}{2} \quad \textit{if} \quad n \ge 4.$$

Let $u \in W_2^1(B)$ be a weak solution to (1.1) in B with some $f \in L_q(B)$, $q > n/2$. Then $u \in L_\infty(B_{1/2})$ and

$$\|u\|_{L_\infty(B_{1/2})} \ \le \ C\ \Big(\|u\|_{W_2^1(B)} + \|f\|_{L_q(B)} \Big)$$

where the constant C depends only on n, p, q and $\|b\|_{L_p(B)}$.

Theorem 3.5 was proved (in the parabolic version) in [**Z**]. For the reader's convenience we present the proof of this theorem in Appendix.

Let us consider the following

EXAMPLE 3.6. Assume $n \ge 4$ and put

$$u(x) = \ln r, \qquad b = (n-3) \left(\frac{1}{r}\, \mathbf{e}_r - (n-3)\, \frac{z}{r^2}\, \mathbf{e}_z \right),$$

where $r^2 = x_1^2 + ... + x_{n-1}^2$, $z = x_n$, and $\mathbf{e}_r$, $\mathbf{e}_z$ are the basis vectors of the corresponding cylindrical coordinate system in $\mathbb{R}^n$. Then $u \in \overset{\circ}{W}{}_2^1(\Omega)$, and

$$-\Delta u + b \cdot \nabla u \ = 0.$$

Next, $\operatorname{div} b = 0$, $b(x) = O(r^{-2})$ near the axis of symmetry, and hence

$$b \in L_p(B) \quad \text{for any} \quad p \ < \ \frac{n-1}{2}.$$

Clearly, the assumption $b \in L_2(\Omega)$ leads to the restriction $n \ge 6$. So, for divergence-free drifts $b \in L_2(\Omega)$ we have the following picture. Assume $u \in W_2^1(\Omega)$ is a weak solution to (1.6) with $f \in L_p(\Omega)$, $p > n/2$. Theorem 3.4 means that

$$\varphi \in L_\infty(\partial\Omega) \cap W_2^{1/2}(\partial\Omega) \qquad \Longrightarrow \qquad u \in L_\infty(\Omega) \quad \text{for any} \ \ n \ge 2.$$

The Example 3.6 shows that for general φ we have

$$\varphi \in W_2^{1/2}(\partial\Omega) \quad \Longrightarrow \quad \begin{cases} \text{if } n \le 3 \text{ then } u \in L_{\infty,loc}(\Omega), \\ \text{if } n \ge 6 \text{ then it is possible } u \notin L_{\infty,loc}(\Omega), \\ \text{if } n = 4, 5 \text{ -- open questions.} \end{cases}$$

Theorem 3.4 and Example 3.6 together establish an interesting phenomenon: for drifts $b \in L_2(\Omega)$, $\operatorname{div} b = 0$, the property of the elliptic operator in (1.1) to improve the "regularity" of weak solutions (in the sense that every weak solution is locally bounded) depends on the behavior of a weak solution on the boundary of the domain. If the values of $\varphi := u|_{\partial\Omega}$ on the boundary are bounded then this weak solution must be bounded as Theorem 3.4 says. On the other hand, if the function φ is unbounded on $\partial\Omega$ then the weak solution can be unbounded even near internal points of the domain Ω as Example 3.6 shows. To our opinion such a behavior of solutions to an elliptic equation is unexpected. Allowing some abuse of language we can say that *non-regularity of the drift can destroy the hypoellipticity of the operator.*

Theorem 3.4 impose some restrictions on the structure of the set of singular points of weak solutions. Namely, let us define a *singular point* of a weak solution as a point for which the weak solution is unbounded in any its neighborhood, and then define the *singular set* of a weak solution as the set of all its singular points. It is clear that the singular set is closed. Theorem 3.4 shows that if for some weak solution its singular set is non-empty then its 1-dimensional Hausdorff measure must be positive.

THEOREM 3.7. *Let $b \in L_2(\Omega)$, $\operatorname{div} b = 0$, and let $u \in W_2^1(\Omega)$ be a weak solution to (1.1) with $f \in L_p(\Omega)$, $p > n/2$. Denote by $\Sigma \subset \bar{\Omega}$ the singular set of u and assume $\Sigma \cap \Omega \neq \emptyset$. Then any point of the set $\Sigma \cap \Omega$ never can be surrounded by any smooth closed $(n-1)$-dimensional surface $S \subset \bar{\Omega}$ such that $u|_S \in L_\infty(S)$. In particular, this means that*

$$\mathcal{H}^1(\Sigma) > 0, \quad \Sigma \cap \partial\Omega \neq \emptyset, \tag{3.5}$$

where $\mathcal{H}^1$ is one-dimensional Hausdorff measure in $\mathbb{R}^n$.

PROOF. The first assertion is clear. Let us prove (3.5). Assume $\Sigma \cap \Omega \neq \emptyset$ and $x_0 \in \Sigma \cap \Omega$. Denote $d := \operatorname{dist}\{x_0, \partial\Omega\}$. Let $z_0 \in \partial\Omega$ be a point such that $|z_0 - x_0| = d$ and denote by $[x_0, z_0]$ the straight line segment connecting x_0 with z_0. Let us take arbitrary $\delta > 0$ and consider any countable covering of Σ by open balls $\{B_{\rho_i}(y_i)\}$ such that $\rho_i \leq \delta$. For any i denote $r_i := |x_0 - y_i|$. If $r_i \leq d$ then denote $z_i := [x_0, z_0] \cap \partial B_{r_i}(x_0)$. By Theorem 3.4 for any $r \leq d$ we have $\Sigma \cap \partial B_r(x_0) \neq \emptyset$. Therefore,

$$[x_0, z_0] \subset \bigcup_{r_i \leq d} B_{\rho_i}(z_i).$$

This inclusion means that

$$\mathcal{H}^1(\Sigma) \geq \mathcal{H}^1([x_0, z_0]) = d > 0.$$

□

Theorem 3.7 in particular implies that no isolated singularity is possible. This exactly what Example 3.6 demonstrates: the singular set in this case is the axis of symmetry.

Note that the divergence free condition brings significant improvements into the local boundedness results. Without the condition $\operatorname{div} b = 0$ one can prove local boundedness of weak solutions to (1.1) only for $b \in L_n(\Omega)$ $(n \geq 3)$, while if $\operatorname{div} b = 0$ the local boundedness is valid for any $b \in L_p(\Omega)$ with $p > \frac{n}{2}$. Note also that for the moment of writing of this paper we can say nothing about analogues of neither Theorem 3.5 nor Example 3.6 if $p \in [\frac{n-1}{2}, \frac{n}{2}]$. *We state this problem as an open question.*

The final issue we need to discuss is the problem of further regularity of solutions to the equation (1.1). The example of a bounded weak solution which is not locally continuous was constructed originally in [**SSSZ**] for $n = 3$ and $b \in L_1(\Omega)$, $\operatorname{div} b = 0$ (actually the method of [**SSSZ**] allowed to extend their example for $b \in L_p$, $p \in [1, 2)$). Later the first author in [**F**] generalized this example for all $n \geq 3$ and for all $p \in [1, n)$.

THEOREM 3.8. *Assume $n \geq 3$, $p < n$. Then there exist $b \in L_p(B)$ satisfying $\operatorname{div} b = 0$ and a weak solution u to (1.1) with $f \equiv 0$ such that $u \in W_2^1(B) \cap L_\infty(B)$ but $u \notin C(\bar{B}_{1/2})$.*

The latter result shows that if one is interested in the local continuity of weak solutions then the assumption $b \in L_n(\Omega)$ can not be weakened in the Lebesgue scale and the structure condition $\operatorname{div} b = 0$ does not help in this situation.

It is not difficult to construct also a weak solution to (1.1) which is continuous but not Hölder continuous.

EXAMPLE 3.9. Assume $n \geq 4$ and take

$$u(x) = \frac{1}{\ln r}, \quad b = \left(\frac{n-3}{r} - \frac{2}{r \ln r} \right) \mathbf{e}_r + \left(\frac{(n-3)^2}{r^2} - \frac{2(n-3)}{r^2 \ln r} - \frac{2}{r^2 \ln^2 r} \right) z\, \mathbf{e}_z.$$

Here $r^2 = x_1^2 + \ldots + x_{n-1}^2$, $z = x_n$, and $\mathbf{e}_r$, $\mathbf{e}_z$ are the basis vectors of the cylindrical coordinate system. Then $u \in W_2^1(B_{1/2}) \cap C(B_{1/2})$, $-\Delta u + b \cdot \nabla u = 0$, $\operatorname{div} b = 0$ in $B_{1/2}$ and $b \in L_p(B_{1/2})$ for any $p < \frac{n-1}{2}$.

Thus, for weak solutions of (1.1) with $b \in L_2(\Omega)$, $\operatorname{div} b = 0$, in large space dimensions (at least for $n \geq 6$) the following sequence of implications can break at any step:

$$u \in W_2^1(\Omega) \quad \not\Longrightarrow \quad u \in L_{\infty,loc}(\Omega) \quad \not\Longrightarrow \quad u \in C_{loc}(\Omega) \quad \not\Longrightarrow \quad u \in C_{loc}^\alpha(\Omega).$$

4. Beyond the L_p–scale

Theorem 3.8 shows that in order to obtain the local continuity of weak solutions to (1.1) for drifts weaker than $b \in L_n(\Omega)$ one needs to go beyond the Lebesgue scale.

We start with the question of the boundedness of the operator T defined by the formula (2.2). The necessary and sufficient condition on b is obtained in [**MV**] in the case $\Omega = \mathbb{R}^n$.

THEOREM 4.1. *The inequality (2.3) holds true if and only if the drift b can be represented as a sum $b = b_0 + b_1$, where the function b_0 is such that*

$$\int\limits_{\mathbb{R}^n} |b_0|^2 |\eta|^2 \, dx \; \leq \; C \int\limits_{\mathbb{R}^n} |\nabla \eta|^2 \, dx, \qquad \forall \, \eta \in C_0^\infty(\mathbb{R}^n), \tag{4.1}$$

b_1 is divergence-free, $\operatorname{div} b_1 = 0$, and $b_1 \in BMO^{-1}(\mathbb{R}^n)$. It means that $b_1(x) = \operatorname{div} A(x)$, $A(x)$ is a skew-symmetric matrix, $A_{ij} = -A_{ji}$, and $A_{ij} \in BMO(\mathbb{R}^n)$.

Here $BMO(\Omega)$ is the space of functions f with *bounded mean oscillation*, i.e.

$$\sup_{\substack{x \in \Omega \\ 0 < r < \infty}} \frac{1}{r^n} \int\limits_{B_r(x) \cap \Omega} |f(y) - (f)_{B_r(x) \cap \Omega}| \, dy < \infty, \quad \text{where} \quad (f)_\omega = \frac{1}{|\omega|} \int\limits_\omega f(y) \, dy.$$

Clearly, each divergence-free vector b_1 can be represented as $b_1 = \operatorname{div} A$ with a skew-symmetric matrix $A(x)$.

This Theorem mentions that the behaviour of the bilinear form $\int\limits_\Omega \nabla u \cdot b\, \eta\, dx$ already distinguish between general drifts and divergence-free drifts. First, let us discuss general drifts. If b satisfies (2.1) then it satisfies the estimate (4.1) too. But we can not use the condition (4.1) instead (2.1) for the regularity theory, as the example of Remark 2.10 shows. Indeed, for functions satisfying

$$|b(x)| \leq \frac{C}{|x|} \tag{4.2}$$

the estimate (4.1) is fulfilled by the Hardy inequality.

On the other hand, the case of the drift b having a one-point singularity (say, at the origin) with the asymptotics which includes homogeneous of degree -1 functions like (4.2), is also interesting. There are several papers, see [**LZ**], [**Sem**], [**SSSZ**] and [**NU**], dealing with different classes of divergence-free drifts which cover (4.2). All these papers contain also the results for parabolic equation (1.8), but we discuss only (simplified) elliptic versions of them. We address the interested readers to the original papers.

The approach of [**SSSZ**] seems to be the most general one. Assume $b \in BMO^{-1}(\Omega)$ and $\operatorname{div} b = 0$. In this case we understand the equation $-\Delta u + b\cdot\nabla u = 0$ in the sense of the integral identity

$$\int_\Omega (\nabla u \cdot \nabla \eta + A\nabla u \cdot \nabla \eta)\, dx = 0 \qquad \forall\, \eta \in C_0^\infty(\Omega), \tag{4.3}$$

where the skew-symmetric matrix $A \in BMO(\Omega)$ is defined via $\operatorname{div} A(x) = b(x)$.

THEOREM 4.2. *Let* $b \in BMO^{-1}(\Omega)$ *and* $\operatorname{div} b = 0$. *Then*

1) *The maximum principle holds. If* $u \in W_2^1(\Omega)$ *satisfies* (4.3) *and* $\varphi := u|_{\partial\Omega}$ *is bounded, then* $\|u\|_{L_\infty(\Omega)} \leq \|\varphi\|_{L_\infty(\partial\Omega)}$. *In particular, the weak solution to* (1.6) *is unique.*

2) *Any weak solution* u *to* (1.1) *is Hölder continuous,* $u \in C^\alpha_{loc}(\Omega)$ *for some* $\alpha > 0$.

For the proof see [**NU**] or [**SSSZ**]. The regularity theory developped in Section 2 is slightly better as it guarantees that weak solutions are locally Hölder continuous with *any* exponent $\alpha < 1$. Nevertheless, Theorem 4.2 means that divergence-free drifts from BMO^{-1} can be also considered as regular ones.

REMARK 4.3. Note that the case $n = 2$, $b \in L_2(\Omega)$, $\operatorname{div} b = 0$, is the particular case of this situation. Indeed, such drifts can be represented as a vector-function with components $b_1 = \partial_2 h$, $b_2 = -\partial_1 h$, where h is a scalar function $h \in W_2^1(\Omega)$. By the imbedding theorem $W_2^1(\Omega) \subset BMO(\Omega)$ we have

$$A(x) = \begin{pmatrix} 0 & -h(x) \\ h(x) & 0 \end{pmatrix} \in BMO(\Omega).$$

5. Appendix

First we prove Theorem 2.9.

PROOF. We present the proof in the case $n \geq 3$ only. The case $n = 2$ differs from it by routine technical details.

1) The statement similar to our estimate (2.5) (for more general equations) can be found in [**St**]. In particular, in [**St**, Theorem 4.2] the following estimate for weak solutions to the problem

$$\left\{ \begin{array}{c} -\Delta u + b\cdot\nabla u = f \quad \text{in} \quad \Omega, \\ u|_{\partial\Omega} = 0, \end{array} \right. \tag{5.1}$$

was proved:

$$\|u\|_{L_\infty(\Omega)} \ \le \ C \ \Big(\ \|f\|_{L_p(\Omega)} + \|u\|_{L_2(\Omega)} \ \Big). \tag{5.2}$$

On the other hand,

$$\|u\|_{W^1_2(\Omega)} \ \le \ C \ \|f\|_{W^{-1}_2(\Omega)} \tag{5.3}$$

due to Theorem 2.8. Hence we can exclude the weak norm of u from the right hand side of (5.2) and obtain the estimate (2.5) in the case $\varphi\equiv 0$. In general case we can split a weak solution u of the problem (1.6) as $u = u_1 + u_2$, where u_1 is a weak solution of (5.1) and u_2 is a weak solution to the problem (1.6) with the boundary data φ and zero right hand side. For u_1 we have (5.3) and for u_2 we have $\|u_2\|_{L_\infty(\Omega)} \ \le \ \|\varphi\|_{L_\infty(\partial\Omega)}$ by Theorem 2.6.

2) As $b\in L_n(\Omega)$ we can complete the integral identity (1.2) up to the test functions $\eta\in \overset{\circ}{W}{}^1_2(\Omega)$. Denote $k_0 := \|\varphi\|_{L_\infty(\partial\Omega)}$ and assume $k\ge k_0$. Take in (1.2) $\eta = (u-k)_+$, where we denote $(u)_+ := \max\{u, 0\}$. As $k\ge k_0$ we have $\eta\in \overset{\circ}{W}{}^1_2(\Omega)$ and $\nabla\eta = \chi_{A_k}\nabla u$ where χ_{A_k} is the characteristic function of the set

$$A_k \ := \ \{ \ x\in\Omega : \ u(x) > k \ \}.$$

We obtain the identity

$$\int\limits_{A_k} |\nabla u|^2 \ dx \ + \ \int\limits_{A_k} b\cdot(u-k)\nabla u \ dx \ = \ \int\limits_{A_k} f(u-k) \ dx.$$

The second term vanishes

$$\int\limits_{A_k} b\cdot(u-k)\nabla u \ dx \ = \ \frac12 \int\limits_{\Omega} b\cdot\nabla|(u-k)_+|^2 \ dx \ = \ 0,$$

as $\operatorname{div} b = 0$, and hence

$$\int\limits_{A_k} |\nabla u|^2 \ dx \ = \ \int\limits_{A_k} f(u-k) \ dx, \qquad \forall\ k\ge k_0.$$

The rest of the proof goes as in the usual elliptic theory. Applying the imbedding theorem we obtain

$$\Big(\int\limits_{A_k} |\nabla u|^2 \ dx \Big)^{\frac12} \ \le \ C(n) \ \Big(\int\limits_{A_k} |f|^{\frac{2n}{n+2}} \ dx \Big)^{\frac{n+2}{2n}},$$

and using the Hölder inequality we get

$$\|f\|_{L_{\frac{2n}{n+2}}(A_k)} \ \le \ |A_k|^{\frac{n+2}{2n}-\frac1p} \ \|f\|_{L_p(A_k)}.$$

So we arrive at

$$\int\limits_{A_k} |\nabla u|^2 \, dx \;\le\; C(n)\, \|f\|^2_{L_p(\Omega)}\, |A_k|^{1-\frac{2}{n}+\varepsilon}, \qquad \forall\, k \ge k_0,$$

where $\varepsilon := 2\left(\frac{2}{n} - \frac{1}{p}\right) > 0$. This inequality yields the following estimate, see [**LU**, Chapter II, Lemma 5.3],

$$\operatorname{esssup}_\Omega (u - k_0)_+ \;\le\; C(n, p, \Omega)\, \|f\|_{L_p(\Omega)}.$$

The estimate of $\operatorname{essinf}_\Omega u$ can be obtained in a similar way if we replace u by $-u$. □

In order to prove Theorem 3.1 we need some auxiliary results.

THEOREM 5.1. *Assume $n \ge 3$, $b \in C^\infty(\bar\Omega)$, $\operatorname{div} b = 0$ in Ω, $f \in L_1(\Omega)$, and assume $u \in \overset{\circ}{W}{}^1_2(\Omega)$ is a weak solution of (1.6) with $\varphi \equiv 0$. Then for any $q \in \left[1, \frac{n}{n-2}\right)$ the following estimate holds:*

$$\|u\|_{L_q(\Omega)} \;\le\; C(n, q, \Omega)\, \|f\|_{L_1(\Omega)}. \tag{5.4}$$

PROOF. Assume $q \in \left(1, \frac{n}{n-2}\right)$. By duality we have

$$\|u\|_{L_q(\Omega)} \;=\; \sup_{g \in L_{q'}(\Omega),\ \|g\|_{L_{q'}(\Omega)} \le 1} \ \int\limits_\Omega ug\, dx,$$

where $q' := \frac{q}{q-1}$, $q' > \frac{n}{2}$. For any $g \in L_{q'}(\Omega)$ denote by $w_g \in W^2_{q'}(\Omega)$ a solution to the problem

$$\left\{ \begin{aligned} -\Delta w_g - b\cdot \nabla w_g \;&=\; g \qquad \text{in} \quad \Omega, \\ w_g|_{\partial\Omega} \;&=\; 0. \end{aligned} \right.$$

From Theorem 2.9 we conclude that for w_g the following estimate holds:

$$\|w_g\|_{L_\infty(\Omega)} \;\le\; C(n, q, \Omega)\, \|g\|_{L_{q'}(\Omega)}.$$

Integrating by parts we obtain

$$\int\limits_\Omega ug\, dx \;=\; \int\limits_\Omega u(-\Delta w_g - b\cdot\nabla w_g)\, dx \;=\; \int\limits_\Omega \nabla u \cdot (\nabla w_g + b w_g)\, dx \;=\; \int\limits_\Omega f w_g\, dx.$$

Then for any $g \in L_{q'}(\Omega)$ such that $\|g\|_{L_{q'}(\Omega)} \le 1$ we get

$$\int\limits_\Omega ug\, dx \;=\; \int\limits_\Omega f w_g\, dx \;\le\; \|f\|_{L_1(\Omega)} \|w_g\|_{L_\infty(\Omega)} \;\le\; C(n, q, \Omega)\, \|f\|_{L_1(\Omega)}.$$

Hence we obtain (5.4). □

Another auxiliary result we need is the following extension theorem.

THEOREM 5.2. *Assume $\Omega \subset \mathbb{R}^n$ is a bounded domain of class C^1. Then there exists a bounded linear extension operator $T : L_\infty(\partial\Omega) \cap W^{1/2}_2(\partial\Omega) \to L_\infty(\Omega) \cap W^1_2(\Omega)$ such that*

$$T\varphi|_{\partial\Omega} \;=\; \varphi, \qquad \forall\, \varphi \in L_\infty(\partial\Omega) \cap W^{1/2}_2(\partial\Omega),$$

$$\|T\varphi\|_{W^1_2(\Omega)} \;\le\; C(\Omega)\, \|\varphi\|_{W^{1/2}_2(\partial\Omega)}, \qquad \|T\varphi\|_{L_\infty(\Omega)} \;\le\; C(\Omega)\, \|\varphi\|_{L_\infty(\partial\Omega)}.$$

PROOF. For the sake of completeness we briefly recall the proof of Theorem 5.2. After the localization and flattening of the boundary it is sufficient to construct the extension operator from $\mathbb{R}^{n-1}$ to $\mathbb{R}^n_+ := \mathbb{R}^{n-1}\times(0,+\infty)$. Then we can take the standard operator

$$(T\varphi)(x',x_n) \ = \ \eta(x_n)\int\limits_{\mathbb{R}^{n-1}}\varphi(x'-x_n\xi')\psi(\xi')\,d\xi', \qquad (x',x_n)\in\mathbb{R}^n_+,$$

where $x' := (x_1,\dots,x_{n-1}) \in \mathbb{R}^{n-1}$, $\eta\in C_0^\infty(\mathbb{R})$, $\eta(0)=1$, $\psi\in C_0^\infty(\mathbb{R}^{n-1})$, $\int\limits_{\mathbb{R}^{n-1}}\psi(\xi')\,d\xi' = 1$. This operator is bounded from $W_2^{1/2}(\mathbb{R}^{n-1})$ to $W_2^1(\mathbb{R}^n_+)$ and also from $L_\infty(\mathbb{R}^{n-1})$ to $L_\infty(\mathbb{R}^n_+)$. More details can be found in [**BIN**]. □

Now we can give an elementary proof of Theorem 3.1.

PROOF. The function $v_k := u_k - u \in \overset{\circ}{W}{}^1_2(\Omega)$ is a weak solution to the problem

$$\left\{\begin{aligned} -\Delta v_k + b_k\cdot\nabla v_k &= f_k \quad \text{in} \quad \Omega,\\ v_k|_{\partial\Omega} &= 0,\end{aligned}\right.$$

where

$$f_k := (b-b_k)\cdot\nabla u, \qquad f_k\in L_1(\Omega), \qquad \|f_k\|_{L_1(\Omega)} \ \to\ 0.$$

Assume $q\in\left[1,\frac{n}{n-2}\right)$. By Theorem 5.1 we have

$$\|v_k\|_{L_q(\Omega)} \ \le\ C(n,\Omega)\ \|f_k\|_{L_1(\Omega)} \ \to\ 0,$$

and hence (3.2) follows.

Now assume additionally $\varphi\in L_\infty(\partial\Omega)$. Denote $\tilde\varphi := T\varphi$ where T is the extension operator from Theorem 5.2. Taking in the integral identity (1.4) for u_k and b_k the test function $\eta = u_k - \tilde\varphi \in \overset{\circ}{W}{}^1_2(\Omega)$ we obtain

$$\int\limits_\Omega |\nabla u_k|^2\,dx \ - \ \int\limits_\Omega u_kb_k\cdot\nabla(u_k-\tilde\varphi)\,dx \ = \ \int\limits_\Omega \nabla u_k\cdot\nabla\tilde\varphi\,dx \ + \ \langle f, u_k-\tilde\varphi\rangle.$$

Using the condition $\operatorname{div} b_k = 0$ we get

$$\int\limits_\Omega u_kb_k\cdot\nabla(u_k-\tilde\varphi)\,dx \ = \ \int\limits_\Omega \tilde\varphi b_k\cdot\nabla(u_k-\tilde\varphi)\,dx.$$

Therefore,

$$\begin{aligned}\|\nabla u_k\|^2_{L_2(\Omega)} \ \le\ \Big(\|\tilde\varphi\|_{L_\infty(\Omega)}\|b_k\|_{L_2(\Omega)} \ + \ \|f\|_{W_2^{-1}(\Omega)}\Big)\Big(\|u_k\|_{W_2^1(\Omega)} + \|\tilde\varphi\|_{W_2^1(\Omega)}\Big) \ + \\ + \ \|\nabla u_k\|_{L_2(\Omega)}\|\nabla\tilde\varphi\|_{L_2(\Omega)}.\end{aligned}$$

Applying Friedrichs' and Young's inequalities we obtain the estimate

$$\|u_k\|_{W_2^1(\Omega)} \ \le\ C, \tag{5.5}$$

with a constant C independent on k. As the convergence (3.2) is already established, from (5.5) we derive (3.3). □

Now we turn to the proof of Theorem 3.3.

PROOF. We take a sequence $b_k \in C^\infty(\bar\Omega)$, $\operatorname{div} b_k = 0$, such that $b_k \to b$ in $L_2(\Omega)$. Let $u_k \in W^1_2(\Omega)$ be a weak solution to the problem (3.1). Repeating the arguments in the proof of Theorem 3.1, we obtain the estimate (5.5) with a constant C independent on k. Using this estimate we can extract a subsequence satisfying (3.3) for some $u \in W^1_2(\Omega)$. The weak convergence (3.3) and the strong convergence $b_k \to b$ in $L_2(\Omega)$ allow us to pass to the limit in the integral identities (1.2) corresponding to u_k and b_k. Therefore, u is a weak solution to (1.6). □

Now we present the proof of Theorem 3.4.

PROOF. Let b_k be smooth divergence-free vector fields such that $b_k \to b$ in $L_2(\Omega)$. Denote by u_k the weak solution to the problem (3.1). By Theorem 2.9

$$\|u_k\|_{L_\infty(\Omega)} \ \le\ \|\varphi\|_{L_\infty(\partial\Omega)} + C\ \|f\|_{L_p(\Omega)} \tag{5.6}$$

with the constant C depending only on n, p and Ω. From Theorem 3.1 we have the convergence $u_k \to u$ in $L_1(\Omega)$ and hence we can extract a subsequence (for which we keep the same notation) such that

$$u_k \to u \quad \text{a.e. in} \quad \Omega.$$

Passing to the limit in (5.6) we obtain (3.4). □

Finally we give the proof of Theorem 3.5.

PROOF. To simplify the presentation we give the proof only in the case $f \equiv 0$. The extension of the result for non-zero right hand side can be done by standard methods, see [**HL**, Theorem 4.1] or [**FSh**]. First we derive the estimate

$$\|u\|_{L_\infty(B_{1/2})} \ \le\ C\left(1 + \|b\|_{L_p(B)}\right)^\mu\ \|u\|_{L_{2p'}(B)}, \qquad p' := \frac{p}{p-1} \tag{5.7}$$

(with some positive constants C and μ depending only on n and p) under additional assumption $u \in C^\infty(B)$. We explore Moser's iteration technique, see [**Mo**]. Assume $\beta \ge 0$ is arbitrary and let $\zeta \in C_0^\infty(B)$ be a cut-off function. Take a test function $\eta = \zeta^2|u|^\beta u$ in the identity (1.2). Denote $w := |u|^{\frac{\beta+2}{2}}$. Then after integration by parts and some routine calculations we obtain the inequality

$$\int\limits_B |\nabla(\zeta w)|^2\, dx \ \le\ C \int\limits_B |w|^2 \left(|\nabla\zeta|^2 + |b|\ |\nabla\zeta|\right) dx \tag{5.8}$$

Applying the imbedding theorem and the Hölder inequality and choosing the test function ζ in an appropriate way we arrive at the inequality

$$\|w\|_{L_{\frac{2n}{n-2}}(B_r)} \ \le\ C\left(\frac{1}{R-r} \ +\ \|b\|_{L_p(B_R)}\right)\ \|w\|_{L_{2p'}(B_R)},$$

which holds for any $\frac12 \le r < R \le 1$. Note that $\frac{2n}{n-2} > 2p'$ as $p > \frac{n}{2}$ if $n \ge 4$ and $p = 2$ if $n = 3$. The latter inequality gives us the estimate

$$\|u\|_{L_{\frac{n\gamma}{n-2}}(B_r)} \ \le\ C^{\frac{2}{\gamma}}\left(\frac{1}{R-r} + \|b\|_{L_p(B_R)}\right)^{\frac{2}{\gamma}}\ \|u\|_{L_{p'\gamma}(B_R)} \tag{5.9}$$

with an arbitrary $\gamma \ge 2$, $\gamma := \beta + 2$. Denote $s_0 = 2p'$, $s_m := \chi s_{m-1}$, where $\chi := \frac{n(p-1)}{p(n-2)}$, and denote also $R_m = \frac12 + \frac{1}{2^{m+1}}$. Taking in (5.9) $r = R_m$, $R = R_{m-1}$,

$\gamma = \frac{s_{m-1}}{p'}$ we obtain

$$\|u\|_{L_{s_m}(B_{R_m})} \leq \left(C\, 2^{m+1} + C\|b\|_{L_p(B)} \right)^{\frac{1}{\chi^{m-1}}} \|u\|_{L_{s_{m-1}}(B_{R_{m-1}})}$$

Iterating this inequality we arrive at (5.7).

Now we need to get rid of the assumption $u \in C^\infty(B)$. Assume $u \in W_2^1(B)$ is an arbitrary weak solution to (1.1). Let $\zeta \in C_0^\infty(B)$ be a cut-off function such that $\zeta \equiv 1$ on $B_{5/6}$ and denote $v := \zeta u$. Then v is a weak solution to the boundary value problem

$$\left\{ \begin{array}{c} -\Delta v + b \cdot \nabla v = g \quad \text{in} \quad B \\ v|_{\partial B} = 0 \end{array} \right.$$

where

$$g := -u\Delta\zeta - 2\nabla u \cdot \nabla\zeta + bu \cdot \nabla\zeta.$$

Note that $g \equiv 0$ and $v \equiv u$ on $B_{5/6}$. As $b \in L_p(B)$ with $p > \frac{n}{2}$ we have $g \in W_2^{-1}(B)$. Now we take a sequence $b_k \in C^\infty(\bar{B})$, $\operatorname{div} b_k = 0$, such that $b_k \to b$ in $L_p(B)$ and let v_k be the weak solution to the problem

$$\left\{ \begin{array}{c} -\Delta v_k + b_k \cdot \nabla v_k = g \quad \text{in} \quad B \\ v_k|_{\partial B} = 0 \end{array} \right.$$

From Theorem 3.1 we have $v_k \rightharpoonup v$ in $W_2^1(B)$ and as $p > \frac{n}{2}$ we can extract a subsequence (for which we keep the same notation) such that $v_k \to v$ a.e. in B and $v_k \to v$ in $L_{2p'}(B)$. As $g \equiv 0$ on $B_{5/6}$ from the usual elliptic theory (see [**LU**]) we conclude that $v_k \in C^\infty(B_{5/6})$. Applying (5.7) (with the obvious modification in radius) we obtain the estimate

$$\|v_k\|_{L_\infty(B_{1/2})} \leq C \left(1 + \|b_k\|_{L_p(B)}\right)^\mu \|v_k\|_{L_{2p'}(B_{3/4})}.$$

Hence v_k are equibounded on $B_{1/2}$. Passing to the limit in the above inequality and taking into account that $v = u$ on $B_{5/6}$ we obtain

$$\|u\|_{L_\infty(B_{1/2})} \leq C \left(1 + \|b\|_{L_p(B)}\right)^\mu \|u\|_{L_{2p'}(B_{3/4})}.$$

To conclude the proof we remark that for $p > \frac{n}{2}$ from the imbedding theorem we have

$$\|u\|_{L_{2p'}(B)} \leq C(n,p)\, \|u\|_{W_2^1(B)}.$$

□

References

[A] D. G. Aronson, *Non-negative solutions of linear parabolic equations*, Ann. Scuola Norm. Sup. Pisa (3) **22** (1968), 607–694. MR0435594

[BIN] O. V. Besov, V. P. Il'in, S. M. Nikol'skii, *Integral representations of functions and imbedding theorems*, Izdat. "Nauka", Moscow, 1975; English translation, Scripta Series in Mathematics, V. H. Winston & Sons, Washington, D.C.; Halsted Press [John Wiley & Sons], New York-Toronto, Ont.-London, 1979.

[F] N. Filonov, *On the regularity of solutions to the equation* $-\Delta u + b \cdot \nabla u = 0$ (English, with English and Russian summaries), Zap. Nauchn. Sem. S.-Peterburg. Otdel. Mat. Inst. Steklov. (POMI) **410** (2013), no. Kraevye Zadachi Matematicheskoĭ Fiziki i Smezhnye Voprosy Teorii Funktsiĭ. 43, 168–186, 189; English transl., J. Math. Sci. (N.Y.) **195** (2013), no. 1, 98–108. MR3048265

[FSh] N. Filonov, T. Shilkin, *On the local boundedness of weak solutions to elliptic equations with divergence-free drifts*, Preprint 2714, Technische Universität Darmstadt, 2017.

[HL] Qing Han and Fanghua Lin, *Elliptic partial differential equations*, Courant Lecture Notes in Mathematics, vol. 1, New York University, Courant Institute of Mathematical Sciences, New York; American Mathematical Society, Providence, RI, 1997. MR1669352

[LU] O. A. Ladyzhenskaya, N. N. Uraltseva, *Linear and quasilinear equations of elliptic type*, Academic Press, 1968.

[LZ] Vitali Liskevich and Qi S. Zhang, *Extra regularity for parabolic equations with drift terms*, Manuscripta Math. **113** (2004), no. 2, 191–209. MR2128546

[MV] V. G. Maz′ya and I. E. Verbitsky, *Form boundedness of the general second-order differential operator*, Comm. Pure Appl. Math. **59** (2006), no. 9, 1286–1329. MR2237288

[Mo] Jürgen Moser, *A new proof of De Giorgi's theorem concerning the regularity problem for elliptic differential equations*, Comm. Pure Appl. Math. **13** (1960), 457–468. MR0170091

[NU] A. I. Nazarov and N. N. Ural′tseva, *The Harnack inequality and related properties of solutions of elliptic and parabolic equations with divergence-free lower-order coefficients* (Russian, with Russian summary), Algebra i Analiz **23** (2011), no. 1, 136–168; English transl., St. Petersburg Math. J. **23** (2012), no. 1, 93–115. MR2760150

[Sem] Yu. A. Semenov, *Regularity theorems for parabolic equations*, J. Funct. Anal. **231** (2006), no. 2, 375–417. MR2195337

[SSSZ] Gregory Seregin, Luis Silvestre, Vladimír Šverák, and Andrej Zlatoš, *On divergence-free drifts*, J. Differential Equations **252** (2012), no. 1, 505–540. MR2852216

[SV] Luis Silvestre and Vlad Vicol, *Hölder continuity for a drift-diffusion equation with pressure*, Ann. Inst. H. Poincaré Anal. Non Linéaire **29** (2012), no. 4, 637–652. MR2948291

[SVZ] Luis Silvestre, Vlad Vicol, and Andrej Zlatoš, *On the loss of continuity for supercritical drift-diffusion equations*, Arch. Ration. Mech. Anal. **207** (2013), no. 3, 845–877. MR3017289

[St] Guido Stampacchia, *Le problème de Dirichlet pour les équations elliptiques du second ordre à coefficients discontinus* (French), Ann. Inst. Fourier (Grenoble) **15** (1965), no. fasc. 1, 189–258. MR0192177

[Su] M. D. Surnachev, *On the uniqueness of a solution to a stationary convection-diffusion equation with a generalized divergence-free drift*, Complex Variables and Elliptic Equations, 2017.

[Z] Qi S. Zhang, *A strong regularity result for parabolic equations*, Comm. Math. Phys. **244** (2004), no. 2, 245–260. MR2031029

[Zhi] V. V. Zhikov, *Remarks on the uniqueness of the solution of the Dirichlet problem for a second-order elliptic equation with lower order terms* (Russian, with Russian summary), Funktsional. Anal. i Prilozhen. **38** (2004), no. 3, 15–28; English transl., Funct. Anal. Appl. **38** (2004), no. 3, 173–183. MR2095131

V.A. Steklov Mathematical Institute, St.-Petersburg, Fontanka 27, 191023, Russia
Email address: `filonov@pdmi.ras.ru`

V.A. Steklov Mathematical Institute, St.-Petersburg, Fontanka 27, 191023, Russia
Email address: `shilkin@pdmi.ras.ru`

Contemporary Mathematics
Volume **710**, 2018
http://dx.doi.org/10.1090/conm/710/14367

Time-periodic solutions to the Navier-Stokes equations in the three-dimensional whole-space with a non-zero drift term: Asymptotic profile at spatial infinity

Giovanni P. Galdi and Mads Kyed

ABSTRACT. An asymptotic expansion at spatial infinity of a weak time-periodic solution to the Navier-Stokes equations with a non-zero drift term in the three-dimensional whole-space is carried out. The asymptotic profile is explicitly identified and expressed in terms of the well-known Oseen fundamental solution. A pointwise estimate is given for the remainder term.

1. Introduction

We investigate the asymptotic structure at spatial infinity of a time-periodic solution to a Navier-Stokes problem in the three-dimensional whole-space. More specifically, we consider a solution $u : \mathbb{R} \times \mathbb{R}^3 \to \mathbb{R}^3$ to the Navier-Stokes equations

$$
\begin{cases}
\partial_t u - \Delta u - \lambda \partial_{x_1} u + \nabla \mathfrak{p} + u \cdot \nabla u = f & \text{in } \mathbb{R} \times \mathbb{R}^3, \\
\operatorname{div} u = 0 & \text{in } \mathbb{R} \times \mathbb{R}^3, \\
\lim_{|x| \to \infty} u(t,x) = 0
\end{cases}
\tag{1.1}
$$

that is time-periodic with period $\mathcal{T} > 0$,

$$
\forall (t,x) \in \mathbb{R} \times \mathbb{R}^3 : \quad u(t,x) = u(t + \mathcal{T}, x),
\tag{1.2}
$$

and corresponds to time-periodic data f of the same period. In this context, $\mathbb{R} \times \mathbb{R}^3$ is a time-space domain. We denote by $t \in \mathbb{R}$ the time variable and by $x \in \mathbb{R}^3$ the spatial variable. The velocity field $u : \mathbb{R} \times \mathbb{R}^3 \to \mathbb{R}^3$ and scalar function $\mathfrak{p} : \mathbb{R} \times \mathbb{R}^3 \to \mathbb{R}$ represent the Eulerian velocity field and pressure term, respectively, of a fluid flow described by the Navier-Stokes equations (1.1). Physically, (1.1) models the flow of an incompressible, viscous, Newtonian fluid past an object that moves with velocity $\lambda \mathrm{e}_1 \in \mathbb{R}^3$. We shall consider the case $\lambda \neq 0$ corresponding to the case of an object moving with non-zero velocity. The motion of the object is then manifested in the so-called drift term $\lambda \partial_{x_1} u$ in (1.1).

2010 *Mathematics Subject Classification.* Primary 35Q30, 35B10, 35C20, 76D05.
Key words and phrases. Navier-Stokes, time-periodic, asymptotic expansion.
Partially supported by NSF-DMS grant 1614011.

The goal in the following is to establish an asymptotic expansion of a time-periodic solution $u(t,x)$ as $|x| \to \infty$, that is, a pointwise identity

$$u(t,x) = \mathcal{G}(t,x) + \mathscr{R}(t,x) \tag{1.3}$$

with $\mathcal{G}$ an explicitly known vector field, which depends on the data f, and $\mathscr{R}(t,x)$ a remainder term that decays faster than $\mathcal{G}(t,x)$ as $|x| \to \infty$. In this case, (1.3) is an asymptotic expansion of u at spatial infinity with *asymptotic profile* $\mathcal{G}(t,x)$.

Significant physical properties of the fluid flow described by a solution to the Navier-Stokes equations are related to the structure of the solution at spatial infinity. In particular, the wake region of a flow past a body is characterized in terms of the asymptotic profile. A celebrated result due to Finn [**9, 10**], Babenko [**1**] and Galdi [**12**] (see also [**8**]) states that an asymptotic expansion of a weak solution $v : \mathbb{R}^3 \to \mathbb{R}^3$ to the three-dimensional *steady-state* Navier-Stokes equations with a non-zero drift term satisfies

$$v(x) = \Gamma_{\mathrm{O}}^{\lambda}(x)\Big(\int_{\mathbb{R}^3} f(x)\,\mathrm{d}x\Big) + O\big(|x|^{-\frac{3}{2}+\varepsilon}\big) \quad \text{for all } \varepsilon > 0, \tag{1.4}$$

where $\Gamma_{\mathrm{O}}^{\lambda}$ denotes the well-known Oseen fundamental solution. Observe that a steady-state solution is, trivially, also time-periodic. In the following, we shall establish for any weak time-periodic solution to (1.1) the asymptotic expansion

$$u(t,x) = \Gamma_{\mathrm{O}}^{\lambda}(x)\Big(\frac{1}{\mathcal{T}}\int_0^{\mathcal{T}}\int_{\mathbb{R}^3} f(t,x)\,\mathrm{d}x\mathrm{d}t\Big) + O\big(|x|^{-\frac{3}{2}+\varepsilon}\big) \quad \text{for all } \varepsilon > 0, \tag{1.5}$$

provided the solution possesses a certain amount of integrability at the outset, but without any restriction on its "size". As a consequence of this result, virtually all the physical properties that can be established for steady states from (1.4) can now also be established for time-periodic fluid flows from (1.5). For example, it is well known (see [**13**, Remark X.8.1]) that (1.4) implies the existence of a wake region in the fluid flow; a property that renders the flow described by v reasonable from a physical point of view. By (1.5), the same is true for time-periodic flows.

Existence of the weak type of time-periodic solutions that we shall consider in the following (Definition 2.1) was shown in [**22**] without any restriction on the "size" of the data. Such a solution is therefore a natural starting point for further investigation. In order to establish (1.5), however, we shall need to assume that the solution possesses a certain amount of additional integrability; see (2.4). Condition (2.4) corresponds to a well-known regularity criteria for the corresponding Navier-Stokes initial-value problem.

Pointwise information such as (1.5) is typically derived from a fundamental solution to an appropriate linearization of the system of partial differential equations to which u is a solution. If we in (1.1) neglect the nonlinear term, we obtain the time-periodic Oseen system. In the following, we shall identify what can be viewed as a fundamental solution to the time-periodic Oseen system. For this purpose, we employ an idea going back to [**22**] and formulate (1.1) as a system of partial differential equations on the locally compact abelian group $G := \mathbb{R}/\mathcal{T}\mathbb{Z} \times \mathbb{R}^3$ and utilize the Fourier transform $\mathscr{F}_G$. We thereby obtain a fundamental solution for which pointwise estimates and integrability properties can be established.

Investigation of time-periodic solutions to Navier-Stokes equations in three-dimensional unbounded domains, which necessarily involves some analysis of their decay at spatial infinity, was initiated by Maremonti [**26–28**]. A representation formula for time-periodic solutions was given by Kozono and Nakao in [**20**], and later utilized by Yamazaki [**29**] to establish estimates in weak–L^3 spaces. Interestingly, the representation formula proposed by Kozono and Nakao can be derived with the help of Poisson's summation formula from the fundamental solution that we shall introduce in the following. Pointwise estimates of time-periodic solutions to the Navier-Stokes equations in a three-dimensional exterior domain were established by Galdi and Sohr [**11**].

An asymptotic expansion of a solution to the three-dimensional steady-state Navier-Stokes equations *without* a drift term ($\lambda = 0$) is known only under restrictions on the magnitude of the data. In this case the asymptotic profile was identified by Korolev and Šverák [**19**] as the so-called Landau solution. It was later shown by Kang, Miura and Tsai [**18**] that a time-periodic solution satisfies a similar expansion. The asymptotic structure of time-periodic solutions in the case $\lambda \neq 0$ was first investigated in [**22**], where an asymptotic expansion similar to (1.5) was established, but only for "small" data and with an inferior decay estimate of the remainder term.

In addition to the classical steady-state Navier-Stokes equations, a substantial amount of research has also gone into the investigation of steady-state solutions to the three-dimensional Navier-Stokes equations in a rotating frame of reference, which models the fluid flow past a rotating body. Also in this case one has to differentiate between the equations with and without a drift term. In the former case ($\lambda \neq 0$) it was shown by Galdi and Kyed [**14**] that weak solutions possess a substantial amount of integrability, which in turn enabled Kyed [**25**] to identify the asymptotic profile. Similar results were obtained with different methods in a series of papers by Deuring, Kračmar and Nečasová; see [**4**] and the references therein. It is easy to verify that a steady-state solution to the Navier-Stokes equations in a rotating frame of reference is time-periodic in an inertial frame. Consequently, the expansion (1.5) furnishes an extension of some of the results established in [**4, 14, 25**]. An asymptotic expansion of steady-state solutions to the Navier-Stokes equations in a rotating frame of reference *without* a drift term ($\lambda = 0$) was obtained by Farwig and Hishida [**7**]; see also [**6**].

Estimates in L^p norms of solutions to both the steady-state and time-periodic Oseen system play an important role in the following. Indeed, we shall spend a lot of effort establishing time-periodic estimates—the steady-state estimates are well-known. We shall employ standard L^p spaces. In the steady-state case, estimates can also be obtained in L^p spaces with anisotropic weights; see [**5, 21**]. It might be possible to establish equivalent estimates for solutions to the time-periodic Oseen system in such spaces.

2. Statement of the main result

The main result in this article is an asymptotic expansion at spatial infinity of a weak solution u to (1.1) with an explicit identification of the asymptotic profile and a pointwise estimate of the remainder term. We shall assume slightly more integrability of the solution than a weak solution possesses at the outset, but the expansion is obtained without any restriction on the "size" of the data f.

In order to state the main theorem, we introduce some function spaces. We first recall the notation $C_{0,\sigma}^{\infty}(\mathbb{R}^3)$ for the space of smooth solenoidal vector fields of compact support, and the notation

$$D^{1,2}_{0,\sigma}(\mathbb{R}^3) := \overline{C^\infty_{0,\sigma}(\mathbb{R}^3)}^{\|\nabla\cdot\|_2} = \{u \in L^6(\mathbb{R}^3)^3 \mid \nabla u \in L^2(\mathbb{R}^3)^{3\times 3},\ \operatorname{div} u = 0\}$$

for the homogeneous Sobolev space of solenoidal vector fields with finite Dirichlet integral (the latter equality above is due to Sobolev embedding). Moreover, we introduce the following spaces

$$\begin{aligned} C^\infty_{\mathrm{per}}(\mathbb{R}\times\mathbb{R}^3) &:= \big\{w \in C^\infty(\mathbb{R}\times\mathbb{R}^3) \mid \forall t\in\mathbb{R}:\ w(t+\mathcal{T},\cdot) = w(t,\cdot)\big\},\\ C^\infty_{0,\mathrm{per}}(\mathbb{R}\times\mathbb{R}^3) &:= \big\{w \in C^\infty_{\mathrm{per}}(\mathbb{R}\times\mathbb{R}^3) \mid w \in C^\infty_0([0,\mathcal{T}]\times\mathbb{R}^3)\big\},\\ C^\infty_{0,\sigma,\mathrm{per}}(\mathbb{R}\times\mathbb{R}^3) &:= \big\{w \in C^\infty_{0,\mathrm{per}}(\mathbb{R}\times\mathbb{R}^3)^3 \mid \operatorname{div}_x w = 0\big\} \end{aligned}$$

of smooth $\mathcal{T}$-time-periodic functions. In addition, we define for sufficiently regular $\mathcal{T}$-time-periodic functions $u : \mathbb{R}\times\mathbb{R}^3 \to \mathbb{R}$ the operators

$$\mathcal{P}u(x) := \frac{1}{\mathcal{T}}\int_0^{\mathcal{T}} u(s,x)\,\mathrm{d}s \quad\text{and}\quad \mathcal{P}_\perp u(t,x) := u(t,x) - \mathcal{P}u(t,x). \tag{2.1}$$

Note that $\mathcal{P}$ and $\mathcal{P}_\perp$ decompose $u = \mathcal{P}u + \mathcal{P}_\perp u$ into a time-independent part $\mathcal{P}u$ and a time-periodic part $\mathcal{P}_\perp u$ with vanishing time-average over the period. Also note that $\mathcal{P}$ and $\mathcal{P}_\perp$ are complementary projections. We shall refer to $\mathcal{P}u$ as the *steady state* part of u, and to $\mathcal{P}_\perp u$ as the *purely oscillatory* part of u.

An asymptotic expansion at spatial infinity will be established for a class of time-periodic weak solutions to (1.1) that we call *physically reasonable.*

DEFINITION 2.1. *Assume* $f \in L^1_{loc}\big(\mathbb{R}\times\mathbb{R}^3\big)^3$ *satisfies* (1.2). *We say that a vector field* $\mathcal{U} \in L^1_{loc}\big(\mathbb{R}\times\mathbb{R}^3\big)^3$ *satisfying* (1.2) *is a* physically reasonable weak time-periodic solution *to* (1.1) *if*

1) $\mathcal{U} \in L^2\big((0,\mathcal{T}); D^{1,2}_{0,\sigma}(\mathbb{R}^3)\big)$,
2) $\mathcal{P}_\perp\mathcal{U} \in L^\infty\big((0,\mathcal{T}); L^2(\mathbb{R}^3)^3\big)$,
3) $\mathcal{U}$ *is a generalized* $\mathcal{T}$*-time-periodic solution to* (1.1) *in the sense that for all test functions* $\Phi \in C^\infty_{0,\sigma,\mathrm{per}}\big(\mathbb{R}\times\mathbb{R}^3\big)$ *holds*

$$\begin{aligned} \int_0^{\mathcal{T}}\int_{\mathbb{R}^3} -\mathcal{U}\cdot\partial_t\Phi + \nabla\mathcal{U} : \nabla\Phi - \lambda\partial_1\mathcal{U}\cdot\Phi + (\mathcal{U}\cdot\nabla\mathcal{U})\cdot\Phi\,\mathrm{d}x\mathrm{d}t&\\ = \int_0^{\mathcal{T}}\int_{\mathbb{R}^3} f\cdot\Phi\,\mathrm{d}x\mathrm{d}t.& \end{aligned} \tag{2.2}$$

The above class of physically reasonable weak solutions was introduced in [**22**], where existence of a such a solution was shown for any $f \in L^2\big((0,\mathcal{T}); D^{-1,2}_0(\mathbb{R}^3)^3\big)$ regardless of the data's "size". The class is therefore a natural starting point for further investigations. The description of the solutions as *physically reasonable* is due to property 2), which in physical terms expresses that the kinetic energy of the purely oscillatory part $\mathcal{P}_\perp\mathcal{U}$ is finite. In [**22**] it was further required of a physically weak solution that it satisfies an energy inequality. We shall not need this property in the following and thus extend the class of physically weak solutions by leaving this condition out in Definition 2.1 above.

The asymptotic profile in the expansion will be given in terms of the well-known Oseen fundamental solution

$$\begin{aligned} &\varGamma_{\mathrm{O}}^{\lambda}:\mathbb{R}^3\setminus\{0\}\to\mathbb{R}^{3\times 3}, \quad \big[\varGamma_{\mathrm{O}}^{\lambda}(x)\big]_{ij} := (\delta_{ij}\Delta-\partial_i\partial_j)\Phi^{\lambda}(x),\\ &\Phi^{\lambda}(x) := \frac{1}{4\pi\lambda}\int_0^{\lambda(|x|+x_1)/2}\frac{1-\mathrm{e}^{-\tau}}{\tau}\,\mathrm{d}\tau. \end{aligned} \tag{2.3}$$

See also [**13**, Chapter VII.3] for a closed-form expression of $\varGamma_{\mathrm{O}}^{\lambda}$.

We are now in a position to state the main theorem of the paper.

THEOREM 2.2. *Let $\lambda\neq 0$ and $f\in C_{0,\mathrm{per}}^{\infty}\big(\mathbb{R}\times\mathbb{R}^3\big)^3$ be a $\mathcal{T}$-time-periodic vector-field. If u is a physically reasonable weak time-periodic solution to* (1.1), *in the sense of Definition* 2.1, *which satisfies*

$$\exists r\in(5,\infty):\quad \mathcal{P}_{\bot}u\in L^r\big((0,\mathcal{T})\times\mathbb{R}^3\big)^3, \tag{2.4}$$

then

$$\forall (t,x)\in\mathbb{R}\times\mathbb{R}^3:\quad u(t,x)=\varGamma_{\mathrm{O}}^{\lambda}(x)\Big(\frac{1}{\mathcal{T}}\int_0^{\mathcal{T}}\int_{\mathbb{R}^3} f\,\mathrm{d}x\mathrm{d}t\Big)+\mathscr{R}(t,x) \tag{2.5}$$

with $\mathscr{R}$ satisfying

$$\forall\varepsilon>0\ \exists C_1>0\ \forall\,|x|\geq 1,\ t\in\mathbb{R}:\quad |\mathscr{R}(t,x)|\leq C_1\,|x|^{-\frac{3}{2}+\varepsilon}. \tag{2.6}$$

The asymptotic structure of $\varGamma_{\mathrm{O}}^{\lambda}$ is anisotropic. Indeed, the explicit expressions for $\varGamma_{\mathrm{O}}^{\lambda}$ reveal that $\varGamma_{\mathrm{O}}^{\lambda}=O\big(|x|^{-1}(|x|+x_1)^{-1}\big)$ as $|x|\to\infty$ and that $\varGamma_{\mathrm{O}}^{\lambda}(x)$ decays like $|x|^{-1}$ in any given paraboloidal region with its axis directed along the negative x_1-axis; see also [**13**, Chapter VII.3]. Theorem 2.2 therefore yields an asymptotic expansion of u with a remainder term that decays faster than the profile in such a so-called wake region. For a steady-state solution, anisotropic refinements of estimate (2.6) for the remainder $\mathscr{R}$ are known; see for example [**10**] and [**2**]. It is likely that similar refinements can be established for time-periodic solutions, but we shall not address this issue here.

The additional integrability (2.4) is imposed for technical reasons only. It is needed to ensure additional regularity of the weak solution. We emphasize that condition (2.4) does *not* translate into additional decay of the solution at spatial infinity. If the data f is sufficiently "small", condition (2.4) is automatically satisfied; see [**23**, Theorem 2.3].

3. Notation

Points in $\mathbb{R}\times\mathbb{R}^3$ are denoted by (t,x) with $t\in\mathbb{R}$ and $x\in\mathbb{R}^3$. We refer to t as the time variable and to x as the spatial variable.

We use $B_R:=\{x\in\mathbb{R}^3\mid |x|<R\}$ to denote balls in $\mathbb{R}^3$. Furthermore, for $R_1<R_2$ we let $B_{R_2,R_1}:=\{x\in\mathbb{R}^3\mid R_1<|x|<R_2\}$. In addition, we put $B^R:=\mathbb{R}^3\setminus B_R$.

For a sufficiently regular function $u:\mathbb{R}\times\mathbb{R}^3\to\mathbb{R}$, we put $\partial_i u:=\partial_{x_i}u$.

For two vectors $a,b\in\mathbb{R}^3$, we let $a\otimes b\in\mathbb{R}^{3\times 3}$ denote the tensor with $(a\otimes b)_{ij}:=a_ib_j$. We denote by I the identity tensor $I\in\mathbb{R}^{3\times 3}$.

Constants in capital letters in the proofs and theorems are global, while constants in small letters are local to the proof in which they appear. Unless otherwise stated, constants are positive.

We make use of Einstein's summation convention and implicitly sum over all repeated indices appearing in a formula.

4. Reformulation in a group setting

We shall take advantage of the formalism developed in [**22**, **24**] to reformulate the time-periodic Navier-Stokes problem as a system of partial differential equation on the group $G := \mathbb{R}/\mathcal{T}\mathbb{Z} \times \mathbb{R}^3$. We endow G with the canonical topology and differentiable structure inherited from $\mathbb{R} \times \mathbb{R}^3$ via the quotient mapping

$$\pi : \mathbb{R} \times \mathbb{R}^3 \to \mathbb{R}/\mathcal{T}\mathbb{Z} \times \mathbb{R}^3, \quad \pi(t,x) := ([t],x).$$

Clearly, G is then a locally compact abelian group. There is a natural correspondence between $\mathcal{T}$-time-periodic functions defined on $\mathbb{R} \times \mathbb{R}^3$ and functions defined on G. We shall take advantage of this correspondence and reformulate (1.1)–(1.2) and the main theorem in a setting of vector fields defined on G. The Fourier transform on G then enables us to establish a representation formula in terms of Fourier multipliers.

The Haar measure $\mathrm{d}g$ on G is unique up-to a normalization factor. Using the restriction $\Pi := \pi_{|[0,\mathcal{T})\times\mathbb{R}^3}$, which is clearly a (continuous) bijection between $[0,\mathcal{T}) \times \mathbb{R}^3$ and G, we choose the normalization factor so that

$$\int_G u(g)\,\mathrm{d}g = \frac{1}{\mathcal{T}} \int_0^{\mathcal{T}} \int_{\mathbb{R}^3} u \circ \Pi(t,x)\,\mathrm{d}x\mathrm{d}t.$$

For the sake of convenience, we will omit the symbol Π in integrals with respect to $\mathrm{d}x\mathrm{d}t$ of G-defined functions, that is, instead of $\frac{1}{\mathcal{T}} \int_0^{\mathcal{T}} \int_{\mathbb{R}^3} u \circ \Pi(t,x)\,\mathrm{d}x\mathrm{d}t$ we simply write $\frac{1}{\mathcal{T}} \int_0^{\mathcal{T}} \int_{\mathbb{R}^3} u(t,x)\,\mathrm{d}x\mathrm{d}t$.

We make use of the Schwartz-Bruhat space of generalized Schwartz functions $\mathscr{S}(G)$ and the corresponding dual space $\mathscr{S}'(G)$ of tempered distributions; see for example [**3**, **22**, **24**] for the exact definition. We identify the dual group $\widehat{G}$ with $\mathbb{Z} \times \mathbb{R}^3$ by associating elements $(k,\xi) \in \mathbb{Z} \times \mathbb{R}^3$ with the characters $(t,x) \to \mathrm{e}^{ix\cdot\xi + ik\frac{2\pi}{\mathcal{T}}t}$. In the following, we repeatedly use (k,ξ) to denote points in $\widehat{G}$. The Fourier transform on G is then given by the expression

$$\mathscr{F}_G : \mathscr{S}(G) \to \mathscr{S}(\widehat{G}), \quad \mathscr{F}_G(u)(k,\xi) := \frac{1}{\mathcal{T}} \int_0^{\mathcal{T}} \int_{\mathbb{R}^3} u(t,x)\,\mathrm{e}^{-ix\cdot\xi - i\frac{2\pi}{\mathcal{T}}kt}\,\mathrm{d}x\mathrm{d}t.$$

Recall that $\mathscr{F}_G : \mathscr{S}(G) \to \mathscr{S}(\widehat{G})$ is a homeomorphism and extends by duality to a homeomorphism $\mathscr{F}_G : \mathscr{S}'(G) \to \mathscr{S}'(\widehat{G})$.

The space of smooth functions on G is given by

$$C^\infty(G) := \{u : G \to \mathbb{R} \mid u \circ \pi \in C^\infty(\mathbb{R} \times \mathbb{R}^3)\}. \tag{4.1}$$

For $u \in C^\infty(G)$, derivatives are defined by

$$\forall(\alpha,\beta) \in \mathbb{N}_0^n \times \mathbb{N}_0 : \quad \partial_t^\beta \partial_x^\alpha u := \left[\partial_t^\beta \partial_x^\alpha (u \circ \pi)\right] \circ \Pi^{-1}. \tag{4.2}$$

By $C_0^\infty(G)$ we denote the subspace of $C^\infty(G)$ of smooth functions with compact support. We further introduce for $r,s\in\mathbb{N}_0$ the Sobolev space

$$
\begin{aligned}
&W^{r,s,q}(G) := \overline{C_0^\infty(G)}^{\|\cdot\|_{r,s,q}},\\
&\|u\|_{r,s,q} := \Big(\|u\|_q^q + \sum_{(\alpha,\beta)\in\mathbb{N}\times\mathbb{N}^3,\ |\alpha|\leq r,|\beta|\leq s} \|\partial_t^\alpha u\|_q^q + \|\partial_x^\beta u\|_q^q\Big)^{1/q}.
\end{aligned}
\tag{4.3}
$$

It is easy to verify that $W^{r,s,q}(G) = \{u\in L^q(G) \mid \|u\|_{r,s,q}<\infty\}$.

The projections $\mathcal{P}$ and $\mathcal{P}_\perp$ are defined on G-defined functions by the same expressions as in (2.1). Indeed, the integral in (2.1) is well-defined for any $u\in\mathscr{S}(G)$, and clearly both $\mathcal{P}$ and $\mathcal{P}_\perp$ map $\mathscr{S}(G)$ into itself. By duality, both projections extend to continuous linear mappings $\mathcal{P}:\mathscr{S}'(G)\to\mathscr{S}'(G)$ and $\mathcal{P}_\perp:\mathscr{S}'(G)\to\mathscr{S}'(G)$. A direct computation shows that

$$
\mathcal{P}f = \mathscr{F}_G^{-1}\big[\delta_\mathbb{Z}(k)\,\mathscr{F}_G[f]\big],\quad \mathcal{P}_\perp f = \mathscr{F}_G^{-1}\big[\big(1-\delta_\mathbb{Z}(k)\big)\,\mathscr{F}_G[f]\big],
\tag{4.4}
$$

where $\delta_\mathbb{Z}$ denotes the delta distribution on $\mathbb{Z}$, that is,

$$
\delta_\mathbb{Z}:\mathbb{Z}\to\mathbb{C},\quad \delta_\mathbb{Z}(k) := \begin{cases} 1 & \text{if } k=0,\\ 0 & \text{if } k\neq 0.\end{cases}
$$

In other words, $\delta_\mathbb{Z}$ and $(1-\delta_\mathbb{Z})$ is the Fourier symbol of the projection $\mathcal{P}$ and $\mathcal{P}_\perp$, respectively.

The Helmholtz-Weyl projection can be extended from the Euclidean setting into a setting of G-defined vector fields. It is convenient to do so by introducing the Helmholtz-Weyl projection in terms of its Fourier symbol:

$$
\mathcal{P}_H : L^2(G)^3\to L^2(G)^3,\quad \mathcal{P}_H f := \mathscr{F}_G^{-1}\Big[\Big(I-\frac{\xi\otimes\xi}{|\xi|^2}\Big)\mathscr{F}_G[f]\Big].
\tag{4.5}
$$

It follows as in the Euclidean setting that $\mathcal{P}_H$ extends uniquely to a continuous projection $\mathcal{P}_H: L^q(G)^3\to L^q(G)^3$ for any $q\in(1,\infty)$.

We now consider for data F the linear problem

$$
\begin{cases} \partial_t W - \Delta W - \lambda\partial_1 W = \mathcal{P}_\perp\mathcal{P}_H F & \text{in } G,\\ \operatorname{div} W = 0 & \text{in } G.\end{cases}
\tag{4.6}
$$

If we consider the system in the realm of tempered distributions $\mathscr{S}'(G)$ and apply the Fourier transform $\mathscr{F}_G$, we find, formally at least, the expression

$$
W = \mathscr{F}_G^{-1}\bigg[\frac{1-\delta_\mathbb{Z}(k)}{|\xi|^2 + i(\frac{2\pi}{\mathcal{T}}k-\lambda\xi_1)}\,\mathscr{F}_G\big[\mathcal{P}_H F\big]\bigg]
\tag{4.7}
$$

for the solution. We shall briefly recall some result from [**22, 24**] concerning the validity of this representation formula. For this purpose, we introduce the linear operator $S_\lambda^\perp:\ \mathscr{S}(G)^3\to\mathscr{S}'(G)^3$ defined by

$$
S_\lambda^\perp\psi := \mathscr{F}_G^{-1}\bigg[\frac{1-\delta_\mathbb{Z}(k)}{|\xi|^2 + i(\frac{2\pi}{\mathcal{T}}k-\lambda\xi_1)}\,\mathscr{F}_G\big[\psi\big]\bigg]
$$

and $S_\lambda^\perp\circ\operatorname{div}:\ \mathscr{S}(G)^{3\times 3}\to\mathscr{S}'(G)^3$ defined by

$$
\big(S_\lambda^\perp\circ\operatorname{div}\psi\big)_h := \mathscr{F}_G^{-1}\bigg[\frac{(1-\delta_\mathbb{Z}(k))\,i\xi_j}{|\xi|^2 + i(\frac{2\pi}{\mathcal{T}}k-\lambda\xi_1)}\mathscr{F}_G\big[\psi_{jh}\big]\bigg].
$$

It is easy to see that $S_\lambda^\perp$ and $S_\lambda^\perp \circ \operatorname{div}$ are well-defined in the setting of tempered distributions above. Their mapping properties in an $L^q(G)$ setting are given in the following lemma. These properties hold for all $\lambda \in \mathbb{R}$, so for the sake of completeness we make no restriction to $\lambda \neq 0$ in the follow lemmas.

LEMMA 4.1. *Let $\lambda \in \mathbb{R}$ and $q \in (1,\infty)$. The operator $S_\lambda^\perp$ extends uniquely to a bounded linear operator*

$$S_\lambda^\perp : L^q(G)^3 \to W^{1,2,q}(G)^3. \tag{4.8}$$

Let $r := \frac{5q}{5-q}$ if $q \in (1,5)$, let $r \in (5,\infty)$ if $q = 5$, and let $r := \infty$ if $q \in (5,\infty)$. Then the operator $S_\lambda^\perp \circ \operatorname{div}$ extends uniquely to a bounded linear operator

$$S_\lambda^\perp \circ \operatorname{div} : L^q(G)^{3\times 3} \to W^{0,1,q}(G)^3 \cap L^r(G)^3. \tag{4.9}$$

PROOF. Property (4.8) was shown in [**24**, Proof of Theorem 4.8]. The proof relies on the fact that the multiplier

$$M_1 : \widehat{G} \to \mathbb{C}, \quad M_1(k,\xi) := \frac{1-\delta_{\mathbb{Z}}(k)}{|\xi|^2 + i(\frac{2\pi}{\mathcal{T}}k - \lambda\xi_1)}$$

that defines $S_\lambda^\perp$ is smooth and possesses sufficient decay as $|\xi| \to \infty$ and $|k| \to \infty$. Observe in particular that the numerator $1-\delta_{\mathbb{Z}}(k)$ vanishes in a neighborhood of the only zero at $(0,0)$ of the denominator of M_1. With the technique from [**24**, Proof of Theorem 4.8], also the multiplier

$$M_2 : \widehat{G} \to \mathbb{C}, \quad M_2(k,\xi) := \frac{\big(1-\delta_{\mathbb{Z}}(k)\big)\, i\xi_j}{|\xi|^2 + i(\frac{2\pi}{\mathcal{T}}k - \lambda\xi_1)}$$

that defines $S_\lambda^\perp \circ \operatorname{div}$ can be analyzed to show

$$\|\nabla S_\lambda^\perp \circ \operatorname{div} \psi\|_q \leq c_0 \|\psi\|_q. \tag{4.10}$$

Finally, since $\mathscr{F}_G = \mathscr{F}_{\mathbb{R}/2\pi\mathbb{Z}} \circ \mathscr{F}_{\mathbb{R}^3}$, it follows that

$$S_\lambda^\perp \circ \operatorname{div} \psi =$$
$$\mathscr{F}^{-1}_{\mathbb{R}/2\pi\mathbb{Z}}\Big[\big(1-\delta_{\mathbb{Z}}(k)\big)|k|^{-\frac{1}{5}}\Big] *_{\mathbb{R}/2\pi\mathbb{Z}} \mathscr{F}^{-1}_{\mathbb{R}^3}\Big[|\xi|^{-\frac{3}{5}}\Big] *_{\mathbb{R}^3} \mathscr{F}^{-1}_{G}\Big[M_3\, \mathscr{F}_G[\psi]\Big],$$

with

$$M_3 : \widehat{G} \to \mathbb{C}, \quad M_3(k,\xi) := \frac{|k|^{\frac{1}{5}}\, |\xi|^{\frac{3}{5}} \big(1-\delta_{\mathbb{Z}}(k)\big)\, i\xi_j}{|\xi|^2 + i(\frac{2\pi}{\mathcal{T}}k - \lambda\xi_1)}.$$

Again with the technique from [**24**, Proof of Theorem 4.8] it can be shown that M_3 is an $L^q(G)$ multiplier. The estimate

$$\|S_\lambda^\perp \circ \operatorname{div} \psi\|_r \leq c_1 \|\psi\|_q \tag{4.11}$$

thus follows from mapping properties of the two Riesz potentials

$$\psi \to \mathscr{F}^{-1}_{\mathbb{R}^3}\Big[|\xi|^{-\frac{3}{5}}\Big] *_{\mathbb{R}^3} \psi, \quad \psi \to \mathscr{F}^{-1}_{\mathbb{R}/2\pi\mathbb{Z}}\Big[\big(1-\delta_{\mathbb{Z}}(k)\big)|k|^{-\frac{1}{5}}\Big] *_{\mathbb{R}/2\pi\mathbb{Z}} \psi.$$

For the details we refer to [**15**, Proof of Theorem 4.1]. By (4.10) and (4.11), we conclude (4.9). □

LEMMA 4.2. *Let $\lambda \in \mathbb{R}$ and $u \in \mathscr{S}'(G)$ with $\mathcal{P}u = 0$. If $\partial_t u - \Delta u - \lambda\partial_1 u = 0$, then $u = 0$.*

PROOF. Applying the Fourier transform $\mathscr{F}_G$ yields $\big(|\xi|^2 + i(\frac{2\pi}{\mathcal{T}}k - \lambda\xi_1)\big)\widehat{u} = 0$. It follows that $\operatorname{supp}\widehat{u} \subset \{(0,0)\}$. However, since $\mathcal{P}u = 0$ we have $(0,0) \notin \operatorname{supp}\widehat{u}$. Consequently $\operatorname{supp}\widehat{u} = \emptyset$ and thus $u = 0$. □

By Lemma 4.1 and Lemma 4.2 we have obtained a statement concerning existence and uniqueness of a solution to (4.6). More specifically, for data $F \in L^q(G)^3$ we find that $W := S_\lambda^\perp \mathcal{P}_H F$ is a solution to (4.6) in $W^{1,2,q}(G)^3$ that satisfies the representation formula (4.7). Moreover, this solution is unique in the subspace $\mathcal{P}_\perp \mathscr{S}'(G)^3$ of tempered distributions. From (4.7) we observe, formally, that $W = \Gamma_\perp^\lambda * F$ with

$$\Gamma_\perp^\lambda \in \mathscr{S}'(G)^{3\times 3}, \quad \Gamma_\perp^\lambda := \mathscr{F}_G^{-1}\left[\frac{1-\delta_\mathbb{Z}(k)}{|\xi|^2 + i(\frac{2\pi}{\mathcal{T}}k - \lambda\xi_1)}\Big(I - \frac{\xi\otimes\xi}{|\xi|^2}\Big)\right]. \tag{4.12}$$

In the following, we shall establish both pointwise estimates and integrability properties of $\Gamma_\perp^\lambda$. As a consequence, we will see that the solution W can be legitimately expressed as $W = \Gamma_\perp^\lambda * F$. Thus, we may view $\Gamma_\perp^\lambda$ as a fundamental solution to (4.6). The rest of this section is devoted to an analysis of $\Gamma_\perp^\lambda$.

LEMMA 4.3. *Let $\lambda \in \mathbb{R}$, $k \in \mathbb{Z}\setminus\{0\}$ and*

$$\gamma_k^\lambda : \mathbb{R}^3\setminus\{0\} \to \mathbb{C}, \quad \gamma_k^\lambda(x) := \frac{1}{4\pi|x|}\,\mathrm{e}^{-\left(i\frac{2\pi}{\mathcal{T}}k + (\frac{\lambda}{2})^2\right)^{\frac12}|x| - \frac{\lambda}{2}x_1}, \tag{4.13}$$

where $\big(i\frac{2\pi}{\mathcal{T}}k + (\frac{\lambda}{2})^2\big)^{\frac12}$ denotes the square root with non-negative real part. Then

$$|\gamma_k^\lambda(x)| \leq \frac{1}{4\pi|x|}\,\mathrm{e}^{-C_2|k|^{\frac12}|x|}, \tag{4.14}$$

$$\big|\partial_j\big[\gamma_k^\lambda(x)\big]\big| \leq C_3\Big(\frac{1}{|x|^2} + \frac{|k|^{\frac12}}{|x|}\Big)\,\mathrm{e}^{-C_2|k|^{\frac12}|x|}, \tag{4.15}$$

$$\big|\partial_j\partial_h\big[\gamma_k^\lambda(x)\big]\big| \leq C_4\Big(\frac{1}{|x|^3} + \frac{|k|^{\frac12}}{|x|^2} + \frac{|k|}{|x|}\Big)\,\mathrm{e}^{-C_2|k|^{\frac12}|x|}, \tag{4.16}$$

with $C_2 = C_2(\lambda,\mathcal{T})$, $C_3 = C_3(\lambda,\mathcal{T})$ and $C_4 = C_4(\lambda,\mathcal{T})$.

PROOF. The case $\lambda = 0$ is trivial. Assume without loss of generality $\lambda > 0$. We first observe that

$$\begin{aligned}
\operatorname{Re}\Big[\big(i\frac{2\pi}{\mathcal{T}}k + (\lambda/2)^2\big)^{\frac12}\Big] &= \big|i\frac{2\pi}{\mathcal{T}}k + (\lambda/2)^2\big|^{\frac12}\cos\Big(\frac12\arctan\Big(\frac{\frac{2\pi}{\mathcal{T}}k}{(\lambda/2)^2}\Big)\Big)\\
&= \big((\frac{2\pi}{\mathcal{T}})^2k^2 + (\lambda/2)^4\big)^{\frac14}\frac{1}{\sqrt2}\Big(1+\cos\Big(\arctan\Big(\frac{\frac{2\pi}{\mathcal{T}}k}{(\lambda/2)^2}\Big)\Big)\Big)^{\frac12}\\
&= \big((\frac{2\pi}{\mathcal{T}})^2k^2 + (\lambda/2)^4\big)^{\frac14}\frac{1}{\sqrt2}\Big(1+\Big(1+\frac{(\frac{2\pi}{\mathcal{T}})^2k^2}{(\lambda/2)^4}\Big)^{-\frac12}\Big)^{\frac12}\\
&= \frac{1}{\sqrt2}(\lambda/2)\Big(\Big(1+\frac{(\frac{2\pi}{\mathcal{T}})^2k^2}{(\lambda/2)^4}\Big)^{\frac12}+1\Big)^{\frac12}.
\end{aligned}$$

It follows that

$$\forall k \in \mathbb{Z}\setminus\{0\}: \ \operatorname{Re}\Big[\big(i\frac{2\pi}{\mathcal{T}}k + (\lambda/2)^2\big)^{\frac12} - (\lambda/2)\Big] > 0$$

and

$$\lim_{|k|\to\infty}\frac{\operatorname{Re}\left[\left(i\frac{2\pi}{\mathcal{T}}k+(\lambda/2)^2\right)^{\frac{1}{2}}-(\lambda/2)\right]}{|k|^{\frac{1}{2}}}=\sqrt{\frac{\frac{2\pi}{\mathcal{T}}}{2}}.$$

Consequently, there is a constant $c_1=c_1(\lambda,\mathcal{T})>0$ such that

$$\forall k\in\mathbb{Z}\setminus\{0\}:\ \operatorname{Re}\left[\left(i\frac{2\pi}{\mathcal{T}}k+(\lambda/2)^2\right)^{\frac{1}{2}}-(\lambda/2)\right]\geq c_1|k|^{\frac{1}{2}},$$

from which we conclude that

$$\begin{aligned}\operatorname{Re}&\left[-\left(i\frac{2\pi}{\mathcal{T}}k+(\lambda/2)^2\right)^{\frac{1}{2}}|x|-(\lambda/2)x_1\right]\\ &\leq-\operatorname{Re}\left[\left(i\frac{2\pi}{\mathcal{T}}k+(\lambda/2)^2\right)^{\frac{1}{2}}-(\lambda/2)\right]|x|\leq -c_1|k|^{\frac{1}{2}}|x|.\end{aligned}$$

We can now estimate

$$\begin{aligned}|\gamma_k^\lambda(x)|&\leq\frac{1}{4\pi|x|}\,\mathrm{e}^{\operatorname{Re}\left[-\left(i\frac{2\pi}{\mathcal{T}}k+(\lambda/2)^2\right)^{\frac{1}{2}}|x|-(\lambda/2)x_1\right]}\\ &\leq\frac{1}{4\pi|x|}\,\mathrm{e}^{-c_1|k|^{\frac{1}{2}}|x|},\end{aligned}\tag{4.17}$$

and conclude (4.14). Similarly, one may verify (4.15) and (4.16) after taking derivatives. □

We can now establish a pointwise estimate of $\varGamma_\perp^\lambda$.

LEMMA 4.4. *Let $\lambda\in\mathbb{R}$ and $\varGamma_\perp^\lambda$ be defined as in* (4.12). *Then $(j,l,h=1,2,3)$*

$$\forall|x|\geq 2:\quad\left(\frac{1}{\mathcal{T}}\int_0^{\mathcal{T}}\left|\left[\varGamma_\perp^\lambda\right]_{jl}(t,x)\right|^2\mathrm{d}t\right)^{\frac{1}{2}}\leq\frac{C_5}{|x|^3},\tag{4.18}$$

$$\forall|x|\geq 2:\quad\left(\frac{1}{\mathcal{T}}\int_0^{\mathcal{T}}\left|\partial_h\left[\varGamma_\perp^\lambda\right]_{jl}(t,x)\right|^2\mathrm{d}t\right)^{\frac{1}{2}}\leq\frac{C_5}{|x|^4},\tag{4.19}$$

with $C_5=C_5(\lambda,\mathcal{T})$.

PROOF. For $k\in\mathbb{Z}\setminus\{0\}$ we put

$$\varGamma_{k,jl}^\perp:=\mathscr{F}_{\mathbb{R}^3}^{-1}\left[\frac{\xi_j\xi_l}{|\xi|^2}\frac{1}{|\xi|^2+i(\frac{2\pi}{\mathcal{T}}k-\lambda\xi_1)}\right]\in\mathscr{S}'(\mathbb{R}^3).$$

A direct computation yields

$$\varGamma_{k,jl}^\perp=\partial_j\partial_l\mathscr{F}_{\mathbb{R}^3}^{-1}\left[\mathscr{F}_{\mathbb{R}^3}\left(\varGamma_{\mathrm{L}}\right)\mathscr{F}_{\mathbb{R}^3}\left(\gamma_k^\lambda\right)\right]\tag{4.20}$$

with

$$\varGamma_{\mathrm{L}}:\mathbb{R}^3\setminus\{0\}\to\mathbb{C},\quad\varGamma_{\mathrm{L}}(x):=\frac{1}{4\pi|x|}$$

and γ_k^λ given by (4.13). In view of (4.14), we see that both $\varGamma_{\mathrm{L}}$ and γ_k^λ are "regular" enough to express the right-hand side of (4.20) in terms of a classical convolution integral. More precisely, we have

$$\varGamma_{k,jl}^\perp(x)=\partial_j\partial_l\int_{\mathbb{R}^3}\varGamma_{\mathrm{L}}(x-y)\,\gamma_k^\lambda(y)\,\mathrm{d}y.\tag{4.21}$$

We introduce at this point a "cut-off" function $\chi \in C_0^\infty(\mathbb{R};\mathbb{R})$ with

$$\chi(x) = \begin{cases} 0 & \text{if } 0 \leq |x| \leq \frac{1}{2} \quad \text{or} \quad 4 \leq |x|, \\ 1 & \text{if } 1 \leq |x| \leq 3, \end{cases}$$

and define for $R > 0$

$$\chi_R \in C_0^\infty(\mathbb{R}^3;\mathbb{R}), \quad \chi_R(x) := \chi(|x|/R).$$

Now consider an $x \in \mathbb{R}^3$ with $|x| > 2$. Put $R := \frac{|x|}{2}$. Note that this implies $x \in B_{3R,R}$. We use χ_R to decompose the convolution integral in (4.21) as follows:

$$\begin{aligned} \Gamma^\perp_{k,jl}(x) = {} & \partial_j\partial_l \int_{B_{4R,R/2}} \Gamma_{\mathrm{L}}(x-y)\,\gamma_k^\lambda(y)\,\chi_R(y)\,\mathrm{d}y \\ & + \partial_j\partial_l \int_{B^{3R}} \Gamma_{\mathrm{L}}(x-y)\,\gamma_k^\lambda(y)\,\big(1-\chi_R(y)\big)\,\mathrm{d}y \\ & + \partial_j\partial_l \int_{B_R} \Gamma_{\mathrm{L}}(x-y)\,\gamma_k^\lambda(y)\,\big(1-\chi_R(y)\big)\,\mathrm{d}y \\ =: {} & I_1(x) + I_2(x) + I_3(x). \end{aligned} \tag{4.22}$$

We can estimate

$$|I_1(x)| \leq \int_{B_{4R,R/2}} \frac{1}{4\pi|x-y|^2}\,\big|\partial_j\big[\gamma_k^\lambda(y)\,\chi_R(y)\big]\big|\,\mathrm{d}y. \tag{4.23}$$

Utilizing (4.14) and (4.15), we see that for $y \in B_{4R,R/2}$ holds

$$\big|\partial_j\big[\gamma_k^\lambda(y)\,\chi_R(y)\big]\big| \leq c_1 \frac{1}{|k|R^4},$$

with $c_1 = c_1(\lambda, \mathcal{T})$. It follows that

$$|I_1(x)| \leq c_2 \frac{1}{|k|R^3}, \tag{4.24}$$

with $c_2 = c_2(\lambda, \mathcal{T})$. Using again (4.14), we further estimate

$$\begin{aligned} |I_2(x)| & \leq c_3 \int_{B^{3R}} \frac{1}{|x-y|^3}\,\big|\gamma_k^\lambda(y)\big|\,\mathrm{d}y \leq c_4 \int_{B^{3R}} \frac{1}{|x-y|^3}\,\frac{1}{|y|}\,\mathrm{e}^{-C_2|k|^{\frac{1}{2}}|y|}\,\mathrm{d}y \\ & \leq c_5 \int_{B^{3R}} \frac{1}{R^3}\,\frac{1}{|y|}\,\frac{1}{|k|^{\frac{3}{2}}|y|^3}\,\mathrm{d}y \leq c_6 \frac{1}{|k|R^3}, \end{aligned} \tag{4.25}$$

with $c_6 = c_6(\lambda, \mathcal{T})$, and

$$\begin{aligned} |I_3(x)| & \leq c_7 \int_{B_R} \frac{1}{|x-y|^3}\,\big|\gamma_k^\lambda(y)\big|\,\mathrm{d}y \leq c_8 \int_{B_R} \frac{1}{R^3}\,\frac{1}{|y|}\,\mathrm{e}^{-C_2|k|^{\frac{1}{2}}|y|}\,\mathrm{d}y \\ & = c_9 \frac{1}{R^3} \int_0^R r\,\mathrm{e}^{-C_2|k|^{\frac{1}{2}}r}\,\mathrm{d}r \leq c_9 \frac{1}{|k|R^3} \int_0^\infty s\,\mathrm{e}^{-C_2 s}\,\mathrm{d}s = c_{10} \frac{1}{|k|R^3}, \end{aligned} \tag{4.26}$$

with $c_{10} = c_{10}(\lambda, \mathcal{T})$. Combining (4.24), (4.25) and (4.26), we deduce

$$\big|\Gamma^\perp_{k,jl}(x)\big| \leq c_{11} \frac{1}{|k||x|^3},$$

with $c_{11} = c_{11}(\lambda, \mathcal{T})$. Denoting by $\mathscr{F}_{\mathbb{R}/\mathcal{T}\mathbb{Z}}$ the Fourier transform on the torus $\mathbb{R}/\mathcal{T}\mathbb{Z}$, we employ Plancherel's theorem to deduce

$$\begin{aligned}
&\left(\frac{1}{\mathcal{T}}\int_0^{\mathcal{T}} \left|\left[\Gamma_\perp^\lambda\right]_{jl}(t,x)\right|^2 \mathrm{d}t\right)^{\frac12}\\
&\quad= \left(\frac{1}{\mathcal{T}}\int_0^{\mathcal{T}} \left|\mathscr{F}^{-1}_{\mathbb{R}/\mathcal{T}\mathbb{Z}}\Big[\big(1-\delta_{\mathbb{Z}}(k)\big)\big(\delta_{jl}\Gamma^\perp_{k,hh}(x) - \Gamma^\perp_{k,jl}(x)\big)\Big]\right|^2 \mathrm{d}t\right)^{\frac12}\\
&\quad= \big\|\big(1-\delta_{\mathbb{Z}}(k)\big)\big(\delta_{jl}\Gamma^\perp_{k,hh}(x) - \Gamma^\perp_{k,jl}(x)\big)\big\|_{\ell^2(\mathbb{Z})}\\
&\quad\leq c_{12}\left(\sum_{k\in\mathbb{Z}\setminus\{0\}} \frac{1}{\big(|k||x|^3\big)^2}\right)^{\frac12} \leq c_{13}\,\frac{1}{|x|^3},
\end{aligned}$$

which yields (4.18). Differentiating both sides in (4.22) we obtain a formula for $\partial_h\Gamma^\perp_{k,jl}$ based on which we can proceed as above to deduce (4.19). □

LEMMA 4.5. *Let $\lambda\in\mathbb{R}$ and $\Gamma_\perp^\lambda$ be defined as in* (4.12). *Then*

$$\forall r\in\left(1,\frac{5}{3}\right):\quad \Gamma_\perp^\lambda \in L^r(G)^{3\times 3}, \tag{4.27}$$

$$\forall r\in\left[1,\frac{4}{3}\right):\quad \partial_j\Gamma_\perp^\lambda \in L^r(G)^{3\times 3}\quad (j=1,2,3). \tag{4.28}$$

PROOF. We recall the definition (4.12) of $\Gamma_\perp^\lambda$ to find that

$$\left[\Gamma_\perp^\lambda\right]_{jl} = \left[\delta_{jl}(\mathfrak{R}_h\mathfrak{R}_h) - \mathfrak{R}_j\mathfrak{R}_l\right]\circ\mathscr{F}_G^{-1}\left[\frac{\big(1-\delta_{\mathbb{Z}}(k)\big)|k|^{\frac25}\big(1+|\xi|^2\big)^{\frac35}}{|\xi|^2 + i(\frac{2\pi}{\mathcal{T}}k - \lambda\xi_1)}\mathscr{F}_G\big[\mathscr{F}_G^{-1}(\mathcal{K})\big]\right],$$

where $\mathfrak{R}_j$ denotes the Riesz transform

$$\mathfrak{R}_j:\mathscr{S}(G)\to\mathscr{S}'(G),\quad \mathfrak{R}_j(f) := \mathscr{F}^{-1}_{\mathbb{R}^3}\left[\frac{\xi_j}{|\xi|}\mathscr{F}_{\mathbb{R}^3}(f)\right],$$

and

$$\mathcal{K}:\mathbb{Z}\times\mathbb{R}^3\to\mathbb{C},\quad \mathcal{K}(k,\xi) := \big(1-\delta_{\mathbb{Z}}(k)\big)|k|^{-\frac25}\big(1+|\xi|^2\big)^{-\frac35}.$$

It is well-known that $\mathfrak{R}_j$ extends to a bounded linear operator $\mathfrak{R}_j : L^q(G)\to L^q(G)$ for all $q\in(1,\infty)$; see for example [**16**, Corollary 4.2.8]. Moreover, one can verify with the technique from [**24**, Proof of Theorem 4.8] that

$$M:\widehat{G}\to\mathbb{C},\quad M(k,\xi) := \frac{\big(1-\delta_{\mathbb{Z}}(k)\big)|k|^{\frac25}\big(1+|\xi|^2\big)^{\frac35}}{|\xi|^2 + i(\frac{2\pi}{\mathcal{T}}k - \lambda\xi_1)}$$

is an $L^q(G)$ multiplier. Thus, (4.27) follows if we can show that $\mathscr{F}_G^{-1}(\mathcal{K})\in L^r(G)$. Choosing $[-\frac12\mathcal{T},\frac12\mathcal{T})$ as a realization of $\mathbb{R}/\mathcal{T}\mathbb{Z}$, it follows that

$$\mathscr{F}^{-1}_{\mathbb{R}/\mathcal{T}\mathbb{Z}}\left[\big(1-\delta_{\mathbb{Z}}(k)\big)|k|^{-\frac25}\right](t) = t^{-\frac35} + h(t), \tag{4.29}$$

where $h\in C^\infty(\mathbb{R}/\mathcal{T}\mathbb{Z})$; see for example [**16**, Example 3.1.19]. Additionally, one can estimate

$$\left|\mathscr{F}^{-1}_{\mathbb{R}^3}\left[\big(1+|\xi|^2\big)^{-\frac35}\right](x)\right| \leq c_1\left(|x|^{-\frac95}\chi_{B_2}(x) + \mathrm{e}^{-\frac{|x|}{2}}\right); \tag{4.30}$$

see for example [**17**, Proposition 6.1.5]. We can therefore deduce

$$\mathscr{F}_G^{-1}(\mathcal{K}) = \mathscr{F}_{\mathbb{R}/\mathcal{T}\mathbb{Z}}^{-1}\left[\left(1-\delta_{\mathbb{Z}}(k)\right)|k|^{-\frac{2}{5}}\right] \otimes \mathscr{F}_{\mathbb{R}^3}^{-1}\left[\left(1+|\xi|^2\right)^{-\frac{3}{5}}\right] \in L^r(G)$$

for all $r \in \left(1, \frac{5}{3}\right)$, which concludes (4.27). To show (4.28), we observe that

$$\partial_j \left[\Gamma_\perp^\lambda\right]_{jl} =$$

$$\left[\delta_{jl}(\mathfrak{R}_h\mathfrak{R}_h) - \mathfrak{R}_j\mathfrak{R}_l\right] \circ \mathscr{F}_G^{-1}\left[\frac{\left(1-\delta_{\mathbb{Z}}(k)\right)|k|^{\frac{1}{4}}\left(1+|\xi|^2\right)^{\frac{3}{8}}\xi_j}{|\xi|^2 + i(\frac{2\pi}{\mathcal{T}}k - \lambda\xi_1)}\mathscr{F}_G\left[\mathscr{F}_G^{-1}(\mathcal{J})\right]\right],$$

where

$$\mathcal{J} : \mathbb{Z}\times\mathbb{R}^3 \to \mathbb{C}, \quad \mathcal{J}(k,\xi) := \left(1-\delta_{\mathbb{Z}}(k)\right)|k|^{-\frac{1}{4}}\left(1+|\xi|^2\right)^{-\frac{3}{8}}.$$

We then obtain (4.28) for all $q \in (1, \frac{4}{3})$ by the same computations as above. It follows in particular that $\partial_j \Gamma_\perp^\lambda \in L^1_{loc}(G)$. The asymptotic decay as $|x| \to \infty$ established in (4.19) in Lemma 4.4 therefore allows us to further conclude $\partial_j \Gamma_\perp^\lambda \in L^1(G)$. □

5. Proof of the main theorem

We shall make use of the framework for the linearized time-periodic Navier-Stokes system developed in [**24**] and [**22**]. For this purpose, we introduce for $q \in (1,2)$ and $s \in (1,\infty)$ the function space

$$\begin{aligned} &X^{q,s}_{\mathrm{Oseen}}(\mathbb{R}^3) := \left\{v \in L^1_{loc}(\mathbb{R}^3)^3 \mid \|v\|_{q,s,\mathrm{Oseen}} < \infty\right\}, \\ &\|v\|_{q,s,\mathrm{Oseen}} := \|v\|_{\frac{2q}{2-q}} + \|\nabla v\|_{\frac{4}{4-q}} + \|\partial_1 v\|_q + \|\nabla^2 v\|_q + \|\nabla^2 v\|_s. \end{aligned} \tag{5.1}$$

The Banach space $X^{q,s}_{\mathrm{Oseen}}(\mathbb{R}^3)$ is intrinsically linked to the three-dimensional Oseen operator. More specifically, the norm $\|v\|_{q,s,\mathrm{Oseen}}$ captures precisely the generic integrability of a solution to the three-dimensional steady-state Oseen system. To describe similar properties for a solution to the time-periodic Oseen system, we introduce for $q \in (1,\infty)$ the space

$$\begin{aligned} &W^{1,2,q}_{\mathrm{per}}(\mathbb{R}\times\mathbb{R}^3) := \overline{C^\infty_{0,\mathrm{per}}(\mathbb{R}\times\mathbb{R}^3)}^{\|\cdot\|_{1,2,q}}, \\ &\|u\|_{1,2,q} := \left(\frac{1}{\mathcal{T}}\int_0^{\mathcal{T}}\int_{\mathbb{R}^3} |\partial_t u|^q \,\mathrm{d}x\mathrm{d}t + \sum_{|\beta|\le 2}\frac{1}{\mathcal{T}}\int_0^{\mathcal{T}}\int_{\mathbb{R}^3} |\partial_x^\beta u|^q \,\mathrm{d}x\mathrm{d}t\right)^{\frac{1}{q}}. \end{aligned} \tag{5.2}$$

For $q \in (1,3)$ and $s \in (1,\infty)$ we further define

$$\begin{aligned} &Y^{q,s}_{\mathrm{per}}(\mathbb{R}\times\mathbb{R}^3) := \overline{C^\infty_{0,\mathrm{per}}(\mathbb{R}\times\mathbb{R}^3)}^{\|\cdot\|_{Y^{q,s}_{\mathrm{per}}}}, \\ &\|\mathfrak{p}\|_{Y^{q,s}_{\mathrm{per}}} := \left(\int_0^{\mathcal{T}} \|\mathfrak{p}(t,\cdot)\|^q_{\frac{3q}{3-q}} + \|\nabla\mathfrak{p}(t,\cdot)\|_q^q \,\mathrm{d}t\right)^{1/q} + \|\nabla\mathfrak{p}\|_s, \end{aligned} \tag{5.3}$$

which is the space we shall use to characterize the corresponding pressure term.

We first show that a physically reasonable weak time-periodic solution, recall Definition 2.1, satisfying the integrability condition (2.4) is in fact a strong solution. We then recall the regularity results for strong solutions from [**23**], which yield that u is smooth when the data is smooth.

LEMMA 5.1. *Let $\lambda \neq 0$, $f \in C^{\infty}_{0,\mathrm{per}}(\mathbb{R}\times\mathbb{R}^3)^3$ and u be a physically reasonable weak $\mathcal{T}$-time-periodic solution to* (1.1) *in the sense of Definition* 2.1. *If u satisfies* (2.4), *then*

$$\forall q \in (1,\infty): \ \mathcal{P}_\perp u \in W^{1,2,q}_{\mathrm{per}}(\mathbb{R}\times\mathbb{R}^3)^3 \tag{5.4}$$

and

$$\forall q \in (1,2),\ s \in (1,\infty): \ \mathcal{P}u \in X^{q,s}_{\mathrm{Oseen}}(\mathbb{R}^3). \tag{5.5}$$

Moreover, there is a pressure term $\mathfrak{p}$ with

$$\forall q \in (1,3),\ s \in (1,\infty): \ \mathfrak{p} \in Y^{q,s}_{\mathrm{per}}(\mathbb{R}\times\mathbb{R}^3) \tag{5.6}$$

such that $(u,\mathfrak{p})$ is a solution to (1.1). *Additionally,*

$$(u,\mathfrak{p}) \in C^\infty(\mathbb{R}\times\mathbb{R}^3)^3 \times C^\infty(\mathbb{R}\times\mathbb{R}^3). \tag{5.7}$$

PROOF. Put $v := \mathcal{P}u$ and $w := \mathcal{P}_\perp u$. From Definition 2.1 and standard Sobolev embedding, we infer

$$v \in L^6(\mathbb{R}^3)^3, \quad \nabla v \in L^2(\mathbb{R}^3)^{3\times 3} \tag{5.8}$$

and

$$\begin{aligned} &w \in L^\infty\big((0,\mathcal{T}); L^2(\mathbb{R}^3)^3\big) \cap L^2\big((0,\mathcal{T}); L^6(\mathbb{R}^3)^3\big),\\ &\nabla w \in L^2\big((0,\mathcal{T}); L^2(\mathbb{R}^3)^{3\times 3}\big). \end{aligned} \tag{5.9}$$

It is easy to verify from (2.2) that v is a generalized solution (sometimes referred to as a *D-solution*) to the steady-state Navier-Stokes problem

$$\begin{cases} -\Delta v - \lambda\partial_1 v + \nabla p + v\cdot\nabla v = \mathcal{P}f - \mathcal{P}\big[w\cdot\nabla w\big] & \text{in } \mathbb{R}^3,\\ \operatorname{div} v = 0 & \text{in } \mathbb{R}^3,\\ \lim_{|x|\to\infty} v = 0. \end{cases} \tag{5.10}$$

Problem (5.10) has been studied extensively over the years. A comprehensive exposition can be found in [**13**, Chapter X]. Although [**13**, Chapter X] deals with the exterior domain case, virtually all techniques found there can be used in the whole-space case as well. A straightforward utilization of the integral Minkowski inequality in combination with (5.9) yields $\mathcal{P}\big[w\cdot\nabla w\big] \in L^1(\mathbb{R}^3)^3 \cap L^{\frac{3}{2}}(\mathbb{R}^3)^3$. From [**13**, Chapter X] it thus follows that

$$v \in L^3(\mathbb{R}^3)^3. \tag{5.11}$$

Now consider the linear operator

$$S_\lambda : \mathscr{S}(\mathbb{R}^3)^3 \to \mathscr{S}'(\mathbb{R}^3)^3, \quad S_\lambda\psi := \mathscr{F}^{-1}_{\mathbb{R}^3}\left[\frac{1}{|\xi|^2 - i\lambda\xi_1}\mathscr{F}_{\mathbb{R}^3}\big[\mathcal{P}_H\psi\big]\right].$$

By a well-known application of Lizorkin's multiplier theorem (see for example [**13**, Theorem VII.4.1]), one can show that S_λ extends uniquely to a homeomorphism

$$\forall q \in (1,2), s \in (1,\infty): \quad S_\lambda : L^q(\mathbb{R}^3)^3 \cap L^s(\mathbb{R}^3)^3 \to X^{q,s}_{\mathrm{Oseen}}(\mathbb{R}^3). \tag{5.12}$$

Additionally, it follows from [**13**, Theorem VII.4.2] that the composition $S_\lambda \circ \operatorname{div}$ extends to a continuous linear operator

$$\begin{aligned} &\forall q \in (1,4): \quad S_\lambda \circ \operatorname{div} : L^q(\mathbb{R}^3)^{3\times 3} \to L^{\frac{4q}{4-q}}(\mathbb{R}^3)^3,\\ &\forall q \in (1,\infty): \quad S_\lambda \circ \operatorname{div} : L^q(\mathbb{R}^3)^{3\times 3} \to D^{1,q}(\mathbb{R}^3)^3. \end{aligned} \tag{5.13}$$

We shall employ (5.12) and (5.13) iteratively to establish increased integrability of v. Observe that

$$\begin{aligned} v &= S_\lambda\big[\mathcal{P}f - \mathcal{P}\big[w\cdot\nabla w\big] - v\cdot\nabla v\big] \\ &= S_\lambda\mathcal{P}f - S_\lambda\circ\operatorname{div}\mathcal{P}\big[w\otimes w\big] - S_\lambda\circ\operatorname{div}\big[v\otimes v\big]. \end{aligned} \tag{5.14}$$

The first equality above follows by a standard uniqueness argument since both sides of the equation is a solution in $L^6(\mathbb{R}^3)^3$ to the same Oseen system. One may verify the second equality by a simple computation. It follows from (5.9) that $\mathcal{P}\big[w\otimes w\big]\in L^1(\mathbb{R}^3)^{3\times3}\cap L^3(\mathbb{R}^3)^{3\times3}$. By (5.12) and (5.13) we thus have

$$\forall q\in(2,12]:\quad S_\lambda\mathcal{P}f - S_\lambda\circ\operatorname{div}\mathcal{P}\big[w\otimes w\big]\in L^q(\mathbb{R}^3). \tag{5.15}$$

By (5.8) and (5.11) we have $v\otimes v\in L^{\frac{3}{2}}(\mathbb{R}^3)^{3\times3}\cap L^3(\mathbb{R}^3)^{3\times3}$, whence (5.13), (5.14) and (5.15) imply $v\in L^{\frac{12}{5}}(\mathbb{R}^3)^3\cap L^{12}(\mathbb{R}^3)^3$. From this, $v\otimes v\in L^{\frac{6}{5}}(\mathbb{R}^3)^{3\times3}\cap L^6(\mathbb{R}^3)^{3\times3}$ follows. Consequently, (5.13), (5.14) and (5.15) yield

$$\forall q\in(2,12]:\quad v\in L^q(\mathbb{R}^3)^3. \tag{5.16}$$

We now turn our attention to the integrability of w. It follows directly from the definitions (4.3) and (5.2) that lifting by π is a homeomorphism between $W^{1,2,q}(G)$ and $W^{1,2,q}_{\mathrm{per}}(\mathbb{R}\times\mathbb{R}^3)$. We may therefore consider w as a function defined on G. It is easy to verify from (2.2) that w is a solution in the sense of $\mathscr{S}'(G)$-distributions to

$$\begin{cases} \partial_t w - \Delta w - \lambda\partial_1 w = \mathcal{P}_\perp\mathcal{P}_H f - \mathcal{P}_\perp\mathcal{P}_H\big[w\cdot\nabla w + w\cdot\nabla v + v\cdot\nabla w\big] & \text{in } G,\\ \operatorname{div} w = 0 & \text{in } G. \end{cases} \tag{5.17}$$

Observe that the integrability of the right-hand side above is sufficient to apply the Helmholtz projection $\mathcal{P}_H$. In fact, due to the assumption that w satisfies (2.4) it is easy to verify that all terms on the right-hand side belong to some $L^q(G)$ space. Consequently, we can employ Lemma 4.1 and Lemma 4.2 to deduce

$$w = S_\lambda^\perp\mathcal{P}_H f - S_\lambda^\perp\mathcal{P}_H[w\cdot\nabla w] - S_\lambda^\perp\mathcal{P}_H[w\cdot\nabla v] - S_\lambda^\perp\mathcal{P}_H[v\cdot\nabla w] \tag{5.18}$$

$$= S_\lambda^\perp\mathcal{P}_H f - S_\lambda^\perp\circ\operatorname{div}\mathcal{P}_H[w\otimes w - v\otimes w - w\otimes v]. \tag{5.19}$$

More precisely, due to (4.8) the right-hand side in (5.18) is well-defined as a solution to (5.17) in $\mathcal{P}_\perp\mathscr{S}'(G)$. From Lemma 4.2 we thus obtain the identity in (5.18), and in turn, by an easy calculation, (5.19). For $q\in(1,10)$, the following implication follows from Lemma 4.1:

$$w\in L^q(G)^3 \Rightarrow \mathcal{P}_H[w\otimes w]\in L^{\frac{q}{2}}(G)^{3\times3} \Rightarrow S_\lambda^\perp\circ\operatorname{div}\mathcal{P}_H[w\otimes w]\in L^{\frac{5q}{10-q}}(G).$$

Recalling (5.16), we also deduce for $q\in(1,60/7)$ validity of the implication

$$w\in L^q(G)^3 \Rightarrow \mathcal{P}_H[v\otimes w]\in L^{\frac{12q}{12+q}}(G)^{3\times3} \Rightarrow S_\lambda^\perp\circ\operatorname{div}\mathcal{P}_H[v\otimes w]\in L^{\frac{60q}{60-7q}}(G).$$

Employing Lemma 4.1 in a similar manner also for $q \in [60/7, \infty)$, we thus deduce from (5.19) the implication

(5.20)
$$\forall q \in (1,\infty): \ w \in L^q(G)^3 \ \Rightarrow \begin{cases} w \in L^{\frac{5q}{10-q}}(G) & \text{if } q \in [5,6], \\ w \in L^{\frac{60q}{60-7q}}(G) & \text{if } q \in [6, 60/7), \\ \forall s \in [q,\infty): \ w \in L^s(G) & \text{if } q = 60/7, \\ w \in L^\infty(G) & \text{if } q \in (60/7, \infty). \end{cases}$$

By assumption (2.4), $w \in L^r(G)^3$. Since $r > 5$, iterating (5.20) with $q = r$ as starting point yields

$$w \in L^\infty(G)^3. \tag{5.21}$$

With this improved integrability of w, we return to (5.14) and improve (5.16) to

$$\forall q \in (2,\infty): \quad v \in L^q(\mathbb{R}^3)^3. \tag{5.22}$$

From (5.13), (5.14), (4.9) and (5.19) it now follows that

$$\forall q \in (1,\infty): \quad \nabla v \in L^q(\mathbb{R}^3)^{3\times 3}, \ \nabla w \in L^q(G)^{3\times 3}. \tag{5.23}$$

Combining (5.21), (5.22), (5.23) and (5.18), we conclude (5.4) by Lemma 4.1. Similarly, recalling (5.13) and (5.14), we deduce (5.5).

The integrability we have established for v and w enables us to employ the Helmholtz-Weyl projection $\mathcal{P}_H$ in (1.1) and obtain the equation

$$\nabla \mathfrak{p} = \big(\mathrm{Id} - \mathcal{P}_H\big)\big[f - w\cdot\nabla w - w\cdot\nabla v - v\cdot\nabla w - v\cdot\nabla v\big] \quad \text{in } G$$

for the pressure term $\mathfrak{p}$. Applying the Fourier transform $\mathscr{F}_{\mathbb{R}^3}$, we thus formally arrive at the expression

$$\mathfrak{p} = \mathscr{F}^{-1}_{\mathbb{R}^3}\Big[\frac{\xi_j}{|\xi|^2}\mathscr{F}_{\mathbb{R}^3}\big[f - w\cdot\nabla w - w\cdot\nabla v - v\cdot\nabla w - v\cdot\nabla v\big]\Big].$$

By well-known mapping properties of the Riesz potential (see for example [**17**, Theorem 6.1.3]) and L^p-boundedness of the Riesz operators, we deduce from the integrability properties of v and w that the expression for $\mathfrak{p}$ is valid as an element of $Y^{q,s}_{\mathrm{per}}(\mathbb{R}\times\mathbb{R}^3)$ for any $q \in (1,3)$ and $s \in (1,\infty)$. This concludes (5.6). Finally, we obtain from [**23**, Corollary 2.5] that both u and $\mathfrak{p}$ are smooth. □

The smoothness of the solution u in Lemma 5.1 enables us to analyze its pointwise behavior. We start by showing a decay estimate of ∇u. The lemma below is a minor modification of [**13**, Lemma X.8.2].

LEMMA 5.2. *Let $f \in C^\infty_{0,\mathrm{per}}\big(\mathbb{R}\times\mathbb{R}^3\big)^3$ and $(u,\mathfrak{p})$ be a $\mathcal{T}$-time-periodic solution to* (1.1) *with*

$$\begin{aligned} &u \in C^2(\mathbb{R}\times\mathbb{R}^3)^3, \quad \mathfrak{p} \in C^1(\mathbb{R}\times\mathbb{R}^3), \\ &\nabla u \in L^2\big(\mathbb{R}\times\mathbb{R}^3\big)^{3\times 3} \cap L^{\frac{3}{2}}\big(\mathbb{R}\times\mathbb{R}^3\big)^{3\times 3}, \\ &\forall s \in (2,3]: \ u \in L^s(G)^3, \quad \exists s' \in [3/2, 2): \ \mathfrak{p} \in L^{s'}(G). \end{aligned} \tag{5.24}$$

Then

$$\forall \varepsilon > 0: \quad \frac{1}{\mathcal{T}}\int_0^{\mathcal{T}}\int_{B^R} |\nabla u|^2 \,\mathrm{d}x\mathrm{d}t \le C_6\, R^{-1+\varepsilon}. \tag{5.25}$$

PROOF. The proof follows that of [**13**, Lemma X.8.2]. We put

$$\mathcal{G}(R) := \frac{1}{\mathcal{T}}\int_0^{\mathcal{T}}\int_{B^R} |\nabla u|^2 \,\mathrm{d}x\mathrm{d}t = \frac{1}{\mathcal{T}}\int_0^{\mathcal{T}}\int_R^\infty\int_{\partial B_r} |\nabla u|^2 \,\mathrm{d}S\mathrm{d}r\mathrm{d}t.$$

By assumption $\nabla u \in C(\mathbb{R}\times\mathbb{R}^3)$, whence $\mathcal{G}\in C^1(0,\infty)$. Multiplying both sides in (1.1) by u and subsequently integrating over $B_{R^*,R}\times[0,\mathcal{T}]$, we obtain

$$\begin{aligned}(5.26)\qquad &\frac{1}{\mathcal{T}}\int_0^{\mathcal{T}}\int_{B_{R^*,R}} |\nabla u|^2 \,\mathrm{d}x\mathrm{d}t = \\ &\qquad\frac{1}{\mathcal{T}}\int_0^{\mathcal{T}}\int_{\partial B_{R^*,R}} \frac{\lambda}{2}|u|^2 n_1 - \frac{1}{2}|u|^2 u\cdot n + u\cdot(\nabla u\cdot n) - \mathfrak{p}\, u\cdot n \,\mathrm{d}S\mathrm{d}t.\end{aligned}$$

From the assumptions in (5.24), it follows that the function

$$(5.27)\qquad r \to \frac{1}{\mathcal{T}}\int_0^{\mathcal{T}}\int_{\partial B_r} |u|^3 + \Big|-\frac{1}{2}|u|^2 u\cdot n + u\cdot(\nabla u\cdot n) - \mathfrak{p}\, u\cdot n\Big| \,\mathrm{d}S\mathrm{d}t$$

belongs to $L^1\big((1,\infty)\big)$. Consequently, there is a sequence $\{R_k\}_{k=1}^\infty \subset (0,\infty)$ with $\lim_{k\to\infty} R_k = \infty$ and

$$\lim_{k\to\infty} R_k \frac{1}{\mathcal{T}}\int_0^{\mathcal{T}}\int_{\partial B_{R_k}} |u|^3 + \Big|-\frac{1}{2}|u|^2 u\cdot n + u\cdot(\nabla u\cdot n) - \mathfrak{p}\, u\cdot n\Big| \,\mathrm{d}S\mathrm{d}t = 0.$$

Employing Hölder's inequality, we observe that

$$\begin{aligned}(5.28)\qquad \frac{1}{\mathcal{T}}\int_0^{\mathcal{T}}\int_{\partial B_R} |u|^2 \,\mathrm{d}S\mathrm{d}t &\leq c_1 \frac{1}{\mathcal{T}}\int_0^{\mathcal{T}}\Big(\int_{\partial B_R} |u|^3 \,\mathrm{d}S\Big)^{\frac{2}{3}} R^{\frac{2}{3}} \,\mathrm{d}t \\ &\leq c_1\Big(R\frac{1}{\mathcal{T}}\int_0^{\mathcal{T}}\int_{\partial B_R} |u|^3 \,\mathrm{d}S\mathrm{d}t\Big)^{\frac{2}{3}}.\end{aligned}$$

It follows that

$$\lim_{k\to\infty}\frac{1}{\mathcal{T}}\int_0^{\mathcal{T}}\int_{\partial B_{R_k}} \frac{\lambda}{2}|u|^2 n_1 - \frac{1}{2}|u|^2 u\cdot n + u\cdot(\nabla u\cdot n) - \mathfrak{p}\, u\cdot n \,\mathrm{d}S\mathrm{d}t = 0.$$

Put

$$F(R) := \frac{1}{\mathcal{T}}\int_0^{\mathcal{T}}\int_{\partial B_R} \frac{\lambda}{2}|u|^2 n_1 - \frac{1}{2}|u|^2 u\cdot n + u\cdot(\nabla u\cdot n) - \mathfrak{p}\, u\cdot n \,\mathrm{d}S\mathrm{d}t.$$

Choosing $R^* = R_k$ in (5.26) and letting $k\to\infty$, we see that

$$(5.29)\qquad \mathcal{G}(R) = F(R).$$

By a similar estimate as in (5.28), we find for $q\in(1,\infty)$

$$\frac{1}{\mathcal{T}}\int_0^{\mathcal{T}}\int_{\partial B_R} |u|^2 \,\mathrm{d}S\mathrm{d}t \leq c_2\, R^{\frac{2(q-1)}{q}}\Big(\frac{1}{\mathcal{T}}\int_0^{\mathcal{T}}\int_{\partial B_R} |u|^{2q} \,\mathrm{d}S\mathrm{d}t\Big)^{\frac{1}{q}}.$$

Employing Young's inequality, we then deduce

$$(5.30)\qquad R^{-\varepsilon}\frac{1}{\mathcal{T}}\int_0^{\mathcal{T}}\int_{\partial B_R} |u|^2 \,\mathrm{d}S\mathrm{d}t \leq c_3\Big(R^{2-\varepsilon\frac{q}{q-1}} + \frac{1}{\mathcal{T}}\int_0^{\mathcal{T}}\int_{\partial B_R} |u|^{2q} \,\mathrm{d}S\mathrm{d}t\Big).$$

Now fix $q \in (1, \frac{3}{3-\varepsilon})$. Then $2 - \varepsilon \frac{q}{q-1} < -1$. Moreover, by assumption (5.24) we have $u \in L^{2q}(\mathbb{R} \times \mathbb{R}^3)^3$. Thus

$$R \to R^{-\varepsilon} \frac{1}{\mathcal{T}} \int_0^{\mathcal{T}} \int_{\partial B_R} |u|^2 \, \mathrm{d}S\mathrm{d}t \in L^1\big((1,\infty)\big).$$

This information together with the integrability of the function in (5.27) implies that $R \to R^{-\varepsilon} F(R) \in L^1\big((1,\infty)\big)$. Recalling (5.29), we deduce

$$R \to R^{-\varepsilon} \mathcal{G}(R) \in L^1\big((1,\infty)\big). \tag{5.31}$$

Combining (5.31) with the fact that

$$\mathcal{G}'(R) = -\frac{1}{\mathcal{T}} \int_0^{\mathcal{T}} \int_{\partial B_R} |\nabla u|^2 \, \mathrm{d}S\mathrm{d}t \leq 0,$$

we finally conclude by [**13**, Lemma X.8.1] that

$$\mathcal{G}(R) R^{1-\varepsilon} \leq c_4,$$

which yields (5.25) □

We observe in the following corollary that the decay property established for u in Lemma 5.2 also holds for $\mathcal{P}u$ and $\mathcal{P}_\perp u$ separately.

COROLLARY 5.3. *Under the same assumptions as in Lemma 5.2, it holds that*

$$\forall \varepsilon > 0 : \quad \frac{1}{\mathcal{T}} \int_0^{\mathcal{T}} \int_{B^R} |\nabla \mathcal{P}_\perp u|^2 \, \mathrm{d}x\mathrm{d}t + \int_{B^R} |\nabla \mathcal{P} u|^2 \, \mathrm{d}x \leq C_7 \, R^{-1+\varepsilon}. \tag{5.32}$$

PROOF. We simply observe that

$$\begin{aligned} &\frac{1}{\mathcal{T}} \int_0^{\mathcal{T}} \int_{B^R} |\nabla u|^2 \, \mathrm{d}x\mathrm{d}t \\ &\quad = \frac{1}{\mathcal{T}} \int_0^{\mathcal{T}} \int_{B^R} |\nabla \mathcal{P}_\perp u|^2 \, \mathrm{d}x\mathrm{d}t + \int_{B^R} |\nabla \mathcal{P} u|^2 \, \mathrm{d}x + 2 \frac{1}{\mathcal{T}} \int_0^{\mathcal{T}} \int_{B^R} \nabla \mathcal{P} u : \nabla \mathcal{P}_\perp u \, \mathrm{d}x\mathrm{d}t \end{aligned}$$

and

$$\frac{1}{\mathcal{T}} \int_0^{\mathcal{T}} \int_{B^R} \nabla \mathcal{P} u : \nabla \mathcal{P}_\perp u \, \mathrm{d}x\mathrm{d}t = \int_{B^R} \nabla \mathcal{P} u : \Big(\frac{1}{\mathcal{T}} \int_0^{\mathcal{T}} \nabla \mathcal{P}_\perp u \, \mathrm{d}t \Big) \, \mathrm{d}x = 0.$$

□

Next, we recall some properties of the Oseen fundamental solution $\varGamma_{\mathrm{O}}^{\lambda}$.

LEMMA 5.4. *Let $\lambda \neq 0$. The Oseen fundamental solution $\varGamma_{\mathrm{O}}^{\lambda}$ satisfies*

$$\forall |x| > 0 : \quad \big|\varGamma_{\mathrm{O}}^{\lambda}(x)\big| \leq C_8 \, |x|^{-1}, \tag{5.33}$$

$$\forall |x| > 1 : \quad \big|\nabla \varGamma_{\mathrm{O}}^{\lambda}(x)\big| \leq C_9 \, |x|^{-\frac{3}{2}}, \tag{5.34}$$

$$\forall r > 0 : \quad \int_{\partial B_r} \big|\nabla \varGamma_{\mathrm{O}}^{\lambda}(x)\big| \, \mathrm{d}S \leq C_{10} \, r^{-\frac{1}{2}}. \tag{5.35}$$

Moreover, $\varGamma_{\mathrm{O}}^{\lambda}$ *enjoys for any* $r > 0$ *the integrability properties*

$$\forall q \in (2,\infty): \qquad \varGamma_{\mathrm{O}}^{\lambda} \in L^q(\mathbb{R}^3 \setminus B_r)^{3\times 3}, \tag{5.36}$$

$$\forall q \in [1,3): \qquad \varGamma_{\mathrm{O}}^{\lambda} \in L^q_{loc}(\mathbb{R}^3)^{3\times 3}, \tag{5.37}$$

$$\forall q \in (4/3,\infty): \qquad \partial_j \varGamma_{\mathrm{O}}^{\lambda} \in L^q(\mathbb{R}^3 \setminus B_r)^{3\times 3} \quad (j=1,2,3), \tag{5.38}$$

$$\forall q \in [1,3/2): \qquad \partial_j \varGamma_{\mathrm{O}}^{\lambda} \in L^q_{loc}(\mathbb{R}^3)^{3\times 3} \quad (j=1,2,3). \tag{5.39}$$

PROOF. We refer to [**13**, Chapter VII: (3.24), (3.28), (3.32), (3.33)] for (5.33), (5.36), (5.34), and (5.38). Estimate (5.35) follows from [**13**, Exercise VII.3.1]. Finally, (5.37) and (5.39) are direct consequences of (5.33) and (5.35), respectively. □

PROOF OF THEOREM 2.2. We start by associating to the weak solution u the pressure term $\mathfrak{p}$ from Lemma 5.1. By Lemma 5.1, $(u, \mathfrak{p})$ is then a smooth strong solution. Additionally, the integrability properties (5.4)–(5.6) obtained in Lemma 5.1 ensure that $(u, \mathfrak{p})$ satisfies (5.24). We can therefore utilize Lemma 5.2 and Corollary 5.3. Further observe that since $\varGamma_{\mathrm{O}}^{\lambda}$ is smooth away from the origin, it is enough to show the estimate in (2.6) for large $|x|$. Consider therefore an arbitrary $x \in \mathbb{R}^3$ with $|x| > 2$. Let $R := \frac{|x|}{2}$. Put $v := \mathcal{P}u$ and $w := \mathcal{P}_\perp u$. We will take advantage of the fact that w satisfies (5.17) and, as in the proof of Lemma 5.1, use Lemma 4.1 and Lemma 4.2 to deduce (5.18). By Lemma 4.5, the componentwise convolution $\varGamma_\perp^\lambda * \psi$ is well-defined in a classical sense for any $\psi \in L^q(G)^3$, $q \in [1,\infty)$. Since clearly $S_\lambda^\perp \mathcal{P}_H \psi = \varGamma_\perp^\lambda * \psi$, we thus have

$$w = \varGamma_\perp^\lambda * \Big(f - w\cdot\nabla w - w\cdot\nabla v - v\cdot\nabla w \Big). \tag{5.40}$$

Put

$$\begin{aligned} I_1(t,x) &:= \varGamma_\perp^\lambda * \big[w\cdot\nabla w\big](t,x),\\ I_2(t,x) &:= \varGamma_\perp^\lambda * \big[w\cdot\nabla v\big](t,x),\\ I_3(t,x) &:= \varGamma_\perp^\lambda * \big[v\cdot\nabla w\big](t,x),\\ I_4(t,x) &:= \varGamma_\perp^\lambda * f(t,x). \end{aligned} \tag{5.41}$$

We shall give a pointwise estimate of I_1, I_2, I_3, and I_4. We first decompose

$$\begin{aligned} I_1(t,x) &= \frac{1}{\mathcal{T}}\int_0^{\mathcal{T}}\int_{B_R} \varGamma_\perp^\lambda(t-s,x-y)\,\big[w\cdot\nabla w\big](s,y)\,\mathrm{d}y\mathrm{d}s\\ &\quad + \frac{1}{\mathcal{T}}\int_0^{\mathcal{T}}\int_{B^R} \varGamma_\perp^\lambda(t-s,x-y)\,\big[w\cdot\nabla w\big](s,y)\,\mathrm{d}y\mathrm{d}s\\ &=: I_{11} + I_{12}. \end{aligned}$$

Employing Hölder's inequality and recalling (4.18), we deduce

$$\begin{aligned} |I_{11}| &\leq \int_{B_R} \Big(\frac{1}{\mathcal{T}}\int_0^{\mathcal{T}} \big|\varGamma_\perp^\lambda(s,x-y)\big|^2\,\mathrm{d}s\Big)^{\frac12} \Big(\frac{1}{\mathcal{T}}\int_0^{\mathcal{T}} \big|w\cdot\nabla w\big|^2\,\mathrm{d}s\Big)^{\frac12}\,\mathrm{d}y\\ &\leq c_1 \frac{1}{R^3}\Big(\int_{B_R} 1\,\mathrm{d}y\Big)^{\frac12} \Big(\frac{1}{\mathcal{T}}\int_0^{\mathcal{T}}\int_{B_R} \big|w\cdot\nabla w\big|^2\,\mathrm{d}y\mathrm{d}s\Big)^{\frac12}\\ &\leq c_2 R^{-\frac32}, \end{aligned}$$

where we in the last inequality use that $w \cdot \nabla w \in L^2(G)^3$, which is a direct consequence of Lemma 5.1. Consider $\alpha \in (1, \frac{5}{3})$ and let α' denote the corresponding Hölder conjugate. Recalling Corollary 5.3, we find that

$$\begin{aligned}|I_{12}| &\leq \Big(\frac{1}{\mathcal{T}}\int_0^{\mathcal{T}}\int_{B^R} \big|\Gamma_\perp^\lambda(s,x-y)\big|^\alpha \,\mathrm{d}y\mathrm{d}s\Big)^{\frac{1}{\alpha}} \Big(\frac{1}{\mathcal{T}}\int_0^{\mathcal{T}}\int_{B^R} |w|^{\alpha'} |\nabla w|^{\alpha'} \,\mathrm{d}y\mathrm{d}s\Big)^{\frac{1}{\alpha'}} \\ &\leq c_3 \,\|\Gamma_\perp^\lambda\|_\alpha \,\|w\|_{L^\infty(\mathbb{R}/\mathcal{T}\mathbb{Z}\times B^R)} \,\|\nabla w\|_\infty^{\frac{\alpha'-2}{\alpha'}} \Big(\frac{1}{\mathcal{T}}\int_0^{\mathcal{T}}\int_{B^R} |\nabla w|^2 \,\mathrm{d}y\mathrm{d}s\Big)^{\frac{1}{\alpha'}} \\ &\leq c_4 \,\|\Gamma_\perp^\lambda\|_\alpha \,\|w\|_{L^\infty(\mathbb{R}/\mathcal{T}\mathbb{Z}\times B^R)} \,\|\nabla w\|_\infty^{\frac{2}{\alpha}-1} \,R^{(-1+\varepsilon)\frac{\alpha-1}{\alpha}}\end{aligned}$$

for all $\varepsilon > 0$. By Sobolev embedding, see [**15**, Theorem 4.1], $\nabla w \in L^\infty(G)$. Recalling (4.27) and letting $\alpha \to \frac{5}{3}$, we thus obtain

$$\forall \varepsilon > 0: \quad |I_{12}| \leq c_5 \,R^{-\frac{2}{5}+\varepsilon} \,\|w\|_{L^\infty(\mathbb{R}/\mathcal{T}\mathbb{Z}\times B^R)}.$$

We conclude

$$\forall \varepsilon > 0: \quad |I_1| \leq c_6 \big(R^{-\frac{3}{2}} + R^{-\frac{2}{5}+\varepsilon} \,\|w\|_{L^\infty(\mathbb{R}/\mathcal{T}\mathbb{Z}\times B^R)}\big). \tag{5.42}$$

Similarly, we estimate

$$\forall \varepsilon > 0: \quad |I_2| \leq c_7 \big(R^{-\frac{3}{2}} + R^{-\frac{2}{5}+\varepsilon} \,\|w\|_{L^\infty(\mathbb{R}/\mathcal{T}\mathbb{Z}\times B^R)}\big), \tag{5.43}$$

$$\forall \varepsilon > 0: \quad |I_3| \leq c_8 \big(R^{-\frac{3}{2}} + R^{-\frac{2}{5}+\varepsilon} \,\|v\|_{L^\infty(B^R)}\big). \tag{5.44}$$

Due to (4.18) and the fact that $\operatorname{supp} f$ is compact, we deduce

$$|I_4| \leq c_9 \int_{\operatorname{supp} f} \Big(\frac{1}{\mathcal{T}}\int_0^{\mathcal{T}} \big|\Gamma_\perp^\lambda(s,x-y)\big|^2 \,\mathrm{d}s\Big)^{\frac{1}{2}} \,\mathrm{d}y \leq c_{10} R^{-3} \tag{5.45}$$

for R sufficiently large. It follows from (5.40) and (5.42)-(5.45) that

$$\forall \varepsilon > 0: \quad |w(t,x)| \leq c_{11} \big(R^{-\frac{3}{2}} + R^{-\frac{2}{5}+\varepsilon}(\|v\|_{L^\infty(B^R)} + \|w\|_{L^\infty(\mathbb{R}/\mathcal{T}\mathbb{Z}\times B^R)})\big). \tag{5.46}$$

We now turn our attention to v. To establish a pointwise estimate of v, we follow essentially the proof of [**13**, Theorem X.8.1]. Recall that v satisfies (5.10). We claim that

$$v = \Gamma_{\mathrm{O}}^\lambda * \big(\mathcal{P}f - \mathcal{P}\big[w\cdot\nabla w\big] - v\cdot\nabla v\big). \tag{5.47}$$

The integrability properties (5.36) and (5.37) of the Oseen fundamental solution $\Gamma_{\mathrm{O}}^\lambda$ combined with the properties obtained for v and w in Lemma 5.1 imply that the convolution on the right-hand side above is well-defined in a classical sense. Since both sides in (5.47) satisfy (5.10), a standard uniqueness argument yields the identity. Recalling (5.33) and that f has compact support, we see for R sufficiently large that

$$\big|\Gamma_{\mathrm{O}}^\lambda * \mathcal{P}f(x)\big| \leq c_{12} \int_{\operatorname{supp}\mathcal{P}f} \frac{1}{|x-y|} \,\mathrm{d}y \leq c_{13} R^{-1}.$$

Moreover, we can estimate

$$\begin{aligned}
\big|\varGamma_{\mathrm{O}}^{\lambda} * \mathcal{P}\big[w\cdot\nabla w\big](x)\big| &= \Big|\int_{\mathbb{R}^3} \varGamma_{\mathrm{O}}^{\lambda}(x-y)\,\frac{1}{\mathcal{T}}\int_0^{\mathcal{T}} w(t,y)\cdot\nabla w(t,y)\,\mathrm{d}t\,\mathrm{d}y\Big| \\
&\leq \frac{1}{\mathcal{T}}\int_0^{\mathcal{T}}\int_{B_R} \frac{c_{14}}{R}|w\cdot\nabla w|\,\mathrm{d}y\mathrm{d}t + \frac{1}{\mathcal{T}}\int_0^{\mathcal{T}}\int_{B^R}\frac{|w(y,t)|}{|x-y|}|\nabla w(t,y)|\,\mathrm{d}y\mathrm{d}t \\
&\leq c_{15}R^{-1} + \frac{1}{\mathcal{T}}\int_0^{\mathcal{T}}\Big(\int_{B^R}\frac{|w(t,y)|^2}{|x-y|^2}\,\mathrm{d}y\Big)^{\frac{1}{2}}\Big(\int_{B^R}|\nabla w(t,y)|^2\,\mathrm{d}y\Big)^{\frac{1}{2}}\,\mathrm{d}t,
\end{aligned}$$

where we use in the last estimate that $w\cdot\nabla w\in L^1(G)$. We can employ the Hardy-type inequality [**13**, Theorem II.6.1] to increase the integral on the right-hand side above by

$$\begin{aligned}
\frac{1}{\mathcal{T}}\int_0^{\mathcal{T}}\Big(\int_{B^R}\frac{|w(t,y)|^2}{|x-y|^2}\,\mathrm{d}y\Big)^{\frac{1}{2}}\Big(\int_{B^R}|\nabla w(t,y)|^2\,\mathrm{d}y\Big)^{\frac{1}{2}}\,\mathrm{d}t \leq \\
c_{16}\frac{1}{\mathcal{T}}\int_0^{\mathcal{T}}\int_{B^R}|\nabla w(t,y)|^2\,\mathrm{d}y\mathrm{d}t,
\end{aligned}$$

whence by Corollary 5.3 we obtain

$$\forall\varepsilon>0:\quad \big|\varGamma_{\mathrm{O}}^{\lambda} * \mathcal{P}\big[w\cdot\nabla w\big](x)\big| \leq c_{17}R^{-1+\varepsilon}.$$

In a similar manner we show

$$\forall\varepsilon>0:\quad \big|\varGamma_{\mathrm{O}}^{\lambda} * \mathcal{P}\big[v\cdot\nabla v\big](x)\big| \leq c_{18}R^{-1+\varepsilon}.$$

We conclude that

$$\forall\varepsilon>0:\quad |v(x)|\leq c_{19}R^{-1+\varepsilon}. \tag{5.48}$$

Inserting (5.48) into (5.46), we then find

$$\forall\varepsilon>0:\quad |w(t,x)|\leq c_{20}\big(R^{-\frac{2}{5}+\varepsilon}\|w\|_{L^\infty(\mathbb{R}/\mathcal{T}\mathbb{Z}\times B^R)} + R^{-\frac{7}{5}+\varepsilon}\big). \tag{5.49}$$

By Lemma 5.1, recall (5.21), $w\in L^\infty(G)$. We can thus iterate (5.49) a sufficient number of times to obtain

$$\forall\varepsilon>0:\quad |w(t,x)|\leq c_{21}\,R^{-\frac{7}{5}+\varepsilon}. \tag{5.50}$$

We now return to the representation formula (5.40) of w. We shall utilize (5.48) and (5.50) to extract even better decay estimates for w. For this purpose, we recall (5.41) and observe that the integrability of v, w and $\varGamma_{\perp}^{\lambda}$ established in Lemma 5.1 and Lemma 4.5, respectively, allows us to integrate by parts and obtain

$$\begin{aligned}
\big[I_1(t,x)\big]_i &= \partial_k\big[\varGamma_{\perp}^{\lambda}\big]_{ij} * \big[w_jw_k\big](t,x), \\
\big[I_2(t,x)\big]_i &= \partial_k\big[\varGamma_{\perp}^{\lambda}\big]_{ij} * \big[v_jw_k\big](t,x), \\
\big[I_3(t,x)\big]_i &= \partial_k\big[\varGamma_{\perp}^{\lambda}\big]_{ij} * \big[w_jv_k\big](t,x).
\end{aligned} \tag{5.51}$$

We again decompose

$$\begin{aligned}
\big[I_1(t,x)\big]_i = &\frac{1}{\mathcal{T}}\int_0^{\mathcal{T}}\int_{B_R}\partial_k\big[\varGamma_{\perp}^{\lambda}\big]_{ij}(t-s,x-y)\,\big[w_jw_k\big](s,y)\,\mathrm{d}y\mathrm{d}s \\
&+\frac{1}{\mathcal{T}}\int_0^{\mathcal{T}}\int_{B^R}\partial_k\big[\varGamma_{\perp}^{\lambda}\big]_{ij}(t-s,x-y)\,\big[w_jw_k\big](s,y)\,\mathrm{d}y\mathrm{d}s \\
=:&\ \tilde{I}_{11}+\tilde{I}_{12}.
\end{aligned}$$

Employing Hölder's inequality and recalling (4.19), we deduce

$$
\begin{aligned}
|\tilde{I}_{11}| &\leq \int_{B_R} \Big(\frac{1}{\mathcal{T}}\int_0^{\mathcal{T}} \big|\partial_k \big[\Gamma_\perp^\lambda\big]_{ij}(s,x-y)\big|^2\,\mathrm{d}s\Big)^{\frac{1}{2}} \Big(\frac{1}{\mathcal{T}}\int_0^{\mathcal{T}} |w|^4\,\mathrm{d}s\Big)^{\frac{1}{2}}\,\mathrm{d}y\\
&\leq c_{22}\frac{1}{R^4}\Big(\int_{B_R} 1\,\mathrm{d}y\Big)^{\frac{1}{2}} \|w\|_4^2 \leq c_{23}R^{-\frac{5}{2}},
\end{aligned}
$$

where we in the last inequality use that $w\in L^4(G)$, which is a direct consequence of Lemma 5.1. We recall from Lemma 4.5 that $\partial_k\Gamma_\perp^\lambda\in L^1(G)$ and infer by (5.50) the estimate

$$
\forall\varepsilon>0:\quad |\tilde{I}_{12}|\leq \|\partial_k\Gamma_\perp^\lambda\|_1\,\|w\|^2_{L^\infty(\mathbb{R}/\mathcal{T}\mathbb{Z}\times B^R)}\leq c_{24}R^{-\frac{14}{5}+\varepsilon}.
$$

Consequently,

$$
\forall\varepsilon>0:\quad |I_1|\leq c_{25}\big(R^{-\frac{5}{2}}+R^{-\frac{14}{5}+\varepsilon}\big).
$$

In a similar manner, we estimate

$$
\forall\varepsilon>0:\quad |I_2|+|I_3|\leq c_{26}\big(R^{-\frac{5}{2}}+R^{-\frac{12}{5}+\varepsilon}\big).
$$

Recalling (5.45), we may thus conclude

$$
\forall\varepsilon>0:\quad |w(t,x)|\leq c_{27}\,|x|^{-\frac{12}{5}+\varepsilon}. \tag{5.52}
$$

We finally return to the representation formula (5.47) for v. Recalling the integrability properties (5.38) and (5.39) of $\nabla\Gamma_{\mathrm{O}}^\lambda$, we estimate

$$
\begin{aligned}
\big|\Gamma_{\mathrm{O}}^\lambda * \mathcal{P}\big[w\cdot\nabla w\big](x)\big| &= \Big|\frac{1}{\mathcal{T}}\int_0^{\mathcal{T}}\int_{\mathbb{R}^3}\Gamma_{\mathrm{O}}^\lambda(x-y)\,\operatorname{div}\big[w\otimes w\big](t,y)\,\mathrm{d}y\mathrm{d}t\Big|\\
&\leq c_{28}\frac{1}{\mathcal{T}}\int_0^{\mathcal{T}}\int_{\mathbb{R}^3}\big|\nabla\Gamma_{\mathrm{O}}^\lambda(x-y)\big|\,|w(t,y)|^2\,\mathrm{d}y\mathrm{d}t.
\end{aligned}
$$

We then use (5.34) to estimate

$$
\begin{aligned}
\frac{1}{\mathcal{T}}\int_0^{\mathcal{T}}\int_{B_R}\big|\nabla\Gamma_{\mathrm{O}}^\lambda(x-y)\big|\,|w(t,y)|^2\,\mathrm{d}y\mathrm{d}t &\leq c_{29}R^{-\frac{3}{2}}\,\frac{1}{\mathcal{T}}\int_0^{\mathcal{T}}\int_{B_R}|w(t,y)|^2\,\mathrm{d}y\mathrm{d}t\\
&\leq c_{30}R^{-\frac{3}{2}},
\end{aligned}
$$

where we in the last inequality recall that $w\in L^2(G)$. Moreover, in view of (5.35) and (5.52), we see that

$$
\begin{aligned}
\frac{1}{\mathcal{T}}\int_0^{\mathcal{T}}\int_{B_{3R,R}}\big|\nabla\Gamma_{\mathrm{O}}^\lambda(x-y)\big|\,|w(t,y)|^2\,\mathrm{d}y\mathrm{d}t &\leq c_{31}R^{-\frac{24}{5}+\varepsilon}\int_{B_{3R,R}}\big|\nabla\Gamma_{\mathrm{O}}^\lambda(x-y)\big|\,\mathrm{d}y\\
\leq c_{31}R^{-\frac{24}{5}+\varepsilon}\int_{B_{6R}}\big|\nabla\Gamma_{\mathrm{O}}^\lambda(y)\big|\,\mathrm{d}y &\leq c_{32}R^{-\frac{24}{5}+\varepsilon}\int_0^{6R}\int_{\partial B_r}\big|\nabla\Gamma_{\mathrm{O}}^\lambda(y)\big|\,\mathrm{d}S\mathrm{d}r\\
\leq c_{33}R^{-\frac{24}{5}+\varepsilon}\int_0^{6R} r^{-\frac{1}{2}}\,\mathrm{d}r &\leq c_{34}R^{-\frac{24}{5}+\frac{1}{2}+\varepsilon}
\end{aligned}
$$

for all $\varepsilon>0$. Finally, employing again (5.34) and the fact that $w\in L^2(G)$, we estimate

$$
\frac{1}{\mathcal{T}}\int_0^{\mathcal{T}}\int_{B^{3R}}\big|\nabla\Gamma_{\mathrm{O}}^\lambda(x-y)\big|\,|w(t,y)|^2\,\mathrm{d}y\mathrm{d}t\leq c_{35}R^{-\frac{3}{2}}.
$$

We thus conclude that

$$\left|\Gamma_{\mathrm{O}}^{\lambda} * \mathcal{P}\big[w \cdot \nabla w\big](x)\right| \leq c_{36}|x|^{-\frac{3}{2}}.$$

It is well-known how to obtain estimates for the other terms on the right-hand side in (5.47). More specifically, in view of (5.48) we can use the arguments from [**13**, Proof of Theorem X.8.1] to obtain

$$\forall \varepsilon > 0: \quad \left|\Gamma_{\mathrm{O}}^{\lambda} * \big[v \cdot \nabla v\big](x)\right| \leq c_{37}|x|^{-\frac{3}{2}+\varepsilon}$$

and

$$\left|\Gamma_{\mathrm{O}}^{\lambda} * \mathcal{P}f(x) - \Gamma_{\mathrm{O}}^{\lambda}(x)\bigg(\int_{\mathbb{R}^3} \mathcal{P}f\bigg)\right| \leq c_{38}|x|^{-\frac{3}{2}}.$$

We therefore finally deduce, recalling (5.47) and (5.52), that

$$u(t,x) = v(x) + w(t,x) = \Gamma_{\mathrm{O}}^{\lambda}(x)\bigg(\int_{\mathbb{R}^3} \mathcal{P}f\bigg) + \mathscr{R}(t,x)$$

with $\mathscr{R}(t,x)$ satisfying (2.6). □

References

[1] K.I. Babenko, *On stationary solutions of the problem of flow past a body of a viscous incompressible fluid*, Math. USSR, Sb. **20** (1973), 1–25.

[2] K. I. Babenko and M. M. Vasil′ev, *On the asymptotic behavior of a steady flow of viscous fluid at some distance from an immersed body* (Russian), Prikl. Mat. Meh. **37** (1973), 690–705; English transl., J. Appl. Math. Mech. **37** (1973), 651–665. MR0347214

[3] François Bruhat, *Distributions sur un groupe localement compact et applications à l'étude des représentations des groupes $\wp$-adiques* (French), Bull. Soc. Math. France **89** (1961), 43–75. MR0140941

[4] Paul Deuring, Stanislav Kračmar, and Šárka Nečasová, *A leading term for the velocity of stationary viscous incompressible flow around a rigid body performing a rotation and a translation*, Discrete Contin. Dyn. Syst. **37** (2017), no. 3, 1389–1409. MR3640557

[5] Reinhard Farwig, *The stationary exterior 3D-problem of Oseen and Navier-Stokes equations in anisotropically weighted Sobolev spaces*, Math. Z. **211** (1992), no. 3, 409–447. MR1190220

[6] Reinhard Farwig, Giovanni P. Galdi, and Mads Kyed, *Asymptotic structure of a Leray solution to the Navier-Stokes flow around a rotating body*, Pacific J. Math. **253** (2011), no. 2, 367–382. MR2878815

[7] Reinhard Farwig and Toshiaki Hishida, *Leading term at infinity of steady Navier-Stokes flow around a rotating obstacle*, Math. Nachr. **284** (2011), no. 16, 2065–2077. MR2844678

[8] Reinhard Farwig and Hermann Sohr, *Weighted estimates for the Oseen equations and the Navier-Stokes equations in exterior domains*, Theory of the Navier-Stokes equations, Ser. Adv. Math. Appl. Sci., vol. 47, World Sci. Publ., River Edge, NJ, 1998, pp. 11–30. MR1643022

[9] Robert Finn, *On the exterior stationary problem for the Navier-Stokes equations, and associated perturbation problems*, Arch. Rational Mech. Anal. **19** (1965), 363–406. MR0182816

[10] Robert Finn, *Estimates at infinity for stationary solutions of the Navier-Stokes equations*, Bull. Math. Soc. Sci. Math. Phys. R. P. Roumaine (N.S.) **3 (51)** (1959), 387–418. MR0166495

[11] Giovanni Galdi and Hermann Sohr, *Existence and uniqueness of time-periodic physically reasonable Navier-Stokes flow past a body*, Arch. Ration. Mech. Anal. **172** (2004), no. 3, 363–406. MR2062429

[12] Giovanni P. Galdi, *On the asymptotic structure of D-solutions to steady Navier-Stokes equations in exterior domains*, Mathematical problems relating to the Navier-Stokes equation, Ser. Adv. Math. Appl. Sci., vol. 11, World Sci. Publ., River Edge, NJ, 1992, pp. 81–104. MR1190730

[13] ———, *An introduction to the mathematical theory of the Navier-Stokes equations. Steady-state problems. 2nd ed.*, New York: Springer, 2011.

[14] Giovanni P. Galdi and Mads Kyed, *Steady-state Navier-Stokes flows past a rotating body: Leray solutions are physically reasonable*, Arch. Ration. Mech. Anal. **200** (2011), no. 1, 21–58. MR2781585

[15] Giovanni P. Galdi and Mads Kyed, *Time-periodic flow of a viscous liquid past a body*, arXiv:1609.09829 (2016).

[16] Loukas Grafakos, *Classical Fourier analysis*, 2nd ed., Graduate Texts in Mathematics, vol. 249, Springer, New York, 2008. MR2445437

[17] Loukas Grafakos, *Modern Fourier analysis*, 2nd ed., Graduate Texts in Mathematics, vol. 250, Springer, New York, 2009. MR2463316

[18] Kyungkuen Kang, Hideyuki Miura, and Tai-Peng Tsai, *Asymptotics of small exterior Navier-Stokes flows with non-decaying boundary data*, Comm. Partial Differential Equations **37** (2012), no. 10, 1717–1753. MR2971204

[19] A. Korolev and V. Šverák, *On the large-distance asymptotics of steady state solutions of the Navier-Stokes equations in 3D exterior domains*, Ann. Inst. H. Poincaré Anal. Non Linéaire **28** (2011), no. 2, 303–313. MR2784073

[20] Hideo Kozono and Mitsuhiro Nakao, *Periodic solutions of the Navier-Stokes equations in unbounded domains*, Tohoku Math. J. (2) **48** (1996), no. 1, 33–50. MR1373173

[21] Stanislav Kračmar, Antonín Novotný, and Milan Pokorný, *Estimates of Oseen kernels in weighted L^p spaces*, J. Math. Soc. Japan **53** (2001), no. 1, 59–111. MR1800524

[22] Mads Kyed, *Time-Periodic Solutions to the Navier-Stokes Equations*, Habilitationsschrift, Technische Universität Darmstadt (2012).

[23] Mads Kyed, *The existence and regularity of time-periodic solutions to the three-dimensional Navier-Stokes equations in the whole space*, Nonlinearity **27** (2014), no. 12, 2909–2935. MR3291136

[24] Mads Kyed, *Maximal regularity of the time-periodic linearized Navier-Stokes system*, J. Math. Fluid Mech. **16** (2014), no. 3, 523–538. MR3247366

[25] Mads Kyed, *On the asymptotic structure of a Navier-Stokes flow past a rotating body*, J. Math. Soc. Japan **66** (2014), no. 1, 1–16. MR3161389

[26] P. Maremonti and M. Padula, *Existence, uniqueness and attainability of periodic solutions of the Navier-Stokes equations in exterior domains* (English, with English and Russian summaries), Zap. Nauchn. Sem. S.-Peterburg. Otdel. Mat. Inst. Steklov. (POMI) **233** (1996), no. Kraev. Zadachi Mat. Fiz. i Smezh. Vopr. Teor. Funkts. 27, 142–182, 257; English transl., J. Math. Sci. (New York) **93** (1999), no. 5, 719–746. MR1699121

[27] Paolo Maremonti, *Existence and stability of time-periodic solutions to the Navier-Stokes equations in the whole space*, Nonlinearity **4** (1991), no. 2, 503–529. MR1107016

[28] Paolo Maremonti, *Some theorems of existence for solutions of the Navier-Stokes equations with slip boundary conditions in half-space*, Ricerche Mat. **40** (1991), no. 1, 81–135. MR1191888

[29] Masao Yamazaki, *The Navier-Stokes equations in the weak-L^n space with time-dependent external force*, Math. Ann. **317** (2000), no. 4, 635–675. MR1777114

Department of Mechanical Engineering and Materials Science, University of Pittsburgh, Pittsburgh, PA 15261

Email address: galdi@pitt.edu

Fachbereich Mathematik, Technische Universität Darmstadt, Schlossgartenstr. 7, 64289 Darmstadt, Germany

Email address: kyed@mathematik.tu-darmstadt.de

Contemporary Mathematics
Volume **710**, 2018
http://dx.doi.org/10.1090/conm/710/14368

Uniqueness and stability for the Vlasov-Poisson system with spatial density in Orlicz spaces

Thomas Holding and Evelyne Miot

ABSTRACT. In this paper, we establish uniqueness of the solution of the Vlasov-Poisson system with spatial density belonging to a certain class of Orlicz spaces. This extends the uniqueness result of Loeper **[11]** (which holds for density in L^∞) and of the paper **[15]**. Uniqueness is a direct consequence of our main result, which provides a quantitative stability estimate for the Wasserstein distance between two weak solutions with spatial density in such Orlicz spaces, in the spirit of Dobrushin's proof of stability for mean-field PDEs. Our proofs are built on the second-order structure of the underlying characteristic system associated to the equation.

1. Introduction

The purpose of this article is to study uniqueness and stability issues for a class of weak solutions of the Vlasov-Poisson system in dimension $d = 2$ or $d = 3$, which reads:

$$(1.1) \qquad \begin{cases} \partial_t f + v \cdot \nabla_x f + E \cdot \nabla_v f = 0, \quad (t, x, v) \in [0, T] \times \mathbb{R}^d \times \mathbb{R}^d \\ E(t, x) = (K *_x \rho)(t, x), \quad K(x) = \gamma \dfrac{x}{|x|^d} \\ \rho(t, x) = \displaystyle\int_{\mathbb{R}^d} f(t, x, v)\, dv. \end{cases}$$

The system (1.1) describes the evolution of a microscopic density $f = f(t, x, v)$ of interacting particles, that are electric particles for $\gamma = 1$ (Coulombian interaction) or stars for $\gamma = -1$ (gravitational interaction). The function ρ is called macroscopic (or spatial) density.

Existence and uniqueness of classical solutions of (1.1) defined on $[0, T]$ for all $T > 0$ were established by Ukai and Okabe **[16]** for $d = 2$ and by Pfaffelmoser **[18]** for $d = 3$. Arsenev **[2]** proved global existence of weak solutions with finite

2010 *Mathematics Subject Classification.* Primary 35Q83, 35A02, 35A24.

Key words and phrases. Uniqueness and stability of solutions, Kinetic equations, Orlicz spaces.

The first author was supported in part by UK Engineering and Physical Sciences Research Council (EPSRC) grant EP#H023348/1 for the University of Cambridge Centre for Doctoral Training, the Cambridge Centre for Analysis.

The second author is or was partly supported by the french National Research Agency projects INFAMIE ANR-#15-CE40-0, SchEq ANR-#12-JS-0005-01 and GEODISP ANR-12-#BS01-0015-01.

energy. Another kind of global solutions, which propagate the velocity moments, was constructed by Lions and Perthame [**10**]. We refer to the articles [**6, 17**], and to references quoted therein, for further related results. On the other hand, part of the literature is devoted to determining sufficient conditions for uniqueness. Loeper [**11**] established uniqueness on $[0,T]$ in the class of weak solutions such that the spatial density ρ is uniformly bounded: [1]

$$f \in C\left([0,T], \mathcal{M}_+(\mathbb{R}^d \times \mathbb{R}^d) - w^*\right) \quad \text{and } \rho \in L^\infty\left([0,T], L^\infty(\mathbb{R}^d)\right). \tag{1.2}$$

This result was extended by the second author in [**15**] to weak solutions satisfying

$$f \in C\left([0,T], \mathcal{M}_+(\mathbb{R}^d \times \mathbb{R}^d) - w^*\right) \quad \text{and } \sup_{[0,T]} \sup_{p\geq 1} \frac{\|\rho(t)\|_{L^p}}{p} < +\infty. \tag{1.3}$$

In Theorem 1.9 below, we establish uniqueness of the solution with spatial density belonging to a certain class of exponential Orlicz spaces defined in (1.7). These spaces interpolate the functional spaces arising in (1.2) and (1.3). Our uniqueness result actually comes as a by-product of the main result of Theorem 1.9, which states a quantitative stability estimate involving the Wasserstein distance[2] between such weak solutions. We obtain this estimate in the spirit of the method of Dobrushin [**5**] to establish stability estimates for mean field PDE with Lipschitz convolution Kernels K.

In the second part of this paper, we look for sufficient conditions on the initial data ensuring that any corresponding solution has spatial density belonging to the exponential Orlicz spaces defined in (1.7) on $[0,T]$. In Proposition 1.14 we prove that this holds for data with finite exponential velocity moment.

1.1. Main results.

1.1.1. *Preliminary definitions on Orlicz spaces and on the Wasserstein distance.* We begin by recalling some standard definitions related to Orlicz spaces. We refer the reader to e.g. [**19**] for a more thorough exposition.

DEFINITION 1.1 (N-function). We say that a function $\phi : [0,\infty) \to [0,\infty)$ is an N-function if it is continuous, convex with $\phi(\tau) > 0$ for $\tau > 0$ and satisfies both $\lim_{\tau\to 0} \phi(\tau)/\tau = 0$ and $\lim_{\tau\to\infty} \phi(\tau)/\tau = \infty$.

DEFINITION 1.2 (Luxemburg norm). Let U be a domain of $\mathbb{R}^d$. For an N-function ϕ we define the Luxemburg norm of a function f defined on U as

$$\|f\|_{L_\phi(U)} = \inf\left\{\lambda > 0 : \int_U \phi(|f(x)|/\lambda)\, dx < 1\right\}. \tag{1.4}$$

REMARK 1.3. If it holds for some constant C' that

$$\int_U \phi(|f(x)|/\gamma)\, dx < C' \tag{1.5}$$

then $\|f\|_{L_\phi(U)} \leq C\gamma$, where C is an absolute constant depending only on C'.

REMARK 1.4. On bounded domains only the asymptotic behaviour as $\tau \to \infty$ of the N-function ϕ is important in defining the space L_ϕ. In particular, if two N-functions $\phi, \tilde\phi$ have the same behaviour at infinity in the sense that there are $K, \tilde K > 0$ such that $\phi(\tau) \leq \tilde\phi(\tilde K\tau)$ and $\tilde\phi(\tau) \leq \phi(K\tau)$ for all sufficiently large τ,

---

[1] $\mathcal{M}_+(\mathbb{R}^d \times \mathbb{R}^d)$ denotes the space of bounded positive measures.

[2] See Definition 1.7 hereafter of the Wasserstein distance.

then the norms $\|\cdot\|_{L_\phi(U)}$ and $\|\cdot\|_{L_{\tilde\phi}(U)}$ are equivalent for any bounded domain $U \subseteq \mathbb{R}^d$.

DEFINITION 1.5 (Complementary N-function). For an N-function ϕ we define its complementary N-function $\bar\phi$ as

$$\bar\phi(\tau) = \int_0^\tau a(s)\, ds$$

where a is the right inverse of the right derivative of ϕ.

For $\alpha \in [1, +\infty)$ we let, for $\tau \geq 0$,

$$\phi_\alpha(\tau) = \exp(\tau^\alpha) - 1. \tag{1.6}$$

The spaces $L_{\phi_\alpha}(\mathbb{R}^d)$ are exponential Orlicz spaces, and can be equivalently characterised as those functions g which lie in L^p for all $p \in [\alpha, \infty)$ and have the following norm finite:

$$\|g\|_{\phi_\alpha} = \sup_{p \geq \alpha} p^{-1/\alpha} \|g\|_{L^p(\mathbb{R}^d)}, \tag{1.7}$$

which is an equivalent norm to the Luxemburg norm $\|\cdot\|_{L_{\phi_\alpha}(\mathbb{R}^d)}$. This equivalence is standard and can be verified by Taylor expansion of the exponential. Note that in the $\alpha \to \infty$ limiting case we obtain the function ϕ_∞ given by $\phi_\infty(\tau) = \infty$ if $\tau > 1$ and 0 otherwise. Although ϕ_∞ is not an N-function, we will use the convention that $L_{\phi_\infty} = L^\infty$. Therefore with this convention L_{ϕ_α} indeed interpolates the functional spaces for ρ that are considered in [**11**] (for $\alpha = +\infty$) and [**15**] (for $\alpha = 1$).

REMARK 1.6. When working with exponential Orlicz spaces, one has the choice between working with the Luxemburg norm (1.2) directly, or working with L^p norms *uniformly* in p and using (1.7), as is done in [**15**] for $\alpha = 1$. We take the former approach in this work.

We next turn to some definitions concerning the notion of transportation distances. Let $n \geq 2$. We let $\mathcal{M}_+(\mathbb{R}^n)$ denote the space of bounded positive measures on $\mathbb{R}^n$.

DEFINITION 1.7 (Wasserstein distance). For two measures $\mu, \nu \in \mathcal{M}_+(\mathbb{R}^n)$ with the same mass and finite first moments, we define the (Monge-Kantorovich-Rubenstein)-Wasserstein distance $W_1(\mu, \nu)$ as[3]

$$W_1(\mu, \nu) = \inf_{\pi \in \Pi(\mu,\nu)} \int_{\mathbb{R}^n \times \mathbb{R}^n} |x - y|\, d\pi(x, y),$$

where, here and throughout, $\Pi(\mu, \nu)$ denotes the set of *couplings* between μ and ν, by which we mean measures in $\mathcal{M}_+(\mathbb{R}^n \times \mathbb{R}^n)$ which have marginals μ and ν respectively.

REMARK 1.8. The Wasserstein distance is usually defined on probability measures (i.e. elements of $\mathcal{M}_+$ with mass 1) and metrisizes the weak* topology on the space of probability measures with finite first moment. In the case of the extension to general bounded positive measures given above, it should be noted that the Wasserstein distance does not metrisize the weak* topology on $\mathcal{M}_+$ with finite first moment. However, given any fixed mass m, the Wasserstein distance metrisizes the weak* topology on measures in $\mathcal{M}_+$ of mass m with first moment finite.

[3]Here and throughout, $|v|$ denotes the euclidean norm of $v \in \mathbb{R}^n$.

1.1.2. *Main results.* We are now in position to state a quantitative estimate on the Wasserstein distance between two weak solutions of (1.1) with spatial density belonging to some exponential Orlicz space:

$$\rho_j \in L^\infty([0,T]; L_{\phi_\alpha}(\mathbb{R}^d)) \text{ for } j=1,2 \text{ and some } \alpha \in [1,\infty]. \tag{1.8}$$

THEOREM 1.9. *Let $\varepsilon > 0$ and $T > 0$. Let $f_1, f_2 \in C([0,T], \mathcal{M}_+(\mathbb{R}^d \times \mathbb{R}^d) - w^*)$ be two weak solutions of the Vlasov-Poisson system* (1.1) *with the same total mass such that* (1.8) *holds. If $(1+T)^{1+\varepsilon} W_1(f_1(0), f_2(0)) < 1/18$, then we have the bound for $t \in [0, T^*]$:*

- *If $\alpha = 1$ then*

$$W_1(f_1(t), f_2(t)) \le C W_1(f_1(0), f_2(0))^{\exp(-Ct)} \left(1 + t|\ln W_1(f_1(0), f_2(0))|^2\right).$$

- *If $\alpha > 1$ then*

$$W_1(f_1(t), f_2(t)) \le C W_1(f_1(0), f_2(0))^{1/\gamma} \exp(Ct^\gamma) \left(1 + t|\ln W_1(f_1(0), f_2(0))|^{1+(1/\alpha)}\right),$$

where

$$\gamma = \frac{2}{1-(1/\alpha)} \in [2, +\infty),$$

and where T^* satisfies the lower bound

$$T^* \ge \begin{cases} C' \ln|\ln W_1(f_1(0), f_2(0))| - C'^{-1} & \text{when } \alpha = 1, \\ C'\gamma |\ln W_1(f_1(0), f_2(0))|^{1/\gamma} - C'^{-1} & \text{when } \alpha \ne 1. \end{cases}$$

(we set $T^* = T$ if the right hand side is larger than T).

The constants C and C' depend only upon the norms of ρ_1, ρ_2 in (1.8) *and on ε.*

REMARK 1.10. The bound is stated in a way that is easy to understand for *large* t and is suboptimal near $t = 0$. In particular the bound does not converge to $W_1(f_1(0), f_2(0))$ as $t \to 0$[4]. Such a bound could be obtained by a careful analysis of the proofs, but we do not present this here.

REMARK 1.11. As will be clear in the proof of Theorem 1.9, the time T^* essentially corresponds to the first time at which the right hand side becomes larger or equal to 1.

REMARK 1.12. For $\alpha = +\infty$, the estimate of Theorem 1.9 reads

$$W_1(f_1(t), f_2(t)) \le C W_1(f_1(0), f_2(0))^{1/2} \exp(Ct^2) \left(1 + t|\ln W_1(f_1(0), f_2(0))|\right),$$

which is valid up to times of order $|\ln W_1(f_1(0), f_2(0))|^{1/2}$.

In [**5**], Dobrushin considered the stability of measure-valued solutions of first order mean-field PDE with Lipschitz convolution Kernels K and obtained the inequality

$$W_1(f_1(t), f_2(t)) \le W_1(f_1(0), f_2(0)) \exp\left(Ct\|\nabla K\|_{L^\infty}\right).$$

The same estimate was derived by Moussa and Sueur [**14**] for a mixed first/second order PDE. Hauray and Jabin [**7**] handled the case of more singular Kernels, see also the recent work by Lazarovici and Pickl [**9**] on cut-off kernels and the references quoted therein.

---

[4]They do, of course, converge to zero as $W_1(f_1(0), f_2(0)) \to 0$.

In the present situation, we are able to address the case of the singular convolution Kernel $K = \gamma x/|x|^d$ because, in contrast with the works mentioned above, the solutions have some additional regularity - the macroscopic density belongs to L_{ϕ_α}. Nevertheless, as a consequence of the singularity of K, the growth of $W_1(f_1(t), f_2(t))$ in Theorem 1.9 is not linearly bounded in terms of $W_1(f_1(0), f_2(0))$.

We mention that although stability estimates are not explicitly done in [**11**], the computations therein involve a log-Lipschitz Grönwall estimate and would yield the inequality

$$W_2(f_1(t), f_2(t)) \leq C W_2(f_1(0), f_2(0))^{\exp(-Ct)}, \tag{1.9}$$

with W_2 denoting

$$W_2(\mu, \nu) = \left(\inf_{\pi \in \Pi(\mu,\nu)} \int_{\mathbb{R}^d \times \mathbb{R}^d} |x-y|^2 \, d\pi(x,y) \right)^{1/2},$$

so the W_2-Wasserstein distance grows in time roughly like an exponential tower $e^{e^{ct}}$. Therefore the estimate of Theorem 1.9 setting $\alpha = +\infty$, which corresponds to the regularity considered in [**11**], improves this to stretched exponential growth of the form e^{ct^2}. This improvement is due to the second-order structure of the characteristic system (2.5) of ODE associated to the Vlasov-Poisson system, which was already exploited in the proof of uniqueness in [**15**].

Finally, we would like to point out that the same technique of exploiting the second-order structure can be applied to general measure solutions (with no regularity assumption on the spatial density), and allows the Dobrushin estimate to be improved slightly from *Lipschitz* kernels to *log^2-Lipschitz* kernels:

THEOREM 1.13. *Let the convolution kernel K be bounded and satisfy the log^2-Lipschitz property:*

$$|K(x) - K(y)| \leq C|x-y|(\ln|x-y|)^2 \text{ for all } |x-y| \leq 1/9. \tag{1.10}$$

Then the Vlasov-Poisson system (1.1) *has a unique solution such that f belongs to $C([0,T]; \mathcal{M}_+(\mathbb{R}^d \times \mathbb{R}^d) - w^*)$ for any initial datum in $\mathcal{M}_+(\mathbb{R}^d \times \mathbb{R}^d)$. Moreover it obeys the stability estimate, for any two solutions $f_1, f_2 \in C([0,T]; \mathcal{M}_+(\mathbb{R}^d \times \mathbb{R}^d) - w^*)$ with the same mass and satisfying $(1+T)W_1(f_1(0), f_2(0)) < 1/9$,*

$$W_1(f_1(t), f_2(t)) \leq C W_1(f_1(0), f_2(0))^{\exp(-Ct)} (1 + t|\ln W_1(f_1(0), f_2(0))|^2).$$

which holds for times $t \in [0, T^*]$ with T^* defined analogously to Theorem 1.9.

We remark that the conventional improvement of the Dobrushin estimate by replacing the Grönwall inequality with a log-Lipschitz inequality only allows one to treat log-Lipschitz kernels K, rather than the slightly weaker assumption (1.10).

In the second part of our analysis, we seek for initial data f_0 for which the macroscopic density indeed belongs to some exponential Orlicz space.

PROPOSITION 1.14. *Let $f_0 \in L^\infty(\mathbb{R}^d \times \mathbb{R}^d)$ be such that*

$$\int_{\mathbb{R}^d \times \mathbb{R}^d} f_0(x,v) e^{c\langle v \rangle^{d\alpha}} \, dx dv < \infty,$$

for some $\alpha \in [1, \infty)$ and $c > 0$, where $\langle v \rangle = \sqrt{1+|v|^2}$. For $T > 0$, let $f \in C([0,T], \mathcal{M}_+(\mathbb{R}^d \times \mathbb{R}^d) - w^*) \cap L^\infty([0,T], L^1 \cap L^\infty(\mathbb{R}^d \times \mathbb{R}^d))$ be any solution to

(1.1), *with this initial datum, provided by* [**10**, Theo. 1][5]. *Then it satisfies*

$$\sup_{t\in[0,T]} \|\rho(t)\|_{L_{\phi_\alpha}} \le C < \infty.$$

In particular, this solution satisfies the uniqueness criterion of Theorem 1.9.

We remark that setting $\alpha = 1$, we retrieve as a special case the condition obtained in [**15**, Theo. 1.2] to ensure that (1.3) holds. For any $\alpha \in [1, \infty)$, an example of initial data that are allowed by Proposition 1.14 are Maxwell-Boltzmann distributions of the form

$$f_0(x, v) = e^{-c|v|^{d\alpha}}(1 + |v|^p)h_0(x, v), \quad c > 0,\ p \ge 0,\ h_0 \in L^1 \cap L^\infty.$$

The plan of the remainder of the paper is as follows. In Section 2 we prove Theorem 1.9. We first establish in Lemma 2.1 a log-Lipschitz like estimate for the force field $E = K * \rho$ associated to a function ρ satisfying (1.8). Then, we introduce in (2.7) a notion of distance between two solutions in terms of the characteristics defined in (2.5), which controls the Wasserstein distance (see (2.8)). This quantity was used in the original proof of Dobrushin and also in [**15**], while the proof of [**11**] uses a slightly different version. Applying similar arguments as in [**5**], we derive a second-order differential inequality for this distance, which eventually leads to Theorem 1.9. In Subsection 2.4 we show how to adapt this technique to prove Theorem 1.13. Finally, the last Section 3 is devoted to the proof of Proposition 1.14.

2. Proof of Theorem 1.9

2.1. An estimate for the Newton kernel. To prove Theorem 1.9 we have need of the following lemma on the Newton kernel. Note that the complementary N-functions of the ϕ_α behave asymptotically (see Remark 1.4) like

$$\bar{\phi}_\alpha(\tau) \sim \tau \ln(\tau)^{1/\alpha} \text{ as } \tau \to \infty. \tag{2.1}$$

Recall that Orlicz spaces obey a form of Hölder's inequality (see e.g. [**19**])

$$\left|\int_{\mathbb{R}^d} fg\,dx\right| \le C\|f\|_{L_\phi}\|g\|_{L_{\bar{\phi}}}$$

for the constant $C > 1$.

Given $\alpha \in [1, \infty]$ we define the constant $\beta \in [1, 2]$ by

$$\beta = \frac{1}{\alpha} + 1. \tag{2.2}$$

In particular, note that $\beta = 1$ for $\alpha = +\infty$ and $\beta = 2$ as $\alpha = 1$.

LEMMA 2.1. *Let $\alpha \in [1, \infty)$, then there exists $C = C(\alpha) > 0$ such that for all $g \in L_{\phi_\alpha} \cap L^1$ we have the estimate*

$$\int_{\mathbb{R}^d} |K(x - z) - K(y - z)|\,|g(z)|\,dz \le C(\|g\|_{L_{\phi_\alpha}} + \|g\|_{L^1})\psi_\alpha(|x - y|), \tag{2.3}$$

[5]The existence of such a solution is ensured by [**10**, Theo. 1] because f_0 has finite velocity moments of sufficiently large order: $\int |v|^m f_0(x, v)\,dx\,dv < +\infty$ for some $m > d^2 - d$. This is proved in [**10**] for $d = 3$. The case $d = 2$ is a straightforward adaptation of the case $d = 3$.

where ψ is defined by

$$\psi_\alpha(\tau) = \begin{cases} \tau|\ln(\tau)|^\beta, & \textit{for } \tau \in \left[0, \frac{1}{9}\right], \\ \frac{1}{9}\ln(9)^\beta, & \textit{for } \tau \geq \frac{1}{9} \end{cases} \tag{2.4}$$

and where β is defined by (2.2).

REMARK 2.2. In the case $\alpha = +\infty$, namely for $g \in L^1 \cap L^\infty$ the estimate of Lemma 2.1 is standard, see e.g. [**12**, Lemma 8.1] for the case $d = 2$: we have

$$\int_{\mathbb{R}^d} |K(x-z) - K(y-z)|\,|g(z)|\,dz \leq C(\|g\|_{L^1\cap L^\infty})|x-y|\,(1 + |\ln|x-y||)\,.$$

REMARK 2.3. For the case $\alpha = 1$, the following variant of Lemma 2.1 was obtained in [**15**, Lemma 2.2]: for all $x, y \in \mathbb{R}^d$ with $|x-y|$ sufficiently small,

$$\int_{\mathbb{R}^d} |K(x-z) - K(y-z)|\,|g(z)|\,dz \leq C\,p\,(\|g\|_{L^1} + \|g\|_{L^p})|x-y|^{1-d/p}, \quad \forall p > d.$$

In particular, recalling (1.7) for $\alpha = 1$, this yields

$$\int_{\mathbb{R}^d} |K(x-z) - K(y-z)|\,|g(z)|\,dz \leq C\|g\|_{L_{\phi_1}}|x-y|\left(p^2|x-y|^{-d/p}\right), \quad \forall p > d,$$

so setting $p = |\ln|x-y||$ we retrieve the estimate of Lemma 2.1. In fact one can also prove the other cases via this method. Nevertheless, we give a direct proof of Lemma 2.1 below for completeness.

PROOF. We set

$$\delta = \frac{1}{d}\left(1 + \frac{1}{\alpha}\right) = \frac{\beta}{d} \in \left(\frac{1}{d}, \frac{2}{d}\right].$$

By standard estimates using Hölder's inequality (see e.g. [**13**]) it is well-known that, fixing some $p_0 > d$,

$$\int_{\mathbb{R}^d} |K(x-z) - K(y-z)|\,|g(z)|\,dz \leq C(\|g\|_{L^{p_0}} + \|g\|_{L^1}) \leq C(\|g\|_{L_{\phi_\alpha}} + \|g\|_{L^1}).$$

Hence, in view of the form of ψ_α, letting $R = |x-y|$ we may assume without loss of generality that $R \leq 1/9$. We introduce $A = (x+y)/2$. Since $R|\ln R|^\delta < 1$, we may split the integral as follows:

$$\begin{aligned} &\int_{\mathbb{R}^d} |K(x-z) - K(y-z)|\,|g(z)|\,dz \\ &= \int_{\mathbb{R}^d\setminus B(A,|\ln R|^{-\delta})} |K(x-z) - K(y-z)|\,|g(z)|\,dz \\ &+ \int_{B(A,|\ln R|^{-\delta})\setminus B(A,R)} |K(x-z) - K(y-z)|\,|g(z)|\,dz \\ &+ \int_{B(A,R)} |K(x-z) - K(y-z)|\,|g(z)|\,dz \\ &= I_1 + I_2 + I_3. \end{aligned}$$

For I_1 we apply the mean value theorem to obtain the bound

$$I_1 \leq C\|g\|_{L^1}\,R \sup_{u\in[x,y],z\in\mathbb{R}^d\setminus B(A,|\ln R|^{-\delta})} \frac{1}{|u-z|^d} \leq C\|g\|_{L^1}\,R|\ln R|^{d\delta}$$

where $[x, y]$ is the line segment joining x and y, and where we have used that

$$|u - z| \geq |\ln R|^{-\delta} - \frac{R}{2} \geq \frac{|\ln R|^{-\delta}}{2}$$

in the considered supremum. Therefore we have obtained

$$I_1 \leq C\|g\|_{L^1} R |\ln R|^{\beta}.$$

For I_3 we apply Hölder's inequality for Orlicz spaces,

$$I_3 \leq C\|g\|_{L_{\phi_\alpha}} \|1_{B(A,R)}|K(x-z) - K(y-z)|\|_{L_{\bar{\phi}_\alpha}} \leq C\|g\|_{L_{\phi_\alpha}} \|1_{B(0,3R/2)} K\|_{L_{\bar{\phi}_\alpha}}$$

where we have used the fact that $\bar{\phi}_\alpha$ is increasing, that

$$|K(x-z) - K(y-z)| \leq |K(x-z)| + |K(y-z)|$$

and that $z \in B(A, R)$ implies that both $x - z$ and $y - z$ lie in $B(0, 3R/2)$.

Now we set $\lambda = R|\ln R|^{1/\alpha}$ and we consider the integral

$$\int_{\mathbb{R}^d} \bar{\phi}_\alpha(1_{B(0,3R/2)}|K(z)|/\lambda)\, dz.$$

By Remark 1.3, to show that $I_3 \leq C\|g\|_{L_{\phi_\alpha}} \lambda$ it is sufficient to show that the integral above is bounded by a constant. Furthermore, by Remark 1.4 using the fact that $|K(z)|/\lambda \geq 1 > 0$ on $B(0, 3R/2)$ we may work with the asymptotic form (2.1).

Thus, we estimate

$$\begin{aligned}\int_{|z|\leq 3R/2} \frac{|K(z)|}{\lambda} \ln\left(\frac{|K(z)|}{\lambda}\right)^{1/\alpha} dz &= \frac{1}{\lambda}\int_{|z|\leq 3R/2} |z|^{1-d} \ln\left(\frac{|z|^{1-d}}{\lambda}\right)^{1/\alpha} dz \\ &\leq \frac{C}{\lambda}\int_0^{3R/2} |\ln r|^{1/\alpha}\, dr\end{aligned}$$

where we have used the inequality

$$0 \leq \ln\left(\frac{|z|^{1-d}}{\lambda}\right) \leq (d-1)|\ln|z|| + |\ln\lambda| \leq C|\ln|z||$$

with this definition of λ.

Thus, noting that for $r \leq 3R/2 \leq 1/6$ we have

$$\frac{1}{\alpha}|\ln r|^{(1/\alpha)-1} \leq \frac{1}{2}|\ln r|^{1/\alpha},$$

so that

$$|\ln r|^{1/\alpha} \leq 2\left(|\ln r|^{1/\alpha} - \frac{1}{\alpha}|\ln r|^{(1/\alpha)-1}\right),$$

we obtain

$$\begin{aligned}\int_{|z|\leq 3R/2} \frac{|K(z)|}{\lambda} \ln\left(\frac{|K(z)|}{\lambda}\right)^{1/\alpha} dz &\leq \frac{C}{\lambda}\int_0^{3R/2} \left(|\ln r|^{1/\alpha} - \frac{1}{\alpha}|\ln r|^{(1/\alpha)-1}\right) dr \\ &= \frac{C}{\lambda}\left[r|\ln r|^{1/\alpha}\right]_0^{3R/2} \leq \frac{C}{\lambda} R|\ln R|^{1/\alpha}.\end{aligned}$$

Thus we have shown that

$$I_3 \leq C\|g\|_{L_{\phi_\alpha}} R|\ln R|^{1/\alpha}.$$

Finally, we bound I_2. In the same way as for I_3, we apply Hölder's inequality for Orlicz spaces to obtain

$$I_2 \leq C\|g\|_{L_{\phi_\alpha}} \|1_{B(A,|\ln R|^{-\delta})\setminus B(A,R)}|K(x-z)-K(y-z)|\|_{L_{\bar{\phi}_\alpha}}.$$

Applying the mean value theorem we obtain for $z \in B(A,|\ln R|^{-\delta}) \setminus B(A,R)$

$$\begin{aligned}|K(x-z)-K(y-z)| &\leq C\,R \sup_{u\in[x,y]} |u-z|^{-d} \\ &\leq C\,R \sup_{u\in[x,y]} \frac{1}{||z-A|-|u-A||^d} \\ &\leq C\,R|z-A|^{-d},\end{aligned}$$

where we have used that $|z-A| \geq R$ to obtain the final inequality. Hence, by a change of variables, and since $\bar{\phi}_\alpha$ is increasing, in order to estimate I_2 it is sufficient to obtain the bound

$$\|1_{B(0,|\ln R|^{-\delta})\setminus B(0,R)} R|z|^{-d}\|_{L_{\bar{\phi}_\alpha}} \leq C\,R|\ln R|^\beta.$$

Therefore, setting $\lambda' = R|\ln R|^\beta$, by Remark 1.3 it is enough to show that

$$\int_{R\leq|z|\leq|\ln R|^{-\delta}} \bar{\phi}_\alpha(R|z|^{-d}/\lambda')\,dz \leq C.$$

Let $|z| \leq |\ln R|^{-\delta}$, then we have by definition of λ'

$$\frac{R|z|^{-d}}{\lambda'} \geq \frac{R|\ln R|^{d\delta}}{\lambda'} = 1,$$

so by Remark 1.4 we may instead bound the asymptotic form (2.1). Therefore, we estimate

$$\begin{aligned}\int_{R\leq|z|\leq|\ln R|^{-\delta}} &\frac{R|z|^{-d}}{\lambda'} \ln\left(\frac{R|z|^{-d}}{\lambda'}\right)^{1/\alpha} dz \\ &= |\ln R|^{-\beta} \int_{R\leq|z|\leq|\ln R|^{-\delta}} |z|^{-d}\ln(|z|^{-d}|\ln R|^{-\beta})^{1/\alpha}\,dz. \\ &= C|\ln R|^{-\beta} \int_R^{|\ln R|^{-\delta}} r^{-1}|\ln(r^{-d}|\ln R|^{-\beta})|^{1/\alpha}\,dr\end{aligned}$$

Since for $r \leq |\ln R|^{-\delta}$ we have

$$|\ln(r^{-d}|\ln R|^{-\beta})| = \ln(r^{-d}|\ln R|^{-\beta}) = d|\ln r| - \beta\ln|\ln R| \leq d|\ln r|,$$

we infer that

$$\begin{aligned}\int_{R\leq|z|\leq|\ln R|^{-\delta}} &\frac{R|z|^{-d}}{\lambda'} \ln\left(\frac{R|z|^{-d}}{\lambda'}\right)^{1/\alpha} dz \\ &\leq C|\ln R|^{-\beta} \int_R^{|\ln R|^{-\delta}} r^{-1}|\ln r|^{1/\alpha}\,dr \\ &\leq C|\ln R|^{-\beta}\left[-|\ln r|^{(1/\alpha)+1}\right]_R^{|\ln R|^{-\delta}} \\ &\leq C,\end{aligned}$$

as we wanted, and hence we obtain

$$I_2 \leq C\|g\|_{L_{\phi_\alpha}(\mathbb{R}^d)} R|\ln R|^\beta.$$

Finally, putting this all together, we conclude that

$$\int_{\mathbb{R}^d} |K(x-z) - K(y-z)|\,|g(z)|\,dz$$
$$\leq I_1 + I_2 + I_3 \leq C(\|g\|_{L^1} R + \|g\|_{L_{\phi_\alpha}} R|\ln R|^\beta + \|g\|_{L_{\phi_\alpha}} R|\ln R|^{1/\alpha})$$

which implies the claim of the lemma. □

2.2. Lagrangian formulation of the Vlasov-Poisson system and the Wasserstein distance. Let $f \in C([0,T], \mathcal{M}_+(\mathbb{R}^d \times \mathbb{R}^d) - w^*)$ be a weak measure-valued solution of the Vlasov-Poisson system (1.1) on $[0,T]$ such that ρ belongs to $L^\infty([0,T], L^1 \cap L^p(\mathbb{R}^d))$ for some $p > d$. By potential estimates it is well-known that E belongs to $L^\infty([0,T] \times \mathbb{R}^d)$. Moreover, by Caldéron-Zygmund inequality (see e.g. see [**4**, Theo. 4.12]) $\nabla E \in L^\infty([0,T], L^p(\mathbb{R}^d))$. By the theory on transport equations (see [**3**, Theo. III2] or [**1**, Theo. 5.7] for more recent results on the theory), there exists a unique Lagrangian flow associated to E, namely a map $(X,V) \in L^1_{\text{loc}}([0,T] \times \mathbb{R}^d \times \mathbb{R}^d; \mathbb{R}^d \times \mathbb{R}^d)$ such that for a.e. $(x,v) \in \mathbb{R}^d \times \mathbb{R}^d$, $t \mapsto (X,V)(t,x,v)$ is an absolutely continuous integral solution of the characteristic system of ODE

$$\begin{cases} \dot{X}(t,x,v) = V(t,x,v), & X(0,x,v) = x \\ \dot{V}(t,x,v) = E(t, X(t,x,v)), & V(0,x,v) = v. \end{cases} \tag{2.5}$$

Moreover, we have the representation[6]

$$\forall t \in [0,T], \quad f(t) = (X,V)(t)_\# f_0. \tag{2.6}$$

Let f_1, f_2 be two weak solutions of the Vlasov-Poisson equation (1.1) as in Theorem 1.9, then $f_j(t) = (X_j(t), V_j(t))_\# f_{j0}$ for $(X_j, V_j)(t,x,v)$ the solutions to the characteristic equations (2.5) associated to E_j.

REMARK 2.4. In fact, under the assumptions of Theorem 1.9, the characteristic flows are Hölder continuous as functions of (x,v). This may be deduced from a similar Grönwall type estimate to the proof of Lemma 2.5 below using that E_i satisfy a $\log^2$-Lipschitz bound of the form (1.10). This will not be needed for the proof of Theorem 1.9.

Given a coupling $\pi_0 \in \Pi(f_{10}, f_{20})$ (as defined in Definition 1.7) we define the following quantities:

$$\begin{aligned} \mathcal{X}(t) = \mathcal{X}_{\pi_0}(t) &= \int_{\mathbb{R}^{2d}\times\mathbb{R}^{2d}} |X_1(t,x,v) - X_2(t,y,w)|\,d\pi_0(x,v,y,w), \\ \mathcal{V}(t) = \mathcal{V}_{\pi_0}(t) &= \int_{\mathbb{R}^{2d}\times\mathbb{R}^{2d}} |V_1(t,x,v) - V_2(t,y,w)|\,d\pi_0(x,v,y,w). \end{aligned} \tag{2.7}$$

By (2.6), the measure $\pi_t = ((X_1(t), V_1(t)); (X_2(t), V_2(t)))_\# \pi_0$ belongs to the space $\Pi(f_1(t), f_2(t))$. Therefore, by the Definition 1.7 of the Wasserstein distance, we have

$$W_1(f_1(t), f_2(t)) \leq \inf_{\pi_0 \in \Pi(f_{10}, f_{20})} \left(\mathcal{X}_{\pi_0}(t) + \mathcal{V}_{\pi_0}(t)\right), \quad t \in [0,T]. \tag{2.8}$$

On the other hand, note the converse estimate:

---

[6]The # notation means that $f(t)(B) = f_0\left(((X,V)(t,\cdot,\cdot)^{-1}(B)\right)$ for all Borel set $B \subset \mathbb{R}^d$.

(2.9)

$$\begin{aligned}\inf_{\pi_0\in\Pi(f_{10},f_{20})}(\mathcal{X}_{\pi_0}(0)+\mathcal{V}_{\pi_0}(0)) &\leq \inf_{\pi_0\in\Pi(f_{10},f_{20})}\int_{\mathbb{R}^{2d}\times\mathbb{R}^{2d}}(|x-y|+|v-w|)\,d\pi_0(x,v,y,w)\\ &\leq \inf_{\pi_0\in\Pi(f_{10},f_{20})}\int_{\mathbb{R}^{2d}\times\mathbb{R}^{2d}}\sqrt{2}|(x,v)-(y,w)|\,d\pi_0(x,v,y,w)\\ &= \sqrt{2}W_1(f_{1,0},f_{2,0}).\end{aligned}$$

We emphasize that the quantity which would lead to (1.9) in [**11**] is instead

$$\left\{\int_{\mathbb{R}^{2d}\times\mathbb{R}^{2d}}\left(|X_1(t,x,v)-X_2(t,y,w)|^2+|V_1(t,x,v)-V_2(t,y,w)|^2\right)\,d\pi_0(x,v,y,w)\right\}^{1/2},$$

since it controls the Wasserstein distance $W_2(f_1(t),f_2(t))$.

2.3. Proof of Theorem 1.9 completed. We will prove Theorem 1.9 proper with the following lemma which controls the distance involving the spatial characteristics, namely the quantity $\mathcal{X}(t)$.

We recall that $\beta = 1+(1/\alpha)$. For a given $0<A<1/9$, we define the function $G_\alpha(t) = G_\alpha(t;A) = G_\alpha(t;A,c)$ as the solution to

$$G'_\alpha(t;A,c) = -cG_\alpha(t;A,c)^{\beta/2}, \qquad G_\alpha(0;A,c) = \ln(1/A) \tag{2.10}$$

for times $t\in[0,T^*]$ where $T^* = T^*(\alpha,A,c)$ is the maximal time such that

$$G_\alpha(\cdot;A,c) > \ln(9) \quad \text{on}[0,T^*)$$

(we set $T^*=T$ if T^* is larger than T). Note that $G_\alpha(\cdot;A,c)$ is decreasing and is explicitly given by

$$G_\alpha(t;A,c) = \begin{cases}|\ln A|\exp(-ct) & \text{when } \alpha=1,\\ \left(|\ln A|^{1/\gamma}-c\gamma^{-1}t\right)^\gamma & \text{when } \alpha\neq 1,\end{cases} \tag{2.11}$$

where γ is given by (1.9). Moreover, we have

$$T^*(\alpha,A,c) = \begin{cases}\dfrac{1}{c}\ln\left(\dfrac{|\ln A|}{\ln 9}\right) & \text{when } \alpha=1,\\ \dfrac{\gamma}{c}\left(|\ln A|^{1/\gamma}-(\ln 9)^{1/\gamma}\right) & \text{when } \alpha\neq 1.\end{cases} \tag{2.12}$$

LEMMA 2.5. *Let $\pi_0\in\Pi(f_{10},f_{20})$ and (1.8) hold. Assume that*

$$0\leq A := (1+T)(\mathcal{X}(0)+\mathcal{V}(0)) < \frac{1}{9}.$$

Then the following estimate holds

$$\mathcal{X}(t)\leq \exp(-G_\alpha(t;A(t),c)), \quad t\leq T^*(\alpha,A,c),$$

where

$$A(t) = (1+t)(\mathcal{X}(0)+\mathcal{V}(0))\leq A$$

and where $c>0$ is a constant depending only on α and the norms in (1.8).

PROOF. By integrating the characteristic ODEs (2.5) twice we have

(2.13)
$$X_1(t,x,v) - X_2(t,y,w)$$
$$= x - y + (v-w)t + \int_0^t \int_0^s [E_1(\tau, X_1(\tau,x,v)) - E_2(\tau, X_2(\tau,y,w))]\, d\tau ds.$$

Since $f_j(t) = (X_j(t), V_j(t))\# f_{j0}$, we can evaluate the fields E_j as follows, where we omit the τ dependence for brevity:

$$\begin{aligned}
&E_1(X_1(x,v)) - E_2(X_2(y,w))\\
&= \int_{\mathbb{R}^{2d}} K(X_1(x,v) - X_1(x_0,v_0)) f_{10}(x_0,v_0)\, dx_0 dv_0\\
&\quad - \int_{\mathbb{R}^{2d}} K(X_2(x,v) - X_2(y_0,w_0)) f_{20}(y_0,w_0)\, dy_0 dw_0\\
&= \int_{\mathbb{R}^{2d}\times\mathbb{R}^{2d}} [K(X_1(x,v) - X_1(x_0,v_0)) - K(X_2(x,v) - X_2(y_0,w_0))]\, d\pi_0(x_0,v_0,y_0,w_0)\\
&= \int_{\mathbb{R}^{4d}} [K(X_1(x,v) - X_1(x_0,v_0)) - K(X_2(x,v) - X_1(x_0,v_0))]\, d\pi_0(x_0,v_0,y_0,w_0)\\
&\quad + \int_{\mathbb{R}^{4d}} [K(X_2(x,v) - X_1(x_0,v_0)) - K(X_2(x,v) - X_2(y_0,w_0))]\, d\pi_0(x_0,v_0,y_0,w_0)\\
&= \int_{\mathbb{R}^{4d}} [K(X_1(x,v) - z) - K(X_2(x,v) - z)]\rho_1(z)\, dz\\
&\quad + \int_{\mathbb{R}^{4d}} [K(X_2(x,v) - X_1(x_0,v_0)) - K(X_2(x,v) - X_2(y_0,w_0))]\, d\pi_0(x_0,v_0,y_0,w_0).
\end{aligned}$$

Thus, by applying Lemma 2.1 we obtain the estimate

$$\begin{aligned}
&|E_1(X_1(x,v)) - E_2(X_2(y,w))| \le C\psi_\alpha(|X_1(x,v) - X_2(x,v)|)\\
&\quad + \int_{\mathbb{R}^{4d}} |K(X_2(x,v) - X_1(x_0,v_0)) - K(X_2(x,v) - X_2(y_0,w_0))|\, d\pi_0(x_0,v_0,y_0,w_0).
\end{aligned}$$

It follows that

$$\begin{aligned}
&\int_{\mathbb{R}^{4d}} |E_1(X_1(x,v)) - E_2(X_2(y,w))| d\pi_0(x,v,y,w)\\
&\le C \int_{\mathbb{R}^{4d}} \psi_\alpha(|X_1(x,v) - X_2(x,v)|)\, d\pi_0(x,v,y,w)\\
&+ \int_{\mathbb{R}^{4d}} d\pi_0(x,v,y,w)\\
&\left(\int_{\mathbb{R}^{4d}} |K(X_2(x,v) - X_1(x_0,v_0)) - K(X_2(x,v) - X_2(y_0,w_0))|\, d\pi_0(x_0,v_0,y_0,w_0)\right)\\
&\le C \int_{\mathbb{R}^{4d}} \psi_\alpha(|X_1(x,v) - X_2(x,v)|)\, d\pi_0(x,v,y,w)\\
&+ \int_{\mathbb{R}^{4d}} \left(\int_{\mathbb{R}^{4d}} |K(z - X_1(x_0,v_0)) - K(z - X_2(y_0,w_0))| \rho_2(\tau,z)\, dz\right) d\pi_0(x_0,v_0,y_0,w_0)
\end{aligned}$$

where we have exchanged the order of integration with $d\pi_0(x_0,v_0,y_0,w_0)$ and used that $f_1(\tau) = (X_1(\tau), V_1(\tau))\# f_{10}$ in the last inequality. Therefore, by integrating

(2.13) against the measure $d\pi_0(x,v,y,w)$ we obtain

$$\begin{aligned}\mathcal{X}(t) &\le \int_{\mathbb{R}^{4d}} |x-y+t(v-w)|\, d\pi_0(x,v,y,w) \\ &+ \int_0^t \int_0^s \int_{\mathbb{R}^{4d}} |E_1(\tau, X_1(\tau,x,v)) - E_2(\tau, X_2(\tau,y,w))| d\pi_0(x,v,y,w)\, d\tau\, ds \\ &\le [\mathcal{X}(0) + t\mathcal{V}(0)] \\ &+ 2C \int_0^t \int_0^s \int_{\mathbb{R}^{4d}} \psi_\alpha(|X_1(\tau,x,v) - X_2(\tau,x,v)|)\, d\pi_0(x,v,y,w)\, d\tau\, ds\end{aligned}$$

where we have applied Lemma 2.1 (noting Remark 2.2 if $\alpha = \infty$) to find the second inequality.

Using that ψ_α is concave we deduce that

$$\mathcal{X}(t) \le (\mathcal{X}(0) + \mathcal{V}(0)(1+t) + C_0 \int_0^t \int_0^s \psi_\alpha(\mathcal{X}(\tau))\, d\tau\, ds$$

for a constant C_0 depending only on α and the norms in (1.8). For a constant c to be determined later on, let $T^*(\alpha, A, c)$ be the corresponding time defined by (2.12). Let $t_0 \in [0, T^*(\alpha, A, c)]$ be fixed and set

$$\mathcal{F}(t) = (\mathcal{X}(0) + \mathcal{V}(0)(1+t_0) + C_0 \int_0^t \int_0^s \psi_\alpha(\mathcal{X}(\tau))\, d\tau\, ds \ge \mathcal{X}(t), \quad t \in [0, t_0].$$

Define $\varphi_\alpha(t) = \int_0^t \psi_\alpha(s)\, ds$ and note that $\varphi_\alpha(\tau) \le C\tau^2 |\ln(\tau)|^\beta$ for $\tau \le 1/9$ and $\varphi(\tau) \le C\tau$ for $\tau \ge 1/9$. Then it holds that

$$\mathcal{F}(0) = (\mathcal{X}(0) + \mathcal{V}(0)(1+t_0), \quad \mathcal{F}'(0) = 0, \quad \mathcal{F}'(t) \ge 0,$$

and

$$\mathcal{F}''(t) = C_0 \psi_\alpha(\mathcal{X}(t)) \le C_0 \psi_\alpha(\mathcal{F}(t)) = C_0 \varphi'_\alpha(\mathcal{F}(t)).$$

Thus

$$[(\mathcal{F}'(t))^2]' = 2\mathcal{F}''(t)\mathcal{F}'(t) \le 2C_0 \varphi'_\alpha(\mathcal{F}(t))\mathcal{F}'(t) = 2C_0[\varphi_\alpha(\mathcal{F}(t))]'$$

and by integrating we deduce that

$$\mathcal{F}'(t) \le \sqrt{2C_0\, \varphi_\alpha(\mathcal{F}(t))}, \quad \text{for } t \le t_0,$$

which by definition of φ_α implies

$$\mathcal{F}'(t) \le C_1\, \mathcal{F}(t) |\ln \mathcal{F}(t)|^{\beta/2}, \quad \text{for } \mathcal{F}(t) \le \frac{1}{9}, \tag{2.14}$$

where C_1 depends only on α and the norms in (1.8). Now let $y(t)$ be the solution to

$$y'(t) = C_1 y(t) |\ln y(t)|^{\beta/2}, \quad y(0) = \mathcal{F}(0) = A(t_0),$$

for $t \in [0, T^*(\alpha, A, c)]$. In view of the definition (2.12), since $A(t_0) \le A$ we have $T^*(\alpha, A(t_0), C_1) \ge T^*(\alpha, A, C_1)$ so that $y \le 1/9$ on $[0, T^*(\alpha, A, C_1)]$. Then (2.14) obeys $\mathcal{F}(t) \le y(t) \le 1/9$ on its domain of definition. By applying the change of variables $y = e^{-G}$ we deduce that $G' = -C_1 G^{\beta/2}$ and that therefore

$$\mathcal{F}(t) \le y(t) = \exp(-G_\alpha(t, A(t_0), C_1)), \quad t \in [0, t_0],$$

and as t_0 was arbitrary the proof of the lemma is complete by setting $c = C_1$. □

Using this lemma we are now able to prove the main result Theorem 1.9.

Proof of Theorem 1.9. By integrating the characteristic equation (2.5) once we obtain
(2.15)

$$V_1(t,x,v) - V_2(t,y,w) = v - w + \int_0^t [E_1(s, X_1(s,x,v)) - E_2(s, X_2(s,y,w))]\, ds.$$

Letting $\pi_0 \in \Pi(f_{10}, f_{20})$ be arbitrary, in the same way as in the proof of Lemma 2.5 we find that

$$\mathcal{V}(t) \leq \mathcal{V}(0) + C\int_0^t \psi_\alpha(\mathcal{X}(s))\, ds.$$

By (2.9), we may consider only couplings π_0 such that $\mathcal{X}(0)+\mathcal{V}(0) \leq 2W_1(f_1(0), f_2(0))$ and, therefore, by assumption on $W_1(f_1(0), f_2(0))$ in Theorem 1.9

$$A := (\mathcal{X}(0) + \mathcal{V}(0))(1+T) < \frac{1}{9}.$$

So we also have $\mathcal{X}(t) \leq \exp(-G_\alpha(t, A(t), c))$ with $A(t) = (\mathcal{X}(0) + \mathcal{V}(0))(1+t)$ by Lemma 2.5 and for $t \leq T^*(\alpha, A, c)$. Note that by definition (2.12) of the time T^*, since $W_1(f_1(0), f_2(0))(1+T) \leq A$ we have

$$T^*(\alpha, A, c) \geq T^* := T^*(\alpha, W_1(f_1(0), f_2(0))(1+T), c).$$

Thus all the subsequent estimates hold for $t \in [0, T^*]$.

Thus, since ψ_α is an increasing function, we obtain, dropping the α, A and c in G for brevity,

$$\begin{aligned}
\int_0^t \psi_\alpha(\mathcal{X}(s))\, ds &\leq \int_0^t \psi_\alpha(\exp(-G(s)))\, ds = \int_0^t \exp(-G(s))G(s)^\beta\, ds \\
&\leq \int_0^t \exp(-G(t))G(0)^\beta\, ds = t\exp(-G(t))G(0)^\beta \\
&= t\exp(-G(t))|\ln A(0)|^\beta,
\end{aligned}$$

where we have used that $s \mapsto G(s) = G_\alpha(s, A(s), c)$ is a decreasing function of s.

Combining the estimates for $\mathcal{X}$ and $\mathcal{V}$ we have

$$\begin{aligned}
&\mathcal{X}(t) + \mathcal{V}(t) \\
&\quad\leq \exp(-G_\alpha(t; A(t), c)) + \mathcal{V}(0) + Ct\exp(-G_\alpha(t; A(t), c))|\ln A(0)|^\beta \\
&\quad\leq \exp(-G_\alpha(t; A(t), c)) + A(t) + Ct\exp(-G_\alpha(t; A(t), c))|\ln A(0)|^\beta \\
&\quad= \exp(-G_\alpha(t; A(t), c)) + \exp(-G_\alpha(0, A(t), c)) + Ct\exp(-G_\alpha(t; A(t), c))|\ln A(0)|^\beta \\
&\quad\leq 2\exp(-G_\alpha(t; A(t), c)) \\
&\quad+ Ct\exp(-G_\alpha(t; A(t), c))|\ln A(0)|^\beta
\end{aligned}$$

where we have used that $t \mapsto G_\alpha(t, \overline{A}, c)$ is decreasing for fixed $\overline{A}$ in the last line. Thus for $t \in [0, T^*]$ we obtain

$$\mathcal{X}(t) + \mathcal{V}(t) \leq \exp(-G_\alpha(t; A(t), c))(2 + Ct|\ln A(0)|^\beta).$$

We set $B = W_1(f_1(0), f_2(0))$, so that $B \leq A(0) \leq 2B$. By taking the infimum over couplings π_0 (recall (2.8) and (2.9)) we obtain

$$W_1(f_1(t), f_2(t)) \leq \exp(-G_\alpha(t; 2(1+t)B))(2 + Ct|\ln B|^\beta).$$

Now suppose $\alpha = 1$, then by the explicit formula (2.11) we have (recalling that $\beta = 2$ in this case)

$$\begin{aligned} W_1(f_1(t), f_2(t)) &\leq \exp(\ln(2(1+t)B)\exp(-ct))(2 + C\,t\,|\ln B|^2) \\ &\leq (2(1+t)B)^{\exp(-ct)}(2 + C\,t|\ln B|^2) \\ &\leq CB^{\exp(-ct)}(1 + t|\ln B|^2) \end{aligned}$$

where we have used that $(2(1+t))^{\exp(-ct)}$ is bounded by a constant uniformly over $t \in [0, \infty)$.

Suppose instead that $\alpha \neq 1$, then

$$\begin{aligned} W_1(f_1(t), f_2(t)) &\leq \exp(-G_\alpha(t; A(t), c))(2 + C\,t|\ln B|^\beta) \\ &\leq \exp(\gamma^{-1}\ln(2B(1+t)) + Ct^\gamma)(2 + C\,t|\ln B|^\beta) \\ &\leq (2B(1+t))^{1/\gamma}\exp(Ct^\gamma)(2 + C\,t|\ln B|^\beta) \\ &\leq CB^{1/\gamma}\exp(Ct^\gamma)(1 + t|\ln B|^\beta) \end{aligned}$$

where on the last line we have used that $e^{Ct^\gamma}t^\delta \leq Ce^{C't^\gamma}$ for a larger constant $C' > C$, and on the second line we have used the lower bound

$$G_\alpha(t; A, c) \geq \gamma^{-1}\ln(1/A) - Ct^\gamma \tag{2.16}$$

for $\alpha \neq 1$, which we will now prove. Indeed, from (2.11) we use convexity (noting $\gamma \geq 2$) to obtain

$$G(t) \geq G(0) - tG'(0) = G(0) - CtG(0)^{\beta/2},$$

and the desired bound follows from an application of Young's inequality, i.e.

$$G(0)^{\beta/2}t \leq \gamma^{-1}t^\gamma + (\beta/2)G(0).$$

Finally, we infer the lower bound for $T^* = T^*(\alpha, B(1+T), c)$ as follows: since $(1+T)^{1+\varepsilon}B \leq 1$, we have

$$|\ln B(1+T)| \geq \frac{\varepsilon}{1+\varepsilon}|\ln B|.$$

So in view of (2.12), we have for $\alpha = 1$

$$T^* \geq \frac{1}{c}\left(\ln|\ln B| + \ln\left(\frac{\varepsilon}{1+\varepsilon}\right) - \ln\ln 9\right) \geq C'\ln|\ln B| - C'^{-1},$$

and for $\alpha > 1$

$$T^* \geq \frac{\gamma}{c}\left(\left(\frac{\varepsilon}{1+\varepsilon}\right)^{1/\gamma}|\ln B|^{1/\gamma} - (\ln 9)^{1/\gamma}\right) \geq C'\gamma|\ln B|^{1/\gamma} - C'^{-1}$$

for sufficiently large constant C'. □

2.4. Proof of Theorem 1.13. To prove Theorem 1.13 we note that we have the following result, analogous to Lemma 2.1. As its proof is immediate we omit it.

LEMMA 2.6. *Let K be bounded and satisfy* (1.10), *then for any $\mu \in \mathcal{M}_+(\mathbb{R}^d)$ with mass m, we have the inequality*

$$\int_{\mathbb{R}^d} |K(x-z) - K(y-z)|d\mu(z) \leq Cm\psi_1(|x-y|)$$

where C is the constant in (1.10) *and ψ_1 is defined by* (2.4).

Furthermore, we note that due to this lemma the vector fields E_i are $\log^2$-Lipschitz, and as noted in Remark 2.4 this is enough to define the characteristic ODEs. The proof of Theorem 1.13 is now entirely analogous to the proof of Theorem 1.9 for $\alpha = 1$, replacing Lemma 2.1 with this lemma. Thus we leave it to the reader.

3. Proof of Proposition 1.14

We first show that it is sufficient to propagate the exponential velocity moment.

LEMMA 3.1. *Let $f \in L^\infty(\mathbb{R}^d \times \mathbb{R}^d)$ and $c > 0$, then*

$$\int_{\mathbb{R}^d} \exp\left(\left|\int_{\mathbb{R}^d} f(x,v)\,dv\right|^\alpha / \lambda\right) dx \leq C \int_{\mathbb{R}^d\times\mathbb{R}^d} f(x,v) e^{c\langle v\rangle^{d\alpha}}\,dx\,dv$$

for constants C, λ depending only upon $\|f\|_{L^\infty}$ and c.

PROOF. We apply the usual 'interpolation' method: let

$$M(x) = \int_{\mathbb{R}^d} f(x,v) e^{c\langle v\rangle^{d\alpha}}\,dv,$$

then for each x we have

$$\rho(x) := \int_{\mathbb{R}^d} f(x,v)\,dv \leq \int_{|v|\geq R} f(x,v)\,dv + C\|f\|_{L^\infty} R^d \leq e^{-cR^{d\alpha}} M(x) + CR^d,$$

by Markov's inequality. We now choose $R = R(x) = c^{-1/(\alpha d)} \ln(1 \vee M(x))^{1/(d\alpha)}$ which gives

$$\rho(x) \leq 1 + C\ln(1 \vee M(x))^{1/\alpha}.$$

Thus,

$$\begin{aligned}\int_{\mathbb{R}^d} \exp(|\rho(x)|^\alpha/\lambda)\,dx &\leq \int_{\mathbb{R}^d} \exp(|1 + C\ln(1\vee M(x))^{1/\alpha}|^\alpha/\lambda)\,dx \\ &\leq \int_{\mathbb{R}^d} \exp((C/\lambda)(1 + \ln(1 \vee M(x))))\,dx\end{aligned}$$

and choosing $\lambda = C$ we have

$$\int_{\mathbb{R}^d} \exp(|\rho(x)|^\alpha/\lambda)\,dx \leq C\int_{\mathbb{R}^d} M(x) \vee 1\,dx \leq C\int_{\mathbb{R}^d} f(x,v)e^{c\langle v\rangle^{d\alpha}}\,dx\,dv.$$

□

We now prove that the exponential moment is propagated. Since f_0 has finite velocity moments of order larger than $d^2 - d$, the solution provided by [**10**, Theo. 1] has bounded velocity moments of order larger than $d^2 - d$ on $[0,T]$. By [**10**, Cor. 2] it follows that

$$\sup_{t\in[0,T]} \|E(t)\|_{L^\infty} \leq C < \infty \tag{3.1}$$

for any finite T.

LEMMA 3.2. *Let $f \in L^\infty(\mathbb{R}^d \times \mathbb{R}^d)$, $\alpha \geq 1$ and $c > 0$. Define*

$$M(t) = \int_{\mathbb{R}^d\times\mathbb{R}^d} f(t,x,v) e^{1+c\langle v\rangle^{d\alpha}}\,dx\,dv.$$

Then we have the differential inequality along the Vlasov-Poisson flow

$$\frac{dM}{dt} \leq C\big[1 + \big(\ln M(t)\big)^{1-\frac{1}{d\alpha}} M(t)\big]$$

for a constant C depending only upon c, α and $\|f\|_{L^\infty}$.

PROOF. We directly compute, using the weak formulation of the Vlasov-Poisson equation

$$\begin{aligned}\frac{dM}{dt} &= -\int_{\mathbb{R}^d\times\mathbb{R}^d} cE(t,x)\cdot\nabla_v[\langle v\rangle^{d\alpha}]f(t,x,v)e^{1+c\langle v\rangle^{d\alpha}}\,dx\,dv \\ &\le C\int_{\mathbb{R}^d\times\mathbb{R}^d} |E(t,x)|\langle v\rangle^{d\alpha-1}f(t,x,v)e^{1+c\langle v\rangle^{d\alpha}}\,dx\,dv \\ &\le C\|E(t)\|_{L^\infty}\int_{\mathbb{R}^d\times\mathbb{R}^d} \ln(e^{1+c\langle v\rangle^{d\alpha}})^{1-\frac{1}{d\alpha}}e^{1+c\langle v\rangle^{d\alpha}}f(t,x,v)\,dx\,dv.\end{aligned}$$

The claim of the lemma now follows from (3.1) and Jensen's inequality, using the convexity of $\tau\mapsto\tau\ln(\tau)^{1-\frac{1}{d\alpha}}$ on $\tau\in(e,\infty)$. ∎

PROOF OF PROPOSITION 1.14. By using Lemma 3.2 and solving the resulting differential inequality we deduce that

$$\sup_{t\in[0,T]}\int_{\mathbb{R}^d\times\mathbb{R}^d} f(t,x,v)e^{c\langle v\rangle^{d\alpha}}\,dx\,dv\le C<\infty.$$

The claim of the proposition now follows from an application of Lemma 3.1. ∎

References

[1] Luigi Ambrosio and Gianluca Crippa, *Existence, uniqueness, stability and differentiability properties of the flow associated to weakly differentiable vector fields*, Transport equations and multi-D hyperbolic conservation laws, Lect. Notes Unione Mat. Ital., vol. 5, Springer, Berlin, 2008, pp. 3–57. MR2409676

[2] A. A. Arsen′ev, *Existence in the large of a weak solution of Vlasov's system of equations* (Russian), Ž. Vyčisl. Mat. i Mat. Fiz. **15** (1975), 136–147, 276. MR0371322

[3] R. J. DiPerna and P.-L. Lions, *Ordinary differential equations, transport theory and Sobolev spaces*, Invent. Math. **98** (1989), no. 3, 511–547. MR1022305

[4] Javier Duoandikoetxea, *Fourier analysis*, Graduate Studies in Mathematics, vol. 29, American Mathematical Society, Providence, RI, 2001. Translated and revised from the 1995 Spanish original by David Cruz-Uribe. MR1800316

[5] R. L. Dobrušin, *Vlasov equations* (Russian), Funktsional. Anal. i Prilozhen. **13** (1979), no. 2, 48–58, 96. MR541637

[6] I. Gasser, P.-E. Jabin, and B. Perthame, *Regularity and propagation of moments in some nonlinear Vlasov systems*, Proc. Roy. Soc. Edinburgh Sect. A **130** (2000), no. 6, 1259–1273. MR1809103

[7] Maxime Hauray and Pierre-Emmanuel Jabin, *Particle approximation of Vlasov equations with singular forces: propagation of chaos* (English, with English and French summaries), Ann. Sci. Éc. Norm. Supér. (4) **48** (2015), no. 4, 891–940. MR3377068

[8] D. Lazarovici, *The Vlasov-Poisson dynamics as the mean-field limit of rigid charges*, preprint, 2015.

[9] D. Lazarovici and P. Pickl, *A mean-field limit for the Vlasov-Poisson system*, preprint, 2015.

[10] P.-L. Lions and B. Perthame, *Propagation of moments and regularity for the 3-dimensional Vlasov-Poisson system* (English, with French summary), Invent. Math. **105** (1991), no. 2, 415–430. MR1115549

[11] Grégoire Loeper, *Uniqueness of the solution to the Vlasov-Poisson system with bounded density* (English, with English and French summaries), J. Math. Pures Appl. (9) **86** (2006), no. 1, 68–79. MR2246357

[12] Andrew J. Majda and Andrea L. Bertozzi, *Vorticity and incompressible flow*, Cambridge Texts in Applied Mathematics, vol. 27, Cambridge University Press, Cambridge, 2002. MR1867882

[13] Carlo Marchioro and Mario Pulvirenti, *Mathematical theory of incompressible nonviscous fluids*, Applied Mathematical Sciences, vol. 96, Springer-Verlag, New York, 1994. MR1245492

[14] Ayman Moussa and Franck Sueur, *On a Vlasov-Euler system for 2D sprays with gyroscopic effects*, Asymptot. Anal. **81** (2013), no. 1, 53–91. MR3060032

[15] Evelyne Miot, *A uniqueness criterion for unbounded solutions to the Vlasov-Poisson system*, Comm. Math. Phys. **346** (2016), no. 2, 469–482. MR3535893

[16] Seiji Ukai and Takayoshi Okabe, *On classical solutions in the large in time of two-dimensional Vlasov's equation*, Osaka J. Math. **15** (1978), no. 2, 245–261. MR504289

[17] Christophe Pallard, *Space moments of the Vlasov-Poisson system: propagation and regularity*, SIAM J. Math. Anal. **46** (2014), no. 3, 1754–1770. MR3200421

[18] K. Pfaffelmoser, *Global classical solutions of the Vlasov-Poisson system in three dimensions for general initial data*, J. Differential Equations **95** (1992), no. 2, 281–303. MR1165424

[19] M. M. Rao and Z. D. Ren, *Applications of Orlicz spaces*, Monographs and Textbooks in Pure and Applied Mathematics, vol. 250, Marcel Dekker, Inc., New York, 2002. MR1890178

Department of Mathematics, South Kensington Campus, Imperial College London, London, SW7 2AZ, UK

Email address: t.holding@imperial.ac.uk

CNRS - Institut Fourier, Université Grenoble-Alpes, 100, rue des mathématiques, 38610 Gières, France

Email address: evelyne.miot@univ-grenoble-alpes.fr

Contemporary Mathematics
Volume **710**, 2018
http://dx.doi.org/10.1090/conm/710/14369

Strong solutions of the Navier-Stokes equations with singular data

Hideo Kozono and Senjo Shimizu

ABSTRACT. We construct strong solutions in the Serrin class of the Navier-Stokes equations with singular data. In 2D case, our results cover the initial vorticity as the Dirac measure and the external force whose support consists of a single point. In 3D case, we can handle the initial vortex sheet supported on the sphere and the singular external force whose support is concentrated on the surface.

1. Introduction

Let us consider the Cauchy problem of the Navier-Stokes equations in $\mathbb{R}^n$, $n \geq 2$;

$$\text{(N-S)} \qquad \begin{cases} \dfrac{\partial u}{\partial t} - \Delta u + u \cdot \nabla u + \nabla \pi = f & \text{in } \mathbb{R}^n \times (0,T), \\ \operatorname{div} u = 0 & \text{in } \mathbb{R}^n \times (0,T), \\ u|_{t=0} = a & \text{in } \mathbb{R}^n, \end{cases}$$

where $u = u(x,t) = (u_1(x,t), \cdots, u_n(x,t))$ and $\pi = \pi(x,t)$ denote the unknown velocity vector and the unknown pressure at the point $x = (x_1, \cdots, x_n) \in \mathbb{R}^n$ and the time $t \in (0,T)$, respectively, while $a = a(x) = (a_1(x), \cdots, a_n(x))$ and $f = f(x,t) = (f_1(x,t), \cdots, f_n(x,t))$ are the given initial velocity vector and the given external force, respectively.

It is well-known that (N-S) is invariant under such a change of scaling as $u_\lambda(x,t) = \lambda u(\lambda x, \lambda^2 t)$ and $\pi_\lambda(x,t) = \lambda^2 \pi(\lambda x, \lambda^2 t)$ for all $\lambda > 0$. The Banach space $\mathcal{Y}$ of functions with the space and time variables with the norm $\|\cdot\|_{\mathcal{Y}}$ is called *scaling invariant* to (N-S) if it holds that $\|u_\lambda\|_{\mathcal{Y}} = \|u\|_{\mathcal{Y}}$ for all $\lambda > 0$. For instance, in the usual Lebesgue spaces $L^p(\mathbb{R}^n)$, we see that the scaling invariant space $\mathcal{Y}$ to (N-S) is so-called the Serrin class $L^\alpha(0,\infty; L^p(\mathbb{R}^n))$ for $2/\alpha + n/p = 1$ with $n \leq p \leq \infty$. Since the corresponding scaling laws to the initial data a and the external force f are like $a_\lambda(x) = \lambda a(\lambda x)$ and $f_\lambda(x,t) = \lambda^3 f(\lambda x, \lambda^2 t)$, it is suitable to solve (N-S) in Banach spaces X for a and Y for f with the properties that

2010 *Mathematics Subject Classification.* Primary 35Q30; Secondary 76D03, 76D05.
Key words and phrases. Navier-Stokes equations, global strong solutions, singular data, Dirac measure, single layer potential.
The work of H.K. is partially supported by JSPS Grant in aid for Scientific Research S #16H06339. The work of S.S. is partially supported by JSPS Grant in aid for Scientific Research B #16H03945.

$\|a_\lambda\|_X = \|a\|_X$ and $\|f_\lambda\|_Y = \|f\|_Y$ for all $\lambda > 0$, respectively. Since the pioneer work of Fujita-Kato [**13**], there have been a number of results to enlarge the space X such as $H^{\frac{n}{2}-1}(\mathbb{R}^n)$, $L^n(\mathbb{R}^n)$, $\dot{B}^{-1+n/p}_{p,\infty}(\mathbb{R}^n)$ with $n < p < \infty$ and BMO^{-1} which are monotonically increasing. See e.g., Kato [**15**], Giga-Miyakawa [**14**], Kozono-Yamazaki [**21**], Cannone [**4**], Cannone-Planchon [**7**] and Koch-Tataru [**16**]. Amann [**1**] has established a systematic treatment of strong solutions in various function spaces such as Lebesgue space $L^p(\Omega)$, Bessel potential space $H^{s,p}(\Omega)$, Besov space $B^s_{p,q}(\Omega)$ and Nikol'skii space $N^{s,p}(\Omega)$ in general domains Ω. In this direction, based on the Littlewood-Paley decomposition, Cannone-Meyer [**5**] showed how to choose the Banach spaces X for a and $\mathcal{Y}$ for u. Besides these results, in terms of the Stokes operator, Farwig-Sohr [**11**] and Farwig-Sohr-Varnhorn [**12**] proved a necessary and sufficient condition on a that weak solutions u belong to the Serrin class. Furthermore, ill-posedness in the sense of breakdown of continuous dependence of the solution mapping $a \in X \to u \in \mathcal{Y}$ in (N-S) was proved by Bourgain-Pavlović [**2**], Yoneda [**24**] and Wang [**23**].

On the other hand, compared with great efforts to find the optimal space for X, less attention has been paid to find a suitable space for Y. For instance, Giga-Miyakawa [**14**] proved existence of strong solutions for

$$Y = \{f \in C((0,\infty); L^n(\mathbb{R}^n)); \|f\|_Y = \sup_{0<t<\infty} t^{1-\delta}\|(-\Delta)^{-\delta}Pf(t)\|_{L^n} < \infty\}$$

for some $\delta > 0$, where P denotes the Helmholtz projection. Cannone-Planchon [**8**] treated the case $n = 3$ and showed that $Y = \{f = \mathrm{div}\, F; F \in L^\alpha(0,\infty; L^p(\mathbb{R}^3))\}$ for $2/\alpha + 3/p = 2$ with $2/3 < p < \infty$ is a suitable space. See also Planchon [**22**]. Introducing the space of pseudo measures $\mathcal{PM}^k = \{a \in \mathcal{S}'; \sup_{\xi\in\mathbb{R}^n} |\xi|^k|\hat{a}(\xi)| < \infty\}$, Cannone-Karch [**6**] showed that the pair of $X = \mathcal{PM}^2$ and $Y = C_w((0,\infty); \mathcal{PM}^0)$ is suitable for $n = 3$. Recently, the authors [**18**] constructed mild solutions for $X = L^{n,\infty}(\mathbb{R}^n)$ and $Y = L^{\alpha,\infty}(0,\infty; L^{p,\infty}(\mathbb{R}^n))$ for $2/\alpha + n/p = 3$ with $\max\{1, n/3\} < p < \infty$. Farwig, Giga and Hsu [**9**] yield a sufficient condition on initial data and external force to guarantee the existence of local strong solution in time weighted space $L^s_\alpha(L^q)$ in bounded domains in $\mathbb{R}^3$ with $2/s + 3/q = 1 - 2\alpha$, $2 < s < \infty$, and they also deal with the limit case $L^\infty_\alpha(L^q)$ in [**10**]. In [**19**] we consider the case of $s = \infty$ of Farwig, Giga and Hsu [**9**], namely

$$X = \dot{B}^{-1+\frac{n}{p}}_{p,\infty}(\mathbb{R}^n), \quad Y = \{f \in C_w(0,\infty); \dot{B}^{s_0}_{p_0,\infty}(\mathbb{R}^n));$$
$$\|f\|_Y \equiv \sup_{0<t<\infty} t^{\frac{n}{2}(\frac{3}{n}-\frac{1}{p_0})+\frac{s_0}{2}}\|f(t)\|_{\dot{B}^{s_0}_{p_0,\infty}} < \infty\}$$

for $n < p < \infty$, $n/3 < p_0 \le p$ and $s_0 < \min\{0, n/p_0 - 1\}$. However, the mild solution does not reach the strong solution.

In the preceding paper [**20**], we prove the existence and uniqueness of local strong solutions to (N-S) for arbitrary initial data $a \in \dot{B}^{-1+\frac{n}{r}}_{r,q}(\mathbb{R}^n)$ and external force $f \in L^{\alpha,q}((0,T); \dot{B}^s_{p,\infty}(\mathbb{R}^n))$, $0 < T \le \infty$ for $2/\alpha + n/p - s = 3$ with $1 < r \le p$ and $s > -1$. Here $\dot{B}^s_{p,q}$ and $L^{\alpha,q}$ denote the homogeneous Besov space and the Lorentz space, respectively. We also show the existence and uniqueness of global solutions for small $a \in \dot{B}^{-1+\frac{n}{r}}_{r,q}(\mathbb{R}^n)$ and small $f \in L^{\alpha,q}((0,\infty); \dot{B}^s_{p,\infty}(\mathbb{R}^n))$. In this article, we construct strong solutions in the Serrin class of (N-S) equations with singular data. In 2D case, our results cover the initial vorticity as the Dirac measure and the external force whose support consists of a point. In 3D case, we can handle the initial vortex sheet supported on the sphere and the singular external force whose support is concentrated on the surface. In the next section, we state our main results. Section 3 is devoted to the proof of theorems based on the argument of our previous paper [**20**].

2. Results

We first recall definitions of the Besov space $B^s_{p,q} = B^s_{p,q}(\mathbb{R}^n)$ and the homogeneous Besov space $\dot{B}^s_{p,q} = \dot{B}^s_{p,q}(\mathbb{R}^n)$. Let $\{\varphi_j\}_{j\in\mathbb{Z}}$ be the Littlewood-Paley decomposition. We take a function $\phi \in C_0^\infty(\mathbb{R}^n)$ with $\operatorname{supp}\phi = \{\xi \in \mathbb{R}^n;\ 1/2 \le |\xi| \le 2\}$ such that $\sum_{j\in\mathbb{Z}} \phi(2^{-j}\xi) = 1$ for all $\xi \neq 0$. The functions φ_j and ψ are defined by

$$\mathcal{F}\varphi_j(\xi) = \phi(2^{-j}\xi), \quad \mathcal{F}\psi(\xi) = 1 - \sum_{j=1}^{\infty} \phi(2^{-j}\xi),$$

where $\mathcal{F}$ denotes the Fourier transform. For $s \in \mathbb{R}$ and $1 \le p, q \le \infty$, the Besov space $B^s_{p,q}$ and the homogeneous Besov space $\dot{B}^s_{p,q}$ are defined by

$$B^s_{p,q} \equiv \{f \in \mathcal{S}';\ \|f\|_{B^s_{p,q}} < \infty\}, \quad \dot{B}^s_{p,q} \equiv \{f \in \mathcal{S}';\ \|f\|_{\dot{B}^s_{p,q}} < \infty\},$$

respectively, where

$$\|f\|_{B^s_{p,q}} \equiv \begin{cases} \|\psi * f\|_{L^p} + \left(\sum_{j=1}^{\infty} (2^{sj}\|\varphi_j * f\|_{L^p})^q\right)^{1/q}, & 1 \le q < \infty, \\ \|\psi * f\|_{L^p} + \sup_{1\le j<\infty} (2^{sj}\|\varphi_j * f\|_{L^p}), & q = \infty, \end{cases}$$

$$\|f\|_{\dot{B}^s_{p,q}} \equiv \begin{cases} \left(\sum_{j\in\mathbb{Z}} (2^{sj}\|\varphi_j * f\|_{L^p})^q\right)^{1/q}, & 1 \le q < \infty, \\ \sup_{j\in\mathbb{Z}} (2^{sj}\|\varphi_j * f\|_{L^p}), & q = \infty. \end{cases}$$

Here and in what follows, we write simply L^p instead of $L^p(\mathbb{R}^n)$. Let us denote by P the Helmholtz projection from L^p, $1 < p < \infty$ onto the subspace of solenoidal vector fields $PL^p = L^p_\sigma \equiv \{f \in L^p; \operatorname{div} f = 0\}$ as a bounded operator. It is well-known that P is expressed as

$$P = (P_{ij})_{1\le i,j\le n}, \quad P_{ij} = \delta_{ij} + R_i R_j, \quad i, j = 1, \ldots, n,$$

where $\{\delta_{ij}\}_{1\le i,j\le n}$ is the Kronecker symbol and $R_i = \frac{\partial}{\partial x_i}(-\Delta)^{-\frac{1}{2}}$, $i = 1, \cdots, n$ are the Riesz transforms. Since the Riesz transforms are bounded in the homogeneous Besov space $\dot{B}^s_{p,q}$ for all $s \in \mathbb{R}$ and all $1 \le p, q \le \infty$, so is the Helmholtz projection P. Then the solenoidal subspace $\dot{\mathcal{B}}^s_{p,q}$ in $\dot{B}^s_{p,q}$ is defined as the range of P, i.e., $\dot{\mathcal{B}}^s_{p,q} \equiv P\dot{B}^s_{p,q}$ (cf. [**17**, Section 1]). It is known that unless $s < \frac{n}{p}$ or $s = \frac{n}{p}$, $q = 1$, the space $\dot{B}^s_{p,q}$ consists of functions modulo polynomials (cf. [**3**, Remark 2.26]). Throughout of the paper, we consider the case $s < \frac{n}{p}$, $\dot{B}^s_{p,q}$ is considered as a usual Banach space. The Stokes operator A on $\dot{\mathcal{B}}^s_{p,q}$ is defined by $A = -P\Delta$. Then the original equations (N-S) can be rewritten to the abstract evolution equation:

$$\begin{cases} \dfrac{du}{dt} + Au + P(u \cdot \nabla u) = Pf \quad \text{on } (0, T), \\ u(0) = a. \end{cases} \tag{2.1}$$

The solution u of (2.1) is called a *strong* solution of (N-S).

We first state a result on the strong solution with singular initial data and an external force on the plane $\mathbb{R}^2$.

THEOREM 2.1. *Let $n=2$. Let $1 < p < 2$ and $1 \le r \le p$. Suppose that a and f are expressed as*

$$a(x) = -\varepsilon(-\Delta)^{-1}\text{rot }\delta(x), \tag{2.2}$$

$$f(x,t) = \varepsilon t^{-\frac{1}{2}}\delta(x), \tag{2.3}$$

by the small parameter $\varepsilon > 0$, where $\delta(x)$ is the Dirac measure in $\mathbb{R}^2$. There exists $\varepsilon_0 = \varepsilon_0(p,r) > 0$ such that if $0 < \varepsilon \le \varepsilon_0$, then there exists a strong solution u on $(0,\infty)$ of

$$\text{(NS)} \qquad \begin{cases} \dfrac{du}{dt} + Au + P(u\cdot\nabla u) = Pf \quad \text{a.e. } t \in (0,\infty) \text{ in } \dot{\mathcal{B}}_{p,\infty}^{-2+\frac{2}{p}}(\mathbb{R}^2), \\ u(0) = a \quad \text{in } \dot{\mathcal{B}}_{r,\infty}^{-1+2/r}(\mathbb{R}^2) \end{cases}$$

in the class

$$u_t,\ Au \in L^{2,\infty}(0,\infty;\dot{\mathcal{B}}_{p,\infty}^{-2+\frac{2}{p}}(\mathbb{R}^2)). \tag{2.4}$$

Moreover, such a solution u satisfies that

$$u \in L^{\alpha_0,\infty}(0,\infty;\dot{\mathcal{B}}_{p_0,1}^{s_0}(\mathbb{R}^2)) \quad \text{for } 2/\alpha_0 + 2/p_0 - s_0 = 1 \tag{2.5}$$

with $p \le p_0$, $2 < \alpha_0$ and $-1 + \frac{2}{r} < s_0$. The solution u also belongs to the Serrin class

$$u \in L^{\alpha_0,\infty}(0,\infty;L^{r_0}(\mathbb{R}^2)) \quad \text{for } 2/\alpha_0 + 2/r_0 = 1. \tag{2.6}$$

Concerning the uniqueness, there is a constant $\kappa = \kappa(p,r) > 0$ such that if the initial data a as in (2.2) and the solution u of (NS) on $(0,\infty)$ in the class (2.4) satisfy

$$\varepsilon + \|u_t\|_{L^{2,\infty}\left(0,\infty;\dot{B}_{p,\infty}^{-2+\frac{2}{p}}\right)} + \|Au\|_{L^{2,\infty}\left(0,\infty;\dot{B}_{p,\infty}^{-2+\frac{2}{p}}\right)} \le \kappa \tag{2.7}$$

then u is unique.

Next we state a result of a strong solution with a singular datum and an external force in the three dimensional space $\mathbb{R}^3$. Let S_R be the sphere $S_R = \{x \in \mathbb{R}^3;\ |x| = R\}$, and let δ_{S_R} be the single layer potential whose support is on S_R. Notice that δ_{S_R} is defined as a distribution such as

$$(\delta_{S_R}, \Phi) = \int_{S_R} \Phi(x)\, d\sigma(x) \quad \text{for } \Phi \in C_0^\infty(\mathbb{R}^3),$$

where $\sigma(x)$ is the surface element of S_R. Our results now reads

THEOREM 2.2. *Let $n = 3$. Let $2 \le p < \frac{5}{2}$. Suppose that a and f are expressed as*

$$a(x) = \varepsilon(-\Delta)^{-1}\text{rot } P(\delta_{S_R}(x)e_i), \tag{2.8}$$

$$f(x,t) = \varepsilon t^{-\frac{1}{p}}\delta_{S_R}(x), \tag{2.9}$$

with small parameter $\varepsilon > 0$, where $\{e_1, e_2, e_3\}$ is the canonical basis in $\mathbb{R}^3$. There exists $\varepsilon_0 = \varepsilon_0(p) > 0$ such that if $0 < \varepsilon \le \varepsilon_0$, then there exists a strong solution u on $(0,\infty)$ of

$$\text{(NS)} \qquad \begin{cases} \dfrac{du}{dt} + Au + P(u\cdot\nabla u) = Pf \quad \text{a.e. } t \in (0,\infty) \text{ in } \dot{\mathcal{B}}_{p,\infty}^{-3+\frac{5}{p}}(\mathbb{R}^3), \\ u(0) = a \quad \text{in } \dot{\mathcal{B}}_{p,\infty}^{-1+3/p}(\mathbb{R}^3) \end{cases}$$

in the class

$$u_t,\ Au \in L^{p,\infty}(0,\infty;\dot{\mathcal{B}}_{p,\infty}^{-3+\frac{5}{p}}(\mathbb{R}^3)). \tag{2.10}$$

Moreover, such a solution u satisfies that

$$u \in L^{\alpha_0,\infty}(0,\infty;\dot{\mathcal{B}}_{p_0,1}^{s_0}) \quad \text{for } 2/\alpha_0 + 3/p_0 - s_0 = 1 \tag{2.11}$$

with $p \le p_0$, $\alpha < \alpha_0$ and $\frac{1}{2} < s_0$. The solution u also belongs to the Serrin class

$$u \in L^{\alpha_0,\infty}(0,\infty;L^{r_0}) \quad \text{for } 2/\alpha_0 + 3/r_0 = 1. \tag{2.12}$$

Concerning the uniqueness, there is a constant $\kappa = \kappa(p) > 0$ such that if the initial data a as in (2.8) and the solution u of (NS) on $(0,\infty)$ in the class (2.10) satisfy

$$\varepsilon + \|u_t\|_{L^{p,\infty}(0,\infty;\dot{B}_{p,\infty}^{-3+\frac{5}{p}})} + \|Au\|_{L^{p,\infty}(0,\infty;\dot{B}_{p,\infty}^{-3+\frac{5}{p}})} \le \kappa \tag{2.13}$$

then u is unique.

3. Proofs

3.1. Global existence theorem. First we recall the results [**20**] on the global existence and uniqueness of strong solutions to (N-S) for small a and f which are based on the maximal Lorentz regularity theorem of the Stokes equations in the homogeneous Besov space.

THEOREM 3.1 (Theorem 3 in [**20**]). *Let $1 < p < \infty$, $1 < \alpha < \infty$ and $s > -1$ satisfy $2/\alpha + n/p - s = 3$. Let $1 \le q \le \infty$. Assume that $1 \le r \le p$ satisfies*

$$\frac{n}{r} < \frac{2}{\alpha} + \frac{n}{p}. \tag{3.1}$$

There exists a constant $\eta = \eta(n,p,\alpha,s,r,q) > 0$ such that if $a \in \dot{\mathcal{B}}_{r,q}^{-1+n/r}(\mathbb{R}^n)$ and $f \in L^{\alpha,q}(0,\infty;\dot{B}_{p,\infty}^{s}(\mathbb{R}^n))$ satisfy

$$\|a\|_{\dot{B}_{r,q}^{-1+n/r}} + \|f\|_{L^{\alpha,q}(0,\infty;\dot{B}_{p,\infty}^{s})} \le \eta, \tag{3.2}$$

then there is a solution u of

$$\text{(NS)} \qquad \begin{cases} \dfrac{du}{dt} + Au + P(u\cdot\nabla u) = Pf \quad \text{a.e. } t\in(0,\infty) \text{ in } \dot{\mathcal{B}}_{p,\infty}^{s}(\mathbb{R}^n), \\ u(0) = a \quad \text{in } \dot{\mathcal{B}}_{r,q}^{-1+n/r}(\mathbb{R}^n) \end{cases}$$

in the class

$$u_t,\ Au \in L^{\alpha,q}(0,\infty;\dot{\mathcal{B}}_{p,\infty}^{s}(\mathbb{R}^n)). \tag{3.3}$$

Moreover, such a solution u satisfies that

$$u \in L^{\alpha_0,q}(0,\infty;\dot{\mathcal{B}}_{p_0,1}^{s_0}(\mathbb{R}^n)) \quad \text{for } 2/\alpha_0 + n/p_0 - s_0 = 1 \tag{3.4}$$

with $p \le p_0$, $\alpha < \alpha_0$ and $\max\{s, n/r-1\} < s_0$.

In the case $1 \le q < \infty$, the solution u of (NS) on $(0,\infty)$ in the class (3.3) is unique. In the case $q = \infty$, the uniqueness holds under the condition

(3.5)

$$\sup_{N\le j<\infty} 2^{(-1+\frac{n}{r})j}\|\varphi_j * a\|_{L^r} + \limsup_{t\to\infty} t\,\mu\{\tau\in(0,\infty); \|u_t(\tau)\|_{\dot{B}_{p,\infty}^{s}} + \|Au(\tau)\|_{\dot{B}_{p,\infty}^{s}} > t\}^{\frac{1}{\alpha}} \le \kappa$$

for some $N \in \mathbb{N}$ provided $n < r \le p$. Here μ denotes the Lebesgue measure on $\mathbb{R}$.

REMARK 3.2. (i) We may take the third exponent of the Besov spaces freely for the given data a and f, i.e., $a \in \dot{\mathcal{B}}_{r,q}^{-1+n/r}$ with $1 \le q \le \infty$ and $f(t) \in \dot{B}_{p,\infty}^{s}$ for a.e. $t \in (0,\infty)$. Nevertheless, for the solution u, we may always take the corresponding exponent one, which means that $u(t) \in \dot{\mathcal{B}}_{p_0,1}^{s_0}$ for a.e. $t \in (0,\infty)$. This may be regarded as gain of regularity of solutions to (N-S).

(ii) By taking $\alpha_0 > 2$ and $s_0 \ge 0$ we see that $p_0 \le r_0$ with $1/r_0 = 1/p_0 - s_0/n = n^{-1}(1-2/\alpha_0)$. Since it holds $\dot{B}_{p_0,1}^{s_0} \hookrightarrow L^{r_0}$, we see from (3.4) that

$$u \in L^{\alpha_0,q}(0,\infty; L^{r_0}(\mathbb{R}^n))$$

for $2/\alpha_0 + n/r_0 = 2/\alpha_0 + n(1/p_0 - s_0/n) = 2/\alpha_0 + n/p_0 - s_0 = 1$. This implies that our solution belongs to the usual Serrin class.

3.2. Proof of Theorem 2.1. The Dirac measure δ is defined by

$$(\delta, \Phi) = \Phi(0)$$

for every $\Phi \in C_0^\infty(\mathbb{R}^n)$. We recall that $\{\varphi_j\}_{j\in\mathbb{Z}}$ is the Littlewood-Paley decomposition and $\phi \in C_0^\infty(\mathbb{R}^n)$ satisfies $\mathcal{F}\varphi_j(\xi) = \phi(2^{-j}\xi)$. Since δ is the identity element of convolution in the distribution sense, we have

$$(\varphi_j * \delta)(x) = \varphi_j(x) = \mathcal{F}^{-1}[\phi(2^{-j}\cdot)](x) = 2^{nj}[\mathcal{F}^{-1}\phi](2^j x),$$

which yields

$$\|\varphi_j * \delta\|_{L^p} = \|2^{nj}[\mathcal{F}^{-1}\phi](2^j\cdot)\|_{L^p} = 2^{jn(1-\frac{1}{p})}\|\mathcal{F}^{-1}\phi\|_{L^p}. \tag{3.6}$$

Hence it holds that

$$\delta(x) \in \dot{B}_{p,\infty}^{-n+\frac{n}{p}}(\mathbb{R}^n), \quad 1 \le p \le \infty,\ n \in \mathbb{N}.$$

Substituting the regularity $s = -n + \frac{n}{p}$ in the condition $2/\alpha + n/p - s = 3$, we obtain $n = 3 - 2/\alpha$. Combining $1 < \alpha < \infty$ with $n = 3 - 2/\alpha$, we have $1 < n < 3$. Therefore the initial data used by the Dirac measure covers only the case when $n = 2$. And so α is uniquely determined as 2. By the condition $s = -2 + 2/p > -1$ and $1 < p < \infty$, it can be taken as $1 < p < 2$. For all r with $1 \le r \le p$, it holds that $2/r < 2/\alpha + 2/p = 1 + 2/p$. Thus, in the case when $n = 2$, we take $1 < p < 2$ and

$$\omega_0(x) = \operatorname{rot} a(x) = \varepsilon\delta(x) \in \dot{\mathcal{B}}_{r,\infty}^{-2+\frac{2}{r}}(\mathbb{R}^2), \quad 1 \le r \le p, \tag{3.7}$$

$$f(x,t) = \varepsilon t^{-\frac{1}{2}}\delta(x) \in L^{2,\infty}(0,\infty;\ \dot{B}_{p,\infty}^{-2+\frac{2}{p}}(\mathbb{R}^2)), \tag{3.8}$$

where ε_0 is taken sufficiently small to satisfy

$$\|a\|_{\dot{B}_{r,\infty}^{-1+2/r}} + \|f\|_{L^{2,\infty}(0,\infty;\ \dot{B}_{p,\infty}^{-2+\frac{2}{p}}(\mathbb{R}^2))} \le \eta$$

in (3.2). Then the initial data a and the external force f satisfy the conditions in Theorem 3.1, we obtain the global solution

$$u \in L^{\alpha_0,\infty}(0,\infty; \dot{\mathcal{B}}_{p_0,1}^{s_0}) \quad \text{for } 2/\alpha_0 + 2/p_0 - s_0 = 1$$

with $p \le p_0$, $2 < \alpha_0$ and $\max\{-2+\frac{2}{p}, -1+\frac{2}{r}\} = -1+\frac{2}{r} < s_0$. Moreover by Remark 3.2 (ii), we know that the solution also belongs to the Serrin class (2.6).

It remains to prove the uniqueness. Let u_1 and u_2 be two solutions of (NS) on $(0,\infty)$ in the class (2.4). Defining $w = u_1 - u_2$, we have

$$(3.9)\quad \begin{cases} \dfrac{\partial w}{\partial t} + Aw = -P(u_1\cdot\nabla w) - P(w\cdot\nabla u_2) \quad \text{a.e. } t\in(0,\infty) \text{ in } \dot{\mathcal{B}}_{p,\infty}^{-2+\frac{2}{p}}, \\ w(0) = 0. \end{cases}$$

By Lemma 3.1 in [**20**], we know that

$$\|e^{-tA}a\|_{L^{\alpha_0,\infty}(0,\infty;\dot{\mathcal{B}}^{s_0}_{p_0,1})} \le C\|a\|_{\dot{B}^{-1+2/r}_{r,\infty}},$$

where C is independent of a. By Lemma 3.2 in [**20**], it holds that

$$\begin{aligned} &\|P(u_1\cdot\nabla w) + P(w\cdot\nabla u_2)\|_{L^{2,\infty}(0,\infty;\,\dot{B}_{p,\infty}^{-2+\frac{2}{p}}(\mathbb{R}^2))} \\ \le& C\Big(2\|a\|_{\dot{B}^{-1+2/r}_{r,\infty}} + \Big\|\frac{\partial u_1}{\partial t},\, Au_1,\, \frac{\partial u_2}{\partial t},\, Au_2\Big\|_{L^{2,\infty}(0,\infty;\,\dot{B}_{p,\infty}^{-2+\frac{2}{p}}(\mathbb{R}^2))}\Big) \\ &\qquad \times \|w_t,\, Aw\|_{L^{2,\infty}(0,\infty;\,\dot{B}_{p,\infty}^{-2+\frac{2}{p}}(\mathbb{R}^2))}, \end{aligned}$$

where C is independent of a and f. Applying Theorem 1 in [**20**] to (3.9) we obtain

$$\|w_t,\, Aw\|_{L^{2,\infty}(0,\infty;\,\dot{B}_{p,\infty}^{-2+\frac{2}{p}}(\mathbb{R}^2))} \le C\|P(u_1\cdot\nabla w) + P(w\cdot\nabla u_2)\|_{L^{2,\infty}(0,\infty;\,\dot{B}_{p,\infty}^{-2+\frac{2}{p}}(\mathbb{R}^2))},$$

where C is independent of u_1, u_2, and w. Therefore, if we take κ in (2.7) as $\kappa < 1/4C$, then we obtain $w = w(x,t) = 0$, which implies the desired uniqueness. This completes the proof of Theorem 1.

REMARK 3.3. The Dirac measure δ is not a measurable function. Therefore the initial data a defined by (3.7) does not belong to the Lorentz space $L^{2,\infty}(\mathbb{R}^2)$ and the external force f defined by (3.8) does not belong to the Lorentz space $L^{\alpha,\infty}(0,\infty;L^{p,\infty}(\mathbb{R}^2))$ with $2/\alpha + 2/p = 3$.

REMARK 3.4. Since the initial data a defined by (3.7) belongs to $\dot{\mathcal{B}}_{r,\infty}^{-2+\frac{2}{r}}(\mathbb{R}^2)$ with $1\le r\le p$ and $1 < p < 2$, the exponent r is less than the dimension 2. Therefore, the sufficient condition on uniqueness (2.7) cannot be replaced by the weaker condition (3.5) in Theorem 3.1. By the same reason, Theorem 2 in [**20**] is not applicable to obtain local solution for the large data.

3.3. Proof of Theorem 2.2. In order to obtain a further example for the three dimensional space $\mathbb{R}^3$, it is suitable to consider a function with weaker singularities in comparison with the Dirac measure. To this end, we introduce the single layer potential δ_{S_R} supported on the sphere S_R.

For a convolution $f*g$ suitably defined in the distribution sense, we have for every $\Phi\in C_0^\infty(\mathbb{R}^n)$,

$$\begin{aligned} (f*g,h) &= \int_{\mathbb{R}^n_x}\left(\int_{\mathbb{R}^n_y} f(x-y)g(y)\,dy\right)h(x)\,dx \\ &= \int_{\mathbb{R}^n_y}\left(\int_{\mathbb{R}^n_x} f(x-y)h(x)\,dx\right)g(y)\,dy \\ &= \int_{\mathbb{R}^n_y}\left(\int_{\mathbb{R}^n_x} f(y-x)h(x)\,dx\right)g(y)\,dy = (g, f_-*h), \end{aligned}$$

where we set $f_-(x) = f(-x)$. Therefore for the Littlewood-Paley decomposition $\{\varphi_j\}_{j\in\mathbb{Z}}$, we have

$$
\begin{aligned}
(\varphi_j * \delta_{S_R}, \Phi) &= (\delta_{S_R}, (\varphi_j)_- * \Phi) = \int_{S_R} (\varphi_j)_- * \Phi(x)\, d\sigma(x) \\
&= \int_{S_R} \left(\int_{\mathbb{R}^n} \varphi_j(y-x)\Phi(y)\, dy \right) d\sigma(x) = \int_{\mathbb{R}^n} \left(\int_{S_R} \varphi_j(y-x)\, d\sigma(x) \right) \Phi(y)\, dy \\
&= (\int_{S_R} \varphi_j(\cdot - x)\, d\sigma(x), \Phi).
\end{aligned}
$$

and

$$
\begin{aligned}
\|\varphi_j * \delta_{S_R}\|_{L_p}^p &= \int_{\mathbb{R}^n} | \int_{S_R} \varphi_j(x-y)\, d\sigma(y)|^p\, dx \\
&\le \int_{\mathbb{R}^n} \int_{S_R} |\varphi_j(x-y)|^p\, d\sigma(y) (\int_{S_R} d\sigma(y))^{\frac{p}{p'}}\, dx \qquad (3.10)\\
&= |S_R|^{\frac{p}{p'}} \int_{S_R} \left(\int_{\mathbb{R}^n} |\varphi_j(x-y)|^p\, dx \right) d\sigma(y) \\
&= |S_R|^{1+\frac{p}{p'}} \|\varphi_j\|_{L_p}^p,
\end{aligned}
$$

which implies

$$
\|\varphi_j * \delta_{S_R}\|_{L_p} \le |S_R| \|\varphi_j\|_{L_p}, \quad 1 \le p \le \infty, \tag{3.11}
$$

where we set $|S_R| = \int_{S_R} d\sigma$. In particular, taking $p = \infty$, it holds that

$$
\|\varphi_j * \delta_{S_R}\|_{L_\infty} \le |S_R| \|\varphi_j\|_{L_\infty} = |S_R| \|\mathcal{F}^{-1}[\phi(2^{-j}\cdot)]\|_{L_\infty} = 2^{jn} C_{n,\phi} |S_R|,
$$

where $C_{n,\phi} = \|[\mathcal{F}^{-1}\phi](2^j\cdot)\|_{L_\infty}$. This shows

$$
\sup_{j\in\mathbb{Z}} 2^{-jn} \|\varphi_j * \delta_{S_R}\|_{L_\infty} < \infty,
$$

namely

$$
\delta_{S_R} \in \dot{B}^{-n}_{\infty,\infty}(\mathbb{R}^n), \quad n \ge 2. \tag{3.12}
$$

We should remark that (3.11) is not a sharp L^p estimate because we use the Hölder inequality at (3.10). Indeed, it is essentially the same estimate (3.6) in the case of the Dirac measure. In order to obtain a more precise L^p estimate, we calculate $\|\varphi_j * \delta_{S_R}\|_{L^2}$ by using the Fourier transform. Since the Fourier transform of the single layer potential in the even dimension is expressed by the Bessel function with order $n/2$, it seems to be difficult to obtain its explicit expression. By this reason hereafter we consider the the single layer potential only in the case $n = 3$. Since the support of δ_{S_R} is compact, its Fourier transform has an expression as

$$
\mathcal{F}[\delta_{S_R}](\xi) = \int_{S_R} \eta(x) e^{ix\cdot\xi}\, dS_x,
$$

where $\eta(x) \in C_0^\infty(\mathbb{R}^3)$ such that $\eta(x) = 1$ on S_R. By using the polar coordinate $S_R = \{x = R\omega\}$, $\omega = (\sin\theta\cos\varphi, \sin\theta\sin\varphi, \cos\theta)$ with the angles $0 \le \theta < \pi$ and

$0 < \varphi < 2\pi$ in such a way that $x \cdot \xi = |x||\xi| \cos\theta$, we have that

$$\begin{aligned}\mathcal{F}[\delta_{S_R}](\xi) &= \int_0^\pi d\theta \int_0^{2\pi} e^{iR|\xi|\cos\theta} R^2 \sin\theta\, d\varphi \\ &= 2\pi R^2 \int_0^\pi e^{iR|\xi|\cos\theta} \sin\theta\, d\theta \\ &= 4\pi R \frac{\sin R|\xi|}{|\xi|}.\end{aligned}$$

Recalling that $\sum_{j\in\mathbb{Z}} \phi(2^{-j}\xi) = 1$ for all $\xi \neq 0$ and that $\operatorname{supp} \mathcal{F}[\varphi_j] = \{\xi \in \mathbb{R}^3;\ 2^{j-1} \le |\xi| \le 2^{j+1}\}$, by the Plancherel theorem we have

$$\begin{aligned}\|\varphi_j * \delta_{S_R}\|_{L^2}^2 = \|\mathcal{F}[\varphi_j]\mathcal{F}[\delta_{S_R}]\|_{L^2}^2 &\le C \int_{2^{j-1}\le|\xi|\le 2^{j+1}} \frac{|\phi(2^{-j}\xi)|^2}{|\xi|^2}\, d\xi \\ &= C\, 2^j \int_{\frac{1}{2}\le|\xi|\le 2} \frac{|\phi(\xi)|^2}{|\xi|^2}\, d\xi \le C\, 2^j,\end{aligned}$$

which yields that

$$2^{-\frac{j}{2}} \|\varphi_j * \delta_{S_R}\|_{L^2} \le C, \quad \text{for all } j \in \mathbb{Z}.$$

This shows that

$$\delta_{S_R} \in \dot{B}^{-\frac{1}{2}}_{2,\infty}(\mathbb{R}^3). \tag{3.13}$$

Interpolating (3.12) with $n = 3$ and (3.13), we have that

$$\delta_{S_R} \in \dot{B}^{-3+\frac{5}{p}}_{p,\infty}(\mathbb{R}^3), \quad 2 \le p \le \infty. \tag{3.14}$$

Substituting the regularity $s = -3 + \frac{5}{p}$ in the condition $2/\alpha + 3/p - s = 3$, we obtain $\alpha = p$. By the condition $s = -3 + \frac{5}{p} > -1$, p is restricted $p < \frac{5}{2}$. Thus, in the case when $n = 3$, by taking $r = 2$ with

$$\omega_0(x) = P(\delta_{S_R} e_i) \in \dot{B}^{-2+\frac{3}{2}}_{2,\infty}(\mathbb{R}^3),$$

$$a(x) = \varepsilon(-\Delta)^{-1}\mathrm{rot}\, \omega_0 \in \dot{B}^{-1+\frac{3}{2}}_{2,\infty}(\mathbb{R}^3), \tag{3.15}$$

$$f(x,t) = \varepsilon t^{-\frac{1}{p}} \delta_{S_R}(x) \in L^{p,\infty}(0,\infty; \dot{B}^{-3+\frac{5}{p}}_{p,\infty}(\mathbb{R}^3)), \quad 2 \le p < \frac{5}{2}, \tag{3.16}$$

we may take $\varepsilon > 0$ so small that

$$\|a\|_{\dot{B}^{-1+3/2}_{2,\infty}} + \|f\|_{L^{p,\infty}(0,\infty;\, \dot{B}^{-3+\frac{5}{p}}_{p,\infty}(\mathbb{R}^3))} \le \eta$$

is fulfilled provided $0 < \varepsilon \le \varepsilon_0$. Then we see that the initial data a and the external force f satisfy the conditions (3.2) in Theorem 3.1 to obtain the global solution

$$u \in L^{\alpha_0,\infty}(0,\infty; \dot{\mathcal{B}}^{s_0}_{p_0,1}) \quad \text{for } 3/\alpha_0 + 2/p_0 - s_0 = 1$$

with $p \le p_0$, $p < \alpha_0$ and $\max\{-3 + \frac{5}{p}, -1 + \frac{3}{2}\} = \frac{1}{2} < s_0$. Moreover by Remark 3.2 (ii), we know that the solution u also belongs to the Serrin class (2.12). The uniqueness is proved in a similar way to the proof of Theorem 2.1.

REMARK 3.5. The single layer potential δ_{S_R} is not a measurable function. Therefore, the initial data a defined by (3.15) does not belong to the Lorentz space $L^{3,\infty}(\mathbb{R}^3)$ and the external force f defined by (3.16) does not belong to the Lorentz space $L^{\alpha,\infty}(0,\infty; L^{p,\infty}(\mathbb{R}^3))$ with $2/\alpha + 3/p = 3$, either.

REMARK 3.6. Since the initial data a defined by (3.15) belongs to $\dot{B}^{-1+\frac{3}{2}}_{2,\infty}(\mathbb{R}^3)$, the exponent $r = 2$ is less than the dimension 3. Therefore, the sufficient condition (2.7) on uniqueness cannot be replaced by the weaker condition (3.5) in Theorem 3.1. By the same reason, Theorem 2 in [**20**] is not applicable to obtain a local solution for the large data.

4. Appendix

In this appendix, we consider the double layer potential $-\partial_\nu \delta_{S_R}$ with ν the unit outer normal to S_R, which is defined by

$$(-\partial_\nu \delta_{S_R}, \Phi) = \int_{S_R} \frac{\partial \Phi}{\partial \nu}(x)\, dS(x)$$

for every $\Phi(x) \in C_0^\infty(\mathbb{R}^n)$. In the same as the calculationa of L^2 and L^∞ norms of the single layer potential, we have for $n = 3$,

$$\mathcal{F}[-\partial_\nu \delta_{S_R}](\xi) = 4\pi R \cos R|\xi| - \frac{4\pi \sin R|\xi|}{|\xi|},$$

which shows that

$$\|\varphi_j * -\partial_\nu \delta_{S_R}\|_{L^2} = \|\hat{\varphi}_j \mathcal{F}[-\partial_\nu \delta_{S_R}]\|_{L^2} \le C_{n,R} 2^{\frac{3}{2}j},$$
$$\|\varphi_j * -\partial_\nu \delta_{S_R}\|_{L^\infty} = C2^j \|(\nabla \mathcal{F}^{-1}[\phi])_{2^{-j}}\|_{L^\infty} \le C_{n,R} 2^{4j}.$$

This yields that

$$-\partial_\nu \delta_{S_R} \in \dot{B}^{-\frac{3}{2}}_{2,\infty}(\mathbb{R}^3) \cap \dot{B}^{-4}_{\infty,\infty}(\mathbb{R}^3).$$

By interpolation, we have that

$$-\partial_\nu \delta_{S_R} \in \dot{B}^{-4+\frac{5}{p}}_{p,\infty}(\mathbb{R}^3), \quad 2 \le p \le \infty. \tag{4.1}$$

Unfortunately, such an investigation does not fit the hypothesis of Theorem 3.1.

Acknowledgments

The authors express their sincere thanks to Professor Nakao Hayashi who made them aware of the importance of examples of functions with higher singularities. The authors express their thanks to Professor Pierre Gilles Lemarié-Rieusset for our valuable discussions in the conference "Complex Fluids and the Issue of Regularity", Luminy, May 8-12, 2017. The authors also express their thanks to the referee who gives them valuable comments for the paper.

References

[1] Herbert Amann, *On the strong solvability of the Navier-Stokes equations*, J. Math. Fluid Mech. **2** (2000), no. 1, 16–98. MR1755865

[2] Jean Bourgain and Nataša Pavlović, *Ill-posedness of the Navier-Stokes equations in a critical space in 3D*, J. Funct. Anal. **255** (2008), no. 9, 2233–2247. MR2473255

[3] Hajer Bahouri, Jean-Yves Chemin, and Raphaël Danchin, *Fourier analysis and nonlinear partial differential equations*, Grundlehren der Mathematischen Wissenschaften [Fundamental Principles of Mathematical Sciences], vol. 343, Springer, Heidelberg, 2011. MR2768550

[4] Marco Cannone, *Ondelettes, paraproduits et Navier-Stokes* (French), Diderot Editeur, Paris, 1995. With a preface by Yves Meyer. MR1688096

[5] Marco Cannone and Yves Meyer, *Littlewood-Paley decomposition and Navier-Stokes equations*, Methods Appl. Anal. **2** (1995), no. 3, 307–319. MR1362019

[6] Marco Cannone and Grzegorz Karch, *Smooth or singular solutions to the Navier-Stokes system?*, J. Differential Equations **197** (2004), no. 2, 247–274. MR2034160

[7] Marco Cannone and Fabrice Planchon, *Self-similar solutions for Navier-Stokes equations in* $\mathbf{R}^3$, Comm. Partial Differential Equations **21** (1996), no. 1-2, 179–193. MR1373769

[8] Marco Cannone and Fabrice Planchon, *On the non-stationary Navier-Stokes equations with an external force*, Adv. Differential Equations **4** (1999), no. 5, 697–730. MR1696350

[9] Reinhard Farwig, Yoshikazu Giga, and Pen-Yuan Hsu, *Initial values for the Navier-Stokes equations in spaces with weights in time*, Funkcial. Ekvac. **59** (2016), no. 2, 199–216. MR3560499

[10] Reinhard Farwig, Yoshikazu Giga, and Pen-Yuan Hsu, *The Navier-Stokes equations with initial values in Besov spaces of type* $B_{q,\infty}^{-1+3/q}$, J. Korean Math. Soc. **54** (2017), no. 5, 1483–1504. MR3691934

[11] Reinhard Farwig and Hermann Sohr, *Optimal initial value conditions for the existence of local strong solutions of the Navier-Stokes equations*, Math. Ann. **345** (2009), no. 3, 631–642. MR2534111

[12] Reinhard Farwig, Hermann Sohr, and Werner Varnhorn, *On optimal initial value conditions for local strong solutions of the Navier-Stokes equations*, Ann. Univ. Ferrara Sez. VII Sci. Mat. **55** (2009), no. 1, 89–110. MR2506065

[13] Hiroshi Fujita and Tosio Kato, *On the Navier-Stokes initial value problem. I*, Arch. Rational Mech. Anal. **16** (1964), 269–315. MR0166499

[14] Yoshikazu Giga and Tetsuro Miyakawa, *Solutions in* L_r *of the Navier-Stokes initial value problem*, Arch. Rational Mech. Anal. **89** (1985), no. 3, 267–281. MR786550

[15] Tosio Kato, *Strong* L^p*-solutions of the Navier-Stokes equation in* $\mathbf{R}^m$*, with applications to weak solutions*, Math. Z. **187** (1984), no. 4, 471–480. MR760047

[16] Herbert Koch and Daniel Tataru, *Well-posedness for the Navier-Stokes equations*, Adv. Math. **157** (2001), no. 1, 22–35. MR1808843

[17] Kenta Kaneko, Hideo Kozono, and Senjo Shimizu, *Stationary solution to the Navier-Stokes equations in the scale invariant Besov space and its regularity.* to appear in Indiana Univ. Math. J..

[18] Hideo Kozono and Senjo Shimizu, *Navier–Stokes equations with external forces in Lorentz spaces and its application to the self-similar solutions*, J. Math. Anal. Appl. **458** (2018), no. 2, 1693–1708. MR3724749

[19] Hideo Kozono and Senjo Shimizu, *Navier-Stokes equations with external forces in time-weighted Besov spaces.* to be submitted.

[20] Hideo Kozono and Senjo Shimizu, *Strong solutions of the Navier-Stokes equations based on maximal Lorentz regularity in Besov spaces.* to be submitted.

[21] Hideo Kozono and Masao Yamazaki, *Semilinear heat equations and the Navier-Stokes equation with distributions in new function spaces as initial data*, Comm. Partial Differential Equations **19** (1994), no. 5-6, 959–1014. MR1274547

[22] Fabrice Planchon, *Asymptotic behavior of global solutions to the Navier-Stokes equations in* $\mathbf{R}^3$, Rev. Mat. Iberoamericana **14** (1998), no. 1, 71–93. MR1639283

[23] Baoxiang Wang, *Ill-posedness for the Navier-Stokes equations in critical Besov spaces* $\dot{B}_{\infty,q}^{-1}$, Adv. Math. **268** (2015), 350–372. MR3276597

[24] Tsuyoshi Yoneda, *Ill-posedness of the 3D-Navier-Stokes equations in a generalized Besov space near* BMO^{-1}, J. Funct. Anal. **258** (2010), no. 10, 3376–3387. MR2601621

Department of Mathematics, Waseda University, 169-8555 Tokyo, Japan
Email address: `kozono@waseda.jp`

Graduate School of Human and Environmental Studies, Kyoto University, 606-8501 Kyoto, Japan
Email address: `shimizu.senjo.5s@kyoto-u.ac.jp`

Contemporary Mathematics
Volume **710**, 2018
http://dx.doi.org/10.1090/conm/710/14370

L^p-L^q maximal regularity for some operators associated with linearized incompressible fluid-rigid body problems

Debayan Maity and Marius Tucsnak

ABSTRACT. We study an unbounded operator arising naturally after linearizing the system modelling the motion of a rigid body in a viscous incompressible fluid. We show that this operator is $\mathcal{R}$ sectorial in L^q for every $q \in (1, \infty)$, thus it has the maximal L^p-L^q regularity property. Moreover, we show that the generated semigroup is exponentially stable with respect to the L^q norm. Finally, we use the results to prove the global existence for small initial data, in an L^p-L^q setting, for the original nonlinear problem.

1. Introduction and main results

The aim of this work is to show that the semigroup associated to the equations obtained by linearizing some systems modelling fluid-structure interactions has the maximal L^p-L^q regularity property. This result can be seen as an improvement of those in [**21**, **22**], where the result was proved in a Hilbert space setting and in [**24**], where it has been shown that the corresponding semigroup is analytic in $L^{\frac{6}{5}}(\mathbb{R}^3) \cap L^q(\mathbb{R}^3)$, for $q \geqslant 2$. We then apply this result in proving a global existence and uniqueness result, for small data, for the original nonlinear problem. Such a result seems new in an L^p-L^q setting. The references [**21**, **22**] and [**11**] contain closely related results and methods which are often used in the present paper.

Let us first remind the original free boundary system which motivates this work. We will come back to this system later on, in order to prove the global existence of solutions for small initial data. The smallness is measured in a Besov space and the solutions lie in function spaces which are L^p with respect to time and L^q with respect to the space variable. As far as we know, global existence results of this type were known only in an L^2 setting.

Consider a rigid body immersed in a viscous incompressible fluid and moving under the action of forces exerted by the fluid only. At time $t \geqslant 0$, this solid occupies a smooth bounded domain $\Omega_S(t)$. The fluid and rigid body are contained in a bounded domain $\Omega \subset \mathbb{R}^3$ with smooth boundary $\partial\Omega$. We assume that there exists a constant α with

$$\operatorname{dist}(\Omega_S(0), \partial\Omega) \geqslant \alpha > 0. \tag{1.1}$$

2010 *Mathematics Subject Classification.* Primary 76D03, 35Q30,76N10.
Key words and phrases. Fluid Structure interaction, Incompressible flow, Maximal L^p regularity.

At any time $t \geqslant 0$, we denote by $\Omega_F(t) = \Omega \setminus \overline{\Omega_S(t)}$ the domain occupied by the fluid. We assume that the motion of the fluid is governed by the incompressible Navier-Stokes equations, whereas the motion of the structure is governed by the balance equation for linear and angular momentum. The full system of equations modelling the rigid body inside the fluid can be written as

$$
\begin{aligned}
&\partial_t u + (u\cdot\nabla)u - \operatorname{div}\sigma(u,\pi) = 0, && t\in(0,\infty),\ x\in\Omega_F(t),\\
&\operatorname{div} u = 0, && t\in(0,\infty),\ x\in\Omega_F(t),\\
&u = 0, && t\in(0,\infty),\ x\in\partial\Omega,\\
&u = a'(t) + \omega(t)\times(x-a(t)), && t\in(0,\infty),\ x\in\partial\Omega_S(t),\\
&ma''(t) = -\int_{\partial\Omega_S(t)} \sigma(u,\pi)n \,\mathrm{d}\gamma, && t\in(0,\infty),\\
&J\omega'(t) = (J\omega)\times\omega - \int_{\partial\Omega_S(t)} (x-a(t))\times\sigma(u,\pi)n\,\mathrm{d}\gamma, && t\in(0,\infty),\\
&u(0,x) = u_0(x) && x\in\Omega_F(0),\\
&a(0) = 0, \quad a'(0) = \ell_0, \quad \omega(0) = \omega_0.
\end{aligned}
\tag{1.2}
$$

In the above equations, $u(t,x)$ denote the velocity of the fluid, $\pi(t,x)$ denote the pressure of the fluid, $a(t)$ denote the position of the centre of the mass and $\omega(t)$ denote the angular velocity of the rigid body. The domain $\Omega_S(t)$ is defined by

$$\Omega_S(t) = a(t) + Q(t)y, \quad \forall y\in\Omega_S(0),\ \forall t>0,$$

where $Q(t)\in\mathbb{M}_{3\times3}(\mathbb{R})$ is the orthogonal matrix giving the orientation of the solid. More precisely, $\omega(t)$ and $Q(t)$ are related to each other through the following relation

$$\dot{Q}(t)Q(t)^{-1}y = A(\omega(t))y = \omega(t)\times y, \quad \forall y\in\mathbb{R}^3, \quad Q(0) = I_3, \tag{1.3}$$

where the skew-symmetric matrix $A(\omega)$ is given by

$$A(\omega) = \begin{pmatrix} 0 & -\omega_3 & \omega_2\\ \omega_3 & 0 & -\omega_1\\ -\omega_2 & \omega_1 & 0\end{pmatrix}, \qquad \omega\in\mathbb{R}^3.$$

The constant $m>0$ denote the mass of the rigid structure and $J(t)\in\mathcal{M}_{3\times3}(\mathbb{R})$ its tensor of inertia at time t. This tensor is given by

$$J(t)a\cdot b = \int_{\Omega_S(0)} \rho_S(y)(a\times Q(t)y)\cdot(b\times Q(t)y)\,dy, \qquad \forall a,b\in\mathbb{R}^3, \tag{1.4}$$

where $\rho_S>0$ is the density of the structure. One can check that

$$J(t)a\cdot a \geqslant C_J|a|^2 > 0, \tag{1.5}$$

where C_J is independent of $t>0$. In the above, we have denoted by $\partial\Omega_S(t)$ the boundary of the rigid structure at time t and by $n(t,x)$ the unit normal to $\partial\Omega_S(t)$ at the point x directed towards the interior of the rigid body. The Cauchy stress tensor $\sigma(u,\pi)$ is given by

$$\sigma(u,\pi) = -\pi I_3 + 2\nu\varepsilon(u), \quad \varepsilon(u) = \frac{1}{2}\left(\nabla u + \nabla u^\top\right), \tag{1.6}$$

where the positive constant ν is the viscosity of the fluid.

Linearizing the above equations around the zero solution we obtain a system coupling Stokes equations in a fixed domain and an ODE system. The corresponding equations read as

$$
(1.7)\quad
\begin{aligned}
&\partial_t u - \nu\Delta u + \nabla\pi = f, \quad \operatorname{div} u = 0, && t\in(0,\infty),\ y\in\Omega_F(0),\\
&u = 0, && t\in(0,\infty),\ y\in\partial\Omega,\\
&u = \ell + \omega\times y, && t\in(0,\infty),\ y\in\partial\Omega_S(0),\\
&m\ell' = -\int_{\partial\Omega_S(0)} \sigma(u,\pi)n\, d\gamma + g_1, && t\in(0,\infty),\\
&J(0)\omega' = -\int_{\partial\Omega_S(0)} y\times\sigma(u,\pi)n\, d\gamma + g_2, && t\in(0,\infty),\\
&u(0,y) = u_0(y), && y\in\Omega_F(0),\\
&\ell(0) = \ell_0, \quad \omega(0) = \omega_0,
\end{aligned}
$$

where n is the unit normal to $\partial\Omega_S(0)$ directed towards the interior of the rigid body.

Let us now define the operator associated with the above linear fluid-structure interaction problem, which was first introduced in [**21**, **22**]. The idea is to extend the fluid velocity u by $\ell(t)+\omega(t)\times y$ in $\Omega_S(0)$. More precisely, for any $1<q<\infty$ we define

(1.8)

$$
\mathbb{H}^q(\Omega) = \left\{\varphi\in L^q(\Omega)^3 \mid \operatorname{div}\varphi = 0 \text{ in } \Omega, \quad \varepsilon(\varphi) = 0 \text{ in } \Omega_S(0), \quad \varphi\cdot n = 0 \text{ on } \partial\Omega\right\}
$$

We define

(1.9)

$$
\mathcal{D}(\mathbb{A}) = \left\{\varphi\in W_0^{1,q}(\Omega)^3 \mid \varphi|_{\Omega_F(0)}\in W^{2,q}(\Omega_F(0))^3,\ \operatorname{div}\varphi = 0 \text{ in } \Omega,\ \varepsilon(\varphi) = 0 \text{ in } \Omega_S(0)\right\}.
$$

For all $v\in\mathcal{D}(\mathbb{A})$ we set

$$
\mathcal{A}v = \begin{cases} -\nu\Delta v & \text{in } \Omega_F(0),\\ 2\nu m^{-1}\displaystyle\int_{\partial\Omega_S(0)} \varepsilon(v)n\, d\gamma + \left(2\nu J(0)^{-1}\int_{\partial\Omega_S(0)} y\times\varepsilon(v)n\, d\gamma\right)\times y & \text{in } \Omega_S(0),\end{cases}
$$

and

$$
(1.10)\qquad \mathbb{A}v = \mathbb{P}\mathcal{A}v,
$$

where $\mathbb{P}$ is the projection from $L^q(\Omega)^3$ onto $\mathbb{H}^q(\Omega)$. The existence of such projector $\mathbb{P}$ can be found in [**24**, Theorem 2.2].

Takahashi and Tucsnak [**22**] proved that the operator $\mathbb{A}$ defined above generates an analytic semigroup on $\mathbb{H}^2(\Omega)$ when $\Omega = \mathbb{R}^2$. When Ω is a smooth bounded domain in $\mathbb{R}^2$ the same result was proved by Takahashi [**21**]. Later, Wang and Xin [**24**] proved that the operator $\mathbb{A}$ generates an analytic semigroup on $\mathbb{H}^{6/5}(\mathbb{R}^3)\cap\mathbb{H}^q(\mathbb{R}^3)$ if $q\geqslant 2$ and when the solid is a ball in $\mathbb{R}^3$ the operator $\mathbb{A}$ generates an analytic semigroup on $\mathbb{H}^2(\mathbb{R}^3)\cap\mathbb{H}^q(\mathbb{R}^3)$ if $q\geqslant 6$. In this article, as a corollary of our main result, we prove that the operator $\mathbb{A}$ generates an analytic semigroup on $\mathbb{H}^q(\Omega)$ for any $1<q<\infty$.

Before we state our main result, we introduce the notion of maximal L^p-regularity. Let us consider the following Cauchy problem:

$$
(1.11)\qquad z'(t) = Az(t) + f(t), \quad z(0) = z_0,
$$

where A is a closed, linear densely defined unbounded operator in a Banach space $\mathcal{X}$ with domain $\mathcal{D}(A)$, $f : \mathbb{R}^+ \mapsto \mathcal{X}$ is a locally integrable function and $z_0 \in \mathcal{X}$.

DEFINITION 1.1. We say A has maximal L^p-regularity property for $1 < p < \infty$, on $[0, T)$, $0 < T \leqslant \infty$, if for $z_0 = 0$ and for every $f \in L^p(0, T; \mathcal{X})$ there exists a unique $z \in W^{1,p}_{\text{loc}}[0, \infty); \mathcal{X})$ satisfying (1.11) almost everywhere and such that z' and Az belong to $L^p(0, T; \mathcal{X})$. We denote the class of all such operators by $\mathcal{MR}_p([0, T); \mathcal{X})$.

REMARK 1.2. In the above definition we do not assume that $z \in L^p(0, T; \mathcal{X})$. In fact, if $T < \infty$ or $0 \in \rho(A)$, where $\rho(A)$ is the resolvent set of A, $z' \in L^p(0, T; \mathcal{X})$ can be replaced by $z \in W^{1,p}(0, T; \mathcal{X})$ ([**6**, Theorem 2.4]).

We now state our first main result:

THEOREM 1.3. *Let $1 < p, q < \infty$ and $T < \infty$. Then $\mathbb{A} \in \mathcal{MR}_p([0, T]; \mathbb{H}^q(\Omega))$. In particular, the operator $\mathbb{A}$ generates an analytic semigroup on $\mathbb{H}^q(\Omega)$ for any $q \in (1, \infty)$.*

To prove the above result we use the characterization of maximal L^p regularity due to Weis ([**25**, Theorem 4.2]), which says that maximal L^p regularity property in a UMD Banach space, in particular for L^q spaces, is equivalent to the $\mathcal{R}$-sectoriality property of the operator (see Section 2 for definition and properties of $\mathcal{R}$-sectorial operators).

The maximal L^p-L^q regularity property in finite time interval of the system (1.7), when $\Omega = \mathbb{R}^3$, was already proved in [**11, 13**]. However, the approach of those papers is different from our approach. In fact, in those papers, fluid and structure equations are treated separately and maximal L^p-L^q regularity property of the linear system (1.7) is proved by a fixed point argument. In our approach, we solve the fluid and structure equations simultaneously. In the study of fluid-structure interactions, this method is known as a monolithic approach. We refer to Maity and Tucsnak [**18**] where a similar approach has been used to prove maximal L^p-L^q regularity for several other fluid structure models.

The main advantage of such approach is that, by studying resolvent of the linear operator $\mathbb{A}$, we can conclude that the operator $\mathbb{A}$ generates a C^0-semigroup of negative type. This allows us to obtain the maximal L^p-L^q regularity of the system (1.7) on $[0, \infty)$. As a consequence, we obtain global existence and uniqueness for the full non-linear system (1.2) under a smallness condition on the initial data.

In order to state global existence and uniqueness result, we introduce some notation. Firstly $W^{s,q}(\Omega)$, with $s \geqslant 0$ and $q > 1$, denote the usual Sobolev spaces. We introduce the space

$$L^q_m(\Omega) = \left\{ f \in L^q(\Omega) \mid \int_\Omega f = 0 \right\}$$

and we set

$$W^{s,q}_m(\Omega) = W^{s,q}(\Omega) \cap L^q_m(\Omega).$$

Let $k \in \mathbb{N}$. For every $0 < s < k$, $1 \leqslant p < \infty$, $1 \leqslant q < \infty$, we define the Besov spaces by real interpolation of Sobolev spaces

$$B^s_{q,p}(\Omega) = (L^q(\Omega), W^{k,q}(\Omega))_{s/k,p}.$$

We refer to [**23**] for more details on Besov spaces. We also need a definition of Sobolev spaces in the time dependent domain $\Omega_F(t)$. Let $\Lambda(t,\cdot)$ be a C^1-diffeomorphism from $\Omega_F(0)$ onto $\Omega_F(t)$ such that all the derivatives up to second order in space variable and all the derivatives up to first order in time variable exist are continuous. For all functions $v(t,\cdot):\Omega_F(t)\mapsto\mathbb{R}$, we denote $\widehat{v}(t,y)=v(t,\Lambda(t,y))$. Then for any $1<p,q<\infty$ we define

$$\begin{aligned}
&L^p(0,T;L^q(\Omega_F(\cdot)))=\{v\mid\widehat{v}\in L^p(0,T;L^q(\Omega_F(0)))\},\\
&L^p(0,T;W^{2,q}(\Omega_F(\cdot)))=\left\{v\mid\widehat{v}\in L^p(0,T;W^{2,q}(\Omega_F(0)))\right\},\\
&W^{1,p}(0,T;L^q(\Omega_F(\cdot)))=\left\{v\mid\widehat{v}\in W^{1,p}(0,T;L^q(\Omega_F(0)))\right\},\\
&C([0,T];B^{2(1-1/p)}_{q,p}(\Omega_F(\cdot)))=\left\{v\mid\widehat{v}\in C([0,T];B^{2(1-1/p)}_{q,p}(\Omega_F(0))\right\}.
\end{aligned}$$

THEOREM 1.4. *Let* $1<p,q<\infty$ *satisfying the conditions* $\dfrac{1}{p}+\dfrac{1}{2q}\neq 1$ *and* $\dfrac{1}{p}+\dfrac{3}{2q}\leqslant\dfrac{3}{2}$. *Let* $\eta\in(0,\eta_0)$, *where* η_0 *is the constant introduced in Theorem 4.1. Then there exist two constants* $\delta_0>0$ *and* $C>0$, *depending on* p,q,η *and* $\Omega_F(0)$, *such that, for all* $\delta\in(0,\delta_0)$ *and for all* $(u_0,\ell_0,\omega_0)\in B^{2(1-1/p)}_{q,p}(\Omega_F(0))^3\times\mathbb{R}^3\times\mathbb{R}^3$ *satisfying the compatibility conditions*

$$\begin{aligned}
&\operatorname{div} u_0=0 \text{ in } \Omega_F(0),\\
&u_0=\ell_0+\omega_0\times y \text{ on } \partial\Omega_S(0),\quad u_0=0 \text{ on } \partial\Omega \text{ if } \frac{1}{p}+\frac{1}{2q}<1\\
\text{and } &u_0\cdot n=(\ell_0+\omega_0\times y)\cdot n \text{ on } \partial\Omega_S(0),\qquad u_0\cdot n=0 \text{ on } \partial\Omega \text{ if } \frac{1}{p}+\frac{1}{2q}>1,
\end{aligned}$$

and

$$\|u_0\|_{B^{2(1-1/p)}_{q,p}(\Omega_F(0))^3}+\|\ell_0\|_{\mathbb{R}^3}+\|\omega_0\|_{\mathbb{R}^3}\leqslant\delta, \tag{1.12}$$

the system (1.2) *admits a unique strong solution* (u,π,ℓ,ω) *in the class of functions satisfying*

$$\begin{aligned}
\|e^{\eta(\cdot)}u\|_{L^p(0,\infty;W^{2,q}(\Omega_F(\cdot)))^3}&+\|e^{\eta(\cdot)}u\|_{W^{1,p}(0,\infty;L^q(\Omega_F(\cdot)))^3}\\
&+\|e^{\eta(\cdot)}u\|_{L^\infty(0,\infty;B^{2(1-1/p)^3}_{q,p}(\Omega_F(\cdot)))}\\
+\|e^{\eta(\cdot)}\pi\|_{L^p(0,\infty;W^{1,q}_m(\Omega_F(\cdot)))}&+\|a\|_{L^\infty(0,\infty;\mathbb{R}^3)}++\|e^{\eta(\cdot)}a'\|_{L^p(0,\infty;\mathbb{R}^3)}\\
&+\|e^{\eta(\cdot)}a''\|_{L^p(0,\infty;\mathbb{R}^3)}+\|e^{\eta(\cdot)}\omega\|_{W^{1,p}(0,\infty;\mathbb{R}^3)}\leqslant C\delta.
\end{aligned}\tag{1.13}$$

Moreover, $\operatorname{dist}(\Omega_S(t),\partial\Omega)\geqslant\alpha/2$ *for all* $t\in[0,\infty)$. *In particular, we have*

$$\|u(t,\cdot)\|_{B^{2(1-1/p)}_{q,p}(\Omega_F(t))}+\|a'(t)\|_{\mathbb{R}^3}+\|\omega(t)\|_{\mathbb{R}^3}\leqslant C\delta e^{-\eta t}.$$

REMARK 1.5. When $p=q=2$, the above result was proved in [**21**, Corollary 9.2].

REMARK 1.6. Our proof of Theorem 1.4 also applies to the 2 dimensional case. In this case, we have to choose $1<p,q<\infty$ such that $\dfrac{1}{p}+\dfrac{1}{2q}\neq 1$ and $\dfrac{1}{p}+\dfrac{1}{2q}\leqslant\dfrac{3}{2}$.

The plan of this paper is as follows. In Section 2, we recall the definition and some basic properties of $\mathcal{R}$-sectorial operators. In Section 3, we prove Theorem 1.3. The stability of the operator $\mathbb{A}$ is proved in Section 4. Maximal L^p-L^q regularity of the linear fluid structure system on $(0,\infty)$ is studied in Section 5. Finally, in Section 6 we prove Theorem 1.4.

2. Some background on $\mathcal{R}$-sectorial operators

In this section we recall some definitions and basic results concerning maximal regularity and $\mathcal{R}$-boundedness in Banach spaces. For detailed information on these subjects we refer to [**5**, **16**, **25**] and references therein. For $\theta \in (0,\pi)$ we define the sector Σ_θ in the complex plane by

$$\Sigma_\theta = \{\lambda \in \mathbb{C} \setminus \{0\} \quad | \quad |\arg\lambda| < \theta\}. \tag{2.1}$$

In order to state Weis' theorem concerning maximal L^p-regularity of the Cauchy problem (1.11) we need to introduce so-called *UMD* spaces.

DEFINITION 2.1. Let $\mathcal{X}$ be a Banach space. The Hilbert transform of a function $f \in \mathcal{S}(\mathbb{R};\mathcal{X})$, the Schwartz space of $\mathcal{X}$-valued rapidly decreasing functions, is defined by

$$Hf(t) = \frac{1}{\pi} \lim_{\epsilon \mapsto 0} \int_{|s|>\epsilon} \frac{f(t-s)}{s}\, ds, \quad t \in \mathbb{R}.$$

A Banach space $\mathcal{X}$ is said to be of class $\mathcal{HT}$, if the Hilbert transform is bounded on $L^p(\mathbb{R};\mathcal{X})$ for some (thus all) $1 < p < \infty$.

These spaces are also called *UMD* Banach spaces, where *UMD* stands for *unconditional martingale differences.* Hilbert spaces, all closed subspaces and quotient spaces of $L^q(\Omega)$ with $1 < q < \infty$ are examples of *UMD* spaces. We refer the reader to [**1**, pp. 141-147] for more information about *UMD* spaces.

We next introduce the notion of $\mathcal{R}$-bounded family of operators and $\mathcal{R}$-sectoriality of a densely defined linear operator.

DEFINITION 2.2 ($\mathcal{R}$ - bounded family of operators). Let $\mathcal{X}$ and $\mathcal{Y}$ be Banach spaces. A family of operators $\mathcal{T} \subset \mathcal{L}(\mathcal{X},\mathcal{Y})$ is called $\mathcal{R}$- bounded on $\mathcal{L}(\mathcal{X},\mathcal{Y})$, if there exist constants $C > 0$ and $p \in [1,\infty)$ such that for every $n \in \mathbb{N}$, $\{T_j\}_{j=1}^n \subset \mathcal{T}$, $\{x_j\}_{j=1}^n \subset \mathcal{X}$ and for all sequences $\{r_j(\cdot)\}_{j=1}^n$ of independent, symmetric, $\{-1,1\}$ valued random variables on $[0,1]$, we have

$$\left\| \sum_{j=1}^n r_j(\cdot) T_j x_j \right\|_{L^p([0,1];\mathcal{Y})} \leqslant C \left\| \sum_{j=1}^n r_j(\cdot) x_j \right\|_{L^p([0,1];\mathcal{X})}.$$

The smallest such C is called $\mathcal{R}$-bound of $\mathcal{T}$ on $\mathcal{L}(\mathcal{X},\mathcal{Y})$ and denoted by $\mathcal{R}_{\mathcal{L}(\mathcal{X},\mathcal{Y})}(\mathcal{T})$.

DEFINITION 2.3 ($\mathcal{R}$-sectorial operator). Let A be a densely defined closed linear operator on a Banach space $\mathcal{X}$ with domain $\mathcal{D}(A)$. Then A is said sectorial of angle $\theta \in (0,\pi)$ if $\sigma(A) \subseteq \overline{\Sigma_\theta}$ and for any $\theta_1 > \theta$ the set $\{\lambda(\lambda I - A)^{-1} \mid \theta_1 \leqslant |\arg(\lambda)| \leqslant \pi\}$ is bounded. If this set is $\mathcal{R}$-bounded then A is $\mathcal{R}$-sectorial of angle θ.

We now state several useful properties concerning $\mathcal{R}$-boundedness, which will be used later on

PROPOSITION 2.4.

(1) *If $\mathcal{T} \subset \mathcal{L}(\mathcal{X}, \mathcal{Y})$ is $\mathcal{R}$-bounded then it is uniformly bounded with*

$$\sup \{\|T\| \mid T \in \mathcal{T}\} \leqslant \mathcal{R}_{\mathcal{L}(\mathcal{X},\mathcal{Y})}(\mathcal{T}).$$

(2) *If $\mathcal{X}$ and $\mathcal{Y}$ are Hilbert spaces, $\mathcal{T} \subset \mathcal{L}(\mathcal{X}, \mathcal{Y})$ is $\mathcal{R}$-bounded if and only if $\mathcal{T}$ is uniformly bounded.*

(3) *Let $\mathcal{X}$ and $\mathcal{Y}$ be Banach spaces and let $\mathcal{T}$ and $\mathcal{S}$ be $\mathcal{R}$-bounded families on $\mathcal{L}(\mathcal{X}, \mathcal{Y})$. Then $\mathcal{T} + \mathcal{S}$ is also $\mathcal{R}$-bounded on $\mathcal{L}(\mathcal{X}, \mathcal{Y})$, and*

$$\mathcal{R}_{\mathcal{L}(\mathcal{X},\mathcal{Y})}(\mathcal{T} + \mathcal{S}) \leqslant \mathcal{R}_{\mathcal{L}(\mathcal{X},\mathcal{Y})}(\mathcal{T}) + \mathcal{R}_{\mathcal{L}(\mathcal{X},\mathcal{Y})}(\mathcal{S}).$$

(4) *Let $\mathcal{X}, \mathcal{Y}$ and $\mathcal{Z}$ be Banach spaces and let $\mathcal{T}$ and $\mathcal{S}$ be $\mathcal{R}$-bounded families on $\mathcal{L}(\mathcal{X}, \mathcal{Y})$ and $\mathcal{L}(\mathcal{Y}, \mathcal{Z})$ respectively. Then $\mathcal{S}\mathcal{T}$ is $\mathcal{R}$-bounded on $\mathcal{L}(\mathcal{X}, \mathcal{Z})$, and*

$$\mathcal{R}_{\mathcal{L}(\mathcal{X},\mathcal{Z})}(\mathcal{S}\mathcal{T}) \leqslant \mathcal{R}_{\mathcal{L}(\mathcal{X},\mathcal{Y})}(\mathcal{T})\mathcal{R}_{\mathcal{L}(\mathcal{Y},\mathcal{Z})}(\mathcal{S}).$$

The following characterization of maximal L^p regularity is due to Weis ([**25**, Theorem 4.2])

THEOREM 2.5. *Let $\mathcal{X}$ be a Banach space of class $\mathcal{HT}$, $1 < p < \infty$ and let A be a closed, densely defined unbounded operator with domain $\mathcal{D}(A)$. Then A has maximal L^p-regularity on $\mathbb{R}^+$ if and only if*

$$\mathcal{R}_{\mathcal{L}(\mathcal{X})}\left\{\lambda(\lambda - A)^{-1} \mid \lambda \in \Sigma_\theta\right\} \leqslant C \text{ for some } \theta > \pi/2. \tag{2.2}$$

In other words, A has maximal L^p-regularity if and only if $-A$ is $\mathcal{R}$-sectorial of angle $\theta < \pi/2$.

Next we state a perturbation result due to Kunstmann and Weis [**15**], which states that $\mathcal{R}$-boundedness is preserved by A small perturbations.

PROPOSITION 2.6. *Let $\mathcal{X}$ be a Banach space let A be a closed, densely defined unbounded operator with domain $\mathcal{D}(A)$. Let us assume that there exist $\gamma_0 \geqslant 0$ and $\theta \in (0, \pi)$ such that*

$$\mathcal{R}_{\mathcal{L}(\mathcal{X})}\left\{\lambda(\lambda - A)^{-1} \mid \lambda \in \gamma_0 + \Sigma_\theta\right\} \leqslant C.$$

Let B be a A-bounded operator with relative bound zero, i.e., for all $\delta > 0$ there exists $C(\delta) > 0$ such that

$$\|Bz\| \leqslant \delta\|Az\| + C(\delta)\|z\| \text{ for all } z \in \mathcal{D}(A). \tag{2.3}$$

Then there there exists $\mu_0 \geqslant \gamma_0$ such that

$$\mathcal{R}_{\mathcal{L}(\mathcal{X})}\left\{\lambda(\lambda - (A + B))^{-1} \mid \lambda \in \mu_0 + \Sigma_\theta\right\} \leqslant C.$$

We conclude this section by stating an existence and uniqueness result for the abstract Cauchy problem (1.11) on $\mathbb{R}^+$, which we will use to prove maximal L^p-L^q regularity of the system (1.7) (see [**6**, Theorem 2.4]).

THEOREM 2.7. *Let $\mathcal{X}$ be a Banach space of class $\mathcal{HT}$, $1 < p < \infty$ and let A be a closed, densely defined unbounded operator with domain $\mathcal{D}(A)$. Let us assume that $A \in \mathcal{MR}_p([0, T]; \mathcal{X})$ and the semigroup generated by A has negative exponential type. Then for every $z_0 \in (\mathcal{X}, \mathcal{D}(A))_{1-1/p,p}$ and for every $f \in L^p(0, \infty; \mathcal{X})$, (1.11) admits a unique strong solution in $L^p(0, \infty; \mathcal{D}(A)) \cap W^{1,p}(0, \infty; \mathcal{X})$. Moreover, there exists a positive constant C such that*

(2.4)

$$\|z\|_{L^p(0,\infty;\mathcal{D}(A))} + \|z\|_{W^{1,p}(0,\infty;\mathcal{X})} \leqslant C\left(\|z_0\|_{(\mathcal{X},\mathcal{D}(A))_{1-1/p,p}} + \|f\|_{L^p(0,\infty;\mathcal{X})}\right).$$

3. $\mathcal{R}$-sectoriality of the operator $\mathbb{A}$.

Let us recall the operator $\mathbb{A}$ introduced in (1.10). The aim of this section is to prove Theorem 1.3. Due to Theorem 2.5, it is enough to prove the following theorem:

THEOREM 3.1. *Let $1 < q < \infty$. There exists $\mu_0 > 0$ and $\theta \in (\pi/2, \pi)$ such that $\mu_0 + \Sigma_\theta \subset \rho(\mathbb{A})$ and*

$$\mathcal{R}_{\mathcal{L}(\mathbb{H}^q(\Omega))}\left\{\lambda(\lambda I - \mathbb{A})^{-1} \mid \lambda \in \mu_0 + \Sigma_\theta\right\} \leqslant C. \tag{3.1}$$

Let us remark that when $q = 2$ the above theorem is already proved in [**22**]. To prove the above theorem we will first obtain an equivalence formulation of the resolvent equation.

3.1. Reformulation of the resolvent equation. Given $\lambda \in \mathbb{C}$, $f \in L^q(\Omega_F(0))^3$ and $(g_1, g_2) \in \mathbb{C}^3 \times \mathbb{C}^3$, we consider the system

$$\begin{aligned}
&\lambda u - \nu\Delta u + \nabla\pi = f, \quad \text{div } u = 0 && \text{in } \Omega_F(0),\\
&u = 0 && \text{on } \partial\Omega,\\
&u = \ell + \omega \times y && y \in \partial\Omega_S(0),\\
&\lambda m\ell = -\int_{\partial\Omega_S(0)} \sigma(u,\pi)n \, d\gamma + g_1, &&\\
&\lambda J(0)\omega = -\int_{\partial\Omega_S(0)} y \times \sigma(u,\pi)n \, d\gamma + g_2, &&
\end{aligned} \tag{3.2}$$

of unknowns (u, π, ℓ, ω). Following [**21, 22**] we have the following equivalence

PROPOSITION 3.2. *Let $1 < q < \infty$. Let us assume that $f \in L^q(\Omega_F(0))$ and $(g_1, g_2) \in \mathbb{C}^3 \times \mathbb{C}^3$. Then $(u, \pi, \ell, \omega) \in W^{2,q}(\Omega_F(0))^3 \times W^{1,q}_m(\Omega_F(0)) \times \mathbb{C}^3 \times \mathbb{C}^3$ is a solution to (3.2) if and only if*

$$(\lambda I - \mathbb{A})v = \mathbb{P}F \tag{3.3}$$

where

$$v = u\mathbb{1}_{\Omega_F(0)} + (\ell + \omega \times y)\mathbb{1}_{\Omega_S(0)}, \quad F = \mathbb{P}\left(f\mathbb{1}_{\Omega_F(0)} + \left(m^{-1}g_1 + J(0)^{-1}y \times g_2\right)\mathbb{1}_{\Omega_S(0)}\right).$$

Next, we derive another equivalent formulation of the resolvent equation (3.2). In this case, we do not extend the fluid velocity by the structure velocity everywhere in the domain Ω, rather we work on the fluid domain $\Omega_F(0)$. The idea is to eliminate the pressure from both the fluid and the structure equations. To eliminate the pressure from the fluid equation we use Leray projector

$$\mathcal{P} : L^q(\Omega_F(0))^3 \mapsto \mathcal{V}^q_n(\Omega_F(0)) := \left\{\varphi \in L^q(\Omega_F(0))^3 \mid \text{div } \varphi = 0, \ \varphi \cdot n = 0 \text{ on } \partial\Omega_F(0)\right\}.$$

Note that the projector $\mathcal{P}$ is different from the projector $\mathbb{P}$ used in (1.10). Following [**19**], first, we decompose the fluid velocity into two parts $\mathcal{P}u$ and $(I - \mathcal{P})u$. Next, we obtain an expression of pressure, which can be broken down into two parts, one which depends on $\mathcal{P}u$ and another part which depends on (ℓ, ω). This will allow us to eliminate the pressure term from the structure equations and rewrite the system (3.2) as an operator equation of $(\mathcal{P}u, \ell, \omega)$.

The advantage of this formulation is that we can prove the $\mathcal{R}$-boundedness of the resolvent operator just by using the fact that Stokes operator with *homogeneous Dirichlet boundary conditions* is $\mathcal{R}$-sectorial and a perturbation argument. This idea has been used in several fluid-solid interaction problems in the Hilbert space setting and when the structure is deformable and located at the boundary (see, for instance, [**17**, **20**] and references therein).

Let $1 < q < \infty$ and q' denote the conjugate of q, i.e., $\dfrac{1}{q} + \dfrac{1}{q'} = 1$. Let n denote the normal to $\partial\Omega_F(0)$ exterior to $\Omega_F(0)$. For $1 < q < \infty$, we first introduce the space

$$W_{q,div}(\Omega) = \left\{\varphi \in L^q(\Omega)^3 \mid \text{div } \varphi \in L^q(\Omega)\right\},$$

equipped with the norm

$$\|\varphi\|_{W_{q,div}(\Omega)} := \|\varphi\|_{L^q(\Omega)^3} + \|\text{div } \varphi\|_{L^q(\Omega)}.$$

It is easy to check that $W_{q,div}(\Omega)$ is a Banach space. We have the following classical lemma:

LEMMA 3.3. *Let Ω be a bounded domain with smooth boundary. The linear mapping*

$$\varphi \mapsto \gamma_n \varphi := \varphi|_{\partial\Omega} \cdot n$$

defined on $C^\infty(\overline{\Omega})^3$ can be extended to a continuous and surjective map from $W_{q,div}(\Omega)$ onto $W^{-1/q,q}(\partial\Omega)$.

PROOF. For proof see [**9**, Lemma 1]. $\square$

Let us set

$$\mathcal{V}_n^q(\Omega_F(0)) = \overline{\{\varphi \in C_c^\infty(\Omega_F(0)) \mid \text{div } \varphi = 0\}}^{\|\cdot\|_{L^q}}.$$

As $\Omega_F(0)$ is bounded, we actually have

$$\mathcal{V}_n^q(\Omega_F(0)) = \left\{\varphi \in L^q(\Omega_F(0))^3 \mid \text{div } \varphi = 0,\ \varphi \cdot n = 0 \text{ on } \partial\Omega_F(0)\right\}.$$

We have the following Helmholtz-Weyl decomposition of $L^q(\Omega_F(0))^3$

PROPOSITION 3.4. *The space $L^q(\Omega_F(0))^3$ admits the following decomposition in a direct sum:*

$$L^q(\Omega_F(0))^3 = \mathcal{V}_n^q(\Omega_F(0)) \oplus G^q(\Omega_F(0)),$$

where

$$G^q(\Omega_F(0)) = \left\{\nabla\varphi \mid \varphi \in W^{1,q}(\Omega_F(0))\right\}.$$

The projection operator from $L^q(\Omega_F(0))^3$ onto $\mathcal{V}_n^q(\Omega_F(0))$ is denoted by $\mathcal{P}$. The projector $\mathcal{P} : L^q(\Omega_F(0))^3 \mapsto \mathcal{V}_n^q(\Omega_F(0))$ is defined by

$$\mathcal{P}u = u - \nabla\varphi,$$

where $\varphi \in W^{1,q}(\Omega_F(0))$ solves the following Neumann problem

$$\Delta\varphi = \text{div } u \text{ in } \Omega_F(0), \quad \frac{\partial\varphi}{\partial n} = \varphi \cdot n \text{ on } \partial\Omega_F(0).$$

PROOF. For the proof of the above result we refer to Section 3 and Theorem 2 of [**9**]. $\square$

Let us denote by $A_0 = \nu\mathcal{P}\Delta$, the Stokes operator in $\mathcal{V}_n^q(\Omega_F(0))$ with domain

$$\mathcal{D}(A_0) = W^{2,q}(\Omega_F(0)) \cap W_0^{1,q}(\Omega_F(0))^3 \cap \mathcal{V}_n^q(\Omega_F(0)).$$

PROPOSITION 3.5. *The Stokes operator* $-A_0$ *is* $\mathcal{R}$*-sectorial in* $\mathcal{V}^q_n(\Omega_F(0))$ *of angle* 0. *In particular, there exists* $\theta \in (\pi/2, \pi)$ *such that*

$$\mathcal{R}_{\mathcal{L}(\mathcal{V}^q_n(\Omega_F(0)))}\left\{\lambda(\lambda I - A_0)^{-1} \mid \lambda \in \Sigma_\theta\right\} \leqslant C. \tag{3.4}$$

Moreover, A_0 *generates a* C^0*-semigroup of negative type.*

PROOF. For proof we refer to Theorem 1.4 and Corollary 1.4 of [**12**]. □

Now we are going to rewrite the first three equations of (3.2) in terms of $\mathcal{P}u$ and $(I - \mathcal{P})u$. Let us consider the following problem

$$\begin{cases} -\nu\Delta w + \nabla\psi = 0, \quad \text{div } w = 0, & y \in \Omega_F(0), \\ w = 0, & y \in \partial\Omega, \\ w = \ell + \omega \times y, & y \in \partial\Omega_S(0), \\ \displaystyle\int_{\Omega_F(0)} \psi \,\mathrm{d}y = 0. \end{cases} \tag{3.5}$$

LEMMA 3.6. *Let* $(\ell, \omega) \in \mathbb{C}^3 \times \mathbb{C}^3$ *and let* $\{e_i\}$ *denote the canonical basis in* $\mathbb{C}^3$. *Then the solution* (w, π) *of* (3.5) *can be expressed as follows*

$$w = \sum_{i=1}^{3} \ell_i W_i + \sum_{i=4}^{6} \omega_{i-3} W_i, \quad \psi = \sum_{i=1}^{3} \ell_i \Psi_i + \sum_{i=4}^{6} \omega_{i-3}\Psi_i, \tag{3.6}$$

where (W_i, Ψ_i), $i = 1, 2, \cdots, 6$ *solves the following system*

$$\begin{cases} -\nu\Delta W_i + \nabla\Psi_i = 0, \quad \text{div} W_i = 0, \quad y \in \Omega_F(0), \\ W_i = 0, \quad y \in \partial\Omega, \\ W_i = e_i, \ \textit{for } i = 1,2,3 \textit{ and } W_i = e_{i-3} \times y, \ \textit{for } i = 4,5,6, \quad y \in \partial\Omega_F(0), \\ \displaystyle\int_{\Omega_F(0)} \Psi \,\mathrm{d}y = 0. \end{cases} \tag{3.7}$$

Moreover,

$$\begin{pmatrix} \displaystyle\int_{\partial\Omega_S(0)} \sigma(w, \psi)n \, d\gamma \\ \displaystyle\int_{\partial\Omega_S(0)} y \times \sigma(w, \psi)n \, d\gamma \end{pmatrix} = \mathbb{B}\begin{pmatrix} \ell \\ \omega \end{pmatrix},$$

where

$$\mathbb{B}_{i,j} = \int_{\Omega_F(0)} DW_i : DW_j. \tag{3.8}$$

and the matrix $\mathbb{B}$ *is invertible.*

PROOF. The expressions of w and π are easy to see. The expression of the matrix $\mathbb{B}$ follows easily by putting the expressions of w and π and by integration by parts. The proof of invertibility of the matrix $\mathbb{B}$ can be found in [**14**, Chapter 5]. □

Let us introduce the following operators:

- The Dirichlet lifting operator $D \in \mathcal{L}(\mathbb{C}^6, W^{2,q}(\Omega_F(0)))$ and $D_{pr} \in \mathcal{L}(\mathbb{C}^6, W^{1,q}_m(\Omega_F(0)))$ defined by

$$D(\ell,\omega) = w, \quad D_{pr}(\ell,\omega) = \psi, \tag{3.9}$$

where (w,ψ) is the solution to the problem (3.5).
- The Neumann operator $N \in \mathcal{L}(W^{1-1/q,q}_m(\partial\Omega_F(0)), W^{2,q}_m(\Omega_F(0)))$ defined by $Nh = \varphi$, where φ is the solution to the Neumann problem

$$\Delta\varphi = 0 \text{ in } \Omega_F(0), \quad \frac{\partial\varphi}{\partial n} = h \text{ on } \partial\Omega_F(0). \tag{3.10}$$

We set

$$N_S h = N(\mathbb{1}_{\partial\Omega_S(0)} h) \text{ for } h \in W^{1-1/q,q}_m(\partial\Omega_S(0)). \tag{3.11}$$

We now rewrite the equations satisfied by u in system (3.2) as a new system of two equations, one satisfied by $\mathcal{P}u$ and another by $(I-\mathcal{P})u$. More precisely, we have the following proposition

PROPOSITION 3.7. *Let $1 < q < \infty$. Let us assume that $(f,\ell,\omega) \in \mathcal{V}^q_n(\Omega_F(0)) \times \mathbb{C}^3 \times \mathbb{C}^3$. A pair $(u,\pi) \in W^{2,q}(\Omega_F(0)) \times W^{1,q}_m(\Omega_F(0))$ satisfies the system*

$$\begin{cases} \lambda u - \nu\Delta u + \nabla\pi = f, \quad \text{div } u = 0, \quad y \in \Omega_F(0), \\ u = 0 \quad y \in \partial\Omega, \\ u = \ell + \omega \times y \quad y \in \partial\Omega_S(0), \end{cases} \tag{3.12}$$

if and only if

$$\begin{cases} \lambda\mathcal{P}u - A_0\mathcal{P}u + A_0\mathcal{P}D(\ell,\omega) = \mathcal{P}f, \\ (I-\mathcal{P})u = (I-\mathcal{P})D(\ell,\omega) \\ \pi = N(\nu\Delta\mathcal{P}u\cdot n) - \lambda N_S((\ell+\omega\times y)\cdot n). \end{cases} \tag{3.13}$$

PROOF. Let $(u,\pi) \in W^{2,q}(\Omega_F(0))^3 \times W^{1,q}_m(\Omega_F(0))$ satisfies the system (3.12). We set

$$\widetilde{u} = u - D(\ell,\omega), \quad \widetilde{\pi} = \pi - D_{pr}(\ell,\omega).$$

The pair $(\widetilde{u},\widetilde{p})$ satisfies the following system

$$\begin{aligned} &\lambda\widetilde{u} + \lambda D(\ell,\omega) - \nu\Delta\widetilde{u} + \nabla p = f, \quad \text{div } \widetilde{u} = 0, \text{ in } \Omega_F(0), \\ &\widetilde{u} = 0 \quad \text{ on } \partial\Omega_F(0). \end{aligned}$$

Note that $\widetilde{u} \in \mathcal{D}(A_0)$ and $\mathcal{P}\widetilde{u} = \widetilde{u}$. Thus applying the projection $\mathcal{P}$ on the above system it is easy to see that $\mathcal{P}u$ satisfies the following

$$\lambda\mathcal{P}u - A_0\mathcal{P}u + A_0\mathcal{P}D(\ell,\omega) = \mathcal{P}f.$$

Since $(I-\mathcal{P})\widetilde{u} = 0$, we obtain

$$(I-\mathcal{P})u = (I-\mathcal{P})(\widetilde{u} + D(\ell,\omega)) = (I-\mathcal{P})D(\ell,\omega).$$

Note that, from the expression of $\mathcal{P}$ in Proposition 3.4, it follows that $\Delta(I-\mathcal{P})u = 0$ in $\Omega_F(0)$. Therefore the first equation of (3.12) can be written as

$$\lambda u - \Delta\mathcal{P}u + \nabla\pi = f \quad \text{in } \Omega_F(0).$$

By applying the divergence and normal trace operators to the above equation, we obtain that π is the solution of the problem

$$\begin{cases} \Delta\pi = 0 & \text{in } \Omega_F(0), \\ \dfrac{\partial \pi}{\partial n} = \nu \Delta \mathcal{P} u \cdot n - \lambda u \cdot n & \text{on } \partial\Omega_F(0). \end{cases} \tag{3.14}$$

Since $\operatorname{div}\Delta\mathcal{P}u = 0$ it follows that $\nu\Delta\mathcal{P}u\cdot n$ belongs to $W^{-1/q,q}(\partial\Omega)$ and satisfies the following condition

$$\langle \nu\Delta\mathcal{P}u\cdot n, 1\rangle_{W^{-1/q,q},W^{1-1/q',q'}} = 0$$

Since $\Omega_F(0)$ is a smooth domain, (3.14) admits a unique solution in $W^{1,q}_m(\Omega_F(0))$ ([**8**, Theorem 9.2]) and the expression of π in (3.13) follows from the definition of the operators N and N_S.

Conversely, let $(u,\pi) \in W^{2,q}(\Omega_F(0))^3 \times W^{1,q}_m(\Omega_F(0))$ satisfies the system (3.13). Since $(I-\mathcal{P})u = (I-\mathcal{P})D(\ell,\omega)$ we get $\widetilde{u} := u - D(\ell,\omega) \in \mathcal{D}(A_0)$. Thus $(3.13)_1$ can be written as

$$\mathcal{P}\left(\lambda\widetilde{u} - A_0\widetilde{u}\right) = \mathcal{P}(f - \lambda D(\ell,\omega)).$$

Therefore, there exists $\widetilde{\pi} \in W^{1,q}_m(\Omega_F(0))$ such that $(\widetilde{u},\widetilde{\pi})$ satisfies

$$\lambda\widetilde{u} - \nu\Delta\widetilde{u} + \nabla\widetilde{\pi} = f - \lambda D(\ell,\omega), \quad \operatorname{div}\widetilde{u} = 0 \text{ in } \Omega_F(0), \quad \widetilde{u} = 0 \text{ on } \partial\Omega_F(0).$$

Then (u,π), with $\pi = \widetilde{\pi} + D_{pr}(\ell,\omega)$, satisfies the system (3.12). □

Using the expression of the pressure π obtained in (3.7), we rewrite the equations satisfied by ℓ and ω in (3.2) in the form

$$\begin{aligned} \lambda m\ell &= -2\nu \int_{\partial\Omega_S(0)} \varepsilon(u)n\, d\gamma + \int_{\partial\Omega_S(0)} \pi n\, d\gamma + g_1 \\ &= -2\nu\left[\int_{\partial\Omega_S(0)} \varepsilon(\mathcal{P}u)n\, d\gamma + \int_{\partial\Omega_S(0)} \varepsilon((I-\mathcal{P})D(\ell,\omega))n\, d\gamma\right] \\ &\quad + \int_{\partial\Omega_S(0)} N(\Delta\mathcal{P}u\cdot n)n\, d\gamma - \lambda\int_{\partial\Omega_S(0)} N_S((\ell+\omega\times y)\cdot n)n\, \mathrm{d}\gamma + g_1, \end{aligned}$$

and

$$\begin{aligned} J(0)\lambda\omega &= -2\nu\left[\int_{\partial\Omega_S(0)} y\times\varepsilon(\mathcal{P}u)n\, d\gamma + \int_{\partial\Omega_S(0)} y\times\varepsilon((I-\mathcal{P})D(\ell,\omega))n\, d\gamma\right] \\ &\quad + \int_{\partial\Omega_S(0)} y\times N(\Delta\mathcal{P}u\cdot n)n\, d\gamma - \lambda\int_{\partial\Omega_S(0)} y\times N_S((\ell+\omega\times y)\cdot n)n\, \mathrm{d}\gamma + g_2. \end{aligned}$$

The above two equations can be written as

$$\lambda\mathbb{K}\begin{pmatrix}\ell\\ \omega\end{pmatrix} = \mathcal{C}_1\mathcal{P}u + \mathcal{C}_2\begin{pmatrix}\ell\\ \omega\end{pmatrix} + \begin{pmatrix}g_1\\ g_2\end{pmatrix}, \tag{3.15}$$

where

(3.16)

$$\mathbb{K} = \begin{pmatrix} mI_3 & 0 \\ 0 & J(0)\end{pmatrix} + \mathbb{M}, \quad \mathbb{M}\begin{pmatrix}\ell\\ \omega\end{pmatrix} = \begin{pmatrix} \displaystyle\int_{\partial\Omega_S(0)} N_S((\ell+\omega\times y)\cdot n)n\, \mathrm{d}\gamma \\ \displaystyle\int_{\partial\Omega_S(0)} y\times N_S((\ell+\omega\times y)\cdot n)n\, \mathrm{d}\gamma \end{pmatrix},$$

$$
(3.17)\quad \mathcal{C}_1\mathcal{P}u = \begin{pmatrix} -2\nu\displaystyle\int_{\partial\Omega_S(0)} \varepsilon(\mathcal{P}u)n\,d\gamma + \int_{\partial\Omega_S(0)} N(\Delta\mathcal{P}u\cdot n)n\,d\gamma \\ -2\nu\displaystyle\int_{\partial\Omega_S(0)} y\times\varepsilon(\mathcal{P}u)\,n\,d\gamma + \int_{\partial\Omega_S(0)} y\times N(\Delta\mathcal{P}u\cdot n)n\,d\gamma \end{pmatrix},
$$

and

$$
(3.18)\qquad \mathcal{C}_2\begin{pmatrix}\ell\\ \omega\end{pmatrix} = \begin{pmatrix} -2\nu\displaystyle\int_{\partial\Omega_S(0)} \varepsilon((I-\mathcal{P})D(\ell,\omega))n\,d\gamma \\ -2\nu\displaystyle\int_{\partial\Omega_S(0)} y\times\varepsilon((I-\mathcal{P})D(\ell,\omega))n\,d\gamma \end{pmatrix}.
$$

In the literature, the matrix $\mathbb{M}$ defined above is known as the added mass operator. We are now going to show that the matrix $\mathbb{K}$ is an invertible matrix.

LEMMA 3.8. *The matrix $\mathbb{K}$ defined as in* (3.16) *is an invertible matrix.*

PROOF. The proof may be adapted from that of [**10**, Lemma 4.6] (see also [**11**, Lemma 4.3]). Let us briefly explain the idea of the proof. We are going to show that the matrix $\mathbb{M}$ is symmetric and semipositive definite. For that, we first derive an representation formula of the matrix $\mathbb{M}$. Let us consider the following problem

$$
\Delta\pi^i = 0 \text{ in } \Omega_F(0), \quad \frac{\partial\pi^i}{\partial n} = 0 \text{ on } \partial\Omega,
$$
$$
\frac{\partial\pi^i}{\partial n} = e_i\cdot n \text{ for } i=1,2,3 \text{ and } \frac{\partial\pi^i}{\partial n} = (e_{i-3}\times y)\cdot n \text{ for } i=4,5,6, \quad y\in\partial\Omega_S(0),
$$

where $\{e_i\}$ denote the canonical basis in $\mathbb{C}^3$. Therefore, it is easy to see that

$$
N_S((\ell+\omega\times y)\cdot n) = \sum_{i=1}^{3}\ell_i\pi^i + \sum_{i=4}^{6}\omega_{i-3}\pi^i.
$$

We define

$$
m_{ij} = \begin{cases} \displaystyle\int_{\partial\Omega_S(0)} \pi^i n^j & \text{for } 1\leqslant i\leqslant 6, 1\leqslant j\leqslant 3, \\ \displaystyle\int_{\partial\Omega_S(0)} \pi^i(e_{j-3}\times y)\cdot n & \text{for } 1\leqslant i\leqslant 6, 4\leqslant j\leqslant 6. \end{cases}
$$

One can easily check that, $\mathbb{M} = (m_{ij})_{1\leqslant i,j\leqslant 6}$. With this representation and Gauss' theorem, we can verify that $\mathbb{M}$ is symmetric and semipositive definite. □

Let us set

$$
(3.19)\qquad \mathcal{X} = \mathcal{V}_n^q(\Omega_F(0))\times\mathbb{C}^3\times\mathbb{C}^3.
$$

and consider the operator $\mathcal{A}_{FS} : \mathcal{D}(\mathcal{A}_{FS}) \mapsto \mathcal{X}$ defined by

$$
\mathcal{D}(\mathcal{A}_{FS}) = \{(\mathcal{P}u,\ell,\omega)\in\mathcal{X} \mid \mathcal{P}u - \mathcal{P}D(\ell,\omega)\in\mathcal{D}(A_0)\},
$$

and

$$
(3.20)\qquad \mathcal{A}_{FS} = \begin{pmatrix} A_0 & -A_0\mathcal{P}D \\ \mathbb{K}^{-1}\mathcal{C}_1 & \mathbb{K}^{-1}\mathcal{C}_2 \end{pmatrix}.
$$

Combining the above results, below we obtain an equivalence formulation of the system (3.2).

PROPOSITION 3.9. *Let $1 < q < \infty$. Let us assume that $(f, g_1, g_2) \in \mathcal{V}^q_n(\Omega_F(0)) \times \mathbb{C}^3 \times \mathbb{C}^3$. Then $(u, p, \ell, \omega) \in W^{2,q}(\Omega_F(0)) \times W^{1,q}_m(\Omega_F(0)) \times \mathbb{C}^3 \times \mathbb{C}^3$ satisfy the system (3.2) if and only if*

$$(\lambda I - \mathcal{A}_{FS}) \begin{pmatrix} \mathcal{P}u \\ \ell \\ \omega \end{pmatrix} = \begin{pmatrix} \mathcal{P}f \\ \widetilde{g}_1 \\ \widetilde{g}_2 \end{pmatrix}, \tag{3.21}$$

$$(I - \mathcal{P})u = (I - \mathcal{P})D(\ell, \omega),$$

$$\pi = N(\nu \Delta \mathcal{P}u \cdot n) - \lambda N_S((\ell + \omega \times y) \cdot n),$$

where $(\widetilde{g}_1, \widetilde{g}_2)^\top = \mathbb{K}^{-1}(g_1, g_2)^\top$.

We end this subsection with the following lemma

LEMMA 3.10. *The map*

$$(\mathcal{P}u, \ell, \omega) \mapsto \|\mathcal{P}u\|_{W^{2,q}(\Omega_F(0))} + \|\ell\|_{\mathbb{C}^3} + \|\omega\|_{\mathbb{C}^3},$$

is a norm on $\mathcal{D}(\mathcal{A}_{FS})$ equivalent to the graph norm.

PROOF. The proof is similar to that of [**20**, Proposition 3.3]. □

3.2. $\mathcal{R}$-boundedness of the resolvent operator. In this subsection we are going to prove Theorem 3.1. In view of Proposition 3.2 and Proposition 3.9, it is enough to prove the following theorem

THEOREM 3.11. *Let $1 < q < \infty$. There exist $\mu_0 > 0$ and $\theta \in (\pi/2, \pi)$ such that $\mu_0 + \Sigma_\theta \subset \rho(\mathcal{A}_{FS})$ and*

$$\mathcal{R}_{\mathcal{L}(\mathcal{X})} \left\{ \lambda(\lambda I - \mathcal{A}_{FS})^{-1} \mid \lambda \in \mu_0 + \Sigma_\theta \right\} \leqslant C. \tag{3.22}$$

PROOF. We write $\mathcal{A}_{FS}$ in the form $\mathcal{A}_{FS} = \widetilde{\mathcal{A}}_{FS} + B_{FS}$ where

$$\widetilde{\mathcal{A}}_{FS} = \begin{pmatrix} A_0 & -A_0 \mathcal{P} D \\ 0 & 0 \end{pmatrix}, \quad B_{FS} = \begin{pmatrix} 0 & 0 \\ \mathbb{K}^{-1}\mathcal{C}_1 & \mathbb{K}^{-1}\mathcal{C}_2 \end{pmatrix}.$$

We first show that $\widetilde{\mathcal{A}}_{FS}$ with $\mathcal{D}(\widetilde{\mathcal{A}}_{FS}) = \mathcal{D}(\mathcal{A}_{FS})$ is a $\mathcal{R}$-sectorial operator on $\mathcal{X}$. Observe that

$$\lambda(\lambda I - \widetilde{\mathcal{A}}_{FS})^{-1} = \begin{pmatrix} \lambda(\lambda I - A_0)^{-1} & -(\lambda I - A_0)^{-1} A_0 \mathcal{P} D \\ 0 & I \end{pmatrix}.$$

Since

$$-(\lambda I - A_0)^{-1} A_0 \mathcal{P} D = -\lambda(\lambda I - A_0)^{-1} \mathcal{P} D + \mathcal{P} D,$$

we get

$$\lambda(\lambda I - \widetilde{\mathcal{A}}_{FS})^{-1} = \begin{pmatrix} \lambda(\lambda I - A_0)^{-1} & -\lambda(\lambda I - A_0)^{-1} \mathcal{P} D + \mathcal{P} D \\ 0 & I \end{pmatrix}.$$

Therefore by Proposition 3.5 and Proposition 2.4, there exists $\theta \in (\pi/2, \pi)$ such that

$$\mathcal{R}_{\mathcal{L}(\mathcal{X})} \left\{ \lambda(\lambda I - \widetilde{\mathcal{A}}_{FS})^{-1} \mid \lambda \in \Sigma_\theta \right\} \leqslant C. \tag{3.23}$$

Let us now show that, $\mathcal{C}_1 \in \mathcal{L}(\mathcal{D}(\mathcal{A}_{FS}), \mathbb{C}^3 \times \mathbb{C}^3)$. By Lemma 3.10, for any $(\mathcal{P}u, \ell, \omega) \in \mathcal{D}(\mathcal{A}_{FS})$ we have $(\mathcal{P}u, \ell, \omega) \in W^{2,q}(\Omega_F(0)) \times \mathbb{C}^3 \times \mathbb{C}^3$. Therefore, by trace theorem $\varepsilon(\mathcal{P}u)n \in W^{1-1/q,q}(\partial\Omega_S(0))$ and hence $\int_{\partial\Omega_S(0)} \varepsilon(\mathcal{P}u)n \, d\gamma \in \mathbb{C}^3$. On the other hand, $\Delta \mathcal{P}u \in L^q(\Omega_F(0))$ and div $\Delta \mathcal{P}u = 0$. Therefore by Lemma 3.3, the term $\Delta \mathcal{P}u \cdot n$ belongs to $W^{-1/q,q}(\partial\Omega_F(0))$ and satisfies the following condition

$$\langle \Delta \mathcal{P}u \cdot n, 1\rangle_{W^{-1/q,q},W^{1-1/q',q'}} = 0.$$

Thus by [**8**, Theorem 9.2], $N(\Delta \mathcal{P}u{\cdot}n) \in W^{1,q}(\Omega_F(0))$ and $\displaystyle\int_{\partial\Omega_S(0)} N(\Delta\mathcal{P}u{\cdot}n)n \, \mathrm{d}\gamma \in \mathbb{C}^3$. Other terms of the operator $\mathcal{C}_1$ can be checked in a similar manner. Thus $\mathcal{C}_1 \in \mathcal{L}(\mathcal{D}(\mathcal{A}_{FS}), \mathbb{C}^3 \times \mathbb{C}^3)$. Similarly, one can easily verify that $\mathcal{C}_2 \in \mathcal{L}(\mathbb{C}^3 \times \mathbb{C}^3, \mathbb{C}^3 \times \mathbb{C}^3)$. Therefore the operator B_{FS} with $\mathcal{D}(B_{FS}) = \mathcal{D}(\mathcal{A}_{FS})$ is a finite rank operator. By [**7**, Chapter III, Lemma 2.16], B_{FS} is a $\widetilde{\mathcal{A}}_{FS}$-bounded operator with relative bound zero. Finally using Proposition 2.6 we conclude the proof of the theorem. □

4. Exponential stability of linear fluid-structure interaction operator

The aim of this section is to show that the operator $\mathbb{A}$ or equivalently the operator $\mathcal{A}_{FS}$ generates an exponentially stable semigroup. More precisely, we prove:

THEOREM 4.1. *Let $1 < q < \infty$. The operator $\mathcal{A}_{FS}$ generates an exponentially stable semigroup $\left(e^{t\mathcal{A}_{FS}}\right)_{t\geqslant 0}$ on $\mathcal{X}$. Equivalently, the operator $\mathbb{A}$ generates an exponentially stable semigroup $\left(e^{t\mathbb{A}}\right)_{t\geqslant 0}$ on $\mathbb{H}^q(\Omega)$. In other words, there exist constants $C > 0$ and $\eta_0 > 0$ such that*

$$\left\| e^{t\mathcal{A}_{FS}}(u_0, \ell_0, \omega_0)^\top \right\|_{\mathcal{X}} \leqslant Ce^{-\eta_0 t} \left\|(u_0, \ell_0, \omega_0)^\top\right\|_{\mathcal{X}}. \tag{4.1}$$

To prove this theorem we first show that the set $\{\lambda \in \mathbb{C} \mid \mathrm{Re}\lambda \geqslant 0\}$, i.e, the entire right half plane is contained in the resolvent set of $\mathcal{A}_{\mathcal{FS}}$.

THEOREM 4.2. *Assume $1 < q < \infty$ and $\lambda \in \mathbb{C}$, with $\mathrm{Re}\lambda \geqslant 0$. Then for any $(f, g_1, g_2) \in \mathcal{X}$, the system (3.2) admits a unique solution satisfying the estimate*

$$\|u\|_{W^{2,q}(\Omega_F(0))^3} + \|p\|_{W^{1,q}_m(\Omega_F(0))} + \|\ell\|_{\mathbb{C}^3} + \|\omega\|_{\mathbb{C}^3} \leqslant C\|(f, g_1, g_2)\|_{\mathcal{X}}. \tag{4.2}$$

PROOF. Let us recall, by Proposition 3.9, the system (3.2) is equivalent to

$$\begin{aligned} &(\lambda I - \mathcal{A}_{FS}) \begin{pmatrix} \mathcal{P}u \\ \ell \\ \omega \end{pmatrix} = \begin{pmatrix} \mathcal{P}f \\ \widetilde{g}_1 \\ \widetilde{g}_2 \end{pmatrix}, \\ &(I - \mathcal{P})u = (I - \mathcal{P})D(\ell, \omega), \\ &\pi = N(\nu \Delta \mathcal{P}u \cdot n) - \lambda N_S((\ell + \omega \times y) \cdot n), \end{aligned} \tag{4.3}$$

where $(\widetilde{g}_1, \widetilde{g}_2)^\top = \mathbb{K}^{-1}(g_1, g_2)^\top$. By Theorem 3.11, there exists $\widetilde{\lambda} > \mu_0$ such that $(\widetilde{\lambda} I - \mathcal{A}_{FS})$ is invertible. Consequently, (4.3) can be written as

$$\begin{aligned} &\begin{pmatrix} \mathcal{P}u \\ \ell \\ \omega \end{pmatrix} = \left[I + (\lambda - \widetilde{\lambda})(\widetilde{\lambda} - \mathcal{A}_{FS})^{-1} \right]^{-1} \left(\widetilde{\lambda} - \mathcal{A}_{FS}\right)^{-1} \begin{pmatrix} \mathcal{P}f \\ \widetilde{g}_1 \\ \widetilde{g}_2 \end{pmatrix}, \\ &(I - \mathcal{P})u = (I - \mathcal{P})D(\ell, \omega), \\ &\pi = N(\nu \Delta \mathcal{P}u \cdot n) - \lambda N_S((\ell + \omega \times y) \cdot n). \end{aligned} \tag{4.4}$$

Since $\left(\widetilde{\lambda} - \mathcal{A}_{FS}\right)^{-1}$ is a compact operator, in view of Fredholm alternative theorem, the existence and uniqueness of system (4.4) are equivalent. Therefore, in the sequel we show the uniqueness of the solutions (3.2). Once we prove the uniqueness, the estimate (4.2) follows easily from (4.4). Let $(u, \pi, \ell, \omega) \in W^{2,q}(\Omega_F(0))^3 \times W^{1,q}_m(\Omega_F(0)) \times \mathbb{C}^3 \times \mathbb{C}^3$ satisfies the homogeneous system

$$(4.5)\qquad \begin{aligned} &\lambda u - \nu\Delta u + \nabla\pi = 0, \quad \operatorname{div} u = 0, && \text{in } \Omega_F(0),\\ &u = 0 && \text{on } \partial\Omega,\\ &u = \ell + \omega\times y && \text{on } \partial\Omega_S(0),\\ &\lambda m\ell = -\int_{\partial\Omega_S(0)} \sigma(u,\pi)n\, d\gamma, &&\\ &\lambda J(0)\omega = -\int_{\partial\Omega_S(0)} y\times\sigma(u,\pi)n\, d\gamma. && \end{aligned}$$

We first show that $(u, \pi) \in W^{2,2}(\Omega_F(0))^3 \times W^{1,2}_m(\Omega_F(0))$. If $q \geqslant 2$, this follows from Hölder's estimate. Assume $1 < q < 2$. In that case, we can rewrite (4.5) as follows

$$(4.6)\qquad \begin{aligned} &(\widetilde{\lambda} I - \mathcal{A}_{FS})\begin{pmatrix}\mathcal{P}u\\ \ell\\ \omega\end{pmatrix} = (\widetilde{\lambda} - \lambda)\begin{pmatrix}\mathcal{P}u\\ \ell\\ \omega\end{pmatrix},\\ &(I - \mathcal{P})u = (I - \mathcal{P})D(\ell, \omega),\\ &\pi = N(\nu\Delta\mathcal{P}u\cdot n) - \lambda N_S((\ell + \omega\times y)\cdot n). \end{aligned}$$

Since $W^{2,q}(\Omega_F(0)) \subset L^2(\Omega_F(0))$ and $(\widetilde{\lambda} I - \mathcal{A}_{FS})$ is invertible, we deduce that $(u, \pi) \in W^{2,2}(\Omega_F(0))^3 \times W^{1,2}_m(\Omega_F(0))$.

Multiplying $(4.5)_1$ by $\overline{u}$, $(4.5)_4$ by $\overline{\ell}$ and $(4.5)_5$ by $\overline{\omega}$, we obtain after integration by parts,

$$\lambda\int_{\Omega_F(0)} |u|^2\, dy + 2\nu\int_{\Omega_F(0)} \varepsilon(u) : \varepsilon(\overline{u})\, dy + \lambda m|\ell|^2 + \lambda J(0)\omega\cdot\overline{\omega} = 0.$$

Taking real part of the above equation we obtain

$$\operatorname{Re}\lambda\int_{\Omega_F(0)} |u|^2\, dy + 2\nu\int_{\Omega_F(0)} |\varepsilon(u)|^2\, dy + \operatorname{Re}\lambda m|\ell|^2 + \operatorname{Re}(\lambda J(0)\omega\cdot\overline{\omega}) = 0.$$

Since $\operatorname{Re}\lambda \geqslant 0$, we have

$$2\nu\int_{\Omega_F(0)} |\varepsilon(u)|^2\, dy = 0.$$

The above estimate and the fact that $u = 0$ on $\partial\Omega$ imply that $u = 0$. Next, using $u = \ell + \omega\times y$ for $y \in \partial\Omega_S(0)$, we get $\ell = \omega = 0$. Finally, as $\pi \in W^{1,q}_m(\Omega_F(0))$, we have $\pi = 0$. □

Proof of Theorem 4.1. From Theorem 4.2, we have

$$\{\lambda \in \mathbb{C} \mid \operatorname{Re}\lambda \geqslant 0\} \subset \rho(\mathcal{A}_{FS}).$$

Also, by Theorem 3.11 we have the existence of a constant $C > 0$ such that for any $\lambda \in \mu_0 + \Sigma_\theta$ with $\theta \in (\pi/2, \pi)$,

$$\left\|(\lambda - \mathcal{A}_{FS})^{-1}\right\|_{\mathcal{L}(\mathcal{X})} \leqslant C.$$

Since $\{\lambda \in \mathbb{C} \mid \operatorname{Re}\lambda \geqslant 0\} \setminus [\mu_0 + \Sigma_\theta]$ is a compact set, we deduce the existence of a constant $C > 0$ such that for any $\lambda \in \mathbb{C}$ with $\operatorname{Re}\lambda \geqslant 0$

$$\left\|(\lambda - \mathcal{A}_{FS})^{-1}\right\|_{\mathcal{L}(\mathcal{X})} \leqslant C.$$

This yields that
$$\{\lambda \in \mathbb{C} \mid \text{Re } \lambda \geqslant -\eta\} \subset \rho(\mathcal{A}_{FS}),$$
for some $\eta > 0$. As $\mathcal{A}_{FS}$ generates an analytic semigroup, applying Proposition 2.9 of [**3**, Part II, Chapter 1, pp 120], we obtain exponential stability of $\mathcal{A}_{FS}$ in $\mathcal{X}$. $\square$

5. Maximal L^p-L^q regularity of the system (1.7)

In this section we prove the maximal L^p-L^q regularity of a version of the system the (1.7) with non zero divergence. Treating a non zero divergence term will be useful in the next section in order to tackle some terms coming from a simple change of variables. More precisely, we consider the system

$$(5.1)\quad \begin{aligned}
&\partial_t u - \nu\Delta u + \nabla\pi = f, \quad \text{div } u = \text{div } h && t \in (0,\infty),\ y \in \Omega_F(0),\\
&u = 0 && t \in (0,\infty),\ y \in \partial\Omega,\\
&u = \ell + \omega \times y && t \in (0,\infty),\ y \in \partial\Omega_S(0),\\
&m\frac{d}{dt}\ell = -\int_{\partial\Omega_S(0)} \sigma(u,\pi)n\, d\gamma + g_1 && t \in (0,\infty),\\
&J(0)\frac{d}{dt}\omega = -\int_{\partial\Omega_S(0)} y \times \sigma(u,\pi)n\, d\gamma + g_2 && t \in (0,\infty),\\
&u(0,y) = u_0(y) && y \in \Omega_F(0),\\
&\ell(0) = \ell_0, \quad \omega(0) = \omega_0.
\end{aligned}$$

We set
$$W^{2,1}_{q,p}(Q^F_\infty) = L^p(0,\infty; W^{2,q}(\Omega_F(0))) \cap W^{1,p}(0,\infty; L^q(\Omega_F(0))),$$
with
$$\|u\|_{W^{2,1}_{q,p}(Q^F_\infty)} := \|u\|_{L^p(0,\infty;W^{2,q}(\Omega_F(0)))} + \|u\|_{W^{1,p}(0,\infty;L^q(\Omega_F(0)))}.$$
We prove the following theorem

THEOREM 5.1. *Let $1 < p, q < \infty$ such that $\dfrac{1}{p} + \dfrac{1}{2q} \neq 1$. Let $\eta \in [0, \eta_0)$, where η_0 is the constant introduced in Theorem 4.1. Let us also assume that $\ell_0 \in \mathbb{R}^3$, $\omega_0 \in \mathbb{R}^3$ and $u_0 \in B^{2(1-1/p)}_{q,p}(\Omega_F(0))$ satisfying the compatibility conditions*
$$\text{div } u_0 = 0 \textit{ in } \Omega_F(0),$$
$$(5.2)\quad u_0 = \ell_0 + \omega_0 \times y \textit{ on } \partial\Omega_S(0), \quad u_0 = 0 \textit{ on } \partial\Omega \textit{ if } \frac{1}{p} + \frac{1}{2q} < 1$$
$$\textit{and } u_0 \cdot n = (\ell_0 + \omega_0 \times y) \cdot n \textit{ on } \partial\Omega_S(0), \qquad u_0 \cdot n = 0 \textit{ on } \partial\Omega \textit{ if } \frac{1}{p} + \frac{1}{2q} > 1.$$
Then for any $e^{\eta t} f \in L^p(0,\infty; L^q(\Omega_F(0)))^3$, $e^{\eta t} h \in W^{2,1}_{q,p}(Q^F_\infty)^3$, $e^{\eta t} g_1 \in L^p(0,\infty;\mathbb{R}^3)$ and $e^{\eta t} g_2 \in L^p(0,\infty;\mathbb{R}^3)$ satisfying
$$h(0,y) = 0 \textit{ for all } (t,y) \in (0,\infty) \times \Omega_F(0) \textit{ and } h|_{\partial\Omega_F(0)} = 0,$$
the system (1.7) admits a unique strong solution
$$e^{\eta t} u \in L^p(0,\infty; W^{2,q}(\Omega_F(0))^3) \cap W^{1,p}(0,\infty; L^q(\Omega_F(0))^3)$$
$$e^{\eta t} \pi \in L^p(0,\infty; W^{1,q}_m(\Omega_F(0)))$$
$$e^{\eta t} \ell \in W^{1,p}(0,\infty;\mathbb{R}^3), \quad e^{\eta t} \omega \in W^{1,p}(0,\infty;\mathbb{R}^3).$$

Moreover, there exists a constant $C_L > 0$ depending only on Ω, p and q such that

$$\begin{aligned}(5.3)\quad &\|e^{\eta(\cdot)}u\|_{W^{2,1}_{q,p}(Q^F_\infty)^3} + \|e^{\eta(\cdot)}\pi\|_{L^p(0,\infty;W^{1,q}(\Omega_F(0)))} + \|e^{\eta(\cdot)}\ell\|_{L^p(0,\infty;\mathbb{R}^3)} \\ &+ \|e^{\eta(\cdot)}\omega\|_{L^p(0,\infty;\mathbb{R}^3)} \leqslant C_L\Big(\|u_0\|_{B^{2(1-1/p)}_{q,p}(\Omega_F(0))} + \|\ell_0\|_{\mathbb{R}^3} + \|\omega_0\|_{\mathbb{R}^3} \\ &+\|e^{\eta(\cdot)}f\|_{L^p(0,\infty;L^q(\Omega_F(0)))} + \|e^{\eta(\cdot)}h\|_{W^{2,1}_{q,p}(Q^F_\infty)^3} + \|e^{\eta(\cdot)}g_1\|_{L^p(0,\infty;\mathbb{R}^3)} + \|e^{\eta(\cdot)}g_2\|_{L^p(0,\infty;\mathbb{R}^3)}\Big).\end{aligned}$$

PROOF. We first consider the case $\eta = 0$. Let us set $v = u - h$. Then (v, π, ℓ, ω) satisfies the following system

$$(5.4)\quad \begin{aligned} &\partial_t v - \nu\Delta v + \nabla\pi = F, \quad \operatorname{div} v = 0 && \text{in } (0,\infty)\times\Omega_F(0), \\ &v = 0 && \text{on } (0,\infty)\times\partial\Omega, \\ &v = \ell + \omega \times y && \text{on } (0,\infty)\times\partial\Omega_S(0), \\ &m\frac{d}{dt}\ell = -\int_{\partial\Omega_S(0)} \sigma(v,\pi)n\, d\gamma + G_1 && t\in(0,\infty), \\ &J(0)\frac{d}{dt}\omega = -\int_{\partial\Omega_S(0)} y\times\sigma(v,\pi)n\, d\gamma + G_2 && t\in(0,\infty), \\ &v(0,y) = u_0(y) && \text{in } \Omega_F(0), \\ &\ell(0) = \ell_0, \quad \omega(0) = \omega_0, \end{aligned}$$

where

$$F = f - \partial_t h + \nu\Delta h,\ G_1 = g_1 - \int_{\partial\Omega_S(0)} \varepsilon(h)n\ \mathrm{d}\gamma,\ G_2 = g_2 - \int_{\partial\Omega_S(0)} y\times\varepsilon(h)n\ \mathrm{d}\gamma.$$

Proceeding as Proposition 3.9, it is easy to see that, the above system is equivalent to

$$(5.5)\quad \begin{cases} \dfrac{d}{dt}\begin{pmatrix}\mathcal{P}v\\ \ell\\ \omega\end{pmatrix} = \mathcal{A}_{FS}\begin{pmatrix}\mathcal{P}v\\ \ell\\ \omega\end{pmatrix} + \begin{pmatrix}\mathcal{P}F\\ \widetilde{G}_1\\ \widetilde{G}_2\end{pmatrix}, \quad \begin{pmatrix}\mathcal{P}v(0)\\ \ell(0)\\ \omega(0)\end{pmatrix} = \begin{pmatrix}\mathcal{P}u_0\\ \ell_0\\ \omega_0\end{pmatrix}, \\ (I-\mathcal{P})v = (I-\mathcal{P})D(\ell,\omega), \end{cases}$$

where

$$\widetilde{G}_1 = \int_{\partial\Omega_S(0)} N((F-\nabla\varphi)\cdot n)n\ \mathrm{d}\gamma + G_1,\ \widetilde{G}_2 = \int_{\partial\Omega_S(0)} y\times N((F-\nabla\varphi)\cdot n)n\ d\gamma + G_2,$$

and φ is the solution of the problem

$$-\Delta\varphi = \operatorname{div} F \text{ in } \Omega_F(0), \quad \varphi = 0 \text{ on } \partial\Omega_F(0).$$

(see also [**20**, Section 4.2] or [**17**, Proposition 3.7]). The operator $\mathcal{A}_{FS}$ is defined as in (3.20). Let us recall that the operator $\mathcal{A}_{FS}$ is an $\mathcal{R}$-sectorial operator in $\mathcal{X}$ (Theorem 3.11). One can easily verify that, under the hypothesis of the theorem, $(\mathcal{P}F, \widetilde{G}_1, \widetilde{G}_2) \in L^p(0,\infty;\mathcal{X})$ and

$$\begin{aligned}\|(\mathcal{P}F, \widetilde{G}_1, \widetilde{G}_2)\|_{L^p(0,\infty;\mathcal{X})} \leqslant C\Big(&\|f\|_{L^p(0,\infty;L^q(\Omega_F(0)))} + \|h\|_{W^{2,1}_{q,p}(Q^F_\infty)^3} \\ &+ \|g_1\|_{L^p(0,\infty;\mathbb{R}^3)} + \|g_2\|_{L^p(0,\infty;\mathbb{R}^3)}\Big).\end{aligned}$$

From [**2**, Theorem 3.4], we obtain $(\mathcal{P}u_0, \ell_0, \omega_0) \in (\mathcal{X}, \mathcal{D}(\mathcal{A}_{FS}))_{1-1/p,p}$. Then by Theorem 2.7, the system (5.5) admits a unique solution $(\mathcal{P}v, \ell, \omega) \in L^p(0,\infty;\mathcal{D}(\mathcal{A}_{FS}))\cap$

$W^{1,p}(0,\infty;\mathcal{X})$. From the expression of $(I-\mathcal{P})v$ together with Lemma 3.10, one can easily check that $v \in W^{2,1}_{q,p}(Q^F_\infty)^3$ and thus $u \in W^{2,1}_{q,p}(Q^F_\infty)^3$. The estimate (5.3) is easy to obtain.

The case $\eta > 0$ can be reduced to the previous case by multiplying all the function by $e^{\eta t}$ and using the fact that $\mathcal{A}_{FS} + \eta$ generates an C^0-semigroup of negative type for all $\eta \in (0, \eta_0)$. □

6. Global in time existence and uniqueness

In this section we are going to prove Theorem 1.4. As the domain of the fluid equation for the full nonlinear problem is also a unknown of the problem, we first rewrite the system in a fixed spatial domain.

6.1. Change of variables. We describe a change of variable to rewrite the system (1.2) in a fixed spatial domain. We follow the approach of [**4**]. Let us assume that (1.1) is satisfied and we also assume

$$\|a\|_{L^\infty(0,\infty;\mathbb{R}^3)} + \|Q - I_3\|_{L^\infty(0,\infty;\mathbb{R}^{3\times 3})}\mathrm{diam}(\Omega_S(0)) \leqslant \frac{\alpha}{2}. \tag{6.1}$$

With the above choice we have dist $(\Omega_S(t), \partial\Omega) \geqslant \alpha/2$ for all $t \in [0,\infty)$. We consider a cut-off function ψ which satisfies

$$\psi \in C^\infty(\overline{\Omega}), \quad \psi = 1 \text{ if } \mathrm{dist}(x,\partial\Omega) > \alpha/4, \quad \psi = 0 \text{ if } \mathrm{dist}(x,\partial\Omega) < \alpha/8. \tag{6.2}$$

We introduce a function ξ defined in $(0,\infty)\times\Omega$ by

$$\xi(t,x) = a'(t) + (x - a(t)) + \frac{|x-a(t)|^2}{2}\omega(t)$$

and Λ in $(0,\infty)\times\Omega$ by

$$\Lambda(t,x) = \psi(x)\ \left(a'(t) + \omega(t)\times(x-a(t))\right) + \begin{pmatrix} \dfrac{\partial\psi}{\partial x_2}(x)\xi_3(t,x) - \dfrac{\partial\psi}{\partial x_3}(x)\xi_2(t,x) \\ \dfrac{\partial\psi}{\partial x_3}(x)\xi_1(t,x) - \dfrac{\partial\psi}{\partial x_1}(x)\xi_3(t,x) \\ \dfrac{\partial\psi}{\partial x_1}(x)\xi_2(t,x) - \dfrac{\partial\psi}{\partial x_2}(x)\xi_1(t,x) \end{pmatrix}.$$

With the above definitions, it is easy to see that Λ satisfies the following lemma

LEMMA 6.1. *Let us assume that $a \in W^{2,p}(0,\infty)$ and $\omega \in W^{1,p}(0,\infty)$. Let Λ be defined as above. Then we have*

- $\Lambda(t,x) = 0$ *for all $t \in [0,\infty)$ and for all x such that* $\mathrm{dist}(x,\partial\Omega) < \alpha/8$.
- $\mathrm{div}\Lambda(t,x) = 0$ *for all $t \in [0,\infty)$ and $x \in \Omega$.*
- $\Lambda(t,x) = a'(t) + \omega(t)\times(x-a(t))$ *for all $t \in [0,\infty)$ and $x \in \Omega_S(t)$.*
- $\Lambda \in C([0,\infty)\times\Omega;\mathbb{R}^3)$. *Moreover, for all $t \in [0,\infty)$, $\Lambda(t,\cdot)$ is a C^∞ function for all $x \in \Omega$, the function $\Lambda(\cdot,x) \in W^{1,p}(0,\infty;\mathbb{R}^3)$.*

Next we consider the characteristic X associated to the flow Λ, that is the solution of the Cauchy problem

$$\begin{aligned} &\partial_t X(t,y) = \Lambda(t, X(t,y)) \quad (t > 0), \\ &X(0,y) = y \in \overline{\Omega}. \end{aligned} \tag{6.3}$$

We have the following lemma

LEMMA 6.2. *For all $y \in \Omega$, the initial value problem (6.3) admits a unique solution $X(\cdot, y) : [0,\infty) \mapsto \mathbb{R}^3$, which is a C^1 function in $[0,\infty)$. Furthermore X satisfies the following properties*

- *For any $t \in [0,\infty)$, $X(t,\cdot)$ is a C^1- diffeomorphism from Ω onto Ω and $\Omega_F(0)$ onto $\Omega_F(t)$.*
- *For all $y \in \Omega$ and $t \in [0,\infty)$, we have*
$$\det \nabla X(t,\cdot) = 1.$$
- *For each $t \geqslant 0$, we denote by $Y(t,\cdot) = [X(t,\cdot)]^{-1}$ the inverse of $X(t,\cdot)$.*

PROOF. See [**4**, Lemma 2.2]. □

We consider the following change of variables

$$\widetilde{u}(t,y) = Q^{-1}(t)u(t,X(t,y)), \qquad \widetilde{\pi}(t,y) = \pi(t,X(t,y)) \tag{6.4}$$

$$\widetilde{\ell}(t) = Q^{-1}(t)\dot{a}(t), \qquad \widetilde{\omega}(t) = Q^{-1}(t)\omega(t), \tag{6.5}$$

for $(t,y) \in (0,\infty) \times \Omega_F(0)$.

Then $(\widetilde{u}, \widetilde{p}, \widetilde{\ell}, \widetilde{\omega})$ satisfies the following system

$$\begin{aligned}
&\partial_t \widetilde{u} - \nu\Delta\widetilde{u} + \nabla\widetilde{p} = \mathcal{F}, \quad \operatorname{div} u = \operatorname{div} \mathcal{H}, && t \in (0,\infty),\ y \in \Omega_F(0),\\
&\widetilde{u} = 0, && t \in (0,\infty), y \in \partial\Omega,\\
&\widetilde{u} = \widetilde{\ell} + \widetilde{\omega} \times y, && t \in (0,\infty),\ y \in \partial\Omega_S(0),\\
&m\widetilde{\ell}' = -\int_{\partial\Omega_S(0)} \sigma(\widetilde{u},\widetilde{p})n \, d\gamma + \mathcal{G}_1, && t \in (0,\infty),\\
&J(0)\widetilde{\omega}' = -\int_{\partial\Omega_S(0)} y \times \sigma(\widetilde{u},\widetilde{p})n \, d\gamma + \mathcal{G}_2, && t \in (0,\infty),\\
&u(0,y) = u_0(y) && y \in \Omega_F(0),\\
&\ell(0) = \ell_0, \quad \omega(0) = \omega_0,
\end{aligned} \tag{6.6}$$

where

$$\dot{Q} = QA(\tilde{\omega}), \quad Q(0) = I_3, \tag{6.7}$$

is the rotation matrix of the solid at instant t,

$$X(t,y) = y + \int_0^t \Lambda(s, X(s,y)) \, \mathrm{d}s, \quad \text{and} \quad \nabla Y(t, X(t,y)) = [\nabla X]^{-1}(t,y), \tag{6.8}$$

for every $y \in \Omega_F(0)$ and $t \geqslant 0$. Using the notation

$$Z(t,y) = (Z_{i,j})_{1 \leqslant i,j \leqslant 3} = [\nabla X]^{-1}(t,y) \qquad (t \geqslant 0,\ y \in \Omega_F(0)), \tag{6.9}$$

the remaining terms in (6.6) are defined by

(6.10)

$$\begin{aligned}
\mathcal{F}_i(\widetilde{u},\widetilde{\pi},\widetilde{\ell},\widetilde{\omega}) = &-[(Q - I_3)\partial_t\widetilde{u}]_i - (\omega \times Q\widetilde{u})_i + \partial_t X \cdot Z^T \nabla(Q\widetilde{u})_i - \widetilde{u} \cdot Z^T \nabla(Q\widetilde{u})_i\\
&+ \nu \sum_{l,j,k} \frac{\partial^2 (Q\tilde{u})_i}{\partial y_l \partial y_k} (Z_{k,j} - \delta_{k,j}) Z_{l,j} + \nu \sum_{l,k} \frac{\partial^2 (Q\tilde{u})_i}{\partial y_l \partial y_k} (Z_{l,k} - \delta_{l,k})\\
&+ \nu \left[(Q - I)\Delta\tilde{u}\right]_i + \nu \sum_{l,j,k} Z_{l,j} \frac{\partial (Q\tilde{u})_i}{\partial y_k} \frac{\partial Z_{k,j}}{\partial y_l} - \left((Z^\top - I_3)\nabla\widetilde{p}\right)_i,
\end{aligned}$$

$$\mathcal{H}(\widetilde{u},\widetilde{\pi},\widetilde{\ell},\widetilde{\omega}) = (I_3 - [ZQ]^T)\widetilde{u}, \tag{6.11}$$

$$\mathcal{G}_1(\widetilde{\ell},\widetilde{\omega}) = -m(\widetilde{\omega}\times\widetilde{\ell}), \qquad \mathcal{G}_2(\widetilde{\ell},\widetilde{\omega}) = J(0)\widetilde{\omega}\times\widetilde{\omega}. \tag{6.12}$$

6.2. Estimate of nonlinear terms. In this section, we are going to estimate the nonlinear terms $\mathcal{F},\mathcal{H},\mathcal{G}_1$ and $\mathcal{G}_2$ defined as in (6.10) - (6.12).

Throughout this section we assume $1 < p, q < \infty$ satisfying the conditions $\frac{1}{p}+\frac{1}{2q}\neq 1$ and $\frac{1}{p}+\frac{3}{2q}\leqslant\frac{3}{2}$. Let p' denote the conjugate of p, i.e., $\frac{1}{p}+\frac{1}{p'}=1$. Let us fix $\eta\in(0,\eta_0)$, where η_0 is the constant introduced in Theorem 4.1 and we introduce the following ball

$$\mathcal{S}_\gamma = \left\{(\widetilde{u},\widetilde{\pi},\widetilde{\ell},\widetilde{\omega}) \mid \widetilde{\rho}(t,y) = \left\|(\widetilde{u},\widetilde{\pi},\widetilde{\ell},\widetilde{\omega})\right\|_{\mathcal{S}} \leqslant \gamma\right\},$$

where

$$\begin{aligned}\left\|(\widetilde{u},\widetilde{p},\widetilde{\ell},\widetilde{\omega})\right\|_{\mathcal{S}} &:= \|e^{\eta(\cdot)}\widetilde{u}\|_{L^p(0,\infty;W^{2,q}(\Omega_F(0)))^3} + \|e^{\eta(\cdot)}\widetilde{u}\|_{W^{1,p}(0,\infty;L^q(\Omega_F(0)))^3} \\ &+ \|e^{\eta(\cdot)}\widetilde{p}\|_{L^p(0,\infty;W^{1,q}(\Omega_F(0)))} + \|e^{\eta(\cdot)}\widetilde{\ell}\|_{W^{1,p}(0,\infty;\mathbb{R}^3)} + \|e^{\eta(\cdot)}\widetilde{\omega}\|_{W^{1,p}(0,\infty;\mathbb{R}^3)}.\end{aligned} \tag{6.13}$$

Our aim is to estimate the nonlinear terms in (6.10) - (6.12).

PROPOSITION 6.3. *Let us assume $1 < p, q < \infty$ satisfying the condition $\frac{1}{p}+\frac{3}{2q}\leqslant\frac{3}{2}$. There exist constants $\gamma_0\in(0,1)$ and $C_N > 0$ both depending only on p, q, η and $\Omega_F(0)$ such that for every $\gamma\in(0,\gamma_0)$ and for every $(\widetilde{u},\widetilde{\pi},\widetilde{\ell},\widetilde{\omega})\in\mathcal{S}_\gamma$, we have*

$$\begin{aligned}&\|e^{\eta(\cdot)}\mathcal{F}\|_{L^p(0,\infty;L^q(\Omega_F(0)))} + \|e^{\eta(\cdot)}\mathcal{H}\|_{W^{2,1}_{q,p}(Q_F^\infty)} \\ &\qquad + \|e^{\eta(\cdot)}\mathcal{G}_1\|_{L^p(0,\infty;\mathbb{R}^3)} + \|e^{\eta(\cdot)}\mathcal{G}_2\|_{L^p(0,\infty;\mathbb{R}^3)} \leqslant C_N\gamma^2.\end{aligned} \tag{6.14}$$

PROOF. The constants appearing in this proof will be denoted by C and depends only on p, q, η and $\Omega_F(0)$. Let us first show that, there exists $\gamma_0\in(0,1)$, such that, for every $\gamma\in(0,\gamma_0)$ and for every $(\widetilde{u},\widetilde{\pi},\widetilde{\ell},\widetilde{\omega})\in\mathcal{S}_\gamma$ the condition (6.1) is verified.

The solution of (6.7) satisfies $Q\in SO(3)$ and thus $|Q(t)|=1$ for all $t\geqslant 0$. We can rewrite Q as follows

$$Q(t) = I + \int_0^t e^{-\eta s}e^{\eta s}\widetilde{\omega}(s)\times Q(s)\,ds.$$

Therefore

$$\begin{aligned}\|Q - I_3\|_{L^\infty(0,\infty;\mathbb{R}^{3\times 3})} &\leqslant \int_0^\infty e^{-\eta s}e^{\eta s}|\widetilde{\omega}(s)|\,ds \\ &\leqslant \left(\int_0^\infty e^{-p'\eta t}\,dt\right)^{1/p'}\|e^{\eta(\cdot)}\widetilde{\omega}\|_{L^p(0,\infty;\mathbb{R}^3)} \leqslant \left(\frac{1}{p'\eta}\right)^{1/p'}\gamma.\end{aligned} \tag{6.15}$$

Similarly,

$$\|a\|_{L^\infty(0,\infty;\mathbb{R}^3)} \leqslant \int_0^\infty e^{-\eta t}e^{\eta t}|Q(s)||\widetilde{\ell}(t)|\,dt \leqslant \left(\frac{1}{p'\eta}\right)^{1/p'}\gamma. \tag{6.16}$$

Combining (6.15) and (6.16), we get

$$\|a\|_{L^\infty(0,\infty;\mathbb{R}^3)}+\|Q-I_3\|_{L^\infty(0,\infty;\mathbb{R}^{3\times 3})}\operatorname{diam}(\Omega_S(0)) \leqslant \gamma\left(\frac{1}{p'\eta}\right)^{1/p'}(1+\operatorname{diam}(\Omega_S(0))).$$

Let us set

$$\gamma_0 = \min\left\{1, \frac{\alpha}{2C_{p,\eta}(1+\operatorname{diam}(\Omega_F(0)))}\right\}, \quad \text{with } C_{p,\eta} = \left(\frac{1}{p'\eta}\right)^{1/p'}. \tag{6.17}$$

With the above choice of γ_0, we can easily verify the condition (6.1).

Let X be defined as in (6.8). Differentiating (6.8) with respect to y we obtain

$$\nabla X(t,y) = I_3 + \int_0^t \nabla\Lambda(s, X(s,y))\nabla X(s,y)\, ds$$

From the definition of Λ and X we obtain

$$\begin{aligned}
&\|\nabla X(t,\cdot)\|_{C^2(\Omega)}\\
&\leqslant 1 + C\int_0^t e^{-\eta s}e^{\eta s}\left(|\widetilde{\omega}(s)| + |\widetilde{\ell}(s)|\right)\|\nabla X(s,\cdot)\|_{C^\infty(\Omega)}\, ds\\
&\leqslant 1 + C\left(\|e^{\eta(\cdot)}\widetilde{\ell}\|_{L^\infty(0,\infty;\mathbb{R}^3)} + \|e^{\eta(\cdot)}\widetilde{\omega}\|_{L^\infty(0,\infty;\mathbb{R}^3)}\right)\int_0^t e^{-\eta s}\|\nabla X(s,\cdot)\|_{C^2(\Omega)}\, ds\\
&\leqslant 1 + C\int_0^t e^{-\eta s}\|\nabla X(s,\cdot)\|_{C^2(\Omega)}\, ds,
\end{aligned}$$

By Gronwall's inequality

$$\|\nabla X(t,\cdot)\|_{C^2(\Omega)} \leqslant \exp\left(C\int_0^t e^{-\eta s}\, ds\right) \leqslant e^{C/\eta} \text{ for all } t\in(0,\infty).$$

With the above estimate we obtain

$$\|\nabla X(t,\cdot) - I_3\|_{L^\infty(0,\infty;C^2(\Omega))} \leqslant C\int_0^\infty e^{-\eta s}e^{\eta s}\left(\|\widetilde{\omega}(s)\|_{\mathbb{R}^3} + \|\widetilde{\ell}(s)\|_{\mathbb{R}^3}\right) ds \leqslant C\gamma. \tag{6.18}$$

It is also easy to see that

$$\|\operatorname{Cof}\nabla X\|_{L^\infty(0,\infty;C^2(\Omega))} \leqslant C.$$

From Lemma 6.2, we have $\det\nabla X(t,y) = 1$ for all $t\geqslant 0$ and $y\in\overline{\Omega}$. Thus from the relation

$$Z = [\nabla X]^{-1} = \frac{1}{\det\nabla X}\operatorname{Cof}\nabla X,$$

we obtain

$$\|Z\|_{L^\infty(0,\infty;C^2(\Omega))} \leqslant C. \tag{6.19}$$

Using the above estimate and (6.18), we get

$$\|Z - I_3\|_{L^\infty(0,\infty;C^2(\Omega))} \leqslant \|Z\|_{L^\infty(0,\infty;C^2(\Omega))}\|\nabla X - I_3\|_{L^\infty(0,\infty;C^2(\Omega))} \leqslant C\gamma. \tag{6.20}$$

In a similar manner we can obtain the following estimates

$$\begin{aligned}
&\|\partial_t X\|_{L^\infty((0,\infty)\times\Omega_F(0))} \leqslant C\gamma, \quad \|\partial_t Z\|_{L^\infty((0,\infty)\times\Omega_F(0))} \leqslant C\gamma,\\
&\|ZQ - I_3\|_{L^\infty(0,\infty;C^2(\Omega))} \leqslant C\gamma.
\end{aligned} \tag{6.21}$$

We are now in a position to estimate the nonlinear terms.

Estimate of $\mathcal{F}$.

$$\|e^{\eta(\cdot)}\mathcal{F}\|_{L^p(0,\infty;L^q(\Omega_F(0)))} \leqslant C\gamma^2. \tag{6.22}$$

•Estimate of first, second and third term of $\mathcal{F}$: Using (6.15), (6.19) and (6.21) we have

$$\begin{aligned}
&\Big\|e^{\eta(\cdot)}\Big(- [(Q-I_3)\partial_t\widetilde{u}]_i - (\omega \times Q\widetilde{u})_i + \partial_t X \cdot Z^T\nabla(Q\widetilde{u})_i\Big)\Big\|_{L^p(0,\infty;L^q(\Omega_F(0)))}\\
&\leqslant C\left(\|Q-I_3\|_{L^\infty(0,\infty;\mathbb{R}^{3\times 3})} + \|\widetilde{\omega}\|_{L^\infty(0,\infty;\mathbb{R}^3)} + \|\partial_t X\|_{L^\infty((0,\infty)\times\Omega_F(0))}\right)\|e^{\eta(\cdot)}\widetilde{u}\|_{W^{2,1}_{q,p}(Q_F^\infty)}\\
&\leqslant C\gamma^2.
\end{aligned}$$

•Estimate of fourth term of $\mathcal{F}$: By Hölder's inequality and using (6.19), we obtain

$$\begin{aligned}
&\|e^{\eta(\cdot)}\widetilde{u}\cdot Z^T\nabla(Q\widetilde{u})_i\|_{L^p(0,\infty;L^q(\Omega_F(0)))}\\
&\leqslant C\|e^{\eta(\cdot)}\widetilde{u}\cdot\nabla\widetilde{u}_i\|_{L^p(0,\infty;L^q(\Omega_F(0)))}\\
&\leqslant C\|e^{\eta(\cdot)}\widetilde{u}\|_{L^{3p}(0,\infty;L^{3q}(\Omega_F(0)))}\|\nabla\widetilde{u}_i\|_{L^{3p/2}(0,\infty;L^{3q/2}(\Omega_F(0)))}.
\end{aligned}$$

Since $\dfrac{1}{p} + \dfrac{3}{2q} \leqslant \dfrac{3}{2}$, one has the following embeddings (see for example [**11**, Proposition 4.3])

$W^{2,1}_{q,p}(Q_F^\infty) \hookrightarrow L^{3p}(0,\infty;L^{3q}(\Omega_F(0)))$ and $W^{2,1}_{q,p}(Q_F^\infty) \hookrightarrow L^{3p/2}(0,\infty;W^{1+3q/2}(\Omega_F(0)))$.

Therefore, using the above embeddings we obtain

$$\|e^{\eta(\cdot)}\widetilde{u}\cdot Z^T\nabla(Q\widetilde{u})_i\|_{L^p(0,\infty;L^q(\Omega_F(0)))} \leqslant C\gamma^2$$

•Estimate of fifth term of $\mathcal{F}$(estimates of remaining terms of $\mathcal{F}$ are similar) : Using (6.19) and (6.20) we have

$$\begin{aligned}
&\left\|\nu e^{\eta(\cdot)}\sum_{l,j,k}\frac{\partial^2(Q\widetilde{u})_i}{\partial y_l\partial y_k}\left(Z_{k,j}-\delta_{k,j}\right)Z_{l,j}\right\|_{L^p(0,\infty;L^q(\Omega_F(0)))}\\
&\leqslant C\|Z-I_3\|_{L^\infty(0,\infty;C^2(\Omega))}\|e^{\eta(\cdot)}\widetilde{u}\|_{W^{2,1}_{q,p}(Q_F^\infty)} \leqslant C\gamma^2.
\end{aligned}$$

Estimate of $\mathcal{H}$.

$$\|e^{\eta(\cdot)}\mathcal{H}\|_{W^{2,1}_{q,p}(Q_F^\infty)} \leqslant C\gamma^2. \tag{6.23}$$

Using (6.20) and (6.21), we obtain

$$\begin{aligned}
&\|e^{\eta(\cdot)}(I_3 - [ZQ]^T)\widetilde{u}\|_{W^{2,1}_{q,p}(Q_F^\infty)}\\
&\leqslant C\Big(\|ZQ-I_3\|_{L^\infty(0,\infty;C^2(\Omega))} + \|\partial_t Z\|_{L^\infty((0,\infty)\times\Omega_F(0))}\\
&\qquad + \|\partial_t Q\|_{L^\infty(0,\infty;\mathbb{R}^3\times\mathbb{R}^3)}\Big)\|e^{\eta(\cdot)}\widetilde{u}\|_{W^{2,1}_{q,p}(Q_F^\infty)}\\
&\leqslant C\gamma^2.
\end{aligned}$$

Estimate of $\mathcal{G}_1$ and $\mathcal{G}_2$: From the expressions of $\mathcal{G}_1$ and $\mathcal{G}_2$ it is easy to see that

$$\|e^{\eta(\cdot)}\mathcal{G}_1\|_{L^p(0,\infty;\mathbb{R}^3)} + \|e^{\eta(\cdot)}\mathcal{G}_2\|_{L^p(0,\infty;\mathbb{R}^3)} \leqslant C\gamma^2. \tag{6.24}$$

Combining (6.22) - (6.24), we obtain (6.14). □

PROPOSITION 6.4. *Let us assume* $1 < p, q < \infty$ *satisfying the condition* $\frac{1}{p} + \frac{3}{2q} \leqslant \frac{3}{2}$. *Let* γ_0 *is defined as in* (6.17). *There exist constant* $C_{lip} > 0$ *depending only on* p, q, η *and* $\Omega_F(0)$ *such that for every* $\gamma \in (0, \gamma_0)$ *and for every* $(\widetilde{u}^j, \widetilde{\pi}^j, \widetilde{\ell}^j, \widetilde{\omega}^j) \in \mathcal{S}_\gamma$, $j = 1, 2$ *we have*

$$\begin{aligned}(6.25)\quad &\|e^{\eta(\cdot)}\mathcal{F}(\widetilde{u}^1, \widetilde{\pi}^1, \widetilde{\ell}^1, \widetilde{\omega}^1) - e^{\eta(\cdot)}\mathcal{F}(\widetilde{u}^2, \widetilde{\pi}^2, \widetilde{\ell}^2, \widetilde{\omega}^2)\|_{L^p(0,\infty;L^q(\Omega_F(0)))} \\ &\quad + \|e^{\eta(\cdot)}\mathcal{H}(\widetilde{u}^1, \widetilde{\pi}^1, \widetilde{\ell}^1, \widetilde{\omega}^1) - e^{\eta(\cdot)}\mathcal{H}(\widetilde{u}^2, \widetilde{\pi}^2, \widetilde{\ell}^2, \widetilde{\omega}^2)\|_{W^{2,1}_{q,p}(Q_F^\infty)} \\ &+ \|e^{\eta(\cdot)}\mathcal{G}_1(\widetilde{\ell}^1, \widetilde{\omega}^1) - e^{\eta(\cdot)}\mathcal{G}_1(\widetilde{\ell}^2, \widetilde{\omega}^2)\|_{L^p(0,\infty;\mathbb{R}^3)} + \|e^{\eta(\cdot)}\mathcal{G}_2(\widetilde{\ell}^1, \widetilde{\omega}^1) - e^{\eta(\cdot)}\mathcal{G}_2(\widetilde{\ell}^2, \widetilde{\omega}^2)\|_{L^p(0,\infty;\mathbb{R}^3)} \\ &\leqslant C_{lip}\gamma \left\|(\widetilde{u}^1, \widetilde{\pi}^1, \widetilde{\ell}^1, \widetilde{\omega}^1) - (\widetilde{u}^2, \widetilde{\pi}^2, \widetilde{\ell}^2, \widetilde{\omega}^2)\right\|_{\mathcal{S}}\end{aligned}$$

PROOF. The proof is similar to the proof of Proposition 6.3. □

6.3. Proof of Theorem 1.4. At first we prove global existence and uniqueness theorem for the transformed system (6.6) -(6.12) under the smallness assumption on the initial data. More precisely we prove the following theorem

THEOREM 6.5. *Let* $1 < p, q < \infty$ *satisfying the conditions* $\frac{1}{p} + \frac{1}{2q} \neq 1$ *and* $\frac{1}{p} + \frac{3}{2q} \leqslant \frac{3}{2}$. *Let* $\eta \in (0, \eta_0)$, *where* η_0 *is the constant introduced in Theorem 4.1. There exist a constant* $\widetilde{\gamma} > 0$ *depending only on* p, q, η *and* $\Omega_F(0)$ *such that, for all* $\gamma \in (0, \widetilde{\gamma})$ *and for all* $(u_0, \ell_0, \omega_0) \in B^{2(1-1/p)}_{q,p}(\Omega_F(0)) \times \mathbb{R}^3 \times \mathbb{R}^3$ *satisfying the compatibility conditions*

$$\begin{aligned}&\operatorname{div} u_0 = 0 \text{ in } \Omega_F(0), \\ &u_0 = \ell_0 + \omega_0 \times y \text{ on } \partial\Omega_S(0), \quad u_0 = 0 \text{ on } \partial\Omega \text{ if } \frac{1}{p} + \frac{1}{2q} < 1 \\ \text{and } &u_0 \cdot n = (\ell_0 + \omega_0 \times y) \cdot n \text{ on } \partial\Omega_S(0), \qquad u_0 \cdot n = 0 \text{ on } \partial\Omega \text{ if } \frac{1}{p} + \frac{1}{2q} > 1,\end{aligned}$$

and

$$\|u_0\|_{B^{2(1-1/p)}_{q,p}(\Omega_F(0))} + \|\ell_0\|_{\mathbb{R}^3} + \|\omega_0\|_{\mathbb{R}^3} \leqslant \frac{\gamma}{2C_L}, \tag{6.26}$$

where C_L *is the continuity constant appear in* (5.3), *the system* (6.6) -(6.12) *admits a unique strong solution* $(\widetilde{u}, \widetilde{\pi}, \widetilde{\ell}, \widetilde{\omega})$ *such that*

$$\left\|(\widetilde{u}, \widetilde{\pi}, \widetilde{\ell}, \widetilde{\omega})\right\|_{\mathcal{S}} \leqslant \gamma. \tag{6.27}$$

PROOF. Let us set

$$\widetilde{\gamma} = \min\left\{\gamma_0, \frac{1}{2C_L C_N}, \frac{1}{2C_L C_{lip}}\right\} \tag{6.28}$$

where γ_0 is defined as in (6.17) and C_L, C_N and C_{lip} are the constants appearing Theorem 5.1, Proposition 6.3 and Proposition 6.4 respectively. Let us choose $\gamma \in$

$(0, \widetilde{\gamma})$ and $(v, \varphi, \kappa, \tau) \in \mathcal{S}_\gamma$. We consider the following problem

$$
(6.29)\quad
\begin{cases}
\partial_t \widetilde{u} - \nu \Delta \widetilde{u} + \nabla \widetilde{\pi} = \mathcal{F}(v, \varphi, \kappa, \tau), \text{ div } u = \text{div } \mathcal{H}(v, \varphi, \kappa, \tau), & t \in (0, \infty), \ y \in \Omega_F(0), \\
\widetilde{u} = 0, & t \in (0, \infty), \ y \in \partial\Omega, \\
\widetilde{u} = \widetilde{\ell} + \widetilde{\omega} \times y, & t \in (0, \infty), \ y \in \partial\Omega_S(0), \\
m\widetilde{\ell}' = -\displaystyle\int_{\partial\Omega_S(0)} \sigma(\widetilde{u}, \widetilde{\pi}) n \, d\gamma + \mathcal{G}_1(\kappa, \tau), & t \in (0, \infty), \\
J(0)\widetilde{\omega}' = -\displaystyle\int_{\partial\Omega_S(0)} y \times \sigma(\widetilde{u}, \widetilde{\pi}) n \, d\gamma + \mathcal{G}_2(\kappa, \tau), & t \in (0, \infty), \\
u(0, y) = u_0(y), & y \in \Omega_F(0), \\
\ell(0) = \ell_0, \ \omega(0) = \omega_0.
\end{cases}
$$

We are going to show the mapping

$$\mathcal{N} : (v, \varphi, \kappa, \tau) \mapsto (\widetilde{u}, \widetilde{\pi}, \widetilde{\ell}, \widetilde{\omega})$$

where $(\widetilde{u}, \widetilde{\pi}, \widetilde{\ell}, \widetilde{\omega})$ is the solution to the system (6.29), is a contraction in $\mathcal{S}_\gamma$. As $(v, \varphi, \kappa, \tau) \in \mathcal{S}_\gamma$, we can apply Theorem 5.1 and Proposition 6.3 to the system (6.29) and using (6.26) and definition of $\widetilde{\gamma}$ we obtain

$$
\begin{aligned}
&\|\mathcal{N}(v, \varphi, \kappa, \tau)\|_{\mathcal{S}} \\
&\leqslant C_L \Big(\|u_0\|_{B^{2(1-1/p)}_{q,p}(\Omega_F(0))} + \|\ell_0\|_{\mathbb{R}^3} + \|\omega_0\|_{\mathbb{R}^3} \Big) + C_L C_N \gamma^2 \\
&\leqslant \gamma.
\end{aligned}
$$

Thus $\mathcal{N}$ is a mapping from $\mathcal{S}_\gamma$ to itself for all $\gamma \in (0, \widetilde{\gamma})$. Next, using Theorem 5.1 and Proposition 6.4, we obtain

$$
\begin{aligned}
&\left\|\mathcal{N}(v^1, \varphi^1, \kappa^1, \tau^1) - \mathcal{N}(v^2, \varphi^2, \kappa^2, \tau^2)\right\|_{\mathcal{S}} \\
&\leqslant C_L C_N \gamma \left\|(v^1, \varphi^1, \kappa^1, \tau^1) - (v^2, \varphi^2, \kappa^2, \tau^2)\right\|_{\mathcal{S}},
\end{aligned}
$$

for all $(v^j, \varphi^j, \kappa^j, \tau^j)$, $j = 1, 2$. Again using the definition of $\widetilde{\gamma}$ one can easily verify that $\mathcal{N}$ is a strict contraction of $\mathcal{S}_\gamma$ for any $\gamma \in (0, \widetilde{\gamma})$, which implies our existence and uniqueness result. □

Proof of Theorem 1.4: Let $(\widetilde{u}, \widetilde{\pi}, \widetilde{\ell}, \widetilde{\omega})$ be the solution of the system (6.6) -(6.12), constructed in Theorem 6.5. Since $\gamma < \widetilde{\gamma}$, (6.1) is verified and $X(t, \cdot)$ is a well defined mapping and it is a C^1-diffeomorphism from $\Omega_F(0)$ into $\Omega_F(t)$. Therefore, there is a unique $Y(t, \cdot)$ from $\Omega_F(t)$ into $\Omega_F(0)$ such that $Y(t, \cdot) = X(t, \cdot)^{-1}$. We set, for all $t \geqslant 0$ and $x \in \Omega_F(t)$

$$
\begin{gathered}
u(t, x) = \widetilde{u}(t, Y(t, x)), \quad \pi(t, x) = \widetilde{\pi}(t, Y(t, x)), \\
a'(t) = Q(t)\widetilde{\ell}(t) \text{ and } \omega(t) = Q(t)\widetilde{\omega}(t).
\end{gathered}
$$

We can easily check that (u, π, a, ω) satisfies the original system (1.2) satisfying (1.13).

References

[1] Herbert Amann, *Linear and quasilinear parabolic problems. Vol. I*, Monographs in Mathematics, vol. 89, Birkhäuser Boston, Inc., Boston, MA, 1995. Abstract linear theory. MR1345385

[2] Herbert Amann, *On the strong solvability of the Navier-Stokes equations*, J. Math. Fluid Mech. **2** (2000), no. 1, 16–98. MR1755865

[3] Alain Bensoussan, Giuseppe Da Prato, Michel C. Delfour, and Sanjoy K. Mitter, *Representation and control of infinite dimensional systems*, 2nd ed., Systems & Control: Foundations & Applications, Birkhäuser Boston, Inc., Boston, MA, 2007. MR2273323

[4] Patricio Cumsille and Takéo Takahashi, *Wellposedness for the system modelling the motion of a rigid body of arbitrary form in an incompressible viscous fluid*, Czechoslovak Math. J. **58(133)** (2008), no. 4, 961–992. MR2471160

[5] Robert Denk, Matthias Hieber, and Jan Prüss, *$\mathcal{R}$-boundedness, Fourier multipliers and problems of elliptic and parabolic type*, Mem. Amer. Math. Soc. **166** (2003), no. 788, viii+114. MR2006641

[6] Giovanni Dore, *L^p regularity for abstract differential equations*, Functional analysis and related topics, 1991 (Kyoto), Lecture Notes in Math., vol. 1540, Springer, Berlin, 1993, pp. 25–38. MR1225809

[7] Klaus-Jochen Engel and Rainer Nagel, *One-parameter semigroups for linear evolution equations*, Graduate Texts in Mathematics, vol. 194, Springer-Verlag, New York, 2000. With contributions by S. Brendle, M. Campiti, T. Hahn, G. Metafune, G. Nickel, D. Pallara, C. Perazzoli, A. Rhandi, S. Romanelli and R. Schnaubelt. MR1721989

[8] Eugene Fabes, Osvaldo Mendez, and Marius Mitrea, *Boundary layers on Sobolev-Besov spaces and Poisson's equation for the Laplacian in Lipschitz domains*, J. Funct. Anal. **159** (1998), no. 2, 323–368. MR1658089

[9] Daisuke Fujiwara and Hiroko Morimoto, *An L_r-theorem of the Helmholtz decomposition of vector fields*, J. Fac. Sci. Univ. Tokyo Sect. IA Math. **24** (1977), no. 3, 685–700. MR0492980

[10] Giovanni P. Galdi, *On the motion of a rigid body in a viscous liquid: a mathematical analysis with applications*, Handbook of mathematical fluid dynamics, Vol. I, North-Holland, Amsterdam, 2002, pp. 653–791. MR1942470

[11] Matthias Geissert, Karoline Götze, and Matthias Hieber, *L^p-theory for strong solutions to fluid-rigid body interaction in Newtonian and generalized Newtonian fluids*, Trans. Amer. Math. Soc. **365** (2013), no. 3, 1393–1439. MR3003269

[12] Matthias Geissert, Matthias Hess, Matthias Hieber, Céline Schwarz, and Kyriakos Stavrakidis, *Maximal L^p-L^q-estimates for the Stokes equation: a short proof of Solonnikov's theorem*, J. Math. Fluid Mech. **12** (2010), no. 1, 47–60. MR2602914

[13] Karoline Götze, *Maximal L^p-regularity for a 2D fluid-solid interaction problem*, Spectral theory, mathematical system theory, evolution equations, differential and difference equations, Oper. Theory Adv. Appl., vol. 221, Birkhäuser/Springer Basel AG, Basel, 2012, pp. 373–384. MR2954075

[14] John Happel and Howard Brenner, *Low Reynolds number hydrodynamics with special applications to particulate media*, Prentice-Hall, Inc., Englewood Cliffs, N.J., 1965. MR0195360

[15] Peer Christian Kunstmann and Lutz Weis, *Perturbation theorems for maximal L_p-regularity*, Ann. Scuola Norm. Sup. Pisa Cl. Sci. (4) **30** (2001), no. 2, 415–435. MR1895717

[16] Peer C. Kunstmann and Lutz Weis, *Maximal L_p-regularity for parabolic equations, Fourier multiplier theorems and H^∞-functional calculus*, Functional analytic methods for evolution equations, Lecture Notes in Math., vol. 1855, Springer, Berlin, 2004, pp. 65–311. MR2108959

[17] Debayan Maity and Jean-Pierre Raymond, *Feedback stabilization of the incompressible Navier-Stokes equations coupled with a damped elastic system in two dimensions*, J. Math. Fluid Mech. **19** (2017), no. 4, 773–805. MR3714503

[18] D. Maity and M. Tucsnak, *A Maximal Regularity Approach to the Analysis of Some Particulate Flows*, Springer International Publishing, Cham, 2017, pp. 1–75.

[19] J.-P. Raymond, *Stokes and Navier-Stokes equations with nonhomogeneous boundary conditions*, Ann. Inst. H. Poincaré Anal. Non Linéaire **24** (2007), no. 6, 921–951. MR2371113

[20] Jean-Pierre Raymond, *Feedback stabilization of a fluid-structure model*, SIAM J. Control Optim. **48** (2010), no. 8, 5398–5443. MR2745779

[21] Takéo Takahashi, *Analysis of strong solutions for the equations modeling the motion of a rigid-fluid system in a bounded domain*, Adv. Differential Equations **8** (2003), no. 12, 1499–1532. MR2029294

[22] Takéo Takahashi and Marius Tucsnak, *Global strong solutions for the two-dimensional motion of an infinite cylinder in a viscous fluid*, J. Math. Fluid Mech. **6** (2004), no. 1, 53–77. MR2027754

[23] Hans Triebel, *Interpolation theory, function spaces, differential operators*, 2nd ed., Johann Ambrosius Barth, Heidelberg, 1995. MR1328645

[24] Yun Wang and Zhouping Xin, *Analyticity of the semigroup associated with the fluid-rigid body problem and local existence of strong solutions*, J. Funct. Anal. **261** (2011), no. 9, 2587–2616. MR2826407

[25] Lutz Weis, *Operator-valued Fourier multiplier theorems and maximal L_p-regularity*, Math. Ann. **319** (2001), no. 4, 735–758. MR1825406

Institut de Mathématiques, Université de Bordeaux, Bordeaux INP, CNRS F-33400 Talence, France

Email address: debayan.maity@u-bordeaux.fr

Institut de Mathématiques, Université de Bordeaux, Bordeaux INP, CNRS F-33400 Talence, France

Email address: marius.tucsnak@u-bordeaux.fr

Contemporary Mathematics
Volume **710**, 2018
http://dx.doi.org/10.1090/conm/710/14371

On an interpolation inequality involving the Stokes operator

Paolo Maremonti

ABSTRACT. We prove an interpolation inequality in a suitable multiplicative form for the L^∞-norm of a function u belonging to J^q and such that $P_p\Delta u \in J^p$. Here J^r, $r \in (1, \infty)$, are the so-called spaces of hydrodynamics.

1. Introduction

In order to introduce the aims of this note it is crucial to make a recall of the so called space of hydrodynamics. For these function spaces and related properties we refer e.g. to [**7**].

In the following considerations $\Omega \subseteq \mathbb{R}^n, n \geq 2$, is assumed simply a C^2-smooth domain. Actually in the statement of our chief theorem $\partial\Omega$ has a suitable regularity. We set:

i. $\mathscr{C}_0(\Omega) := \{\varphi : \varphi \in C_0^\infty(\Omega) \text{ and } \nabla \cdot \varphi = 0\}$;
ii. $J^p(\Omega) :=$completion of $\mathscr{C}_0(\Omega)$ in the $L^p(\Omega)$-norm;
iii. for all $p \in (1, \infty)$, $L^p(\Omega) = J^p(\Omega) \oplus G^p(\Omega)$, with $G^p(\Omega) := \{\psi : \psi := \nabla h \in L^p(\Omega)\}$, and, for p and p' complementary exponents,
$$\text{for all } (u, \psi) \in J^p(\Omega) \times G^{p'}(\Omega),\ (u, \psi) = 0 \text{ holds;} \tag{$*$}$$
iv. P_p denotes the projector from L^p onto J^p; when confusion is not likely to occur simply by P; we call Stokes operator the composition $P\Delta$;
v. finally, we set $H_{q,p}(\Omega) := \{u : u \in J^q(\Omega),\ \Delta u \in L^p(\Omega) \text{ and } \nabla u \in L^p_{\ell oc}(\overline{\Omega})\}$, and $\overset{\circ}{H}_{q,p}(\Omega) := \{u : u \in H_{q,p}(\Omega) \text{ and } \gamma_{tr}(u) = 0 \text{ on } \partial\Omega\}$.

In paper [**9**] we prove that, for suitable r, $\overset{\circ}{H}_{q,p}(\Omega) \subset J^r(\Omega)$. More precisely

$$\|u\|_r \leq c\|P\Delta u\|_p^a \|u\|_q^{1-a}, \text{ for all } u \in \overset{\circ}{H}_{q,p}(\Omega), \tag{1.1}$$

provided that, for $a \in [0, 1)$, the dimensional balance holds:

$$\tfrac{1}{r} = a\Big(\tfrac{1}{p} - \tfrac{2}{n}\Big) + (1-a)\tfrac{1}{q}, \tag{1.2}$$

and with a constant c independent of u.

2010 *Mathematics Subject Classification.* Primary 35Q30, 35R45, Secondary 46N20.

Key words and phrases. Interpolation, Stokes operator, hydrodynamic spaces.

This research was partly supported by GNFM-INdAM, and by MIUR via the PRIN 2012 "Nonlinear Hyperbolic Partial Differential Equations, Dispersive and Transport Equations: Theoretical and Applicative Aspects".

The above result, contained in [**9**], holds if $r \neq \infty$. G.P. Galdi pointed out to the author that the technique employed for the J^r spaces, $r \neq \infty$, does not work in the case of $r = \infty$.

Motivated by this question the author produced the result of paper [**10**]. In [**10**], the semigroup properties of the Stokes operator are considered for initial data in the Lebesgue space $L^s(\Omega)$, $s \in [1, \infty)$, see Theorem 2.1 below.

Since the result related to estimate (1.1) is the intersection of questions at first glance independent, to the end of making clear our aims, we procede as follows:

1) first of all we point out the importance of an estimate of the kind (1.1) for $r = \infty$, and the difficulties in the classical framework of the interpolation and embedding of Sobolev spaces;
2) we motivate the role of the semigroup theory in the question;
3) we point out the importance of the role of Theorem 2.1.

1) As is well known, an interpolation inequality of Gagliardo-Nirenberg kind states properties of a function belonging to some Sobolev space. On the right hand side of an interpolation inequality, derivatives of max order in some L^p-Lebesgue space appear in a multiplicative form with the function evaluated in some L^q-norm. However, in some questions related to the partial differential equations (already for equations), instead of the max order of derivatives it is better to involve a suitable differential operator in the right hand side of the interpolation inequality. This is the case of the Navier-Stokes equations where, instead of all second spatial derivatives, one looks for estimates with the Stokes operator. Therefore our interest in estimate (1.1) is achieved. However, the problem becomes involved by the nature of the domain Ω that one considers. Indeed, if the domain Ω is such that, for $|\alpha| = 2$, the estimate

$$\|D^\alpha u\|_p \leq c_S \|P\Delta u\|_p, \; u \in \overset{\circ}{H}_{q,p}(\Omega),$$

holds with a constant c_S independent of u, then, by means of Gagliardo-Nirenberg's interpolation inequality, estimate (1.1) easily holds:

$$\|u\|_\infty \leq c_{GN} \|D^\alpha u\|_p^a \|u\|_q^{1-a} \leq c_{GN} c_S \|P\Delta u\|_p^a \|u\|_q^{1-a},$$

with ∞, p, q and a satisfying (1.2). The above arguments are possible in the case of a bounded domain Ω (see [**5**], for the special case $p = q = 2$), with c_S depending on the size of Ω (unsatisfactory factor), and in the cases of the whole space $\mathbb{R}^n$ and the half-space $\mathbb{R}^n_+$. In the case of an exterior domain the above approach does not work, since, for $|\alpha| = 2$ and $p \geq \frac{n}{2}$, the following inequality fails to hold

$$\|D^\alpha u\|_p \leq c \|P\Delta u\|_p.$$

In this connection see references [**7**] and [**13**]. From one side this difficulty introduces the need of a new proof of inequality (1.1) and, from another side, if the investigation reaches the goal, it lets understand that the second derivatives are overmuch for the L^∞-estimate of a function.

These difficulties lead to point 2).

2) The study developed in paper [**9**] is based on the semigroup properties of the unsteady Stokes operator. As kindly pointed out to the author by Y. Giga, this technique was not new. Actually, there is a wide literature concerning this technique from the Nash paper [**15**] to date with the contributes of [**8**] (see [**3, 4, 6, 16, 18**]). However, all the quoted papers are concerned with interpolation by means of first derivatives and the functions are related to the Laplace operator as well as to the

semigroup properties of the solutions of the heat equation. Instead, we are in a different context, in the sense that we consider interpolation by means of second order derivatives, functions belonging to the space of the hydrodynamic and, finally, the approach to the result is given by means of a duality argument.

These last aspects make necessary to prove the semigroup properties of the Stokes operator associating a suitable initial boundary value problem. These considerations lead to point 3).

3) The decomposition $L^r(\Omega) = J^r(\Omega) \oplus G^r(\Omega)$ and relation $(*)$ of iii. ensure that if a functional u' on $L^r(\Omega)$ belongs to $J^{r'}(\Omega)$, then, its kernel $\mathrm{Ker}(u')$ contains $G^r(\Omega)$ and $G^r(\Omega) = \bigcap\limits_{u' \in J^{r'}(\Omega)} \mathrm{Ker}(u')$. However, the norm of the functional u' achieves a maximum absolute value on the unit ball of $L^r(\Omega)$ and not of $J^r(\Omega)$. As a consequence of this last fact in our computations developed by duality, arising from a sort of reciprocity formula, the following kind of bound holds:

$$|(u, \frac{\varphi_0}{\|P_{r'}\varphi_0\|_{r'}})| \leq c\|P_p\Delta u\|_p^a \|u\|_q^{1-a}, \text{ for all } \varphi_0 \in L^{r'}(\Omega),$$

which implies

$$|(u, \frac{\varphi_0}{\|\varphi_0\|_{r'}})| \leq c(r)\|P_p\Delta u\|_p^a \|u\|_q^{1-a}, \text{ for all } \varphi_0 \in L^{r'}(\Omega), \tag{1.3}$$

provided that r, p, q and a satisfy the dimensional balance (1.2). The presence of $c(r)$ in (1.3) is a consequence of the Helmholtz decomposition, that it is employed to extend the variational formulation to the elements of $L^{r'}(\Omega)$ and not only of $J^{r'}(\Omega)$. Hence, we find a bound of the L^r-norm of u which depends on r, in particular $c(r) \to \infty$ for $r \to \infty$. This fact motivates the analysis proposed in [**10**] for an initial boundary value problem in a weak form, that is with an initial data in $L^r(\Omega)$. The initial data is not assumed by continuity in J^r-norm, but just in a weak form, as well as, the divergence equation fails to be a continuous function of $t \geq 0$. Nevertheless, by means of semigroup properties, we can deduce estimate (1.3) with a constant c independent of r, in particular for $r = \infty$.

In this note we prove the following result:

THEOREM 1.1. *Let Ω be a C^m-smooth, $2m > n \geq 2$, domain, and $p > \frac{n}{2}$. Then, $\overset{\circ}{H}_{q,p}(\Omega) \subset L^\infty(\Omega)$ with*

$$\|u\|_\infty \leq c\|P\Delta u\|_p^a \|u\|_q^{1-a}, \textit{ for all } u \in \overset{\circ}{H}_{q,p}(\Omega),\ a\big(\tfrac{1}{p} - \tfrac{2}{n}\big) + (1-a)\tfrac{1}{q} = 0, \tag{1.4}$$

provided that $a \in [0, 1)$, where c is a constant independent of u.

Employing the convexity theorem for L^p-norm, as corollary we get

COROLLARY 1.2. *In the hypotheses of Theorem 1.1 we get $\overset{\circ}{H}_{q,p}(\Omega) \subset L^r(\Omega)$ with*

$$\|u\|_r \leq c\|P\Delta u\|_p^a \|u\|_q^{1-a}, \textit{ for all } u \in \overset{\circ}{H}_{q,p}(\Omega),\ \tfrac{1}{r} = a\big(\tfrac{1}{p} - \tfrac{2}{n}\big) + (1-a)\tfrac{1}{q}, \tag{1.5}$$

provided that $a \in [0, 1)$, where c is a constant independent of u and r.

The results of Theorem 1.1 and of Corollary 1.2 are an improvement of the ones contained in Theorem 2.1 of [**9**]. Both for the fact that we state the result for $n \geq 2$ either for the fact that constant c in (1.5) is independent of r.

We remark that in papers [**14**, **17**] inequality (1.1) is proved, for $\Omega \subset \mathbb{R}^3$, in the special form: $\|u\|_\infty \le c\|\Delta u\|_2^{\frac{1}{2}}\|\nabla u\|_2^{\frac{1}{2}}$. In [**14**] it is proved for exterior domains, and in [**17**] for arbitrary domains giving the best constant c independent of regularity of Ω.

We refer to paper [**9**] for further comments on the question and to chap. 6 of [**8**] for the interpolation of the first order.

2. Preliminary results

In the following section we assume that $\Omega \subseteq \mathbb{R}^n$ is an exterior domain whose boundary is assumed of class C^m with $2m > n$. This assumption is made for stating the result of Theorem 2.1 below.

Let X be a Banach space. For all $T > 0$, by the symbol $C(0, T; X)$ we mean the set of functions $u : (0, T) \to X$, which are continuous from $[0, T]$ into X. Analogously, with $L^s(0, T; X)$ we mean the set of functions u such that $\int_0^T \|u(\tau)\|_X^s d\tau < \infty$.

We consider the following initial boundary value problem:

$$
\begin{aligned}
&w_t - \Delta w = -\nabla \pi_w,\ \nabla \cdot w = 0, \text{ in } (0, T) \times \Omega,\\
&w(t, x) = 0 \text{ on } (0, T) \times \partial\Omega,\\
&(w(0), \varphi) = (w_0, \varphi) \text{ for any } \varphi \in \mathscr{C}_0(\Omega).
\end{aligned}
\tag{2.1}
$$

Due to condition $(2.1)_3$, problem (2.1) is a weak form of the usual initial boundary value problem for the Stokes equations. This weak formulation allows to consider initial data in the Lebesgue spaces L^p and not in the space of the hydrodynamics J^p. It was introduced in [**10**]. Its interest is connected with the possibility of deducing estimates in L^r-Lebesgue spaces with $r \in (1, \infty]$ by means of duality arguments. Of course, for an initial data in J^p we come back to the classical Stokes solutions.

For each $T > 0$, for q' conjugate exponent of q, we set $W_{q'} := \{\varphi(t, x) : \varphi \in C^1([0, T] \times \overline{\Omega}) \cap C(0, T; J^{1,q'}(\Omega)) \text{ and } \varphi_t \in C([0, T]; L^{q'}(\Omega))\}$.

For problem (2.1) the following result holds:

THEOREM 2.1. *Let $w_0 \in L^p(\Omega)$. If $p = 1$, then, there exists a unique solution (w, π_w) to problem (2.1) such that*

i. $\eta > 0, q > 1, w \in C(\eta, T; J^q(\Omega)) \cap L^\infty(\eta, T; J^{1,q}(\Omega)), D^2 w(t, x), \nabla\pi_w(t, x) \in L^\infty(\eta, T; L^q(\Omega))$;

ii. $q \in (1, \infty]$,

$$\|w(t)\|_q \le c\|w_0\|_1 t^{-\mu}, \qquad \mu = \tfrac{n}{2}\big(1 - \tfrac{1}{q}\big),\ t > 0;$$

$$\|\nabla w(t)\|_q \le c\|w_0\|_1 t^{-\mu_1}, \quad \mu_1 = \begin{cases} \frac{1}{2} + \mu & \textit{if } t \in (0, 1],\\ \frac{1}{2} + \mu & \textit{if } t > 0 \textit{ and } q \in (1, n],\\ \frac{n}{2} & \textit{if } t > 1 \textit{ and } q > n;\end{cases}$$

$$\|w_t(t)\|_q \le c\|w_0\|_1 t^{-\mu_2}, \qquad \mu_2 = 1 + \mu,\ t > 0;$$

where the constant c is independent of w_0 and the exponent μ_1 is sharp;

iii. $\displaystyle\int_0^t \big[(w(\tau), \varphi_\tau(\tau) - (\nabla w(\tau), \nabla\varphi(\tau))\big] d\tau = (w(t), \varphi(t)) - (w_0, \varphi(0))$, *for each* $\varphi \in W_{r'}$ *provided that* $\frac{1}{2} + \frac{n}{2}(1 - \frac{1}{r}) < 1$; *finally,* $\lim_{t\to 0}(w(t), \varphi) = (w_0, \varphi)$, *for all* $\varphi \in \mathscr{C}_0(\Omega)$.

If $p \in (1,\infty)$, *then, there exists a unique solution* (w,π_w) *to problem* (2.1) *such that*

iv. $\eta > 0, q \geq p,\ w \in C([0,T);J^p(\Omega)),\ w \in L^\infty(\eta,T;J^{1,q}(\Omega)),\ D^2w(t,x),$ $\nabla\pi_w(t,x) \in L^\infty(\eta,T;L^q(\Omega))$;

v. $q \in [p,\infty]$,

$$\|w(t)\|_q \leq c\|w_0\|_p t^{-\mu}, \qquad \mu = \tfrac{n}{2}\big(\tfrac{1}{p}-\tfrac{1}{q}\big),\ t>0;$$

$$\|\nabla w(t)\|_q \leq c\|w_0\|_p t^{-\mu_1}, \quad \mu_1 = \begin{cases} \frac{1}{2}+\mu & \textit{if } t \in (0,1], \\ \frac{1}{2}+\mu & \textit{if } t > 0 \textit{ and } q \in [p,n], \\ \frac{n}{2p} & \textit{if } t > 1 \textit{ and } q \geq n; \end{cases}$$

$$\|w_t(t)\|_q \leq c\|w_0\|_p t^{-\mu_2}, \qquad \mu_2 = 1+\mu,\ t>0;$$

where c *is independent of* w_0*;*

vi. *finally,* $\lim\limits_{t\to 0}(w(t)-w_0,\varphi) = 0$, *for any* $\varphi \in \mathscr{C}_0(\Omega)$.

Remark 2.2. For the proof we refer the reader to paper [**10**] Theorem 2.1. However in [**10**] the proof of Theorem 2.1 is developed only for the case of $\Omega \subseteq \mathbb{R}^n$ $n \geq 3$. The same proof works in the two dimensional case provided that Theorem 3.1 of [**10**] is stated in 2-dimensional case too. We realize this employing the results of the two dimensional case obtained in [**2**]. Hence we consider as achieved the proof of Theorem 2.1 for all $n \geq 2$.

We stress that in item iv. the continuity property in $t=0$ is meant in the sense that $\lim\limits_{t\to 0}\|w(t)-P_p(w_0)\|_p = 0$. For the data $w_0 \in L^p(\Omega)$ we just have the limit property of item iii. and vi. Of course, if $w_0 \in J^p(\Omega)$ the requirement $(2.1)_3$ is the usual one for the initial data of the Stokes problem and, since $w \in C(0,T;J^p(\Omega))$, the data w_0 is continuously assumed as function of $J^p(\Omega)$.

Lemma 2.3. *Let be* $v \in C(0,T;J^p(\Omega))$, $w \in C(0,T;J^{p'}(\Omega))$ *and* $w_0 \in L^{p'}(\Omega)$. *Assume also that* $\lim\limits_{t\to 0}(w(t),\varphi) = (w_0,\varphi)$, *for all* $\varphi \in \mathscr{C}_0(\Omega)$. *Then, for all* $t \in (0,T)$, *we get*

$$\lim_{\delta\to 0}(v(t-\delta),w(\delta)) = (v(t),w_0)\,.$$

Proof. For the proof see Lemma 3.5 in [**11**] or Lemma 5.1 in [**1**]. □

3. Proof of Theorem 1.1

Proof. By assumption we have $\Delta u \in L^p(\Omega)$ and the Helmholtz decomposition $\Delta u = P\Delta u + \nabla\pi_u$ holds, with $P\Delta u,\ \nabla\pi_u \in L^p(\Omega)$. Since $u \in J^q(\Omega)$, we get that the pair (u,π_u) satisfies the system

$$\Delta u - \nabla\pi_u = P\Delta u,\ \nabla\cdot u = 0 \text{ in } \Omega \text{ with } u = 0 \text{ on } \partial\Omega,$$

where the last equation is satisfyed via the assumption $u \in \overset{\circ}{H}_{q,p}(\Omega)$. Now, we consider a solution (w,π_w) of problem (2.1) with $w_0 \in L^1(\Omega)$, whose existence is ensured by Theorem 2.1. Since $u \in W^{1,p}_{\ell oc}(\overline{\Omega})$ and $\Delta u \in L^p(\Omega)$ with $u \in J^q(\Omega)$, and since, for $t>0$, $w \in J^r(\Omega)$, for all $r \in (1,\infty)$, and enjoys properties i)-ii) of Theorem 2.1, then, an integration by parts furnishes

$$(P_p\Delta u, w(t)) = (\Delta u - \nabla\pi_u, w(t)) = (u, P_{q'}\Delta w(t)) = (u, w_t(t)), \text{ a.e. in } t>0.$$

Hence, integrating on (s,t), $s>0$, we get

$$(u,w(s)) = (u,w(t)) - \int_s^t (P\Delta u, w(\tau))d\tau. \tag{3.1}$$

Since $u\in J^q(\Omega)$, by virtue of Lemma 2.3, we get

$$\lim_{s\to 0}(u,w(s)) = (u,w_0).$$

The above limit property and (3.1) give

$$(u,w_0) = (u,w(t)) - \int_0^t (P\Delta u, w(\tau))d\tau.$$

Applying Hölder's inequality, we deduce the estimate

$$|(u,w_0)| \le \|u\|_q\|w(t)\|_{q'} + \|P\Delta u\|_p \int_0^t \|w(\tau)\|_{p'}d\tau.$$

By employing the semigroup properties of w we also deduce

$$|(u,w_0)| \le c(\|u\|_q t^{-\frac{n}{2}\frac{1}{q}} + \|P\Delta u\|_p t^{1-\frac{n}{2}\frac{1}{p}})\|w_0\|_1, \text{ for all } w_0\in L^1(\Omega).$$

We have proved the estimate

$$\|u\|_\infty \le c(\|u\|_q t^{-\frac{n}{2}\frac{1}{q}} + \|P\Delta u\|_p t^{1-\frac{n}{2}\frac{1}{p}}).$$

Hence setting $t=\|u\|_q^\alpha\|P\Delta u\|_p^{-\alpha}$, we get the result for $1-\alpha\frac{n}{2}\frac{1}{q} = \alpha(1-\frac{n}{2}\frac{1}{p})$, that is $\alpha = 2a\frac{q}{n}$. □

References

[1] F. Crispo and P. Maremonti, *On the spatial asymptotic decay of a suitable weak solution to the Navier-Stokes Cauchy problem*, Nonlinearity **29** (2016), no. 4, 1355–1383. MR3476511

[2] Wakako Dan and Yoshihiro Shibata, *Remark on the L_q-L_∞ estimate of the Stokes semigroup in a 2-dimensional exterior domain*, Pacific J. Math. **189** (1999), no. 2, 223–239. MR1696121

[3] E. B. Davies, *Heat kernels and spectral theory*, Cambridge Tracts in Mathematics, vol. 92, Cambridge University Press, Cambridge, 1989. MR990239

[4] E. B. Fabes and D. W. Stroock, *A new proof of Moser's parabolic Harnack inequality using the old ideas of Nash*, Arch. Rational Mech. Anal. **96** (1986), no. 4, 327–338. MR855753

[5] Hiroshi Fujita and Tosio Kato, *On the Navier-Stokes initial value problem. I*, Arch. Rational Mech. Anal. **16** (1964), 269–315. MR0166499

[6] Masatoshi Fukushima, Yōichi Ōshima, and Masayoshi Takeda, *Dirichlet forms and symmetric Markov processes*, De Gruyter Studies in Mathematics, vol. 19, Walter de Gruyter & Co., Berlin, 1994. MR1303354

[7] Giovanni P. Galdi, *An introduction to the mathematical theory of the Navier-Stokes equations. Vol. I*, Springer Tracts in Natural Philosophy, vol. 38, Springer-Verlag, New York, 1994. Linearized steady problems. MR1284205

[8] Mi-Ho Giga, Yoshikazu Giga, and Jürgen Saal, *Nonlinear partial differential equations*, Progress in Nonlinear Differential Equations and their Applications, vol. 79, Birkhäuser Boston, Inc., Boston, MA, 2010. Asymptotic behavior of solutions and self-similar solutions. MR2656972

[9] Paolo Maremonti, *Some interpolation inequalities involving Stokes operator and first order derivatives*, Ann. Mat. Pura Appl. (4) **175** (1998), 59–91. MR1748216

[10] P. Maremonti, *A remark on the Stokes problem with initial data in L^1*, J. Math. Fluid Mech. **13** (2011), no. 4, 469–480. MR2847277

[11] Paolo Maremonti, *On the Stokes problem in exterior domains: the maximum modulus theorems*, Discrete Contin. Dyn. Syst. **34** (2014), no. 5, 2135–2171. MR3124730

[12] P. Maremonti, *Stokes and Navier-Stokes initial boundary value problem with nondecaying data*, Waseda Univ., Japanese-German Int. Workshop on Math. Fluid Dynamics, Lecture Notes, http://www.japan-germany.sci.waseda.ac.jp/event/201206/c04a.pdf

[13] P. Maremonti and V. A. Solonnikov, *An estimate for the solutions of a Stokes system in exterior domains* (Russian, with English summary), Zap. Nauchn. Sem. Leningrad. Otdel. Mat. Inst. Steklov. (LOMI) **180** (1990), no. Voprosy Kvant. Teor. Polya i Statist. Fiz. 9, 105–120, 181; English transl., J. Math. Sci. **68** (1994), no. 2, 229–239. MR1067337

[14] Kyûya Masuda, *On the stability of incompressible viscous fluid motions past objects*, J. Math. Soc. Japan **27** (1975), 294–327. MR0440224

[15] J. Nash, *Continuity of solutions of parabolic and elliptic equations*, Amer. J. Math. **80** (1958), 931–954. MR0100158

[16] N. Th. Varopoulos, L. Saloff-Coste, and T. Coulhon, *Analysis and geometry on groups*, Cambridge Tracts in Mathematics, vol. 100, Cambridge University Press, Cambridge, 1992. MR1218884

[17] Wenzheng Xie, *A sharp pointwise bound for functions with L^2-Laplacians and zero boundary values of arbitrary three-dimensional domains*, Indiana Univ. Math. J. **40** (1991), no. 4, 1185–1192. MR1142710

[18] Atsushi Yoshikawa, *Fractional powers of operators, interpolation theory and imbedding theorems*, J. Fac. Sci. Univ. Tokyo Sect. IA Math. **18** (1971), 335–362. MR0317107

Dipartimento di Matematica e Fisica, Università degli Studi della Campania "L. Vanvitelli", via Vivaldi 43, 81100 Caserta - Italy

Email address: paolo.maremonti@unicampania.it

Contemporary Mathematics
Volume **710**, 2018
http://dx.doi.org/10.1090/conm/710/14372

Brezis-Gallouet-Wainger type inequality and its application to the Navier-Stokes equations

Kohei Nakao and Yasushi Taniuchi

ABSTRACT. In this paper, we give a new Serrin type regularity criterion of solutions to the 3-D Navier-Stokes equations. To this end, we use a logarithmic inequality of Brezis-Gallouet-Wainger type. We also briefly introduce our recent results about logarithmic inequalities of Brezis-Gallouet-Wainger type.

1. Introduction

Let Ω be $\mathbb{R}^n$, $\mathbb{R}^n_+$, a bounded domain, or an exterior domain with $\partial\Omega \in C^\infty$. The motion of a viscous incompressible fluid in $\Omega \subset \mathbb{R}^n$ is governed by the Navier-Stokes equations:

$$\text{(N-S)}\begin{cases} \partial_t u - \Delta u + u\cdot\nabla u + \nabla\pi = 0, & \nabla\cdot u = 0 \qquad t\in(0,T), \quad x\in\Omega, \\ \qquad\qquad\qquad u|_{\partial\Omega} = 0, & u|_{t=0} = u_0, \end{cases}$$

where $u = (u^1(x,t), u^2(x,t), \cdots, u^n(x,t))$ and $\pi = \pi(x,t)$ denote the velocity vector and the pressure, respectively, of the fluid at the point $(x,t) \in \Omega\times(0,T)$ and u_0 is a given initial velocity. In this paper, we establish a new Serrin type regularity criterion of solutions to the 3-D Navier-Stokes equations. Many researchers proved that if Leray-Hopf weak solution u satisfies

$$(Se) \qquad u \in L^s(0,T;L^r) \text{ for some } 3 < r \le \infty, 2 \le s < \infty \text{ with } \frac{3}{r} + \frac{2}{s} \le 1,$$

then u is smooth. See e.g. [**9, 15, 21, 22, 26, 37, 40, 42–44**]. The limiting case $s = \infty, r = 3$ was proven in [**8**]. See also [**31**]. Let $p > 3$. It is also known that the condition (Se) also guarantees the time-extension of strong L^p-solution. That is, if strong L^p-solutions u of the Navier-Stokes equations on $(0,T)$ satisfies (Se), then u can be continued to the strong L^p-solution on $(0,T')$ for some $T' > T$, see [**15**]. In the present paper, we shall slightly relax the condition (Se) in the case $r = \infty$:

$$u \in L^2(0,T;L^\infty(\Omega)),$$

by replacing $L^\infty(\Omega)$ with some Banach space X. For this purpose, we use the Brezis-Gallouet-Wainger type inequality:

$$(BGW)_\beta \qquad \|u\|_{L^\infty} \le C(1 + \|f\|_X \log^\beta(e + \|f\|_Y)).$$

2010 *Mathematics Subject Classification.* Primary 35Q30, 35Q35; Secondary 76D05.

Key words and phrases. Navier-Stokes equations, regularity, logarithmic inequalities.

The second author was supported in part by JSPS KAKENHI Grant Number No.16K05228.

When $\Omega = \mathbb{R}^n$, by using the Fourier transform, Brezis-Gallouet-Wainger [**4, 5**] proved $(BGW)_\beta$ in the case

$$\beta = 1 - 1/p, \quad X = W^{n/p,p}(\mathbb{R}^n), \quad Y = W^{n/q+\alpha,q}(\mathbb{R}^n)(\subset \dot{C}^\alpha)(\alpha > 0).$$

Engler [**7**] proved the same inequality for general domains Ω without using the Fourier transform. Ozawa [**35**] proved the Gagliardo-Nirenberg type inequality

$$\|f\|_{L^q(\mathbb{R}^n)} \le C(p,n) q^{1-1/p} \|(-\Delta)^{n/2p} f\|_{L^p(\mathbb{R}^n)}^{1-p/q} \|f\|_{L^p(\mathbb{R}^n)}^{p/q} \quad \text{for all } q \in [p, \infty)$$

with the explicit growth rate with respect to q and that this estimate directly yields $(BGW)_\beta$ with $\beta = 1 - 1/p$. When $\Omega = \mathbb{R}^n$, Chemin [**6**] proved $(BGW)_\beta$ for $\beta = 1$, $X = B^0_{\infty,\infty}(\mathbb{R}^n)$ and $Y = C^\alpha(\mathbb{R}^n)$. Kozono, Ogawa and the second author [**21, 34**] proved $(BGW)_\beta$ for $0 \le \beta \le 1$, $X = \dot{B}^0_{\infty,1/(1-\beta)}(\mathbb{R}^n)$ and $Y = \dot{C}^\alpha(\mathbb{R}^n) \cap \dot{B}^{-\alpha}_{\infty,\infty}$. When Ω is a bounded domain, in [**33**], $(BGW)_\beta$ was proven in the cases

$$\beta = 1, \quad X = bmo(\Omega), \quad Y = \dot{C}^\alpha(\Omega), \text{ or}$$

$$\beta = 1, \quad X = B(\Omega), \quad Y = \dot{C}^\alpha(\Omega),$$

where $B(\Omega)$ is introduced by the norm $\|f\|_{B(\Omega)} := \sup_{q \ge 1} \frac{\|f\|_{L^q(\Omega)}}{q}$. We note that in [**1, 6, 7, 12, 17, 20, 21, 23, 24, 27, 29, 32, 33, 35, 36, 38, 47**] several inequalities of Brezis-Gallouet-Wainger type were established in the case $Y = \dot{C}^\alpha$ or $Y \subset \dot{C}^\alpha$. On the other hand, there are many choices of X, as mentioned above. Then, we have one question.

What is the largest normed space X that satisfies $(BGW)_\beta$ with $Y = \dot{C}^\alpha(\Omega)$?

In this paper, we also briefly introduce our recent results about this problem.

Throughout this paper we impose the following assumption on the domain.

Assumption 1 *$\Omega \subset \mathbb{R}^n$ is the half-space $\mathbb{R}^n_+$, the whole space $\mathbb{R}^n$, a bounded domain, an exterior domain, a perturbed half-space, or an aperture domain with $\partial\Omega \in C^\infty$.*

For the definitions of perturbed half-spaces and aperture domains, see e.g. Kubo-Shibata [**25**] and Farwig-Sohr [**10**].

We introduce Banach spaces of Morrey type and Besov type which are wider than L^∞.

DEFINITION 1.1. (1) (Morrey type space) Let $\beta > 0$.

- $M_\beta(\Omega) := \{f \in L^1_{loc}(\bar{\Omega}); \|f\|_{M_\beta} < \infty\}$ is introduced by the norm

$$\|f\|_{M_\beta(\Omega)} := \sup_{x \in \Omega,\ 0<t<1} \frac{1}{|B(x,t)| \log^\beta(e + \frac{1}{t})} \int_{B(x,t)\cap\Omega} |f(y)| dy.$$

- $\tilde{M}_\beta(\Omega)$ is defined by

$$\tilde{M}_\beta(\Omega) := \overline{BC(\bar{\Omega})}^{\|\cdot\|_{M_\beta(\Omega)}}.$$

(2) (Modified Vishik's space). Let $\beta > 0$ and ψ be a spherical symmetric smooth function on $\mathbb{R}^n$ with $\hat{\psi}(\xi) = 1$ in $B(0, 1/2)$ and $\hat{\psi}(\xi) = 0$ in $B(0,1)^c$. Then,

- $V_\beta(\mathbb{R}^n) = \{f \in \mathcal{S}'(\mathbb{R}^n); \|f\|_{V_\beta} < \infty\}$ is introduced by the norm

$$\|f\|_{V_\beta} := \sup_{N=1,2,\cdots} \frac{\|\psi_N * f\|_\infty}{N^\beta}, \quad \text{where } \psi_N(x) := 2^{nN}\psi(2^N x).$$

- $\tilde{V}_\beta$ is defined by

$$\tilde{V}_\beta := \overline{BUC(\mathbb{R}^n)}^{\|\cdot\|_{V_\beta}}.$$

REMARK 1.2. (a) V_β is a modified version of spaces introduced by Vishik [**45**].

(b) If $\beta > 0$, then $L^\infty(\mathbb{R}^n) \hookrightarrow M_\beta(\mathbb{R}^n) \hookrightarrow V_\beta(\mathbb{R}^n)$ and $L^\infty(\Omega) \hookrightarrow M_\beta(\Omega)$.

(c) $\tilde{V}_\beta(\mathbb{R}^n)$ and $\tilde{M}_\beta(\Omega)$ satisfy $(BGW)_\beta$. That is, if $\alpha \in (0,1)$ and $\beta > 0$, then there are constants $C_1, C_2 > 0$ such that

$$\|f\|_{L^\infty(\Omega)} \le C_1\Big\{1 + \|f\|_{M_\beta(\Omega)} \log^\beta\left(e + \|f\|_{\dot{C}^\alpha(\Omega)}\right)\Big\} \tag{1.1}$$

for all $f \in \dot{C}^\alpha(\Omega) \cap \tilde{M}_\beta(\Omega)$ and

$$\|f\|_{L^\infty(\mathbb{R}^n)} \le C_2\Big\{1 + \|f\|_{V_\beta(\mathbb{R}^n)} \log^\beta\left(e + \|f\|_{\dot{C}^\alpha(\mathbb{R}^n)}\right)\Big\} \tag{1.2}$$

for all $f \in \dot{C}^\alpha(\mathbb{R}^n) \cap \tilde{V}_\beta(\mathbb{R}^n)$.

We will give proofs of (1.1) and (1.2) in Section 3 in the present paper. See also our previous paper [**30**].

In [**30**] we proved the following two theorems. Roughly speaking, $\tilde{M}_\beta(\Omega)$ (or $\tilde{V}_\beta(\mathbb{R}^n)$) is the largest normed space that satisfies $(BGW)_\beta$ in the following sense.

THEOREM 1.3 ([**30**]). *Let $\beta > 0$ and $X(\Omega)$ be a normed space. Assume that $X(\Omega)$ satisfies the following conditions:*

$(A1)$ $BC(\bar{\Omega}) \hookrightarrow X(\Omega) \subset L^1_{uloc}(\bar{\Omega})$ *and* $BC(\bar{\Omega})$ *is dense in* X,

$(A2)$ $\|\cdot\|_{X(\Omega)}$ *has a translation invariant property in the following sense* $\big\|f(\cdot + y)|_\Omega\big\|_{X(\Omega)} \le \|f\|_{X(\Omega)}$ *for all* $y \in \mathbb{R}^n$ *and all* $f \in C_0(\Omega)$,

$(A3)$ $\|f\|_{X(\Omega)} \le \|g\|_{X(\Omega)}$ *if* $f, g \in BC(\bar{\Omega})$ *and* $|f(x)| \le |g(x)|$ *for all* $x \in \Omega$,

$(A4)$ *there exist constants* $\alpha \in (0,1)$ *and* $C > 0$ *such that*

$$\|f\|_{L^\infty(\Omega)} \le C\Big\{1 + \|f\|_{X(\Omega)} \log^\beta\left(e + \|f\|_{\dot{C}^\alpha(\Omega)}\right)\Big\}$$

for all $f \in \dot{C}^\alpha(\Omega) \cap X(\Omega)$.

Then $X(\Omega)$ is continuously embedded in $\tilde{M}_\beta(\Omega)$.

REMARK 1.4. (i) Since $\tilde{M}_\beta$ satisfies $(A1)$-$(A4)$, Theorem 1.3 implies that $\tilde{M}_\beta$ is the largest normed space that satisfies conditions $(A1)$-$(A4)$.

(ii) Since $f(\cdot) = f(\cdot + y - y)$, (A2) implies that $\|f\|_{X(\Omega)} = \|f(\cdot + y)\|_{X(\Omega)}$, if both of f and $f(\cdot + y)$ belong to $C_0(\Omega)$.

If $\Omega = \mathbb{R}^n$, without the condition (A3), we have a similar result as follows.

THEOREM 1.5 ([**30**]). *Let $\beta > 0$ and X be a normed space. Assume that X satisfies the following conditions:*

$(B1)$ $BUC(\mathbb{R}^n) \hookrightarrow X \hookrightarrow \mathcal{S}'(\mathbb{R}^n)$ *and* BUC *is dense in* X,

$(B2)$ $\|\cdot\|_X$ *has a translation invariant property i.e.,* $\|f(\cdot + y)\|_X \le \|f\|_X$ *for all* $y \in \mathbb{R}^n$ *and all* $f \in BUC(\mathbb{R}^n)$,

$(B3)$ *there exist constants* $\alpha \in (0,1)$ *and* $C > 0$ *such that*

$$\|f\|_{L^\infty(\mathbb{R}^n)} \le C\Big\{1 + \|f\|_X \log^\beta\left(e + \|f\|_{\dot{C}^\alpha(\mathbb{R}^n)}\right)\Big\}$$

for all $f \in BC^\infty(\mathbb{R}^n)$.

Then X is continuously embedded in $\tilde{V}_\beta$. Here, $BC^\infty(\mathbb{R}^n) = \{g \in C^\infty(\mathbb{R}^n)\ ;\ \partial^\alpha g \in BC(\mathbb{R}^n)$ for all multi-indices $\alpha\}$.

REMARK 1.6. Since $\tilde{V}_\beta(\mathbb{R}^n)$ satisfies $(B1)$-$(B3)$, Theorem 1.5 implies that $\tilde{V}_\beta(\mathbb{R}^n)$ is the largest normed space that satisfies conditions $(B1)$-$(B3)$.

2. Preliminaries and Main Result

In this section, as an application of $(BGW)_\beta$, we will establish a new Serrin type regularity criterion for solutions to the 3-D Navier-Stokes equations. We first introduce some notations and function spaces.

We use the notations 1_Ω for the characteristic function on Ω and $B(x,t)$ for the ball centered at x and of radius t. Let $C_0^\infty(\Omega)$ denote the set of all C^∞ functions with compact support in Ω and $C_{0,\sigma}^\infty(\Omega) = C_{0,\sigma}^\infty := \{\phi \in (C_0^\infty(\Omega))^n \ ; \ \nabla \cdot \phi = 0\}$. Then L_σ^r, $1 < r < \infty$, is the closure of $C_{0,\sigma}^\infty$ with respect to the L^r-norm $\| \cdot \|_r$. Concerning Sobolev spaces we use the notations $W^{k,p}(\Omega)$ and $W_0^{k,p}(\Omega)$, $k \in \mathbb{N}$, $1 \le p \le \infty$. Note that very often we will simply write L^r and $W^{k,p}$ instead of $L^r(\Omega)$ and $W^{k,p}(\Omega)$, respectively. The symbol $(\cdot,\cdot)$ denotes the L^2- inner product and the duality pairing between L^p and $L^{p'}$, where $1/p + 1/p' = 1$. Let

$$\dot{W}_{0,\sigma}^{1,2}(\Omega) := \overline{C_{0,\sigma}^\infty}^{\|\nabla \cdot\|_2}.$$

Let us recall the Helmholtz decomposition: $L^r(\Omega) = L_\sigma^r \oplus G_r$ $(1 < r < \infty)$, where $G_r = \{\nabla p \in L^r ; p \in L_{loc}^r(\overline{\Omega})\}$, see Fujiwara-Morimoto [**13**], Miyakawa [**28**], Simader-Sohr [**39**], Borchers-Miyakawa [**2**] and Farwig-Sohr [**11**]; P_r denotes the projection operator from L^r onto L_σ^r along G_r. The Stokes operator A_r on L_σ^r is defined by $A_r = -P_r\Delta$ with domain

$$D(A_r) = W^{2,r} \cap W_0^{1,r} \cap L_\sigma^r.$$

It is known that $(L_\sigma^r)^*$ (the dual space of L_σ^r) $= L_\sigma^{r'}$ and A_r^* (the adjoint operator of A_r) $= A_{r'}$, where $1/r + 1/r' = 1$. It is shown by Giga [**14**], Borchers-Sohr [**3**], Borchers-Miyakawa [**2**], Iwashita [**18**] and Farwig-Sohr [**11**] that $-A_r$ generates a holomorphic semigroup $\{e^{-tA_r} ; t \ge 0\}$ of class C_0 in L_σ^r. Since $P_r u = P_q u$ for all $u \in L^r \cap L^q$ $(1 < r, q < \infty)$ and since $A_r u = A_q u$ for all $u \in D(A_r) \cap D(A_q)$, for simplicity, we shall abbreviate $P_r u, P_q u$ as Pu for $u \in L^r \cap L^q$ and $A_r u, A_q u$ as Au for $u \in D(A_r) \cap D(A_q)$, respectively.

Now our main result reads as follows:

THEOREM 2.1. *Let $n = 3$, $p \in [3, \infty)$, Ω satisfy Assumption 1, $u_0 \in L_\sigma^p(\Omega) \cap \dot{W}_{0,\sigma}^{1,2}(\Omega)$ and u be a solution to (N-S) on $(0,T)$ in the class*

$$S_p(0,T) := C([0,T); L_\sigma^p) \cap C^1((0,T); L_\sigma^p) \cap C((0,T); D(A_p)).$$

Assume that $T < \infty$ and

$$\int_s^T \|u(\tau)\|_{M_{1/2}(\Omega)}^2 d\tau < \infty \quad \text{for some } s \in (0,T). \tag{2.1}$$

Then, u can be continued to the solution in the class $S_p(0,T')$ for some $T' > T$.

REMARK 2.2. (i) Solutions in the class $S_p(0,T)$ are called strong L^p solutions on $(0,T)$. For $p \ge 3$, the existence of strong L^p solutions to (N-S) was proven in [**16, 18, 19, 46**].

(ii) When $\Omega = \mathbb{R}^3$, Condition (2.1) can be replaced by

$$\int_s^T \|u(\tau)\|_{V_{1/2}(\mathbb{R}^3)}^2 d\tau < \infty$$

for some $s \in (0,T)$.

(iii) When Ω is a bounded domain, we can remove the condition $u_0 \in \dot{W}_{0,\sigma}^{1,2}$.

(iv) By using Theorem 2.1 and the standard argument, we can also show the following regularity criterion for weak solutions. If $n = 3$, u is a Leray-Hopf weak solution with the energy inequality in the strong form and if u satisfies $\int_0^T \|u(\tau)\|^2_{M_{1/2}(\Omega)} d\tau < \infty$, then u is smooth in $(0, T] \times \Omega$. Here, for the definitions of Leray-Hopf weak solutions and the energy inequality in the strong form, see e.g. [**42**, Definition 2.1].

(iv) In [**30**], we established Beale-Kato-Majda type criteria by means of

$$\int_s^T \|\text{rot } u\|_{M_1(\Omega)} d\tau \quad \text{and} \quad \int_s^T \|\text{rot } u\|_{V_1(\mathbb{R}^3)} d\tau.$$

3. Proofs of (1.1) and (1.2)

Before presenting the proof of Theorem 2.1, we give proofs of (1.1) and (1.2).

PROOF OF (1.1). We use arguments given in Engler [**7**] and Ozawa [**35**]. See also [**33**]. For the sake of simplicity, we assume $n = 3$. By Assumption 1 we see that $\partial\Omega$ satisfies the interior cone condition. Namely there are $\delta \in (0, 1)$ and $\theta \in (\pi/2, \pi)$ depending only on Ω with the following property: For any point $x \in \Omega$, there exists a spherical sector $C_\delta^\theta(x) = \{x + \xi \in \mathbb{R}^3; 0 < |\xi| < \delta,\ -|\xi| \le \kappa(x) \cdot \xi < |\xi| \cos\theta\}$ having a vertex at x such that $C_\delta^\theta(x) \subset \Omega$, where $\kappa(x)$ is an appropriate unit vector from x. We note that for each $x \in \Omega$, $C_\delta^\theta(x)$ is congruent to $C_\delta^\theta \equiv \{\xi \in \mathbb{R}^3; 0 < |\xi| < \delta,\ -|\xi| \le \xi_3 < |\xi| \cos\theta\}$. In particular, for any boundary point $x \in \partial\Omega$, $C_\delta^\theta(x)$ can be expressed as $C_\delta^\theta(x) \equiv \{x + \xi \in \mathbb{R}^3; 0 < |\xi| < \delta,\ -|\xi| \le \xi \cdot \nu(x) < |\xi| \cos\theta\}$, where $\nu(x)$ denotes the unit outward normal at x.

Let $0 < t \le \delta$ and $C_t^\theta(x) := C_\delta^\theta(x) \cap B(x, t)$. For any fixed $x \in \Omega$ and $y \in C_t^\theta(x) \subset \Omega$,

$$|f(x)| \le |f(x) - f(y)| + |f(y)| \le \|f\|_{\dot{C}^\alpha} |x - y|^\alpha + |f(y)| \le \|f\|_{\dot{C}^\alpha} t^\alpha + |f(y)|.$$

Integrating both sides of the above inequality with respect to y over $C_t^\theta(x)$,

$$\begin{aligned} |f(x)||C_t^\theta(x)| &\le t^\alpha \|f\|_{\dot{C}^\alpha(\Omega)} |C_t^\theta(x)| + \int_{y \in C_t^\theta(x)} |f(y)| dy \\ &\le t^\alpha \|f\|_{\dot{C}^\alpha(\Omega)} |C_t^\theta(x)| + \int_{y \in B(x,t) \cap \Omega} |f(y)| dy \qquad (3.1) \\ &\le t^\alpha \|f\|_{\dot{C}^\alpha(\Omega)} |C_t^\theta(x)| + |B(x,t)| \log^\beta(\frac{1}{t} + e) \|f\|_{M_\beta(\Omega)}. \end{aligned}$$

Since $|B(x,t)|/|C_t^\theta(x)| (< \infty)$ depends only on θ, we have

$$|f(x)| \le t^\alpha \|f\|_{\dot{C}^\alpha(\Omega)} + C \log^\beta(\frac{1}{t} + e) \|f\|_{M_\beta(\Omega)} \qquad (3.2)$$

for all $0 < t \le \delta$. Then we optimize t by letting $t = (1/\|f\|_{\dot{C}^\alpha(\Omega)})^{1/\alpha}$ if $\|f\|_{\dot{C}^\alpha(\Omega)} \ge \delta^{-\alpha}$ and letting $t = \delta$ if $\|f\|_{\dot{C}^\alpha(\Omega)} \le \delta^{-\alpha}$ to obtain (1.1).

□

PROOF OF (1.2). We first recall the Littlewood-Paley decomposition. Let ψ be the function given in Definition 1.1 (2) and let $\phi_j \in \mathcal{S}$ be the functions defined by

$$\hat{\phi}(\xi) := \hat{\psi}(\xi) - \hat{\psi}(2\xi) \text{ and } \hat{\phi}_j(\xi) := \hat{\phi}(\xi/2^j) \text{ for } \xi \in \mathbb{R}^3.$$

Then, supp $\hat{\phi}_j \subset \{2^{j-2} \le |\xi| \le 2^j\}$ and

(3.3)

$$1 = \hat{\psi}(\xi/2^N) + \sum_{j=N+1}^{\infty} \hat{\phi}(\xi/2^j) = \hat{\psi}_N(\xi) + \sum_{j=N+1}^{\infty} \hat{\phi}_j(\xi) \text{ for } \xi \in \mathbb{R}^3, N = 1, 2, \cdots .$$

Using (3.3), we decompose f into two parts such as

$$f(x) = \psi_N * f(x) + \sum_{j=N+1}^{\infty} \phi_j * f(x). \tag{3.4}$$

We have by Definition 1.1 (2)

$$\|\psi_N * f\|_\infty \le N^\beta \|f\|_{V_\beta}. \tag{3.5}$$

Since $\dot{B}^\alpha_{\infty,\infty} = \dot{C}^\alpha$ for $0 < \alpha < 1$, we have

$$\begin{aligned} \sum_{j=N+1}^{\infty} \|\phi_j * f\|_\infty &= \sum_{j=N+1}^{\infty} 2^{\alpha j}\|\phi_j * f\|_\infty 2^{-\alpha j} \\ &\le \|f\|_{\dot{B}^\alpha_{\infty,\infty}} \sum_{j=N+1}^{\infty} 2^{-\alpha j} \le C\|f\|_{\dot{C}^\alpha} 2^{-\alpha N}. \end{aligned} \tag{3.6}$$

Gathering (3.5) and (3.6) with (3.4), we obtain

$$\|f\|_\infty \le C(2^{-\alpha N}\|f\|_{\dot{C}^\alpha} + N^\beta \|f\|_{V_\beta}). \tag{3.7}$$

Now we take $N = \left[\frac{\log(\|f\|_{\dot{C}^\alpha} + e)}{\alpha \log 2}\right] + 1$, where $[\cdot]$ denotes Gauss symbol. Then we have the desired estimate (1.2). □

4. Proof of Theorem 2.1

PROOF OF THEOREM 2.1. Since $u \in C((0,T); D(A_6) \cap \dot{W}^{1,2}_{0,\sigma}(\Omega))$, without loss of generality, we may assume that $u_0 \in D(A_6) \cap \dot{W}^{1,2}_{0,\sigma}(\Omega)$. Since the local existence time of L^p-strong solutions T_* can be estimated from below as

$$T_* > C(\Omega)/\|u_0\|^4_{L^6},$$

it suffices to show that

$$\sup_{0<\tau<T} \|u(\tau)\|_{L^6} \le C(u_0,T) \exp\left(C(u_0,T) \exp\left(C(T) \int_0^T \|u(\tau)\|^2_{M_{1/2}} d\tau\right)\right). \tag{4.1}$$

Recall that u satisfies

$$u(t) = e^{-tA}u_0 - \int_0^t e^{-(t-s)A} P\nabla \cdot (u \otimes u)(s) ds \tag{I.E.}$$

for all $t \in (0,T)$.

From (I.E.), the duality argument and the Gronwall lemma we have

$$\sup_{0\le\tau\le t} \|u(\tau)\|_6 \le C \sup_{0\le\tau\le t} \|\nabla u(\tau)\|_2 \le C\|\nabla u_0\|_2 \exp(C \int_0^t \|u(s)\|^2_\infty ds) \tag{4.2}$$

for all $t \in [0,T)$. See Lemma 5.2 (5.1) in Appendix. Let

$$h(t) := \sup_{0\le\tau\le t} \|u(\tau)\|_6,$$
$$g(t) := \int_0^t \|u(\tau)\|_\infty^2 d\tau$$

for $0 \le t < T$. Then, we have

$$h(t) \le C\|\nabla u_0\|_2 \exp(Cg(t)) \text{ for all } t \in [0,T). \tag{4.3}$$

Letting $0 < \alpha < 1/2$ and substituting $f = \frac{u(s)}{\epsilon\|u(s)\|_{\dot{C}^\alpha}}$ into the Brezis-Gallouet-Wainger type inequality (1.1) with $\beta = 1/2$, we obtain

$$\begin{aligned} \|u(s)\|_\infty &\le C\Big(\epsilon\|u(s)\|_{\dot{C}^\alpha} + \log^{1/2}(e+\frac{1}{\epsilon})\|u(s)\|_{M_{1/2}}\Big), \text{ which means} \\ \|u(s)\|_\infty^2 &\le C\Big(\epsilon^2\|u(s)\|_{\dot{C}^\alpha}^2 + \log(e+\frac{1}{\epsilon})\|u(s)\|_{M_{1/2}}^2\Big) \qquad (4.4) \end{aligned}$$

for all $\epsilon > 0$, where C is a constant independent of s and ϵ. Then, by (4.4), for any positive function $\epsilon(s)$ on $(0,T)$, we have

$$\begin{aligned} g(t) &\le C\int_0^t \epsilon^2(s)\|u(s)\|_{\dot{C}^\alpha}^2 ds + C\int_0^t \log(e+\frac{1}{\epsilon(s)})\|u(s)\|_{M_{1/2}}^2 ds \\ &=: I_1(t) + I_2(t). \end{aligned} \tag{4.5}$$

By the Gagliardo-Nirenberg inequality

$$\|f\|_{\dot{C}^\alpha(\Omega)} \le C\|f\|_{W^{2,6}(\Omega)}^\theta \|f\|_{L^6(\Omega)}^{1-\theta},$$

where $\theta = \frac{1}{4} + \frac{\alpha}{2}$, we have

$$\begin{aligned} &\|e^{-tA}P\nabla\cdot f\|_{\dot{C}^\alpha} \\ &\le C\|(1+A)e^{-(t/2)A}e^{-(t/2)A}P\nabla\cdot f\|_{L^6}^\theta \|e^{-(t/2)A}P\nabla\cdot f\|_{L^6}^{1-\theta} \\ &\le C(T)\Big((1+(t/2)^{-1})\|e^{-(t/2)A}P\nabla\cdot f\|_{L^6}\Big)^\theta \|e^{-(t/2)A}P\nabla\cdot f\|_{L^6}^{1-\theta} \\ &\le C(T)(1+(t/2)^{-\theta})\|e^{-(t/2)A}P\nabla\cdot f\|_{L^6} \end{aligned} \tag{4.6}$$

for all $0 < t < T$ and all $f \in (L^6(\Omega))^{3\times 3}$. Since the duality argument yields $\|e^{-(t/2)A}P\nabla\cdot f\|_{L^6} \le C(t/2)^{-\frac{1}{2}}\|f\|_{L^6}$, by (4.6) we obtain

$$\|e^{-tA}P\nabla\cdot f\|_{\dot{C}^\alpha} \le C(T)(1+t^{-\frac{3}{4}-\frac{\alpha}{2}})\|f\|_6$$

for all $0 < t < T$ and all $f \in (L^6(\Omega))^{3\times 3}$. Thus, from (I.E.) we obtain

$$\begin{aligned} \|u(s)\|_{\dot{C}^\alpha} &\le C\|u_0\|_{D(A_6)} + C(T)\int_0^s (1+(s-\tau)^{-\frac{3}{4}-\frac{\alpha}{2}})\|u\otimes u(\tau)\|_6 d\tau \\ &\le C\|u_0\|_{D(A_6)} + C(T)h(s)\int_0^s (1+(s-\tau)^{-\frac{3}{4}-\frac{\alpha}{2}})\|u(\tau)\|_\infty d\tau, \end{aligned}$$

which yields

$$\|u(s)\|_{\dot{C}^\alpha}^2 \le C\|u_0\|_{D(A_6)}^2 + C(T)h^2(s)\Big(\int_0^s (1+(s-\tau)^{-\frac{3}{4}-\frac{\alpha}{2}})\|u(\tau)\|_\infty d\tau\Big)^2. \tag{4.7}$$

Hence, for $0 < t < T$ we have

$$\begin{aligned} I_1(t) \leq & C\|u_0\|^2_{D(A_6)} T \sup_{0<s<T} \epsilon^2(s) \\ & + C(T) \int_0^t h^2(s)\epsilon^2(s) \Big(\int_0^s (1 + (s-\tau)^{-\frac{3}{4}-\frac{\alpha}{2}}) \|u(\tau)\|_\infty d\tau \Big)^2 ds. \end{aligned}$$

Now we choose $\epsilon(s)$ such as

$$\epsilon(s) := \frac{\delta}{h(s)+1},$$

where $\delta = \delta(T) \in (0,1)$ is a constant to be chosen suitably small later on. Then, since $h^2(s)\epsilon^2(s) < \delta^2$ and $\epsilon^2(s) < \delta^2$, we have

$$I_1(t) \leq C\|u_0\|^2_{D(A_6)} T\delta^2 + C(T)\delta^2 \int_0^t \Big(\int_0^s (1 + (s-\tau)^{-\frac{3}{4}-\frac{\alpha}{2}}) \|u(\tau)\|_\infty d\tau \Big)^2 ds.$$

Since $\frac{3}{4} + \frac{\alpha}{2} < 1$, the Hardy-Littlewood-Sobolev inequality yields

$$\begin{aligned} \int_0^t \Big(\int_0^s (s-\tau)^{-\frac{3}{4}-\frac{\alpha}{2}} \|u(\tau)\|_\infty d\tau \Big)^2 ds & \leq \int_{\mathbb{R}} \Big(\int_{\mathbb{R}} (s-\tau)^{-\frac{3}{4}-\frac{\alpha}{2}} \|u(\tau)\|_\infty 1_{[0,t]}(\tau) d\tau \Big)^2 ds \\ & \leq C \Big\| \|u(\cdot)\|_\infty 1_{[0,t]} \Big\|^2_{L^{\frac{4}{3-2\alpha}}(\mathbb{R})} \\ & \leq C t^{\frac{1}{2}-\alpha} \Big\| \|u(\cdot)\|_\infty \Big\|^2_{L^2(0,t)}. \end{aligned}$$

Hence,

$$\begin{aligned} I_1(t) & \leq C\|u_0\|^2_{D(A_6)} T\delta^2 + C(T)\,(t^2 + t^{\frac{1}{2}-\alpha})\delta^2 \int_0^t \|u\|^2_\infty d\tau \\ & \leq C(T, \|u_0\|_{D(A_6)}) + C_1(T)\delta^2\, g(t). \end{aligned} \tag{4.8}$$

Since (4.3) yields

$$\log(e + \frac{1}{\epsilon(s)}) = \log(e + \frac{h(s)+1}{\delta}) \leq C(\delta)(1 + \|\nabla u_0\|_2 + g(s)),$$

we have

$$I_2(t) \leq C(\delta) \int_0^t \|u(s)\|^2_{M_{1/2}} (1 + \|\nabla u_0\|_2 + g(s)) ds. \tag{4.9}$$

Gathering (4.8) and (4.9) with (4.5), we obtain

$$g(t) \leq C(T, \|u_0\|_{D(A_6)}) + C_1(T)\delta^2 g(t) + C(\delta) \int_0^t \|u(s)\|^2_{M_{1/2}} (1 + \|\nabla u_0\|_2 + g(s)) ds.$$

Thus, letting $\delta^2 = 1/(2C_1(T)+1)$, by the Gronwall lemma, we have

$$g(t) \leq C(\|u_0\|_{D(A_6)}, T, \|\nabla u_0\|_2) \exp\Big(C(T) \int_0^T \|u(s)\|^2_{M_{1/2}} ds \Big)$$

for all $0 < t < T$. Then, this estimate and (4.3) yield the desired estimate (4.1) □

5. Appendix

Here, we prove (4.2).

PROPOSITION 5.1. Let $n=3$ and Ω satisfy Assumption 1.
(i) If $a \in \dot{W}^{1,2}_{0,\sigma}$, then $e^{-tA}a \in C([0,\infty);\dot{W}^{1,2}_{0,\sigma})$ and $\|\nabla e^{-tA}a\|_2 \le \|\nabla a\|_2$.
(ii) If $v \in C([0,T];\dot{W}^{1,2}_{0,\sigma})$, then

$$F(t) := \int_0^t e^{-(t-s)A}P(v\cdot\nabla v)(s)ds \in C([0,T];\dot{W}^{1,2}_{0,\sigma}).$$

PROOF. (i) We first recall $C^\infty_{0,\sigma} \subset D(A_2) \subset \dot{W}^{1,2}_{0,\sigma}$, see Sohr [**41**, Chap. III, Sect. 2.1]. Hence

$$\dot{W}^{1,2}_{0,\sigma} = \overline{D(A_2)}^{\|\nabla\cdot\|_2}.$$

By the definition of $\dot{W}^{1,2}_{0,\sigma}$, there exists $a_n \in C^\infty_{0,\sigma}(\Omega)$ such that $a_n \to a$ in $\dot{W}^{1,2}_{0,\sigma}$. Since $e^{-tA}a_n \in C([0,\infty);D(A_2))$ and since

$$\begin{aligned}\sup_{t\ge0}\|\nabla e^{-tA}a_n - \nabla e^{-tA}a_m\|_2 &= \sup_{t\ge0}\|A^{1/2}e^{-tA}a_n - A^{1/2}e^{-tA}a_m\|_2\\ &\le \|A^{1/2}(a_n-a_m)\|_2 = \|\nabla a_m - \nabla a_n\|_2 \to 0\end{aligned}$$

as $m,n\to\infty$, we see that $e^{-tA}a \in C([0,\infty);\overline{D(A_2)}^{\|\nabla\cdot\|_2}) = C([0,\infty);\dot{W}^{1,2}_{0,\sigma})$. Since $\|\nabla e^{-tA}a_n\|_2 = \|e^{-tA}A_2^{1/2}a_n\|_2 \le \|A_2^{1/2}a_n\|_2 = \|\nabla a_n\|_2$, we have $\|\nabla e^{-tA}a\|_2 \le \|\nabla a\|_2$.

(ii) Let $t \in (0,T]$ be fixed and $0<\epsilon<t$. Since $\|P(v\cdot\nabla v)(s)\|_{3/2} \le C\sup_{0\le s\le T}\|\nabla v(s)\|_2^2$, it is straightforward to see that $\int_0^{t-\epsilon} e^{-(t-s)A}P(v\cdot\nabla v)(s)ds \in D(A_2)$. Since

$$\|\nabla\int_0^{t-\epsilon} e^{-(t-s)A}P(v\cdot\nabla v)(s)ds - \nabla F(t)\|_2 \to 0$$

as $\epsilon\to0+$, we have $F(t)\in\dot{W}^{1,2}_{0,\sigma}$. By the direct calculation, we can also show the continuity of $\nabla F(t)$ in L^2 with respect to $t\in[0,T]$, which proves Proposition 5.1. □

LEMMA 5.2. *(a) Let $n=3$, $3\le p<\infty$ and Ω satisfy Assumption 1. If $v_0 \in L^p_\sigma(\Omega)\cap\dot{W}^{1,2}_{0,\sigma}(\Omega)$, then there exist $T^*>0$ and a unique solution v on $(0,T^*)$ to (N-S) with initial data $v(0)=v_0$ in the class*

$$v \in S_p(0,T^*)\cap S_6(0,T^*)\cap C([0,T^*);\dot{W}^{1,2}_{0,\sigma}(\Omega)).$$

Moreover, it holds that

$$T^* \ge C\|v_0\|_{L^6}^{-4},$$

where C is a constant depending only on Ω.

(b) Let $n=3$ and Ω satisfy Assumption 1. Assume that u is a solution to (N-S) in the class

$$u \in S_6(0,T)\cap C([0,T);\dot{W}^{1,2}_{0,\sigma}(\Omega)).$$

Then, it holds that

$$\sup_{0\le\tau\le t}\|\nabla u(\tau)\|_{L^2} \le 2\|\nabla u_0\|_{L^2}\exp 2\int_0^t \|u(s)\|_\infty^2 ds \tag{5.1}$$

for all $0<t<T$.

PROOF. (a) Since Assertion (a) is proven by the standard iteration argument and Proposition 5.1, we omit the proof of (a).

(b) Let $\phi \in C_{0,\sigma}^\infty$. Since $w(\tau) = e^{-\tau A}\phi$ is a solution to the Stokes equation on $(0,\infty)$ with the initial data $w(0) = \phi$, the energy calculation yields

$$\int_0^t \|\nabla e^{-\tau A}\phi\|_2^2 d\tau \le \|\phi\|_2^2 \quad \text{for all } t > 0,$$

which implies

$$\int_0^t \|A^{1/2}e^{-(t-s)A}\phi\|_2^2 ds \le \|\phi\|_2^2 \quad \text{for all } t > 0.$$

Since

$$(u(t),\psi) = (e^{-tA}u(0),\psi) + \int_0^t (u\cdot\nabla u, e^{-(t-s)A}\psi)ds \quad \text{for all } \psi \in L_\sigma^{6/5},$$

letting $\psi = A^{1/2}\phi$, we have

$$\begin{aligned}
&|(A^{1/2}u(t),\phi)| \\
&\le |(e^{-tA}A^{1/2}u(0),\phi)| + \int_0^t |(u\cdot\nabla u, A^{1/2}e^{-(t-s)A}\phi)|ds \\
&\le \|A^{1/2}u(0)\|_2\|\phi\|_2 + \int_0^t \|u\|_\infty\|\nabla u\|_2\|A^{1/2}e^{-(t-s)A}\phi\|_2 ds \\
&\le \|A^{1/2}u(0)\|_2\|\phi\|_2 + \Big(\int_0^t \|u\|_\infty^2\|\nabla u\|_2^2 ds\Big)^{\frac{1}{2}}\Big(\int_0^t \|A^{1/2}e^{-(t-s)A}\phi\|_2^2 ds\Big)^{\frac{1}{2}} \\
&\le \|A^{1/2}u(0)\|_2\|\phi\|_2 + \Big(\int_0^t \|u\|_\infty^2\|\nabla u\|_2^2 ds\Big)^{\frac{1}{2}}\|\phi\|_2.
\end{aligned}$$

Thus, the duality yields

$$\|\nabla u(t)\|_2 \le \|\nabla u_0\|_2 + \Big(\int_0^t \|u\|_\infty^2\|\nabla u\|_2^2 ds\Big)^{\frac{1}{2}}$$

and consequently

$$\|\nabla u(t)\|_2^2 \le 2\|\nabla u_0\|_2^2 + 2\int_0^t \|u\|_\infty^2\|\nabla u\|_2^2 ds.$$

Therefore from the Gronwall lemma we obtain the desired estimate (5.1), which clearly implies (4.2).

□

References

[1] J. T. Beale, T. Kato, and A. Majda, *Remarks on the breakdown of smooth solutions for the 3-D Euler equations*, Comm. Math. Phys. **94** (1984), no. 1, 61–66. MR763762

[2] Wolfgang Borchers and Tetsuro Miyakawa, *L^2 decay for the Navier-Stokes flow in halfspaces*, Math. Ann. **282** (1988), no. 1, 139–155. MR960838

[3] Wolfgang Borchers and Hermann Sohr, *On the semigroup of the Stokes operator for exterior domains in L^q-spaces*, Math. Z. **196** (1987), no. 3, 415–425. MR913666

[4] H. Brézis and T. Gallouet, *Nonlinear Schrödinger evolution equations*, Nonlinear Anal. **4** (1980), no. 4, 677–681. MR582536

[5] Haïm Brézis and Stephen Wainger, *A note on limiting cases of Sobolev embeddings and convolution inequalities*, Comm. Partial Differential Equations **5** (1980), no. 7, 773–789. MR579997

[6] Jean-Yves Chemin, *Perfect incompressible fluids*, Oxford Lecture Series in Mathematics and its Applications, vol. 14, The Clarendon Press, Oxford University Press, New York, 1998. Translated from the 1995 French original by Isabelle Gallagher and Dragos Iftimie. MR1688875

[7] Hans Engler, *An alternative proof of the Brezis-Wainger inequality*, Comm. Partial Differential Equations **14** (1989), no. 4, 541–544. MR989669

[8] L. Iskauriaza, G. A. Serëgin, and V. Shverak, *$L_{3,\infty}$-solutions of Navier-Stokes equations and backward uniqueness* (Russian, with Russian summary), Uspekhi Mat. Nauk **58** (2003), no. 2(350), 3–44; English transl., Russian Math. Surveys **58** (2003), no. 2, 211–250. MR1992563

[9] E. B. Fabes, B. F. Jones, and N. M. Rivière, *The initial value problem for the Navier-Stokes equations with data in L^p*, Arch. Rational Mech. Anal. **45** (1972), 222–240. MR0316915

[10] Reinhard Farwig and Hermann Sohr, *On the Stokes and Navier-Stokes system for domains with noncompact boundary in L^q-spaces*, Math. Nachr. **170** (1994), 53–77. MR1302366

[11] Reinhard Farwig and Hermann Sohr, *Helmholtz decomposition and Stokes resolvent system for aperture domains in L^q-spaces*, Analysis **16** (1996), no. 1, 1–26. MR1384351

[12] Andrew B. Ferrari, *On the blow-up of solutions of the 3-D Euler equations in a bounded domain*, Comm. Math. Phys. **155** (1993), no. 2, 277–294. MR1230029

[13] Daisuke Fujiwara and Hiroko Morimoto, *An L_r-theorem of the Helmholtz decomposition of vector fields*, J. Fac. Sci. Univ. Tokyo Sect. IA Math. **24** (1977), no. 3, 685–700. MR0492980

[14] Yoshikazu Giga, *Analyticity of the semigroup generated by the Stokes operator in L_r spaces*, Math. Z. **178** (1981), no. 3, 297–329. MR635201

[15] Yoshikazu Giga, *Solutions for semilinear parabolic equations in L^p and regularity of weak solutions of the Navier-Stokes system*, J. Differential Equations **62** (1986), no. 2, 186–212. MR833416

[16] Yoshikazu Giga and Tetsuro Miyakawa, *Solutions in L_r of the Navier-Stokes initial value problem*, Arch. Rational Mech. Anal. **89** (1985), no. 3, 267–281. MR786550

[17] S. Ibrahim, M. Majdoub, and N. Masmoudi, *Double logarithmic inequality with a sharp constant*, Proc. Amer. Math. Soc. **135** (2007), no. 1, 87–97. MR2280178

[18] Hirokazu Iwashita, *L_q-L_r estimates for solutions of the nonstationary Stokes equations in an exterior domain and the Navier-Stokes initial value problems in L_q spaces*, Math. Ann. **285** (1989), no. 2, 265–288. MR1016094

[19] Tosio Kato, *Strong L^p-solutions of the Navier-Stokes equation in $\mathbf{R}^m$, with applications to weak solutions*, Math. Z. **187** (1984), no. 4, 471–480. MR760047

[20] Tosio Kato and Gustavo Ponce, *Commutator estimates and the Euler and Navier-Stokes equations*, Comm. Pure Appl. Math. **41** (1988), no. 7, 891–907. MR951744

[21] Hideo Kozono, Takayoshi Ogawa, and Yasushi Taniuchi, *The critical Sobolev inequalities in Besov spaces and regularity criterion to some semi-linear evolution equations*, Math. Z. **242** (2002), no. 2, 251–278. MR1980623

[22] Hideo Kozono and Yukihiro Shimada, *Bilinear estimates in homogeneous Triebel-Lizorkin spaces and the Navier-Stokes equations*, Math. Nachr. **276** (2004), 63–74. MR2100048

[23] Hideo Kozono and Yasushi Taniuchi, *Limiting case of the Sobolev inequality in BMO, with application to the Euler equations*, Comm. Math. Phys. **214** (2000), no. 1, 191–200. MR1794270

[24] Hideo Kozono and Hidemitsu Wadade, *Remarks on Gagliardo-Nirenberg type inequality with critical Sobolev space and BMO*, Math. Z. **259** (2008), no. 4, 935–950. MR2403750

[25] Takayuki Kubo and Yoshihiro Shibata, *On some properties of solutions to the Stokes equation in the half-space and perturbed half-space*, Dispersive nonlinear problems in mathematical physics, Quad. Mat., vol. 15, Dept. Math., Seconda Univ. Napoli, Caserta, 2004, pp. 149–220. MR2231330

[26] O. A. Ladyženskaja, *Uniqueness and smoothness of generalized solutions of Navier-Stokes equations* (Russian), Zap. Naučn. Sem. Leningrad. Otdel. Mat. Inst. Steklov. (LOMI) **5** (1967), 169–185. MR0236541

[27] Toshitaka Matsumoto and Takayoshi Ogawa, *Interpolation inequality of logarithmic type in abstract Besov spaces and an application to semilinear evolution equations*, Math. Nachr. **283** (2010), no. 12, 1810–1828. MR2560670

[28] Tetsuro Miyakawa, *On nonstationary solutions of the Navier-Stokes equations in an exterior domain*, Hiroshima Math. J. **12** (1982), no. 1, 115–140. MR647234

[29] Kei Morii, Tokushi Sato, and Hidemitsu Wadade, *Brézis-Gallouët-Wainger type inequality with a double logarithmic term in the Hölder space: its sharp constants and extremal functions*, Nonlinear Anal. **73** (2010), no. 6, 1747–1766. MR2661357

[30] Nakao, K., Taniuchi, Y., *Brezis-Gallouet-Wainger type inequalities and blow-up criteria for Navier-Stokes equations in unbounded domains*, Preprint.

[31] Jiří Neustupa, *Partial regularity of weak solutions to the Navier-Stokes equations in the class* $L^\infty(0,T;L^3(\Omega)^3)$, J. Math. Fluid Mech. **1** (1999), no. 4, 309–325. MR1738173

[32] Takayoshi Ogawa, *Sharp Sobolev inequality of logarithmic type and the limiting regularity condition to the harmonic heat flow*, SIAM J. Math. Anal. **34** (2003), no. 6, 1318–1330. MR2000973

[33] Takayoshi Ogawa and Yasushi Taniuchi, *On blow-up criteria of smooth solutions to the 3-D Euler equations in a bounded domain*, J. Differential Equations **190** (2003), no. 1, 39–63. MR1970955

[34] Takayoshi Ogawa and Yasushi Taniuchi, *The limiting uniqueness criterion by vorticity for Navier-Stokes equations in Besov spaces*, Tohoku Math. J. (2) **56** (2004), no. 1, 65–77. MR2028918

[35] T. Ozawa, *On critical cases of Sobolev's inequalities*, J. Funct. Anal. **127** (1995), no. 2, 259–269. MR1317718

[36] Yoshihiro Sawano and Hidemitsu Wadade, *On the Gagliardo-Nirenberg type inequality in the critical Sobolev-Morrey space*, J. Fourier Anal. Appl. **19** (2013), no. 1, 20–47. MR3019769

[37] James Serrin, *On the interior regularity of weak solutions of the Navier-Stokes equations*, Arch. Rational Mech. Anal. **9** (1962), 187–195. MR0136885

[38] Taira Shirota and Taku Yanagisawa, *A continuation principle for the 3-D Euler equations for incompressible fluids in a bounded domain*, Proc. Japan Acad. Ser. A Math. Sci. **69** (1993), no. 3, 77–82. MR1222827

[39] Christian G. Simader and Hermann Sohr, *A new approach to the Helmholtz decomposition and the Neumann problem in* L^q*-spaces for bounded and exterior domains*, Mathematical problems relating to the Navier-Stokes equation, Ser. Adv. Math. Appl. Sci., vol. 11, World Sci. Publ., River Edge, NJ, 1992, pp. 1–35. MR1190728

[40] Hermann Sohr, *Zur Regularitätstheorie der instationären Gleichungen von Navier-Stokes* (German), Math. Z. **184** (1983), no. 3, 359–375. MR716283

[41] Hermann Sohr, *The Navier-Stokes equations*, Birkhäuser Advanced Texts: Basler Lehrbücher. [Birkhäuser Advanced Texts: Basel Textbooks], Birkhäuser Verlag, Basel, 2001. An elementary functional analytic approach. MR1928881

[42] Hermann Sohr, *A regularity class for the Navier-Stokes equations in Lorentz spaces*, J. Evol. Equ. **1** (2001), no. 4, 441–467. Dedicated to the memory of Tosio Kato. MR1877269

[43] Michael Struwe, *On partial regularity results for the Navier-Stokes equations*, Comm. Pure Appl. Math. **41** (1988), no. 4, 437–458. MR933230

[44] Shuji Takahashi, *On interior regularity criteria for weak solutions of the Navier-Stokes equations*, Manuscripta Math. **69** (1990), no. 3, 237–254. MR1078355

[45] Misha Vishik, *Incompressible flows of an ideal fluid with vorticity in borderline spaces of Besov type* (English, with English and French summaries), Ann. Sci. École Norm. Sup. (4) **32** (1999), no. 6, 769–812. MR1717576

[46] Fred B. Weissler, *The Navier-Stokes initial value problem in* L^p, Arch. Rational Mech. Anal. **74** (1980), no. 3, 219–230. MR591222

[47] Wojciech M. Zajączkowski, *Remarks on the breakdown of smooth solutions for the 3-d Euler equations in a bounded domain* (English, with Russian summary), Bull. Polish Acad. Sci. Math. **37** (1989), no. 1-6, 169–181 (1990). MR1101466

Interdisciplinary Graduate School of Science and Technology, Shinshu University, Matsumoto 390-8621, Japan

Department of Mathematical Sciences, Shinshu University, Matsumoto 390-8621, Japan

Email address: taniuchi@math.shinshu-u.ac.jp

Contemporary Mathematics
Volume **710**, 2018
http://dx.doi.org/10.1090/conm/710/14373

On steady solutions to a model of chemically reacting heat conducting compressible mixture with slip boundary conditions

Tomasz Piasecki and Milan Pokorný

Abstract. We consider a model of chemically reacting heat conducting compressible mixture. We investigate the corresponding system of partial differential equations in the steady regime with slip boundary conditions for the velocity and, in dependence on the model parameters, we establish existence of either weak or variational entropy solutions. The results extend the range of parameters for which the existence of weak solutions is known in the case of homogeneous Dirichlet boundary conditions for the velocity.

1. Introduction

The aim of this note is to extend results on a model of chemically reacting heat conducting compressible gaseous mixture based on the model considered e.g. in [**3**]. Results dealing with steady solutions for this model appeared recently in [**4**] and [**13**] for the case of Dirichlet boundary conditions; see also [**15**] which can be considered actually as a first result in this direction (dealing, however, with a slightly simplified model).

The common feature of these three papers is the fact that the weak solutions (and also variational entropy solutions in [**13**]) were obtained for any relatively rough data, without any assumption on their size or on the distance to a known (possibly regular) solution.

This paper is devoted to the proof of existence of weak and variational entropy solutions to the model introduced in [**4**]. Due to the slip boundary conditions for the velocity we are able to extend the range of parameters for which the weak solutions exist. This corresponds to the fact which has been observed several times for the compressible Navier–Stokes or the Navier–Stokes–Fourier system; the slip boundary conditions allow for better density (and sometimes also velocity) estimates which leads to stronger results in this case, see e.g. [**7**], [**14**], [**5**], [**8**], [**6**] or also [**10**].

2010 *Mathematics Subject Classification.* Primary 76N10 Secondary 35Q30 .

Key words and phrases. Steady compressible Navier–Stokes–Fourier system; weak solution; variational entropy solution; multicomponent diffusion flux; Fick's law; entropy inequality.

The work of the first author (TP) was supported by the Polish NCN grant No. UMO-2014/14/M/ST1/00108.

The work of the second author (MP) was supported by the Czech Science Foundation (grant No. 16-03230S).

In what follows, we use standard notation for the Lebesgue, Sobolev and other standard function spaces as well as for norms in these spaces. The scalar valued functions will be denoted by the standard font (e.g., ϱ and ϑ for the density and temperature, respectively), the vector valued functions will be printed in bold face (e.g., $\mathbf{u}$ for the velocity) and the tensor valued functions using a special font (e.g., $\mathbb{S}$ for the viscous part of the stress tensor). The generic constants are denoted by C and their value may change from line to line or even in the same formula.

1.1. The model. We consider the following system of partial differential equations

$$\begin{gathered}\operatorname{div}(\varrho\mathbf{u}) = 0,\\ \operatorname{div}(\varrho\mathbf{u}\otimes\mathbf{u}) - \operatorname{div}\mathbb{S} + \nabla\pi = \varrho\mathbf{f},\\ \operatorname{div}(\varrho E\mathbf{u}) + \operatorname{div}(\pi\mathbf{u}) + \operatorname{div}\mathbf{Q} - \operatorname{div}(\mathbb{S}\mathbf{u}) = \varrho\mathbf{f}\cdot\mathbf{u},\\ \operatorname{div}(\varrho Y_k\mathbf{u}) + \operatorname{div}\mathbf{F}_k = m_k\omega_k, \quad k\in\{1,\dots,n\},\end{gathered}\tag{1.1}$$

where the unknown quantities are the total density ϱ, the velocity field $\mathbf{u}$, the temperature ϑ (appearing in (1.1) implicitly, see below) and the mass fractions $Y_k = \varrho_k/\varrho$, where $\{\varrho_k\}_{k=1}^n$ are the densities of the constituents. As $\sum_{k=1}^n \varrho_k = \varrho$, we have $\sum_{k=1}^n Y_k = 1$. The other functions, i.e. the stress tensor $\mathbb{S}$, the pressure π, the total energy E, the heat flux $\mathbf{Q}$, the diffusion fluxes $\mathbf{F}_k$ and the molar production rates ω_k are given functions of these unknows and will be introduced below. Furthermore, $\mathbf{f}$ is the given field of the external forces (e.g., the gravity force) and m_k denotes the molar masses of the kth constituent, $k = 1, 2, \dots, n$.

System (1.1) is completed by the boundary conditions on $\partial\Omega$

$$\begin{aligned}\mathbf{F}_k\cdot\mathbf{n} &= 0,\\ -\mathbf{Q}\cdot\mathbf{n} + L(\vartheta-\vartheta_0) &= 0,\end{aligned}\tag{1.2}$$

and for the velocity we assume the Navier boundary condition (the slip b.c.)

$$\mathbf{u}\cdot\mathbf{n} = 0, \qquad (\mathbb{S}\mathbf{n} + f\mathbf{u})\times\mathbf{n} = \mathbf{0}.\tag{1.3}$$

Above, the boundary condition for the temperature means that the heat flux is proportional to the difference of the temperature inside and outside the boundary. The coefficient f (assumed to be constant in what follows) denotes the friction. We also prescribe the total mass

$$\int_\Omega \varrho\,\mathrm{d}x = M > 0.\tag{1.4}$$

1.1.1. *The stress tensor and the pressure.* We assume the stress tensor $\mathbb{S}$ to be a given linear function of the symmetric part of the velocity gradient

$$\mathbb{S} = \mathbb{S}(\vartheta, \widetilde{\mathbb{D}}(\mathbf{u})) = \mu\Big[\nabla\mathbf{u} + (\nabla\mathbf{u})^T - \frac{2}{3}\operatorname{div}\mathbf{u}\mathbb{I}\Big] + \nu(\operatorname{div}\mathbf{u})\mathbb{I},\tag{1.5}$$

where $\widetilde{\mathbb{D}}(\mathbf{u}) = \frac{1}{2}(\nabla\mathbf{u} + (\nabla\mathbf{u})^T)$, the coefficients $\mu = \mu(\vartheta) > 0$ (Lipschitz continuous in $\mathbb{R}^+$) $\nu = \nu(\vartheta) \geq 0$ (continuous in $\mathbb{R}^+$), are the shear and bulk viscosity coefficients, respectively. We assume

$$\underline{\mu}(1+\vartheta) \leq \mu(\vartheta) \leq \overline{\mu}(1+\vartheta), \quad 0 \leq \nu(\vartheta) \leq \overline{\nu}(1+\vartheta)\tag{1.6}$$

for positive constants $\underline{\mu}, \overline{\mu}, \overline{\nu}$. Furthermore, $\mathbb{I}$ is the identity matrix.

The pressure

$$\pi = \pi(\varrho,\vartheta) = \pi_c(\varrho) + \pi_m(\varrho,\vartheta),\tag{1.7}$$

where the cold pressure is assumed in the form

$$\pi_c = \varrho^\gamma, \quad \gamma > 1. \tag{1.8}$$

A more general pressure form can be assumed, as in the case of the steady compressible Navier–Stokes–Fourier system, see e.g. [**11**] or [**2**] in the context of the chemically reacting flows. We, however, prefer to keep its form as simple as possible. The molecular pressure π_m, according to the Boyle law, satisfies

$$\pi_m = \pi_m(\varrho, \vartheta) = \sum_{k=1}^n p_k(\varrho, \vartheta) = \sum_{k=1}^n \frac{\varrho Y_k}{m_k}\vartheta, \tag{1.9}$$

where, for simplicity, the gas constant is taken to be equal to one.

1.1.2. *The energy and the heat flux.* The specific total energy E is a sum of the specific kinetic and specific internal energies (we denote $\vec{Y} = (Y_1, Y_2, \dots, Y_n)$)

$$E = E(\varrho, \mathbf{u}, \vartheta, \vec{Y}) = \frac{1}{2}|\mathbf{u}|^2 + e(\varrho, \vartheta, \vec{Y}). \tag{1.10}$$

Due to the form of the pressure the internal energy consists of two components

$$e = e_c(\varrho) + e_m(\vartheta, \vec{Y}), \tag{1.11}$$

where the cold energy e_c and the molecular internal energy e_m are given by

$$e_c = \frac{1}{\gamma - 1}\varrho^{\gamma-1}, \qquad e_m = \sum_{k=1}^n Y_k e_k = \vartheta \sum_{k=1}^n c_{vk} Y_k. \tag{1.12}$$

Above, c_{vk} are the constant-volume specific heats and can be different for different species. They are related to the constant-pressure specific heats by

$$c_{pk} = c_{vk} + \frac{1}{m_k} \tag{1.13}$$

and both c_{vk} and c_{pk} are assumed to be constant.

The heat flux $\mathbf{Q}$ consists of two terms

$$\mathbf{Q} = \sum_{k=1}^n h_k \mathbf{F}_k + \mathbf{q}, \tag{1.14}$$

where the first term represents transfer of energy due to the species molecular diffusion (and h_k, defined below, are the enthalpies) and the second one the Fourier law

$$\mathbf{q} = -\kappa \nabla \vartheta. \tag{1.15}$$

The coefficient $\kappa = \kappa(\vartheta)$ is the thermal conductivity coefficient and we assume

$$\underline{\kappa}(1 + \vartheta^m) \le \kappa(\vartheta) \le \overline{\kappa}(1 + \vartheta^m) \tag{1.16}$$

for some constants $m, \underline{\kappa}, \overline{\kappa} > 0$.

1.1.3. *Diffusion flux and species production rates.* The form of the diffusion flux is the most important part modeling the interaction between the species. Following [**3**] we assume that

$$\mathbf{F}_k = -\sum_{l=1}^{n} C_{kl}\mathbf{d}_l, \quad k = 1, 2, \dots, n, \tag{1.17}$$

where

$$\mathbf{d}_k = \nabla\Big(\frac{p_k}{\pi_m}\Big) + \Big(\frac{p_k}{\pi_m} - \frac{\varrho_k}{\varrho}\Big)\nabla \log \pi_m = \frac{\nabla p_k}{\pi_m} - Y_k\frac{\nabla \pi_m}{\pi_m}. \tag{1.18}$$

Furthermore, we introduce another matrix $\mathbb{D}$

$$C_{kl} = Y_k D_{kl},$$

where the diffusion matrix $\mathbb{D} = \mathbb{D}(\vartheta, \vec{Y})$ has the following properties

$$\begin{aligned} &\mathbb{D} = \mathbb{D}^T, \quad N(\mathbb{D}) = \mathbb{R}\vec{Y}, \quad R(\mathbb{D}) = \vec{U}^{\perp}, \\ &\mathbb{D} \quad \text{is positive semidefinite over } \mathbb{R}^n, \end{aligned} \tag{1.19}$$

with $\vec{Y} = (Y_1, \dots, Y_n)^T > 0$ and $\vec{U} = (1, \dots, 1)^T$. Above $N(\mathbb{D})$ denotes the nullspace of matrix $\mathbb{D}$, $R(\mathbb{D})$ denotes its range, and $\vec{U}^{\perp}$ denotes the orthogonal complement of $\vec{U}$. Moreover, the matrix $\mathbb{D}$ is positively definite over $\vec{U}^{\perp}$ and there exists $\delta > 0$ such that

$$\delta\langle \mathbb{Y}^{-1}\vec{x}, \vec{x}\rangle \le \langle \mathbb{D}\vec{x}, \vec{x}\rangle \quad \forall \vec{x} \in \vec{U}^{\perp}, \tag{1.20}$$

where $\mathbb{Y} = \operatorname{diag}(Y_1, \dots, Y_n)$ and $\langle\cdot,\cdot\rangle$ denotes the scalar product in $\mathbb{R}^n$.

Furthermore, D_{ij} are differentiable functions of $\vartheta, Y_1, \dots, Y_n$ for any $i, j \in \{1, \dots, n\}$ such that

$$|Y_i D_{ij}(\vartheta, \vec{Y})| \le C(\vec{Y})(1 + \vartheta^a) \tag{1.21}$$

for some $a \ge 0$, and $C(\vec{Y})$ is bounded in $[0,1]^n$. Finally,

$$\sum_{k=1}^{n} \mathbf{F}_k = \mathbf{0}. \tag{1.22}$$

Note that we can also consider the Fick law in the form

$$\mathbf{F}_k = D(\vartheta, \vec{Y})\nabla Y_k, \quad k = 1, 2, \dots, n, \tag{1.23}$$

and the function $D(\cdot,\cdot)$ is a differentiable function fulfilling a similar estimate as (1.21), i.e.

$$0 < D_0 \le D(\vartheta, \vec{Y}) \le C(\vec{Y})(1 + \vartheta^a) \tag{1.24}$$

with $a \ge 0$ and $C(\cdot)$ bounded in $[0,1]^n$. Condition (1.22) is indeed fulfilled provided $\sum_{k=1}^n Y_k = 1$.

Concerning the species production rates, we assume that $\{\omega_k\}_{k=1}^n$ are differentiable functions of $\varrho, \vartheta, \vec{Y}$ which are bounded, and such that

$$\omega_k \ge -CY_k^r \quad \text{for some} \quad C, r > 0, \tag{1.25}$$

which means that a species cannot decrease faster than proportionally to some positive power of its fraction (a possible natural choice is $r = 1$). Moreover, this condition clearly implies the compatibility condition $\omega_k \geq 0$ if $Y_k = 0$. Furthermore,

$$\sum_{k=1}^{n} m_k \omega_k = 0. \tag{1.26}$$

Note that e.g. in [**2**], instead of $m_k\omega_k$ it is assumed that the species production rate is modeled as $\varrho m_k \omega_k$. We may easily treat here this version and in a sense (see comments to the entropy inequality below) it is in fact simpler.

1.1.4. *Entropy and other thermodynamic potentials; entropy production rate.* Since our main thermodynamic quantities are the internal energy and the pressure, the other thermodynamic potentials are assumed to be given functions of them. In what follows, we assume the thermodynamics connected with a mixture of ideal gases with addition of the cold pressure term and the corresponding term in the internal energy.

The specific enthalpy (of each constituent) has the form

$$h_k = c_{pk}\vartheta, \qquad h = \sum_{k=1}^{n} Y_k h_k, \tag{1.27}$$

where c_{pk} fulfills (1.13). The specific entropy

$$s_k = c_{vk} \log \vartheta - \frac{1}{m_k} \log\left(\frac{\varrho Y_k}{m_k}\right), \qquad s = \sum_{k=1}^{n} Y_k s_k, \tag{1.28}$$

and the Gibbs function (Gibbs free energy)

$$g_k = h_k - \vartheta s_k, \qquad g = \sum_{k=1}^{n} Y_k g_k. \tag{1.29}$$

Moreover, the Gibbs formula has the form

$$\vartheta \mathbf{D} s = \mathbf{D} e + \pi \mathbf{D}\left(\frac{1}{\varrho}\right) - \sum_{k=1}^{n} g_k \mathbf{D} Y_k. \tag{1.30}$$

Using (1.30) it is possible to derive an equation for the specific entropy s

$$\operatorname{div}(\varrho s \mathbf{u}) + \operatorname{div}\left(\frac{\mathbf{Q}}{\vartheta} - \sum_{k=1}^{n} \frac{g_k}{\vartheta} \mathbf{F}_k\right) = \sigma, \tag{1.31}$$

where the entropy production rate

$$\sigma = \frac{\mathbb{S} : \nabla \mathbf{u}}{\vartheta} - \frac{\mathbf{Q} \cdot \nabla \vartheta}{\vartheta^2} - \sum_{k=1}^{n} \mathbf{F}_k \cdot \nabla\left(\frac{g_k}{\vartheta}\right) - \frac{\sum_{k=1}^{n} m_k g_k \omega_k}{\vartheta}. \tag{1.32}$$

Note that the entropy production rate can be expressed in the form

$$\sigma = \frac{\mathbb{S} : \nabla \mathbf{u}}{\vartheta} + \frac{\kappa |\nabla \vartheta|^2}{\vartheta^2} - \sum_{k=1}^{n} \frac{\mathbf{F}_k}{m_k} \cdot \nabla \log p_k - \frac{\sum_{k=1}^{n} m_k g_k \omega_k}{\vartheta}. \tag{1.33}$$

Then we easily see that the first two terms are non-negative due to the form of the stress tensor and the positivity of κ. Moreover, we assume that

$$\sum_{k=1}^{n} m_k g_k \omega_k \leq 0, \tag{1.34}$$

which implies that also the fourth term is non-negative. Finally,

$$
\begin{aligned}
-\sum_{k=1}^{n} \frac{\mathbf{F}_k}{m_k} \cdot \nabla \log p_k &= \sum_{k,l=1}^{n} \frac{Y_k D_{kl}}{m_k} \Big[\nabla\Big(\frac{p_l}{\pi_m}\Big) + \Big(\frac{p_l}{\pi_m} - \frac{\varrho_l}{\varrho}\Big)\nabla \log \pi_m\Big] \nabla \log p_k \\
&= \sum_{k,l=1}^{n} \frac{Y_k D_{kl}}{m_k} \Big[\frac{\nabla p_l}{\pi_m} - \frac{\varrho_l}{\varrho}\frac{\nabla \pi_m}{\pi_m}\Big] \frac{\nabla p_k}{p_k} \\
&= \frac{\pi_m}{\varrho\vartheta} \sum_{k,l=1}^{n} D_{kl} \Big[\frac{\nabla p_l}{\pi_m} - Y_l \frac{\nabla \pi_m}{\pi_m}\Big]\Big[\frac{\nabla p_k}{\pi_m} - Y_k \frac{\nabla \pi_m}{\pi_m}\Big] \\
+\frac{\pi_m}{\varrho\vartheta} \sum_{k,l=1}^{n} D_{kl} \Big[\frac{\nabla p_l}{\pi_m} - Y_l \frac{\nabla \pi_m}{\pi_m}\Big] Y_k \frac{\nabla \pi_m}{\pi_m} \geq 0
\end{aligned}
$$

due to the properties of the matrix $\mathbb{D}$, as $\sum_{k=1}^n \mathbf{F}_k = \mathbf{0}$ implies

$$
\sum_{k=1}^{n} Y_k D_{kl} = 0 \quad \forall l = 1, 2, \ldots, n. \tag{1.35}
$$

Notice in particular that Fick law (1.23) is not a special case of (1.17)–(1.19).

1.2. Formulation of the problem. We now formulate the problem we treat in this paper.

1.2.1. *Non-diagonal diffusion matrix.* Unfortunately, the general non-diagonal form of the diffusion flux is too complex to be considered in the full generality. In particular, the above deduced lower bound of the corresponding term in the entropy production rate does not allow us to control the gradient of the mass fractions. A certain attempt has been done in the evolutionary case, see [**9**], however, it leads to the necessity to control the (total) density gradient, which can be obtained for the fluids with density dependent viscosities satisfying the Bresch–Desjardins identity. The same idea does not work in the steady problem and therefore we must restrict ourselves (cf. [**4**] or [**13**]) to the case when all molar masses are comparable. We therefore assume that $m_1 = m_2 = \cdots = m_n$ and without loss of generality we set this common value to be equal to one.

Then $\pi_m = \sum_{k=1}^n \varrho_k \vartheta = \varrho\vartheta$, $\frac{\nabla p_l}{\pi_m} - Y_l \frac{\nabla \pi_m}{\pi_m} = \nabla Y_l + Y_l \frac{\nabla(\varrho\vartheta)}{\varrho\vartheta} - Y_l \frac{\nabla(\varrho\vartheta)}{\varrho\vartheta} = \nabla Y_l$. Therefore

$$
\mathbf{F}_k = -\sum_{l=1}^{n} Y_k D_{kl} \nabla Y_l
$$

and

$$
-\sum_{k=1}^{n} \frac{\mathbf{F}_k}{m_k} \cdot \nabla \log p_k = -\sum_{k=1}^{n} \mathbf{F}_k \cdot \Big(\frac{\nabla Y_k}{Y_k} + \frac{\nabla(\varrho\vartheta)}{\varrho\vartheta}\Big) = \sum_{k=1}^{n} D_{kl} \nabla Y_l \nabla Y_k \geq c|\nabla \vec{Y}|^2,
$$

provided $\vec{Y} \geq 0$ and $\sum_{k=1}^n Y_k = 1$. Exactly this estimate allows to obtain the existence of a solution in this case. We may therefore consider

Problem P (Non-diagonal diffusion matrix)

We consider system (1.1) with boundary conditions (1.2), (1.3), given total mass (1.4), and (1.5)–(1.22), (1.25)–(1.32) with equal molar masses $m_1 = m_2 = \cdots = m_n = 1$.

1.2.2. *Fick's law.* A similar estimate as above we get also in the case of the Fick law (1.23). We have for the molar masses being the same (and for notational simplicity, equal to 1)

$$-\sum_{k=1}^{n} \frac{\mathbf{F}_k}{m_k} \cdot \nabla \log p_k = -\sum_{k=1}^{n} \mathbf{F}_k \cdot \nabla \log p_k$$
$$= \sum_{k=1}^{n} \frac{D(\vartheta, \vec{Y})}{Y_k} \nabla Y_k \cdot \nabla Y_k + \sum_{k=1}^{n} D(\vartheta, \vec{Y}) \nabla Y_k \cdot \frac{\nabla(\varrho\vartheta)}{\varrho\vartheta} \geq D_0 |\nabla \vec{Y}|^2.$$

Note that in the case of the same molar masses the Fick law behaves exactly in the same way as Problem P for the same molar masses and therefore we do not consider it separately.

2. Definitions of solutions. Existence results

Problem P with the Dirichlet boundary conditions for the velocity has been studied in [**4**] and [**13**], we therefore present in analogy to this case the definitions of weak and variational entropy solutions, in the spirit of paper [**6**]. We introduce

$$C^1_{\mathbf{n}}(\Omega) = \{\mathbf{w} \in C^1(\overline{\Omega}); \mathbf{w} \cdot \mathbf{n} = 0 \text{ on } \partial\Omega\}.$$

DEFINITION 2.1. We say the set of functions $(\varrho, \mathbf{u}, \vartheta, \vec{Y})$ is a weak solution to system (1.1) with boundary conditions (1.2), (1.3), given total mass (1.4), and (1.5)–(1.22), (1.25)–(1.32) with equal molar masses $m_1 = m_2 = \cdots = m_n = 1$ provided

- $\varrho \geq 0$ a.e. in Ω, $\varrho \in L^{6\gamma/5}(\Omega)$, $\int_\Omega \varrho \, \mathrm{d}x = M$
- $\mathbf{u} \in W^{1,2}(\Omega)$, $\mathbf{u} \cdot \mathbf{n} = 0$ a.e. on $\partial\Omega$, $\varrho|\mathbf{u}|$ and $\varrho|\mathbf{u}|^2 \in L^{\frac{6}{5}}(\Omega)$
- $\vartheta \in W^{1,2}(\Omega) \cap L^{3m}(\Omega)$, $\varrho\vartheta, \varrho\vartheta|\mathbf{u}|, \mathbb{S}\mathbf{u}, \kappa|\nabla\vartheta| \in L^1(\Omega)$
- $\vec{Y} \in W^{1,2}(\Omega)$, $Y_k \geq 0$ a.e. in Ω, $\sum_{k=1}^n Y_k = 1$ a.e. in Ω, $\mathbf{F}_k \cdot \mathbf{n}|_{\partial\Omega} = 0$

and the following integral equalities hold
• the weak formulation of the continuity equation

$$\int_\Omega \varrho \mathbf{u} \cdot \nabla \psi \, \mathrm{d}x = 0 \tag{2.1}$$

holds for any test function $\psi \in C^1(\overline{\Omega})$;
• the weak formulation of the momentum equation

$$-\int_\Omega \big(\varrho\,(\mathbf{u} \otimes \mathbf{u}) : \nabla\boldsymbol{\varphi} - \mathbb{S} : \nabla\boldsymbol{\varphi}\big) \, \mathrm{d}x + f \int_{\partial\Omega} \mathbf{u} \cdot \boldsymbol{\varphi} \, \mathrm{d}S - \int_\Omega \pi \mathrm{div}\, \boldsymbol{\varphi} \, \mathrm{d}x = \int_\Omega \varrho \mathbf{f} \cdot \boldsymbol{\varphi} \, \mathrm{d}x \tag{2.2}$$

holds for any test function $\boldsymbol{\varphi} \in C^1_{\mathbf{n}}(\Omega)$;
• the weak formulation of the species equations

$$-\int_\Omega Y_k \varrho \mathbf{u} \cdot \nabla \psi \, \mathrm{d}x - \int_\Omega \mathbf{F}_k \cdot \nabla \psi \, \mathrm{d}x = \int_\Omega \omega_k \psi \, \mathrm{d}x \tag{2.3}$$

holds for any test function $\psi \in C^1(\overline{\Omega})$ and for all $k = 1, \dots, n$;
• the weak formulation of the total energy balance
(2.4)

$$-\int_\Omega \Big(\frac{1}{2}\varrho|\mathbf{u}|^2 + \varrho e\Big)\mathbf{u}\cdot\nabla\psi\,\mathrm{d}x + \int_\Omega \kappa\nabla\vartheta\cdot\nabla\psi\,\mathrm{d}x - \int_\Omega \Big(\sum_{k=1}^n h_k\mathbf{F}_k\Big)\cdot\nabla\psi\,\mathrm{d}x$$
$$= \int_\Omega \varrho\mathbf{f}\cdot\mathbf{u}\psi\,\mathrm{d}x - \int_\Omega (\mathbb{S}\mathbf{u})\cdot\nabla\psi\,\mathrm{d}x + \int_\Omega \pi\mathbf{u}\cdot\nabla\psi\,\mathrm{d}x$$
$$- \int_{\partial\Omega} L(\vartheta-\vartheta_0)\psi\,\mathrm{d}S - f\int_{\partial\Omega}|\mathbf{u}|^2\psi\,\mathrm{d}S$$

holds for any test function $\psi \in C^1(\overline{\Omega})$.

Indeed, the total energy balance which contains the term behaving as $\varrho|\mathbf{u}|^3$ limits the range for γ and m for which we are able to prove existence of a weak solution. Following a similar situation for the compressible Navier–Stokes–Fourier system (both steady and evolutionary, see [**1**] or [**10**]) we introduce another type of solution, where the total energy balance is replaced by the entropy inequality.

DEFINITION 2.2. We say the set of functions $(\varrho, \mathbf{u}, \vartheta, \vec{Y})$ is a variational entropy solution to problem (1.1) with boundary conditions (1.2), (1.3), given total mass (1.4), and (1.5)–(1.22), (1.25)–(1.32) with equal molar masses $m_1 = m_2 = \cdots = m_n = 1$ provided

- $\varrho \geq 0$ a.e. in Ω, $\varrho \in L^{s\gamma}(\Omega)$ for some $s > 1$, $\int_\Omega \varrho\,\mathrm{d}x = M$
- $\mathbf{u} \in W^{1,2}(\Omega)$, $\mathbf{u}\cdot\mathbf{n} = 0$ a.e. on $\partial\Omega$, $\varrho\mathbf{u} \in L^{\frac{6}{5}}(\Omega)$
- $\vartheta \in W^{1,r}(\Omega) \cap L^{3m}(\Omega)$, $r > 1$, $\varrho\vartheta, \mathbb{S} : \frac{\nabla\mathbf{u}}{\vartheta}, \kappa\frac{|\nabla\vartheta|^2}{\vartheta^2}, \kappa\frac{\nabla\vartheta}{\vartheta} \in L^1(\Omega)$, $\frac{1}{\vartheta} \in L^1(\partial\Omega)$
- $\vec{Y} \in W^{1,2}(\Omega)$, $Y_k \geq 0$ a.e. in Ω, $\sum_{k=1}^n Y_k = 1$ a.e. in Ω, $\mathbf{F}_k\cdot\mathbf{n}|_{\partial\Omega} = 0$

satisfy equations (2.1–2.3), the following entropy inequality

$$\int_\Omega \frac{\mathbb{S}:\nabla\mathbf{u}}{\vartheta}\psi\,\mathrm{d}x + \int \kappa\frac{|\nabla\vartheta|^2}{\vartheta^2}\psi\,\mathrm{d}x - \int_\Omega \sum_{k=1}^n \omega_k(c_{pk} - c_{vk}\log\vartheta + \log Y_k)\psi\,\mathrm{d}x \tag{2.5}$$
$$+ \int_\Omega \psi\sum_{k,l=1}^n D_{kl}\nabla Y_k\cdot\nabla Y_l\,\mathrm{d}x + \int_{\partial\Omega}\frac{L}{\vartheta}\vartheta_0\psi\,\mathrm{d}S \leq \int \frac{\kappa\nabla\vartheta\cdot\nabla\psi}{\vartheta}\,\mathrm{d}x - \int_\Omega \varrho s\mathbf{u}\cdot\nabla\psi\,\mathrm{d}x$$
$$- \int_\Omega \log\vartheta\Big(\sum_{k=1}^n \mathbf{F}_k c_{vk}\Big)\cdot\nabla\psi\,\mathrm{d}x + \int_\Omega \Big(\sum_{k=1}^n \mathbf{F}_k\log Y_k\Big)\cdot\nabla\psi\,\mathrm{d}x + \int_{\partial\Omega} L\psi\,\mathrm{d}S$$

for all non-negative $\psi \in C^1(\overline{\Omega})$ and the global total energy balance (i.e. (2.4) with $\psi \equiv 1$)

$$f\int_{\partial\Omega}|\mathbf{u}|^2\,\mathrm{d}S + \int_{\partial\Omega} L(\vartheta-\vartheta_0)\,\mathrm{d}S = \int_\Omega \varrho\mathbf{f}\cdot\mathbf{u}\,\mathrm{d}x. \tag{2.6}$$

Formally, the entropy inequality (2.5) is nothing but a weak formulation of the entropy inequality (1.31). However, some modifications are required. First of all, we are not able to keep equality, but due to the technique used to prove existence of such solutions we face the problem that in several terms we are not able to pass to the limit directly and we have to apply the weak lower semicontinuity here. Note further that (2.5) does not contain all terms from (1.31), some of them are missing.

These terms are formally equal to zero due to assumptions that ω_k and $\mathbf{F}_k$ sum up to zero. We removed them from the formulation of the entropy inequality due to the fact that we cannot exclude the situation that $\varrho = 0$ in some large portions of Ω (with positive Lebesgue measure), thus $\log \varrho$ is not well defined there. However, the variational entropy solution still has the property that any sufficiently smooth variational entropy solution in the sense above is a classical solution to our problem, provided the density is strictly positive in Ω. Replacing the form of the source terms in the species balance equations by $\varrho\omega_k$ we even do not face this problem.

We are now in position to formulate our main result.

THEOREM 2.3. *Let* $\gamma > 1$, $M > 0$, $m > \max\{\frac{2}{3}, \frac{2}{3(\gamma-1)}\}$, $a < \frac{3m}{2}$, $\vartheta_0 \in L^1(\partial\Omega)$, $\vartheta_0 \geq K_0 > 0$ *a.e. on* $\partial\Omega$. *Let* $\Omega \in C^2$ *be not axially symmetric. Then there exists at least one variational entropy solution to Problem P in the sense of Definition 2.2. Moreover,* $(\varrho, \mathbf{u})$ *is the renormalized solution to the continuity equation.*

In addition, if $m > 1$, $\gamma > \frac{5}{4}$, $a < \frac{3m-2}{2}$, *then the solution is a weak solution in the sense of Definition 2.1.*

If Ω *is axially symmetric, let* $f > 0$. *Then there exists at least one variational entropy solution to Problem P. In addition, if* $\gamma > \frac{5}{4}$, $m > 1$, $m > \frac{16\gamma}{15\gamma-16}$ *(if* $\gamma \in (\frac{5}{4}, \frac{4}{3}]$*) or* $m > \frac{18-6\gamma}{9\gamma-7}$ *(if* $\gamma \in (\frac{4}{3}, \frac{5}{3})$*) then the solution is a weak solution.*

REMARK 2.4. Recall that the pair $(\varrho, \mathbf{u})$ is a renormalized solution to the continuity equation provided $\mathbf{u} \in W^{1,2}(\Omega)$, $\varrho \in L^{\frac{6}{5}}(\Omega)$ and for any $b \in C^1(0,\infty) \cap C([0,\infty))$, $b'(z) = 0$ for $z \geq M$ for some $M > 0$

$$\int_\Omega \Big(b(\varrho)\operatorname{div}\psi + (b(\varrho) - b'(\varrho)\varrho)\operatorname{div}\mathbf{u}\psi\Big)\,\mathrm{d}x = 0$$

for all $\psi \in C^1(\overline{\Omega})$.

3. Proof of the existence results

As explained above, it is enough to prove Theorem 2.3 for the case of the generally nondiagonal diffusion matrix. Indeed, for the Fick law, the proof could be simplified due to the special structure of the diffusion flux, but we prefer not to deal with the modification and indicate only one place which is slightly different. The result is exactly the same as in Theorem 2.3.

PROOF. (of Theorem 2.3). First, we define for positive parameters $\delta > \varepsilon > \lambda > \eta > 0$ the following approximations of different quantities appearing in the formulation of Problem I. We start with

$$\mathbf{J}_k = -\sum_{l=1}^n Y_k Y_l \widehat{D}_{kl}(\vartheta, \vec{Y})\nabla Y_l / Y_l - \big(\varepsilon(\varrho+1)Y_k + \lambda\big)\nabla Y_k / Y_k, \tag{3.1}$$

with

$$\widehat{D}_{kl}(\vartheta, \vec{Y}) = \frac{1}{(\sigma_Y + \varepsilon)^r} D_{kl}(\vartheta, \vec{Y}) \tag{3.2}$$

for suitably chosen $r \geq 0$, where $\sigma_Y = \sum_{k=1}^n Y_k$. The reason for this notation is that, unless we let $\lambda \to 0^+$, it is not clear whether $\sigma_Y = 1$. We will only know that $Y_k \geq 0$. For the case of the Fick law this regularization can be simplified. However, in order to keep the unified approach, we only slightly modify this step. Instead of (3.1) we set

$$\mathbf{J}_k = -\sum_{l=1}^{n} \widehat{D}(\vartheta, \vec{Y})\nabla Y_k - \big(\varepsilon(\varrho+1)Y_k + \lambda\big)\nabla Y_k/Y_k, \tag{3.3}$$

where $\widehat{D}$ is defined similarly as $\widehat{\mathbb{D}}$ in (3.2).

Furthermore, we introduce a regularization of the stress tensor

$$\mathbb{S}_\eta = \frac{\mu_\eta(\vartheta)}{1+\eta\vartheta}\left[\nabla\mathbf{u} + (\nabla\mathbf{u})^T - \frac{2}{3}\operatorname{div}\mathbf{u}\,\mathbb{I}\right] + \frac{\nu_\eta(\vartheta)}{1+\eta\vartheta}(\operatorname{div}\mathbf{u})\,\mathbb{I}, \tag{3.4}$$

where μ_η, ν_η are standard mollifications of the viscosity functions. Next,

$$\kappa_{\delta,\eta} = \kappa^\eta + \delta\vartheta^B + \delta\vartheta^{-1} \tag{3.5}$$

is a regularization of heat conductivity coefficient, where the exponent $B > 0$ sufficiently large will be determined later, and κ^η is the standard mollification of the heat conductivity.

We take the following approximation of the specific entropy

$$s_k^\lambda = c_{vk}\log\vartheta - \log Y_k - \log(\varrho + \sqrt{\lambda}), \tag{3.6}$$

and, similarly

$$g_k^\lambda = c_{pk}\vartheta - \vartheta s_k^\lambda, \quad s^\lambda = \sum_{k=1}^{n} Y_k s_k^\lambda. \tag{3.7}$$

In what follows, we present only the main steps of the existence proof, pointing always out the specific paper, where more details can be found.

Step I: Formulation of the approximate problem. We consider additionally one more parameter, $N \in \mathbf{N}$ denoting the dimension for the Galerkin approximation of the velocity. Let $\{\mathbf{w}_n\}_{n=1}^\infty$ be an orthogonal basis of $W^{1,2}(\Omega)$ such that $\mathbf{w}\cdot\mathbf{n} = 0$ on $\partial\Omega$ such that $\mathbf{w}_i \in W^{2,q}(\Omega)$ for $q < \infty$ (we can take for example eigenfunctions of the Lamé system with slip boundary conditions). We look for $(\varrho_{N,\eta,\lambda,\varepsilon,\delta}, \mathbf{u}_{N,\eta,\lambda,\varepsilon,\delta}, \vec{Y}_{N,\eta,\lambda,\varepsilon,\delta}, \vartheta_{N,\eta,\lambda,\varepsilon,\delta})$ (from now on we skip the indices) such that

• the approximate continuity equation

$$\begin{aligned} \varepsilon\varrho + \operatorname{div}(\varrho\mathbf{u}) &= \varepsilon\Delta\varrho + \varepsilon\overline{\varrho}, \\ \nabla\varrho\cdot\mathbf{n}|_{\partial\Omega} &= 0, \end{aligned} \tag{3.8}$$

where $\bar{\varrho} = \frac{M}{|\Omega|}$, is satisfied pointwisely

• the Galerkin approximation for the momentum equation (note that the convective term reduces to the standard form provided $\operatorname{div}(\varrho\mathbf{u}) = 0$, even in the weak sense)

$$\begin{aligned} &\int_\Omega \left(\frac{1}{2}\varrho\mathbf{u}\cdot\nabla\mathbf{u}\cdot\mathbf{w} - \frac{1}{2}\varrho\,(\mathbf{u}\otimes\mathbf{u}) : \nabla\mathbf{w} + \mathbb{S}_\eta : \nabla\mathbf{w}\right)\,\mathrm{d}x \\ &\qquad + f\int_{\partial\Omega}\mathbf{u}\cdot\mathbf{w}\,\mathrm{d}S - \int_\Omega(\pi + \delta\varrho^\beta + \delta\varrho^2)\operatorname{div}\mathbf{w}\,\mathrm{d}x = \int_\Omega \varrho\mathbf{f}\cdot\mathbf{w}\,\mathrm{d}x \end{aligned} \tag{3.9}$$

is satisfied for each test function $\mathbf{w} \in X_N$, where $\mathbf{u} \in X_N$, $X_N = \operatorname{span}\{\mathbf{w}_i\}_{i=1}^N$, and $\beta > 0$ is large enough

• the approximate species mass balance equations

$$
\begin{aligned}
\operatorname{div} \mathbf{J}_k &= \omega_k + \varepsilon\overline{\varrho}_k - \varepsilon Y_k\varrho - \operatorname{div}(Y_k\varrho\mathbf{u}) + \varepsilon\operatorname{div}(Y_k\nabla\varrho) - \sqrt{\lambda}\log Y_k, \\
\mathbf{J}_k \cdot \mathbf{n}|_{\partial\Omega} &= 0
\end{aligned}
\tag{3.10}
$$

are satisfied pointwisely, where $\sum_{k=1}^n \bar{\varrho}_k = \bar{\varrho}$, for example we take $\bar{\varrho}_k = \frac{\bar{\varrho}}{n}$
• the approximate internal energy balance

$$
\begin{aligned}
-\operatorname{div}\left(\kappa_{\delta,\eta}\frac{\varepsilon+\vartheta}{\vartheta}\nabla\vartheta\right) &= -\operatorname{div}(\varrho e\mathbf{u}) - \pi\operatorname{div}\mathbf{u} + \frac{\delta}{\vartheta} + \mathbb{S}_\eta : \nabla\mathbf{u} \\
&\quad + \delta\varepsilon(\beta\varrho^{\beta-2}+2)|\nabla\varrho|^2 - \operatorname{div}\left(\vartheta\sum_{k=1}^n c_{vk}\mathbf{J}_k\right)
\end{aligned}
\tag{3.11}
$$

with the boundary condition

$$
\kappa_{\delta,\eta}\frac{\varepsilon+\vartheta}{\vartheta}\nabla\vartheta\cdot\mathbf{n}|_{\partial\Omega} + (L+\delta\vartheta^{B-1})(\vartheta-\vartheta_0^\eta) + \varepsilon\log\vartheta + \lambda\vartheta^{\frac{B}{2}}\log\vartheta = 0
\tag{3.12}
$$

is satisfied pointwisely, where ϑ_0^η is a smooth, strictly positive approximation of ϑ_0 and $\kappa_{\delta,\eta}$ is as above.

Next, we write down the entropy equality for the approximate system. Note that it is not an additional assumption, but a consequence of the approximate relations above and it is possible to deduce its form (see [**13**] for more details in the case of the Dirichlet boundary conditions) under the regularity assumptions which correspond to the regularity of solutions to the approximate problem stated above.
Step II: Solvability of the approximate system. Following [**4**] and [**13**] we can prove

PROPOSITION 3.1. *Let δ, ε, λ and η be positive numbers and N be a positive integer. Under the assumptions of Theorem 2.3 there exists a solution to system (3.8–3.12) such that $\varrho \in W^{2,q}(\Omega)$ $\forall q < \infty$, $\varrho \geq 0$ in Ω, $\int_\Omega \varrho\,\mathrm{d}x = M$, $\mathbf{u} \in X_N$, $\vec{Y} \in W^{1,2}(\Omega)$ with $\log Y_k \in W^{2,q}(\Omega)$ $\forall q < \infty$, $Y_k > 0$ a.e. in Ω and $\vartheta \in W^{2,q}(\Omega)$*

$\forall q < \infty$, $\vartheta \geq C(N) > 0$. *Moreover, this solution satisfies the entropy equation*

$$
\begin{aligned}
(3.13)\quad & \int_\Omega \frac{\psi \mathbb{S}_\eta : \nabla \mathbf{u}}{\vartheta}\,\mathrm{d}x + \int_\Omega \kappa_{\delta,\eta} \frac{(\varepsilon+\vartheta)}{\vartheta}\frac{|\nabla\vartheta|^2}{\vartheta^2}\psi\,\mathrm{d}x \\
& - \int_\Omega \omega_k(c_{pk} - c_{vk}\log\vartheta + \log Y_k)\psi\,\mathrm{d}x + \int_\Omega \frac{\delta\psi}{\vartheta^2}\,\mathrm{d}x \\
& - \int_\Omega \psi \sum_{k=1}^n \widehat{\mathbf{F}}_k \cdot \nabla \log Y_k\,\mathrm{d}x + \int_{\partial\Omega} \frac{\psi}{\vartheta}(L+\delta\vartheta^{B-1})\vartheta_0^\eta\,\mathrm{d}S \\
& + \int_\Omega \frac{\delta\varepsilon(\beta\varrho^{\beta-2}+2)|\nabla\varrho|^2\psi}{\vartheta}\,\mathrm{d}x + \int_\Omega \psi\sum_{k=1}^n (\varepsilon(\varrho+1)Y_k+\lambda)\Big|\frac{\nabla Y_k}{Y_k}\Big|^2\,\mathrm{d}x \\
& = \int_\Omega \frac{\kappa_{\delta,\eta}(\varepsilon+\vartheta)\nabla\vartheta\cdot\nabla\psi}{\vartheta^2}\,\mathrm{d}x - \int_\Omega \varrho s^\lambda \mathbf{u}\cdot\nabla\psi\,\mathrm{d}x \\
& - \int_\Omega \sum_{k=1}^n (c_{vk}\log\vartheta - \log Y_k)\widehat{\mathbf{F}}_k\cdot\nabla\psi\,\mathrm{d}x - \varepsilon\int_\Omega \psi\sum_{k=1}^n Y_k c_{pk}(\Delta\varrho + \bar{\varrho} - \varrho)\,\mathrm{d}x \\
& + \int_\Omega \psi\varrho\mathbf{u}\cdot\Big(\Big(\sum_{k=1}^n Y_k\Big)\nabla\log(\varrho+\sqrt{\lambda}) - \nabla\log\varrho\Big)\,\mathrm{d}x \\
& + \int_{\partial\Omega} \frac{\psi}{\vartheta}\big((L+\delta\vartheta^{B-1})\vartheta + \varepsilon\log\vartheta + \lambda\vartheta^{B/2}\log\vartheta\big)\,\mathrm{d}S \\
& - \varepsilon\int_\Omega \sum_{k=1}^n Y_k\nabla\varrho\cdot\nabla\Big(\frac{g_k^\lambda\psi}{\vartheta}\Big)\,\mathrm{d}x - \sqrt{\lambda}\int_\Omega \Big(\sum_{k=1}^n g_k^\lambda \log Y_k\Big)\frac{\psi}{\vartheta}\,\mathrm{d}x \\
& + \int_\Omega \varepsilon(\Delta\varrho+\bar{\varrho}-\varrho)(\varrho^{\gamma-1}+e+\theta)\frac{\psi}{\vartheta}\,\mathrm{d}x \\
& + \varepsilon\int_\Omega \sum_{k=1}^n (\bar{\varrho}_k - Y_k\varrho)\frac{g_k^\lambda\psi}{\vartheta}\,\mathrm{d}x - \int_\Omega \sum_{k=1}^n (\varepsilon(\varrho+1)Y_k+\lambda)\frac{\nabla Y_k}{Y_k}\cdot\nabla\psi\,\mathrm{d}x \\
& + \int_\Omega \sum_{k=1}^n (\varepsilon(\varrho+1)Y_k+\lambda)s_k^\lambda\frac{\nabla Y_k}{Y_k}\cdot\nabla\psi\,\mathrm{d}x \\
& - \int_\Omega \psi\sum_{k=1}^n (\varepsilon(\varrho+1)Y_k+\lambda)\frac{\nabla Y_k}{Y_k}\cdot\nabla\log(\varrho+\sqrt{\lambda})\,\mathrm{d}x
\end{aligned}
$$

and the following estimate

$$
\begin{aligned}
(3.14)\quad & \\
& \sqrt{\lambda}\sum_{k=1}^n\Big(\|Y_k\|_{1,2} + \Big\|\frac{\nabla Y_k}{Y_k}\Big\|_2 + \lambda^{-1/4}\|\log Y_k\|_2\Big) + \sum_{k=1}^n\Big\|\frac{|\nabla Y_k|^2}{Y_k}\Big\|_1 + \|\nabla\vartheta^{B/2}\|_2 \\
& + \Big\|\frac{\nabla\vartheta}{\vartheta^2}\Big\|_2 + \Big\|\frac{\nabla\varrho}{\sqrt{\varrho+\sqrt{\lambda}}}\Big\|_2 + \|\vartheta^{-2}\|_1 + \|\vartheta\|_{B,\partial\Omega} + \Big\|\frac{\log\vartheta}{\vartheta}\Big\|_{1,\partial\Omega} + \|\nabla^2\varrho\|_2 \\
& + \|\mathbf{u}\|_{1,2} + \|\nabla\varrho\|_6 \leq C,
\end{aligned}
$$

where C *is independent of* N, η *and* λ.

Note that if Ω is axially symmetric (and f is thus positive), the estimate of the $\|\mathbf{u}\|_2$ (or, more precisely, of $\|\mathbf{u}\|_{2,\partial\Omega}$) must be deduced from the momentum equation.

We now let subsequently $N \to +\infty$, $\eta \to 0^+$, $\lambda \to 0^+$, $\varepsilon \to 0^+$, and $\delta \to 0^+$.

Step III: Limit passage $N \to +\infty$. Using the bounds from Proposition 3.1, weak lower semicontinuity of several terms in the entropy inequality and the fact that for both Galerkin approximation and the limit momentum balance we can use the corresponding velocity as test function (i.e., we have energy equality in both cases), we may let $N \to +\infty$ in the system (3.8)–(3.13) above. Note that instead of entropy equality we get entropy inequality.

Step IV: Limit passage $\eta \to 0^+$. As we cannot ensure strong convergence of the quadratic term on the rhs of the approximate internal energy balance, before starting with the limit passage we must replace it by the approximate total energy balance, i.e. we add the kinetic energy balance to the limit version of (3.11). We get

$$
\begin{aligned}
(3.15)\qquad & -\int_\Omega \Big[\varrho e + \frac{1}{2}\varrho|\mathbf{u}|^2 + (\pi + \delta\varrho^\beta + \delta\varrho^2)\Big]\mathbf{u}\cdot\nabla\psi\,\mathrm{d}x \\
& \quad - \int_\Omega \Big(\mathbb{S}_\eta\mathbf{u}\cdot\nabla\psi + \delta\vartheta^{-1}\psi\Big)\,\mathrm{d}x + \int_\Omega \kappa_{\delta,\eta}\frac{\varepsilon+\vartheta}{\vartheta}\nabla\vartheta\cdot\nabla\psi\,\mathrm{d}x \\
& + \int_{\partial\Omega} \big[(L + \delta\vartheta^{B-1})(\vartheta - \vartheta_0^\eta) + \varepsilon\log\vartheta + \lambda\vartheta^{\frac{B}{2}}\log\vartheta\big]\psi\,\mathrm{d}S + f\int_{\partial\Omega}|\mathbf{u}|^2\psi\,\mathrm{d}S \\
& + \sum_{k=1}^n c_{vk}\int_\Omega \Big[\vartheta\sum_{l=1}^n Y_k\widehat{D}_{kl}\nabla Y_l\cdot\nabla\psi + \vartheta(\varepsilon(\varrho+1)Y_k + \lambda)\frac{\nabla Y_k}{Y_k}\cdot\nabla\psi\Big]\,\mathrm{d}x \\
& = \int_\Omega \varrho\mathbf{f}\cdot\mathbf{u}\psi\,\mathrm{d}x + \frac{\delta}{\beta-1}\int_\Omega (\varepsilon\beta\overline{\varrho}\varrho^{\beta-1}\psi + \varrho^\beta\mathbf{u}\cdot\nabla\psi - \varepsilon\beta\varrho^\beta\psi)\,\mathrm{d}x \\
& \qquad\qquad + \delta\int_\Omega (2\varepsilon\overline{\varrho}\varrho\psi + \varrho^2\mathbf{u}\cdot\nabla\psi - 2\varepsilon\varrho^2\psi)\,\mathrm{d}x.
\end{aligned}
$$

Recalling that the bounds in (3.14) are independent of η, it is not difficult to see that we may now let $\eta \to 0^+$ and pass to the limit in our system of equations. Now, if Ω is axially symmetric, we read the estimate of $\|\mathbf{u}\|_{2,\partial\Omega}$ from the total energy balance with ψ constant. The same holds also for all subsequent limit passages.

Step V: Limit passage $\lambda \to 0^+$. Recall that at this moment it is not yet true that $\varrho = \sum_{k=1}^n \varrho_k$. However, we have at least (see (6.12) in [**4**])

$$
(3.16)\qquad \|\sum_{k=1}^n \nabla Y_k\|_2 + \|(\sum_{k=1}^n Y_k) - 1\|_6 \le C(\lambda) \sim \sqrt{\lambda} \to 0 \quad \text{for} \quad \lambda \to 0.
$$

This bound, together with (3.14), implies

$$
(3.17)\qquad \sum_{k=1}^n \|\nabla Y_k\|_{\frac{12}{7}} \le C
$$

with C independent of λ. We may therefore let $\lambda \to 0^+$ and pass to the limit in our problem (for the details see [**13**]). Due to (3.16) we see that after the limit passage we have

$$
\sum_{k=1}^n Y_k = 1, \qquad \text{i.e.} \ \sum_{k=1}^n \varrho_k = \varrho.
$$

Step VI: Limit passage $\varepsilon \to 0^+$. The last two limit passages are nowadays standard in the theory of compressible Navier–Stokes–Fourier system. First, to let $\varepsilon \to 0^+$, we need additional estimates of the total density. We may use the technique of Bogovskii type estimates (see e.g. [**12**] for more details) to get for $\beta \gg 1$

$$\|\varrho\|_{\frac{5}{3}\beta} \leq C(\delta). \tag{3.18}$$

This can be achieved by testing the approximate momentum equation on the level $\varepsilon > 0$ by solution to

$$\begin{aligned} \operatorname{div} \boldsymbol{\varphi} &= \varrho^{\frac{2}{3}\beta} - \frac{1}{|\Omega|}\int_\Omega \varrho^{\frac{2}{3}\beta}\,\mathrm{d}x, \\ \boldsymbol{\varphi} &= \mathbf{0} \qquad \text{on } \partial\Omega. \end{aligned}$$

Indeed, this estimate does not imply the compactness of the density sequence and further work must be done: we have to combine the *effective viscous flux identity* and the *renormalized continuity equation*, see [**9**] or [**12**] for more details. After letting $\varepsilon \to 0^+$ we have

- the continuity equation

$$\int_\Omega \varrho\mathbf{u}\cdot\nabla\psi = 0 \tag{3.19}$$

for all $\psi \in C^1(\overline{\Omega})$

- the weak formulation of the approximate momentum equation

$$\begin{aligned} &\int_\Omega \Big(- \varrho\,(\mathbf{u}\otimes\mathbf{u}) : \nabla\boldsymbol{\varphi} - \mathbb{S} : \nabla\boldsymbol{\varphi}\Big)\,\mathrm{d}x + f\int_{\partial\Omega} \mathbf{u}\cdot\boldsymbol{\varphi}\,\mathrm{d}S \\ &\qquad - \int_\Omega (\pi + \delta\varrho^\beta + \delta\varrho^2)\operatorname{div}\boldsymbol{\varphi}\,\mathrm{d}x = \int_\Omega \varrho\mathbf{f}\cdot\boldsymbol{\varphi}\,\mathrm{d}x \end{aligned} \tag{3.20}$$

for all $\boldsymbol{\varphi} \in C^1_{\mathbf{n}}(\Omega)$

- the weak formulation of the approximate species balance equations

$$\int_\Omega \Big(- Y_k\varrho\mathbf{u}\cdot\nabla\psi + \sum_{l=1}^n Y_k D_{kl}\nabla Y_l\cdot\nabla\psi\Big)\,\mathrm{d}x = \int_\Omega \omega_k\psi\,\mathrm{d}x \tag{3.21}$$

for all $\psi \in C^1(\overline{\Omega})$ $(k = 1, 2, \ldots, n)$

- the weak formulation of the approximate total energy equation

$$\begin{aligned} &-\int_\Omega \Big[\varrho e + \frac{1}{2}\varrho|\mathbf{u}|^2 + (\pi + \delta\varrho^\beta + \delta\varrho^2)\Big]\mathbf{u}\cdot\nabla\psi\,\mathrm{d}x \\ &- \int_\Omega \Big(\mathbb{S}\mathbf{u}\cdot\nabla\psi + \delta\vartheta^{-1}\psi\Big)\,\mathrm{d}x + \int_\Omega \kappa_\delta\nabla\vartheta\cdot\nabla\psi\,\mathrm{d}x + f\int_{\partial\Omega} |\mathbf{u}|^2\psi\,\mathrm{d}S \\ &+ \int_{\partial\Omega} \big[(L + \delta\vartheta^{B-1})(\vartheta - \vartheta_0)\big]\psi\,\mathrm{d}S + \int_\Omega \vartheta \sum_{k,l=1}^n c_{vk}Y_k D_{kl}\nabla Y_l\cdot\nabla\psi\,\mathrm{d}x \\ &= \int_\Omega \varrho\mathbf{f}\cdot\mathbf{u}\psi\,\mathrm{d}x + \frac{\delta}{\beta-1}\int_\Omega \varrho^\beta\mathbf{u}\cdot\nabla\psi\,\mathrm{d}x + \delta\int_\Omega \varrho^2\mathbf{u}\cdot\nabla\psi\,\mathrm{d}x \end{aligned} \tag{3.22}$$

for all $\psi \in C^1(\overline{\Omega})$

- the weak formulation of the entropy inequality

(3.23)
$$\int_\Omega \frac{\psi \mathbb{S} : \nabla \mathbf{u}}{\vartheta}\,\mathrm{d}x + \int \kappa_\delta \frac{|\nabla\vartheta|^2}{\vartheta^2}\psi\,\mathrm{d}x - \int_\Omega \sum_{k=1}^n \omega_k (c_{pk} - c_{vk}\log\vartheta + \log Y_k)\psi\,\mathrm{d}x$$
$$+ \int_\Omega \frac{\delta\psi}{\vartheta^2}\,\mathrm{d}x + \int_\Omega \psi \sum_{k=1}^n\sum_{l=1}^n D_{kl}\nabla Y_l \nabla Y_k\,\mathrm{d}x + \int_{\partial\Omega} \frac{\psi}{\vartheta}(L + \delta\vartheta^{B-1})\vartheta_0\,\mathrm{d}S$$
$$\le \int \frac{\kappa_\delta \nabla\vartheta\cdot\nabla\psi}{\vartheta}\,\mathrm{d}x - \int_\Omega \varrho s \mathbf{u}\cdot\nabla\psi\,\mathrm{d}x$$
$$- \int_\Omega \sum_{k=1}^n (c_{vk}\log\vartheta - \log Y_k)\mathbf{F}_k\cdot\nabla\psi\,\mathrm{d}x + \int_{\partial\Omega}(L + \delta\vartheta^{B-1})\psi\,\mathrm{d}S$$

for all $\psi \in C^1(\overline{\Omega})$, nonnegative. More details can be found in [**13**].

Step VII: Estimates independent of δ. We denote the solution corresponding to $\delta > 0$ as $(\varrho_\delta, \mathbf{u}_\delta, \vartheta_\delta, \vec{Y}_\delta)$. The entropy inequality and the total energy balance, both with a constant test function, yield the following estimates for Ω not axially symmetric

$$\|\vartheta_\delta\|_{1,\partial\Omega} + \delta\|\vartheta_\delta^B\|_{1,\partial\Omega} \le C\Big(1 + \Big|\int_\Omega \varrho_\delta \mathbf{u}_\delta\cdot\mathbf{f}\,\mathrm{d}x\Big| + \delta\|\vartheta_\delta^{-1}\|_1\Big). \tag{3.24}$$

(3.25)
$$\|\nabla\vec{Y}_\delta\|_2^2 + \|\nabla\vartheta_\delta^{\frac{m}{2}}\|_2^2 + \|\mathbf{u}_\delta\|_{1,2}^2 + \|\vartheta_\delta^{-1}\|_{1,\partial\Omega}$$
$$+ \delta\big(\|\nabla\vartheta_\delta^{\frac{B}{2}}\|_2^2 + \|\nabla\vartheta_\delta^{-\frac{1}{2}}\|_2^2 + \|\vartheta_\delta^{-2}\|_1 + \|\vartheta_\delta^{B-2}\|_{1,\partial\Omega}\big) \le C(1 + \delta\|\vartheta_\delta^{B-1}\|_{1,\partial\Omega}).$$

If Ω is axially symmetric (and $f > 0$), then we must add to the left-hand side of (3.24) $\|\mathbf{u}_\delta\|_{2,\partial\Omega}^2$ and replace in the left-hand side of (3.25) $\|\mathbf{u}_\delta\|_{1,2}^2$ by

$$\int_\Omega \frac{\mathbb{S}(\vartheta_\delta, \widetilde{\mathbb{D}}(\mathbf{u}_\delta)) : \nabla\mathbf{u}_\delta}{\vartheta_\delta}\,\mathrm{d}x.$$

Recall also that we know $0 \le (Y_k)_\delta \le 1$, $k = 1, 2, \ldots, n$. It is not difficult to bound the δ-dependent terms on the right-hand sides to get for Ω not axially symmetric

$$\begin{aligned} &\|\nabla\vec{Y}_\delta\|_2 + \|\vec{Y}_\delta\|_\infty + \|\nabla\vartheta_\delta^{\frac{m}{2}}\|_2 + \|\mathbf{u}_\delta\|_{1,2} + \|\vartheta_\delta^{-1}\|_{1,\partial\Omega} \\ &+ \delta(\|\nabla\vartheta_\delta^{\frac{B}{2}}\|_2^2 + \|\nabla\vartheta_\delta^{-\frac{1}{2}}\|_2^2 + \|\vartheta_\delta^{-2}\|_1 + \|\vartheta_\delta^{B-2}\|_{1,\partial\Omega}) \le C \end{aligned} \tag{3.26}$$

and

$$\|\vartheta_\delta\|_{3m} \le C\Big(1 + \Big|\int_\Omega \varrho_\delta\mathbf{u}_\delta\cdot\mathbf{f}\,\mathrm{d}x\Big|\Big), \tag{3.27}$$

while for Ω axially symmetric we remove from the left-hand side of (3.26) $\|\mathbf{u}_\delta\|_{1,2}$ and add to the left-hand side of (3.27) $\|\mathbf{u}_\delta\|_{1,2}^2$. The main issue are now the density estimates. We aim at obtaining

$$\sup_{x_0\in\overline{\Omega}} \int_\Omega \frac{\pi(\varrho_\delta, \vartheta_\delta) + (1-\alpha)\varrho_\delta|\mathbf{u}_\delta|^2}{|x - x_0|^\alpha}\,\mathrm{d}x \le C, \tag{3.28}$$

for some $\alpha > 0$ as large as possible. We distinguish three cases. If x_0 is sufficiently far from $\partial\Omega$, then it is possible to use as test function in (3.20)

$$\varphi(x) = \frac{x - x_0}{|x - x_0|^\alpha}\tau^2, \tag{3.29}$$

where τ is a suitable cut-off function. Next, we study the remaining two cases, i.e. $x_0 \in \partial\Omega$ and $x_0 \in \Omega$, but close to $\partial\Omega$. To simplify the idea, let us assume that we deal with the part of boundary of Ω which is flat and is described by $x_3 = 0$, i.e. $z(x') = 0$, $x' \in \mathcal{O} \subset \mathbb{R}^2$ with the normal vector $\mathbf{n} = (0, 0, -1)$ and $\boldsymbol{\tau}_1 = (1, 0, 0)$, $\boldsymbol{\tau}_2 = (0, 1, 0)$ the tangent vectors. The general case can be studied using the standard technique of flattening the boundary, see e.g. [**6**]. Consider first that x_0 lies on the boundary of Ω, i.e. $(x_0)_3 = 0$. Then it is possible to use as the test function in the approximate momentum equation

$$\mathbf{w}(x) = \mathbf{v}(x - x_0),$$

where

$$\mathbf{v}(x) = \frac{1}{|x|^\alpha}(x_1, x_2, x_3) = (x \cdot \boldsymbol{\tau}_1)\boldsymbol{\tau}_1 + (x \cdot \boldsymbol{\tau}_2)\boldsymbol{\tau}_2 + ((0, 0, x_3 - z(x')) \cdot \mathbf{n})\mathbf{n}, \quad x_3 \geq 0.$$

Note that if $(x_0)_3 = 0$ we get precisely what we need, i.e. estimate (3.28) (but with $\sup_{x_0 \in \partial\Omega}$ instead of $\sup_{x_0 \in \overline{\Omega}}$.

However, if x_0 is close to the boundary but not on the boundary, i.e. $(x_0)_3 > 0$, but small, we lose control of some terms for $0 < x_3 < (x_0)_3$. In this case, as for the Dirichlet boundary conditions, we must modify the test functions. We first consider

$$\mathbf{v}^1(x) = \begin{cases} \frac{1}{|x-x_0|^\alpha}\big((x - x_0)_1, (x - x_0)_2, (x - x_0)_3\big), & x_3 \geq \frac{(x_0)_3}{2}, \\ \frac{1}{|x-x_0|^\alpha}\Big((x - x_0)_1, (x - x_0)_2, 4(x - x_0)_3 \frac{x_3^2}{|(x-x_0)_3|^2}\Big), & 0 < x_3 < \frac{(x_0)_3}{2}. \end{cases}$$

Nonetheless, using $\mathbf{v}^1$ as test function we would still miss control of some terms from the convective term, more precisely of those, which contain at least one velocity component u_3, however, only close to the boundary, i.e. for $x_3 < (x_0)_3/2$. Hence we further consider

$$\mathbf{v}^2(x) = \begin{cases} \dfrac{(0, 0, x_3)}{(x_3 + |x - x_0||\ln|x - x_0||^{-1})^\alpha}, & |x - x_0| \leq 1/K, \\ \dfrac{(0, 0, x_3)}{(x_3 + 1/K|\ln K|^{-1})^\alpha}, & |x - x_0| > 1/K \end{cases}$$

for K sufficiently large (but fixed, independently of the distance of x_0 from $\partial\Omega$). Note that both functions have zero normal trace, belong to $W^{1,q}(\Omega; \mathbb{R}^3)$ and their norms are bounded uniformly (with respect to the distance of x_0 from $\partial\Omega$) provided $1 \leq q < \frac{3}{\alpha}$. Thus we finally use as the test function in the approximate momentum balance

$$\varphi = \mathbf{v}^1(x) + K_1\mathbf{v}^2(x) \tag{3.30}$$

with K_1 suitably chosen (large). Note that the choice of K and K_1 is done in such a way that the unpleasant terms from both functions are controlled by those from the other one which provide us a positive information. This is possible due to the fact that the unpleasant terms from $\mathbf{v}^2$ are multiplied by $|\ln|x - x_0||^{-1} \leq |\ln K|^{-1} \ll 1$.

We can therefore verify that

$$\sup_{x_0\in\overline{\Omega}}\int_\Omega \frac{p(\varrho_\delta,\vartheta_\delta)+\delta(\varrho_\delta^\beta+\varrho_\delta^2)+(1-\alpha)\varrho_\delta|\mathbf{u}_\delta|^2}{|x-x_0|^\alpha}\,\mathrm{d}x \tag{3.31}$$
$$\leq C(1+\delta\|\varrho_\delta\|_\beta^\beta+\|p(\varrho_\delta,\vartheta_\delta)\|_1+(1+\|\vartheta_\delta\|_{3m})\|\mathbf{u}_\delta\|_{1,2}+\|\varrho_\delta|\mathbf{u}_\delta|^2\|_1),$$

provided $0<\alpha<\max\{1,\frac{3m-2}{2m}\}$, $m>\frac{2}{3}$, and, moreover, the test function (3.30) belongs to $W^{1,p}(\Omega;\mathbb{R}^3)$ for $1\leq p<\frac{3}{\alpha}$ with the norm bounded independently of the distance of x_0 from $\partial\Omega$.

We exploit the estimates in the following way. We define now for $1\leq a\leq\gamma$ and $0<b<1$

$$\mathcal{B}=\int_\Omega\left(\varrho_\delta^a|\mathbf{u}|^2+\varrho_\delta^b|\mathbf{u}|^{2b+2}\right)\mathrm{d}x. \tag{3.32}$$

Then we have

$$\|\varrho_\delta\mathbf{u}_\delta\|_1\leq C\mathcal{B}^{\frac{a-b}{2(ab+a-2b)}}. \tag{3.33}$$

and for $1<s<\frac{1}{2-a}$ (if $a<2$), $0<(s-1)\frac{a}{a-1}<b<1$

$$\|\varrho_\delta|\mathbf{u}_\delta|^2\|_s\leq C\mathcal{B}^{\frac{a-b/s}{ab+a-2b}}. \tag{3.34}$$

Next, we use the Bogovskii-type estimate and get for $1<s<\frac{1}{2-a}$ (if $a<2$), $0<(s-1)\frac{a}{a-1}<b<1$, $s\leq\frac{6m}{3m+2}$, $m>\frac{2}{3}$

$$\int_\Omega\left(\varrho_\delta^{s\gamma}+\varrho_\delta^{(s-1)\gamma}p(\varrho_\delta,\vartheta_\delta)+(\varrho_\delta|\mathbf{u}_\delta|^2)^s+\delta\varrho_\delta^{\beta+(s-1)\gamma}\right)\mathrm{d}x\leq C(1+\mathcal{B}^{\frac{sa-b}{ab+a-2b}}). \tag{3.35}$$

We distinguish two cases. First, for $m\geq 2$ the only restriction on α is actually $\alpha<1$. In the other case, if $m\in(\frac{2}{3},2)$, we have the restriction $\alpha<\frac{3m-2}{2m}$. Therefore, if $m\geq 2$, we set $a=\gamma$ and using Fatou's lemma and Hölder inequality we show for $b\in((s-1)\frac{\gamma}{\gamma-1},1)$, $1<s<\frac{2}{2-\gamma}$ and $s\leq\frac{6m}{3m+2}$

$$\sup_{x_0\in\overline{\Omega}}\int_\Omega\frac{p(\varrho_\delta,\vartheta_\delta)+(\varrho_\delta|\mathbf{u}_\delta|^2)^b}{|x-x_0|}\,\mathrm{d}x \tag{3.36}$$
$$\leq C\big(1+\delta\|\varrho_\delta\|_\beta^\beta+\|p(\varrho_\delta,\vartheta_\delta)\|_1+(1+\|\vartheta_\delta\|_{3m})\|\mathbf{u}_\delta\|_{1,2}+\|\varrho_\delta|\mathbf{u}_\delta|^2\|_1\big).$$

If $m<2$, we keep $1\leq a<\gamma$ and using Hölder inequality we end up with

$$\begin{aligned}&\sup_{x_0\in\overline{\Omega}}\int_\Omega\frac{\varrho_\delta^a+(\varrho_\delta|\mathbf{u}_\delta|^2)^b}{|x-x_0|}\,\mathrm{d}x\\ &\leq C\big(1+\delta\|\varrho_\delta\|_\beta^\beta+\|p(\varrho_\delta,\vartheta_\delta)\|_1+(1+\|\vartheta_\delta\|_{3m})\|\mathbf{u}_\delta\|_{1,2}+\|\varrho_\delta|\mathbf{u}_\delta|^2\|_1\big)^{\frac{a}{\gamma}}\\ &\quad+C\big(1+\delta\|\varrho_\delta\|_\beta^\beta+\|p(\varrho_\delta,\vartheta_\delta)\|_1+(1+\|\vartheta_\delta\|_{3m})\|\mathbf{u}_\delta\|_{1,2}+\|\varrho_\delta|\mathbf{u}_\delta|^2\|_1\big)^b\end{aligned} \tag{3.37}$$

which holds for $b\in((s-1)\frac{\gamma}{\gamma-1},1)$, $1<s<\frac{2}{2-\gamma}$, $\alpha>\max\{\frac{3a-2\gamma}{a},\frac{3b-2}{b}\}$.

Let us consider now

$$\begin{aligned}-\Delta h&=\varrho_\delta^a+\varrho_\delta^b|\mathbf{u}_\delta|^{2b}-\frac{1}{|\Omega|}\int_\Omega(\varrho_\delta^a+\varrho_\delta^b|\mathbf{u}_\delta|^{2b})\,\mathrm{d}x,\\ \frac{\partial h}{\partial\mathbf{n}}|_{\partial\Omega}&=0.\end{aligned} \tag{3.38}$$

It is well-known that the unique strong solution admits the following representation

$$h(x)=\int_\Omega G(x,y)(\varrho_\delta^a+\varrho_\delta^b|\mathbf{u}_\delta|^{2b})\,\mathrm{d}y-\frac{1}{|\Omega|}\int_\Omega G(x,y)\,\mathrm{d}y\int_\Omega(\varrho_\delta^a+\varrho_\delta^b|\mathbf{u}_\delta|^{2b})\,\mathrm{d}x; \tag{3.39}$$

since $G(x,y) \le C|x-y|^{-1}$, we get due to estimates above

$$\|h\|_\infty \le C(1+\mathcal{B}^{\frac{\gamma-b/s}{b\gamma+\gamma-2b}}), \tag{3.40}$$

provided

$$1 < s < \frac{1}{2-\gamma}, \quad 0 < (s-1)\frac{\gamma}{\gamma-1} < b < 1, \quad s \le \frac{6m}{3m+2}, \quad m \ge 2, \tag{3.41}$$

and

$$\|h\|_\infty \le C(1+\mathcal{B}^{\frac{a-b/s}{ab+a-2b}\frac{a}{\gamma}} + \mathcal{B}^{\frac{a-b/s}{ab+a-2b}b}), \tag{3.42}$$

provided

$$\begin{aligned} &1 < s < \frac{1}{2-a}, \quad 0 < (s-1)\frac{a}{a-1} < b < 1, \quad s \le \frac{6m}{3m+2}, \\ &\alpha > \frac{3a-2\gamma}{a}, \quad \alpha > \frac{3b-2}{b}, \quad \alpha < \frac{3m-2}{2m}, \quad \frac{2}{3}m \le 2. \end{aligned} \tag{3.43}$$

Now, from (3.32) and (3.38), we have

$$\begin{aligned} \mathcal{B} &= \int_\Omega -\Delta h \mathbf{u}_\delta^2 \,\mathrm{d}x + \frac{1}{|\Omega|}\int_\Omega \mathbf{u}_\delta^2 \,\mathrm{d}x \int_\Omega (\varrho_\delta^a + \varrho_\delta^b |\mathbf{u}_\delta|^{2b}) \,\mathrm{d}x \\ &\le 2\|\nabla \mathbf{u}_\delta\|_2 D^{\frac{1}{2}} + C(\varepsilon)\|\mathbf{u}_\delta\|_{1,2}^2 (1+\mathcal{B}^{\Gamma+\varepsilon}) \end{aligned} \tag{3.44}$$

for any $\varepsilon > 0$, where

$$\Gamma = \begin{cases} \dfrac{\gamma-b}{b\gamma+\gamma-2b} & \text{if } m \ge 2 \\ \max\left\{\dfrac{a-b}{ab+a-2b}\dfrac{a}{\gamma}, \dfrac{a-b}{ab+a-2b}b\right\} & \text{if } \frac{2}{3} < m < 2, \end{cases}$$

and

$$\begin{aligned} D = \int_\Omega |\nabla h \otimes \mathbf{u}_\delta|^2 \,\mathrm{d}x - \int_\Omega h\Delta h |\mathbf{u}_\delta|^2 \,\mathrm{d}x - \int_\Omega h \nabla h \cdot \nabla \mathbf{u}_\delta \cdot \mathbf{u}_\delta \,\mathrm{d}x \\ \le \|h\|_\infty (\mathcal{B} + C(\varepsilon)\|\mathbf{u}_\delta\|_{1,2}^2 \mathcal{B}^{\Gamma+\varepsilon} + \|\nabla \mathbf{u}_\delta\|_2 D^{\frac{1}{2}}). \end{aligned}$$

Therefore, due to (3.40),

$$\begin{aligned} \mathcal{B} &\le C\big(1+\mathcal{B}^{\frac{\gamma-b/s}{b\gamma+\gamma-2b}}\big) \quad \text{if } m \ge 2, \\ \mathcal{B} &\le C\big(1+\mathcal{B}^{\frac{a-b/s}{ab+a-2b}\frac{a}{\gamma}} + \mathcal{B}^{\frac{a-b/s}{ab+a-2b}b}\big) \quad \text{if } \tfrac{2}{3} < m < 2. \end{aligned} \tag{3.45}$$

Checking carefully all conditions above we end up for Ω not axially symmetric, $\gamma > 1$ and $m > \frac{2}{4\gamma-3}$, $m > \frac{2}{3}$, and for Ω axially symmetric ($f > 0$), $\gamma > 1$ and $m > \frac{6-2\gamma}{3\gamma-1}$, $m > \frac{2}{3}$ that there exists $s > 1$ such that

$$\begin{aligned} \sup_{\delta>0} \|\varrho_\delta\|_{\gamma s} &< \infty, \\ \sup_{\delta>0} \|\varrho_\delta \mathbf{u}_\delta\|_s &< \infty, \\ \sup_{\delta>0} \|\varrho_\delta |\mathbf{u}_\delta|^2\|_s &< \infty, \\ \sup_{\delta>0} \|\mathbf{u}_\delta\|_{1,2} &< \infty, \\ \sup_{\delta>0} \|\vartheta_\delta\|_{3m} &< \infty, \\ \sup_{\delta>0} \|\vartheta_\delta^{m/2}\|_{1,2} &< \infty, \\ \sup_{\delta>0} \delta\|\varrho_\delta^{\beta+(s-1)\gamma}\|_1 &< \infty. \end{aligned} \tag{3.46}$$

Moreover, we can take $s > \frac{6}{5}$ provided $\gamma > \frac{5}{4}$, $m > \max\{1, \frac{2\gamma+10}{17\gamma-15}\}$ (if Ω is not axially symmetric), or $\gamma > \frac{5}{4}$, $m > \frac{16\gamma}{15\gamma-16}$ (if $\gamma \in (\frac{5}{4}, \frac{4}{3}]$) or $m > \frac{18-6\gamma}{9\gamma-7}$ (if $\gamma \in (\frac{4}{3}, \frac{5}{3})$) (if Ω is axially symmetric).

Step VIII: Compactness of the density sequence. Here, the standard tools from the theory of compressible Navier–Stokes equations can be applied. We have namely the *effective viscous flux identity*

$$\begin{aligned} &\overline{p(\varrho,\vartheta)T_k(\varrho)} - \Big(\frac{4}{3}\mu(\vartheta) + \xi(\vartheta)\Big)\overline{T_k(\varrho)\operatorname{div}\mathbf{u}} \\ &= \overline{p(\varrho,\vartheta)}\;\overline{T_k(\varrho)} - \Big(\frac{4}{3}\mu(\vartheta) + \xi(\vartheta)\Big)\overline{T_k(\varrho)}\operatorname{div}\mathbf{u}, \end{aligned} \tag{3.47}$$

where

$$T_k(z) = kT\Big(\frac{z}{k}\Big), \qquad T(z) = \begin{cases} z \text{ for } 0 \le z \le 1, \\ \text{concave on } (0,\infty), \\ 2 \text{ for } z \ge 3. \end{cases}$$

Next, for the *oscillation defect measure*

$$\mathbf{osc_q}[\varrho_\delta \to \varrho](Q) = \sup_{k>1}\Big(\limsup_{\delta\to 0^+}\int_Q |T_k(\varrho_\delta) - T_k(\varrho)|^q \,\mathrm{d}x\Big) \tag{3.48}$$

we can show that there exists $q > 2$ such that

$$\mathbf{osc_q}[\varrho_\delta \to \varrho](\Omega) < \infty, \tag{3.49}$$

provided $m > \max\{\frac{2}{3(\gamma-1)}, \frac{2}{3}\}$. This implies the validity of the renormalized continuity equation even in the case when the density sequence is not bounded in $L^2(\Omega)$. Then it is not difficult to conclude that

$$\lim_{k\to\infty}\limsup_{\delta\to 0^+}\int_\Omega |T_k(\varrho_\delta) - T_k(\varrho)|^q \,\mathrm{d}x = 0$$

with q as in the estimate of the oscillation defect measure. Hence

$$\|\varrho_\delta - \varrho\|_1 \le \|\varrho_\delta - T_k(\varrho_\delta)\|_1 + \|T_k(\varrho_\delta) - T_k(\varrho)\|_1 + \|T_k(\varrho) - \varrho\|_1,$$

which yields strong convergence of the density in $L^1(\Omega)$, and also in $L^p(\Omega)$ for $1 \le p < s\gamma$. Note that if $s > \frac{6}{5}$ we may pass to the limit in the total energy balance while if $s > 1$ solely, we may pass to the limit only in the entropy inequality and in the total energy balance with a constant test function. This finishes the proof of Theorem 2.3. □

References

[1] Eduard Feireisl and Antonín Novotný, *Singular limits in thermodynamics of viscous fluids*, Advances in Mathematical Fluid Mechanics, Birkhäuser Verlag, Basel, 2009. MR2499296

[2] Eduard Feireisl, Hana Petzeltová, and Konstantina Trivisa, *Multicomponent reactive flows: global-in-time existence for large data*, Commun. Pure Appl. Anal. **7** (2008), no. 5, 1017–1047. MR2410865

[3] Vincent Giovangigli, *Multicomponent flow modeling*, Modeling and Simulation in Science, Engineering and Technology, Birkhäuser Boston, Inc., Boston, MA, 1999. MR1713516

[4] Vincent Giovangigli, Milan Pokorný, and Ewelina Zatorska, *On the steady flow of reactive gaseous mixture*, Analysis (Berlin) **35** (2015), no. 4, 319–341. MR3420315

[5] Didier Jesslé and Antonín Novotný, *Existence of renormalized weak solutions to the steady equations describing compressible fluids in barotropic regime* (English, with English and French summaries), J. Math. Pures Appl. (9) **99** (2013), no. 3, 280–296. MR3017990

[6] Didier Jesslé, Antonín Novotný, and Milan Pokorný, *Steady Navier-Stokes-Fourier system with slip boundary conditions*, Math. Models Methods Appl. Sci. **24** (2014), no. 4, 751–781. MR3163399

[7] P. B. Mucha and M. Pokorný, *On a new approach to the issue of existence and regularity for the steady compressible Navier-Stokes equations*, Nonlinearity **19** (2006), no. 8, 1747–1768. MR2250794

[8] Piotr B. Mucha and Milan Pokorný, *On the steady compressible Navier-Stokes-Fourier system*, Comm. Math. Phys. **288** (2009), no. 1, 349–377. MR2491627

[9] P. B. Mucha, M. Pokorný, and E. Zatorska, *Heat-conducting, compressible mixtures with multicomponent diffusion: construction of a weak solution*, SIAM J. Math. Anal. **47** (2015), no. 5, 3747–3797. MR3403138

[10] P.B. Mucha, M. Pokorný, E. Zatorska, *Existence of stationary weak solutions for the heat conducting flows*, in: Handbook of Mathematical Analysis in Mechanics of Viscous Fluids, eds. Y. Giga and A. Novotný, Springer Verlag, 2018.

[11] A. Novotný and M. Pokorný, *Steady compressible Navier-Stokes-Fourier system for monoatomic gas and its generalizations*, J. Differential Equations **251** (2011), no. 2, 270–315. MR2800154

[12] A. Novotný and I. Straškraba, *Introduction to the mathematical theory of compressible flow*, Oxford Lecture Series in Mathematics and its Applications, vol. 27, Oxford University Press, Oxford, 2004. MR2084891

[13] Tomasz Piasecki and Milan Pokorný, *Weak and variational entropy solutions to the system describing steady flow of a compressible reactive mixture*, Nonlinear Anal. **159** (2017), 365–392. MR3659836

[14] Milan Pokorný and Piotr B. Mucha, *3D steady compressible Navier-Stokes equations*, Discrete Contin. Dyn. Syst. Ser. S **1** (2008), no. 1, 151–163. MR2375591

[15] Ewelina Zatorska, *On the steady flow of a multicomponent, compressible, chemically reacting gas*, Nonlinearity **24** (2011), no. 11, 3267–3278. MR2846926

Institute of Applied Mathematics and Mechanics, University of Warsaw, ul. Banacha 2, 02-097 Warsaw, Poland

Email address: tpiasecki@mimuw.edu.pl

Charles University, Faculty of Mathematics and Physics, Sokolovská 83, 18675 Praha 8, Czech Republic

Email address: pokorny@karlin.mff.cuni.cz

Selected Published Titles in This Series

710 **Raphaël Danchin, Reinhard Farwig, Jiří Neustupa, and Patrick Penel, Editors,** Mathematical Analysis in Fluid Mechanics, 2018

706 **Yeonhyang Kim, Sivaram K. Narayan, Gabriel Picioroaga, and Eric S. Weber, Editors,** Frames and Harmonic Analysis, 2018

705 **Graham J. Leuschke, Frauke Bleher, Ralf Schiffler, and Dan Zacharia, Editors,** Representations of Algebras, 2018

704 **Alain Escassut, Cristina Perez-Garcia, and Khodr Shamseddine, Editors,** Advances in Ultrametric Analysis, 2018

703 **Andreas Malmendier and Tony Shaska, Editors,** Higher Genus Curves in Mathematical Physics and Arithmetic Geometry, 2018

702 **Mark Grant, Gregory Lupton, and Lucile Vandembroucq, Editors,** Topological Complexity and Related Topics, 2018

701 **Joan-Carles Lario and V. Kumar Murty, Editors,** Number Theory Related to Modular Curves, 2018

700 **Alexandre Girouard, Dmitry Jakobson, Michael Levitin, Nilima Nigam, Iosif Polterovich, and Frédéric Rochon, Editors,** Geometric and Computational Spectral Theory, 2017

699 **Mark L. Agranovsky, Matania Ben-Artzi, Catherine Bénéteau, Lavi Karp, Dmitry Khavinson, Simeon Reich, David Shoikhet, Gilbert Weinstein, and Lawrence Zalcman, Editors,** Complex Analysis and Dynamical Systems VII, 2017

698 **Alexander M. Blokh, Leonid A. Bunimovich, Paul H. Jung, Lex G. Oversteegen, and Yakov G. Sinai, Editors,** Dynamical Systems, Ergodic Theory, and Probability: in Memory of Kolya Chernov, 2017

697 **Fabrizio Broglia, Françoise Delon, Max Dickmann, Danielle Gondard-Cozette, and Victoria Ann Powers, Editors,** Ordered Algebraic Structures and Related Topics, 2017

696 **Ara S. Basmajian, Yair N. Minsky, and Alan W. Reid, Editors,** In the Tradition of Ahlfors–Bers, VII, 2017

695 **Katrina Barron, Elizabeth Jurisich, Antun Milas, and Kailash Misra, Editors,** Lie Algebras, Vertex Operator Algebras, and Related Topics, 2017

694 **Manjul Bhargava, Robert Guralnick, Gerhard Hiss, Klaus Lux, and Pham Huu Tiep, Editors,** Finite Simple Groups: Thirty Years of the Atlas and Beyond, 2017

693 **Michael Cwikel and Mario Milman, Editors,** Functional Analysis, Harmonic Analysis, and Image Processing, 2017

692 **Anatole Katok, Yakov Pesin, and Federico Rodriguez Hertz, Editors,** Modern Theory of Dynamical Systems, 2017

691 **Farrell Brumley, Maria Paula Gómez Aparicio, and Alberto Mínguez, Editors,** Around Langlands Correspondences, 2017

690 **Andrés Eduardo Caicedo, James Cummings, Peter Koellner, and Paul B. Larson, Editors,** Foundations of Mathematics, 2017

689 **Erica Flapan, Allison Henrich, Aaron Kaestner, and Sam Nelson, Editors,** Knots, Links, Spatial Graphs, and Algebraic Invariants, 2017

688 **Jeffrey Bergen, Stefan Catoiu, and William Chin, Editors,** Groups, Rings, Group Rings, and Hopf Algebras, 2017

687 **Fernanda Botelho, Raena King, and T. S. S. R. K. Rao, Editors,** Problems and Recent Methods in Operator Theory, 2017

686 **Alp Bassa, Alain Couvreur, and David Kohel, Editors,** Arithmetic, Geometry, Cryptography and Coding Theory, 2017

For a complete list of titles in this series, visit the
AMS Bookstore at **www.ams.org/bookstore/conmseries/**.